The Verilog PLI Handbook

A User's Guide
and
Comprehensive Reference
on the
Verilog Programming Language Interface

Second Edition

The Verilog PLI Handbook

A User's Guide
and
Comprehensive Reference
on the
Verilog Programming Language Interface

Second Edition

by

Stuart Sutherland

Sutherland HDL, Inc.

 Springer

Library of Congress Cataloging-in-Publication Data

Sutherland, Stuart, 1953-
    The Verilog PLI handbook : a user's guide and comprehensive reference on the Verilog
programming language interface / by Stuart Sutherland.—2nd ed.
        p. cm. – (The Springer international series in engineering and computer science; 666)
    Includes bibliographical references and index.
    ISBN: 0-7923-7658-7

    1. Verilog (Computer hardware description language) I. Title. II. Series.

TK7885.7 .S88
621.39'2—dc21   2001058306

Printed in the United States of America.

9 8 7 6 5 4 3 2   SPIN 11583783

springer.com

Dedication

*To my wonderful wife, LeeAnn, and
my children: Ammon, Tamara,
Hannah, Seth and Samuel*

Table of Contents

Part One: The VPI Portion of the Verilog PLI Standard

Appendices

About the Author

*M*r. Stuart Sutherland is a member of the IEEE Verilog standards committee, where he is co-chair of the PLI standards task force and technical editor for the PLI sections of the IEEE 1364 Verilog Language Reference Manual.

Mr. Sutherland has more than 18 years of experience in hardware design and over 14 years of experience with Verilog. He is the founder of *Sutherland HDL Inc.*, located in Portland Oregon. Sutherland HDL provides expert Verilog HDL and Verilog PLI design services, including training, modeling, design verification and software tool evaluation. Verilog PLI training is one of the specialties of Sutherland HDL. Prior to founding Sutherland HDL in 1992, Mr. Sutherland was as an engineer at Sanders Display Products Division in New Hampshire, where he worked on high speed graphics systems for the defense industry. In 1988, he became a senior applications engineer for Gateway Design Automation, the founding company of Verilog. At Gateway, which was acquired by Cadence Design Systems in 1989, Mr. Sutherland specialized in training and support for logic simulation, timing analysis, fault simulation, and the Verilog PLI. Mr. Sutherland has also worked closely with several EDA vendors to specify, test and bring to market Verilog simulation products.

Mr. Sutherland holds a Bachelor of Science in Computer Science, with an emphasis in Electronic Engineering Technology, from Weber State University (Ogden, Utah) and Franklin Pierce College (Nashua, New Hampshire). He has taught Verilog engineering courses at the University of California, Santa Cruz (Santa Clara extension), and has authored the *"Verilog 2001: A Guide to the New Features in Verilog"*, *"Verilog HDL Quick Reference Guide"* and *"Verilog PLI Quick Reference Guide"*. He has presented tutorials and papers at the International Verilog Conference, EDA-Front-to-Back Conference, DesignCon Conference, Synopsys User's Group Conference, and at the International Cadence User's Group Conference.

List of Examples

The source code files for the major examples used in this book are provided on a CD located inside the back of this book. The examples can also be downloaded from the Author's web page. Go to *http://www.sutherland-hdl.com*, and navigate the links to: *"The Verilog PLI Handbook"* — *"PLI Book Examples"*.

CHAPTER 5: Reading and Modifying Values Using VPI Routines

CHAPTER 6: Synchronizing to Simulations Using VPI Callbacks

CHAPTER 7: Interfacing to C Models Using VPI Routines

CHAPTER 8: Creating PLI Applications Using TF and ACC Routines

CHAPTER 9: Interfacing TF/ACC Applications to Verilog Simulators

CHAPTER 15: Details on the ACC Routine Library

CHAPTER 16: Reading and Modifying Values Using ACC Routines

CHAPTER 17: Using the Value Change Link (VCL)

CHAPTER 18: Interfacing to C Models Using ACC Routines

APPENDIX A: Linking PLI Applications to Verilog Simulators

Foreword

by Maq Mannan

President and CEO, DSM Technologies
Chairman of the IEEE 1364 Verilog Standards Group
Past Chairman of Open Verilog International

One of the major strengths of the Verilog language is the Programming Language Interface (PLI), which allows users and Verilog application developers to infinitely extend the capabilities of the Verilog language and the Verilog simulator. In fact, the overwhelming success of the Verilog language can be partly attributed to the existence of its PLI.

Using the PLI, add-on products, such as graphical waveform displays or pre and post simulation analysis tools, can be easily developed. These products can then be used with any Verilog simulator that supports the Verilog PLI. This ability to create third-party add-on products for Verilog simulators has created new markets and provided the Verilog user base with multiple sources of software tools.

Hardware design engineers can, and should, use the Verilog PLI to customize their Verilog simulation environment. A Company that designs graphics chips, for example, may wish to see the simulation results of a new design in some custom graphical display. The Verilog PLI makes it possible, and even trivial, to integrate custom software, such as a graphical display program, into a Verilog simulator. The simulation results can then dynamically be displayed in the custom format during simulation. And, if the company uses Verilog simulators from multiple simulator vendors, this integrated graphical display will work with all the simulators.

As another example, a company designing communication products can benefit from simulating more than just the hardware that makes up one piece of equipment. An internet router, for instance, can be better verified if the simulation models of the router have real data flowing into the router, and pass the outputs to a real environment. The Verilog PLI makes it possible to place an entire Verilog simulation into a larger infrastructure, thus allowing real information to dynamically flow into and out of the simulation.

The possibilities are endless for how the Verilog Programming Language Interface can be utilized in design verification. If an engineer can conceive of a design verification task, and can implement the idea as a program, then the PLI will provide a way to integrate the program into the Verilog simulation environment.

Every engineer working with Verilog logic simulators can benefit from knowing the capabilities of the Verilog PLI. Knowing how to use the PLI for design verification tasks will enable an engineer to be more productive, and for designs to be of the highest quality.

Maq Mannan

Acknowledgments

I would like to express my gratitude for the many people who helped with this book. Thank you **David Roberts** and **Charles Dawson** of *Cadence Design Systems*, **Debi Dalio** of *Model Technology* and **Dhiraj Raj** of *Synopsys* for taking the time to answer questions as I prepared the second edition of this book. Your recommendations have helped to make this book more useful and accurate. I also wish to express my appreciation to *Cadence*, *Model Technology* and *Synopsys* for providing me access to the latest versions of their Verilog simulators to testing the examples in this book.

Thank you once again, all those who reviewed the first edition of this book and provided valuable feedback on how to make the book usable and accurate. The reviewers include: **Drew Lynch** of *SureFire Verification*, **Charles Dawson** and **David Roberts** of *Cadence Design Systems*, **Steve Meyer** of *Pragmatic C Software*, **Chris Spear** of *Synopsys*, and **Marcelo Siero**, an independent software development contractor.

A special thank you to my wife, **LeeAnn Sutherland**, for painstakingly wading through all the hardware engineering jargon and C programming terms in order to review the grammar, punctuation and spelling of the book.

Finally, thank you all others who have helped with the complex process of bringing this book from concept to print, including **Carl Harris** and **Ellie Kerrissey** of *Kluwer Academic Publishers*.

Stuart Sutherland

Introduction

T he Verilog Programming Language Interface, commonly called the Verilog PLI, is one of the more powerful features of Verilog. The PLI provides a means for both hardware designers and software engineers to interface their own programs to commercial Verilog simulators. Through this interface, a Verilog simulator can be customized to perform virtually any engineering task desired. Just a few of the common uses of the PLI include interfacing Verilog simulations to C language models, adding custom graphical tools to a simulator, reading and writing proprietary file formats from within a simulation, performing test coverage analysis during simulation, and so forth. The applications possible with the Verilog PLI are endless.

Intended Audience:

This book is written for digital design engineers with a background in the Verilog Hardware Description Language and a fundamental knowledge of the C programming language. It is expected that the reader:

- Has a basic knowledge of hardware engineering, specifically digital design of ASIC and FPGA technologies.

- Is familiar with the Verilog Hardware Description Language (HDL), and can write models of hardware circuits in Verilog, can write simulation test fixtures in Verilog, and can run at least one Verilog logic simulator.

- Knows basic C-language programming, including the use of functions, pointers, structures and file I/O.

Explanations of the concepts and terminology of digital engineering, the Verilog language or C programming are not included in this book.

The IEEE 1364 Verilog PLI standard

The Institute of Electrical and Electronics Engineers, Inc. (IEEE) Verilog standard encapsulates two older, public domain versions of the Verilog PLI standard. In the *IEEE 1364 Language Reference Manual* (the Verilog "LRM"), these older standards are referred to as:

- **PLI 1.0**, Open Verilog International's 1990 Verilog PLI standard. In the IEEE library, the *TF/ACC routines* in the IEEE standard come from this standard.

- **PLI 2.0**, Open Verilog International's 1990 Verilog PLI standard. In the IEEE library, the *VPI routines* in the IEEE standard come from this standard

The VPI routines offer much greater capabilities and opportunities than the older TF/ACC generation of the PLI. The TF/ACC routines comprise an older generation of the Verilog PLI. These routines have been in use for many years, and are used by hundreds—possibly thousands—of PLI applications.

 As the Verilog Hardware Description Language evolves to meet new engineering methods and requirements, the IEEE Verilog Standards Group will enhance the VPI routines to meet these new requirements. The IEEE 1364 standard includes the TF/ACC routines in order to provide backward compatibility and portability for older PLI applications. However, the IEEE 1364 Verilog Standards Group is only maintaining the older TF/ACC routines—just the VPI routines will be enhanced as the Verilog language and engineering methods evolve.

The history of the Verilog PLI

To understand some of the terms used in the PLI standard, it is helpful to know a little about the origin and evolution of the PLI.

1985: The TF routines

The Verilog PLI was first introduced in 1985 as part of a digital simulator called Verilog-XL, which was developed by Gateway Design Automation (Gateway later merged into Cadence Design Systems). Both the Verilog language (the HDL) and the PLI were proprietary to Gateway—they could only be used with Gateway's simulator products. Verilog-XL quickly grew from an obscure product to one of the most popular digital simulators used by hardware engineers. The powerful capabilities of the PLI were one of the reasons for the growing popularity of Verilog-XL. Gateway recognized the potential of the PLI This earliest generation of the PLI provided a procedural interface to Verilog simulation through constructs known as a *system tasks* and *system functions*. The library of C routines in this generation of the PLI mostly begin with the letters "**tf_**", where the "tf" stands for "*task/function*". This first generation of the PLI routines is sometimes referred to as "utility" routines.

1988: The ACC routines

In 1988 and 1989, Gateway Design Automation added several major new features to the Verilog-XL product, and the accompanying Verilog HDL and PLI. For the PLI, the enhancements provided users a procedural interface that could directly access many kinds of information within a simulation, instead of just the system tasks and functions that the TF routines could access. The ACC library of C functions all begin with the letters "**acc_**", which stands for "access" routines.

The combined TF and ACC libraries are treated as the first generation of the PLI in this book.

1990: The OVI PLI 1.0 standard

The widespread acceptance of Verilog-XL, along with a number of other market factors, encouraged Gateway to make the Verilog language public domain. This would allow companies to share Verilog models and expertise, and for other software products to use the Verilog language. Gateway began the process of separating the Verilog HDL and PLI from their proprietary Verilog-XL simulator in 1989. In late 1989, Gateway Design Automation merged with Cadence Design Systems. Just a few months later, early in 1990, Cadence released the Verilog HDL and PLI to the public domain. That same year, a Verilog user organization was formed, called Open Verilog International (OVI), to manage the public domain Verilog language. OVI labeled the TF and ACC libraries of PLI routines which Cadence had released to the public domain as the *"PLI 1.0"* standard. This term is still prevalently used, and almost every commercial Verilog simulator available today supports the 1990 OVI PLI 1.0 standard.

1993: The OVI PLI 2.0 standard

OVI identified three major shortcomings in the PLI 1.0 standard. First, the standard had evolved over nearly 10 years with no controlling specification. This evolution had led to a very large library of C routines, with many inconsistencies and redundancies. Second, Cadence was continuing to evolve the PLI in the proprietary Verilog-XL simulator, but these new features were not part of the public domain standard. And third, while the TF and ACC routines were very useful, they did not provide a complete procedural interface to everything within a simulation.

In 1993, OVI released a *"PLI 2.0"* standard. The PLI 2.0 standard was well thought out, very simple to use, and corrected many of the shortcomings of the older PLI 1.0 standard. *OVI intended that the PLI 2.0 standard would completely replace the aging PLI 1.0 standard, making the older TF and ACC libraries obsolete.*

1995: The IEEE 1364-1995 PLI standard and VPI routines

In late 1993, OVI submitted the Verilog HDL and the Verilog PLI to the Institute of Electrical and Electronics Engineers, Inc. (the IEEE) for standardization. The standardization process was completed in 1995.

OVI had intended that its PLI 2.0 standard would totally replace the older PLI 1.0 standard. The IEEE 1364 Verilog Standards Group, however, felt that maintaining backward compatibility was important in a standard, and chose to include both the older OVI PLI 1.0 and the newer OVI PLI 2.0 in the IEEE 1364 Verilog standard.

In the IEEE 1364 standard, the terms PLI 1.0 and PLI 2.0 do not exist. The 1364 Verilog standard instead refers to the three primary generations of the PLI as *"TF routines"*, *"ACC routines"* and *"VPI routines"*. The IEEE TF and ACC routines are derived from the OVI PLI 1.0 standard, and the VPI routines are derived from the OVI PLI 2.0 standard.

The VPI routines are a super set of the older TF and ACC routines, making the VPI redundant with the older TF/ACC routines. This means that most applications can be written with either TF and ACC routines, or with just the VPI routines. Note, however, that because the VPI routines are a super set of the older TF/ACC routines, there are a number of powerful applications that can only be written using the VPI routines. Many of the PLI application examples in this book are provided in two forms, using the newer VPI routines and also using older TF and ACC routines.

2001: The IEEE 1364-2001 PLI standard

Hardware design technology and engineering methods continuously change, and the IEEE 1364 Verilog Standards Group is dynamically evolving the Verilog HDL and PLI standard to meet the changing needs of hardware engineering. In 2000, the IEEE Verilog Standards Group submitted a proposed Verilog standard with a number of substantial enhancements for vote by the IEEE membership. These proposed enhancements were officially ratified in March, 2001, as the IEEE 1364-2001 Verilog standard.

NOTE The IEEE 1364-2001 Hardware Description Language, "Verilog-2001", contains a number of major enhancements for modeling and verifying hardware. Only the VPI routines were enhanced to support these new HDL capabilities. The older ACC and TF routines are maintained in the IEEE standard to provide backward compatibility, but no new ACC and TF routines were added. It should be noted, however, that the IEEE 1364-2001 standard does contain several corrections and clarifications to the 1364-1995 ACC and TF routines.

This second edition of *The Verilog PLI Handbook* includes all the enhancements and clarifications to the PLI that are in the 1364-2001 standard.

Ambiguities in the Verilog PLI standard

This 13 year history of the PLI has led to several generations of the PLI standard. As a whole, the evolving generations of the PLI have been good for Verilog users, because each generation has added significant new features and capabilities to the PLI. However, each major generation of the PLI was developed by different engineers and sometimes different organizations. This has resulted in a number of inconsistent conventions within the standard, as well as a number of redundancies. In order to maintain an accurate and backward compatible PLI standard, the IEEE 1364 Verilog Standards Group chose to allow these inconsistencies and redundancies to remain in the IEEE standard. At times, this may make learning the PLI standard a little more difficult, but it is necessary for backward compatibility.

Organization of this book

The PLI is not a trivial topic to explain or to comprehend. It combines in a very unique way digital electronic engineering, digital modeling and simulation, and advanced C programming. To add to the complexity, the PLI has a long and complex history, having been in use—and evolving—since 1985.

This book is structured to serve two specific needs:

1. A tutorial on how to write PLI applications.

   Towards this goal, the book provides explanations of the extensive PLI library from a usage point of view, with many examples showing how the PLI routines are applied.

2. A reference book on the IEEE 1364 Verilog PLI standard.

   The appendices of this book contain a full description of the IEEE standard for the PLI. These appendices are intended to serve as a quick and convenient reference whenever you need to see the syntax or semantics of a specific PLI routine. These reference appendices do not explain the usage of each routine. That is covered in the more generalized "how-to" chapters.

This book is divided into two major parts, in order to effectively cover both generations of the PLI. Each part of the book is completely self contained, allowing a reader to focus on either or both of the generations of the Verilog PLI.

Part One of this book, comprising chapters 1 through 7, presents the *VPI* portion of the PLI standard, which is sometimes referred to as *"PLI 2.0"*. The VPI is the newest of generations of the PLI, and is by far the most versatile of all the PLI generations. The first chapters in Part One explain how to use the VPI routines. A large number of small, but useful examples illustrate the concepts presented. The later chapters in Part One contain a more comprehensive description of the VPI library of routines, and how each routine is used. More extensive examples of PLI applications are presented in these chapters. Appendix D contains a quick reference of the IEEE 1364 Verilog PLI standard for the VPI portion of the PLI.

Part Two of the book, comprising chapters 8 through 18, presents the *TF* portions and the *ACC* portions of the PLI standard. This is the older generation of the PLI, and is still very widely used in PLI applications. The TF and ACC portions of the standard are often referred to as *"PLI 1.0"*. The first chapters in Part Two explain how to use the TF and ACC routines, and include many useful examples. The later chapters present the TF library of routines in much greater detail, followed by a detailed presentation of the ACC library of routines. Appendices B and C contain a quick reference of the IEEE 1364 Verilog PLI standard for the TF and ACC portion of the standard.

About the PLI examples in this book

This book covers using the PLI for simple to intermediate applications. A large number of examples are provided, which are realistic and useful programs for engineers designing with Verilog. The examples are necessarily kept reasonably short and simplified in capability. It is expected that readers will build larger and more advanced PLI applications based on the examples provided in this book.

Portability is a primary focus in this book. The IEEE has standardized the Verilog PLI, but some Verilog simulators do not strictly adhere to this standard, and sometimes ambiguities in the complex IEEE standard are interpreted differently. Many Verilog simulators do not support the entire IEEE standard, and other simulators may add proprietary extensions to the IEEE standard. The examples presented within this book will work with any Verilog simulator that is fully IEEE compliant. Many of the examples will also work with simulators that support a subset of the IEEE standard. This book does not use any PLI extensions that are proprietary to a specific simulator.

The PLI examples in this book were tested using the *NC-Verilog™* simulator, from *Cadence Design Systems, Inc*. NC-Verilog runs on most Unix operating systems, Linux, and on the Windows-2000 operating system. For this book, all examples were tested using NC-Verilog version 3.2, on a Pentium III laptop system with the Windows-2000 operating system. The examples were compiled with the Microsoft Visual C++ compiler.

Many of the examples were also tested using the *Cadence Design Systems Verilog-XL™* simulator with the Linux operating system, the *Synopsys VCS™* simulator with the Solaris operating system, and the *Model Technology ModelSim™* simulator with the Windows-2000 operating system.

The source code files for the major examples used in this book are provided on a CD located in the back of this book. The examples can also be downloaded from the Author's web page. Go to *http://www.sutherland-hdl.com*, and navigate the links to: *"The Verilog PLI Handbook"* — *"PLI Book Examples"*.

Other sources of information

There are many excellent books on the C programming language. Readers of this book who are novice C programmers will want to have their personal favorite C reference book handy.

There are also number of books on the Verilog HDL, any of which can serve as an excellent companion to this book. Some suggested books are:

IEEE Std 1364-2001, Language Reference Manual (LRM)—IEEE Standard Hardware Description Language based on the Verilog Hardware Description Language.

Copyright 2001, IEEE, Inc., New York, NY. ISBN 0-7381-2827-9.

This is the official Verilog HDL and PLI standard. The book is a syntax and semantics reference, not a tutorial for learning Verilog. Softcover, 665 pages. For information on ordering, visit the web site: *http://shop.ieee.org/store/product.asp?prodno=SH94921*, or call 1-800-678-4333 (US and Canada), 1-908-981-9667 (elsewhere).

Verilog 2001: A Guide to the New Features of the Verilog Hardware Description Language by Stuart Sutherland

Copyright 2002, Kluwer Academic Publishers, Norwell MA.
ISBN: 0-7923-7568-8

An overview of the many enhancements added as part of the IEEE 1364-2001 standard, 136 pages. For more information, refer to the web site *www.wkap.nl/ book.htm/0-7923-7568-8.*

The Verilog Hardware Description Language, 4th Edition by Donald E. Thomas and Philip R. Moorby

Copyright 1998, Kluwer Academic Publishers, Norwell MA.
ISBN: 0-7923-8166-1

A complete book on Verilog, covering RTL modeling, behavioral modeling and gate level modeling. The book has more detail on the gate, switch and strength level aspects of Verilog than many other books. Hardcover, 336 pages. For more information, refer to the web site *www.wkap.nl/book.htm/0-7923-8166-1.*

Verilog Quickstart, A Practical Guide to Simulation and Synthesis, 2nd Edition by James M. Lee

Copyright 1999, Kluwer Academic Publishers, Norwell MA.
ISBN: 0-7923-8515-2.

An excellent book for learning the Verilog HDL. The book teaches the basics of Verilog modeling, without getting into the more obscure or advanced aspects of the Verilog language. Hardcover, 328 pages. For more information, refer to the web site *www.wkap.nl/book.htm/0-7923-8515-2.*

Verilog HDL Quick Reference Guide, based on the Verilog-2001 standard by Stuart Sutherland

Copyright 1992, 2001, Sutherland HDL, Inc., Portland, OR.

A small, pocket-sized quick reference on the syntax and semantics of the complete Verilog language. Softcover, 50 pages. For more information, refer to the web site *www.sutherland-hdl.com.*

Verilog PLI Quick Reference Guide, based on the Verilog-2001 standard by Stuart Sutherland

Copyright 1996, 2001, Sutherland HDL, Inc., Portland, OR.

A valuable quick reference on the syntax and semantics of the complete Verilog PLI standard. The main contents of this quick reference are reprinted with permission as the Appendices in this book. Softcover, 114 pages. For more information, refer to the web site *www.sutherland-hdl.com.*

Part One:

The VPI Portion of the Verilog PLI Standard

CHAPTER 1 *Creating PLI Applications Using VPI Routines*

*T*his chapter uses two short examples to introduce how PLI applications are created using the VPI routines. The first example will allow a Verilog simulator to execute the ubiquitous *Hello World* C program. The second example uses the PLI to access specific activity within a Verilog simulation, by listing the name and current logic value of a Verilog signal. The purpose of these examples is to introduce how to write a PLI application using the library of VPI routines in the PLI standard. Subsequent chapters in this part of the book build on the concepts presented in this chapter and show how to write more complex PLI applications.

1.1 The capabilities of the Verilog PLI

The Verilog **Programming Language Interface**, commonly referred to as the Verilog **PLI**, is a user-programmable procedural interface for Verilog digital logic simulators. The PLI gives Verilog users a way to extend the capabilities of a Verilog simulator by providing a means for the simulator to invoke other software programs, and then exchange information and synchronize activity with those programs.

The creators of the Verilog Hardware Description Language (HDL), which was first introduced in 1984, intended that the Verilog language should be focused on modeling digital logic. They felt that design verification tasks not directly related to modeling the design itself, such as reading and writing disk files, were already capably handled in other languages and should not be replicated in the Verilog HDL. Instead, the creators of Verilog added a procedural interface—the PLI—that provides a way for end-users of Verilog to write design verification tasks in the C programming language, and then have Verilog logic simulators execute those programs.

The capabilities of the PLI extend far beyond simple file I/O operations. The Verilog PLI allows Verilog designers to interface virtually any program to Verilog simulators. Some common engineering tasks that are required in the design of hardware, and for which the PLI is aptly fitted, include:

- C language bus-functional models

  Abstract bus-functional hardware models are sometimes represented in the C language instead of the Verilog HDL, perhaps to protect intellectual property or to optimize the performance of simulation. The PLI provides a number of ways to pass data from Verilog simulation to a C language model and from the C language model back into Verilog simulation.

- Access to programming language libraries

  The PLI allows Verilog models and test programs to access the libraries of the C programming language, such as the C math library. Through the PLI, Verilog can pass arguments to a C math function, and pass the return of the math function back to the Verilog simulator.

- Reading test vector files

  Test vectors are a common method of representing stimulus for simulation. A file reader program in C is easy to write, and, using the PLI, it is easy to pass values read by the C program to a Verilog simulation. The IEEE 1364-2001 Verilog standard adds significant file I/O enhancements to the Verilog HDL. Many file operations previously performed through the PLI can now be performed directly in the Verilog HDL. However, the PLI still provides greater and more versatile control of file formatting and file operations.

- Delay calculation

  Through the PLI, a Verilog simulator can call an ASIC or FPGA vendor's delay calculator program. The delay calculator can estimate the effects that fanout, voltage, temperature and other variances will have on the delays of ASIC cells or gates. Through the PLI, the delay calculator program can modify the Verilog simulation data structure so that simulation is using the more accurate delays. This usage of the PLI for dynamic delay calculation was common from the late 1980's to the mid 1990's, but has largely been replaced by static, pre-simulation delay estimation and SDF files.

- Custom output displays

  In order to verify design functionality, a logic simulator must generate outputs. The Verilog HDL provides formatted text output, and most simulators provide waveform displays and other graphical output displays. The PLI can be used to extend these output format capabilities. A video controller, for example, might generate video output for a CRT or other type of display panel. Using the PLI, an engineer can take the outputs from simulation of the video controller, and, while simulation is running, pass the data to a program which can display the data in the same form as the real display output. The PLI allows custom output programs to dynamically read simulation values while simulation is running.

- Co-simulation

  Complex designs often require several types of logic simulation, such as a mix of analog and digital simulations, or a mix of Verilog and VHDL simulations. Each simulator type could be run independently on its own regions of the design, but it is far more effective to simulate all regions of the design at the same time, using multiple simulators together. The PLI can be used as a communication channel for Verilog simulators to transfer data to and from other types of simulators.

- Design debug utilities

  The Verilog logic simulators on the market vary a great deal in what they provide for debugging design problems. There may be some type of information about a design that is needed, which cannot be accessed by the simulator's debugger. The Verilog PLI can access information deep inside a simulation data structure. It is often a very trivial task to write a short C program that uses the PLI to extract some specific information needed for debugging a design.

- Simulation analysis

  Most Verilog logic simulators are intended to simply apply input stimulus to a model and show the resulting model outputs. The PLI can be used to greatly enhance simulation by analyzing what had to happen inside a design to generate the output values. The PLI can be used to generate toggle check reports (how many times each node in a design changed value), code coverage reports (what Verilog HDL statements were exercised by the input tests), power usage (how much energy the chip consumed), etc.

The preceding examples are just a few of the tasks that can be performed using the Verilog PLI. The PLI provides full access to anything that is happening in a Verilog simulation, and allows external programs to modify the simulation. This open access enables truly unlimited possibilities. If an engineer can conceive of an application that requires interaction with Verilog simulations, and can write a program to perform the task, the PLI can be used to interface that application to a Verilog simulator.

1.2 General steps to create a PLI application

A PLI application is a user-defined C language application which can be executed by a Verilog simulator. The PLI application can interact with the simulation by both reading and modifying the simulation logic and delay values.

The general steps to create a PLI application are:

1. Define a *system task* or *system function* name for the application.

2. Write a C language *calltf routine* which will be executed by the simulator when-
    ever simulation encounters the system task name or the system function name.

    Optionally, additional C language routines can be written which will be executed
    by the simulator for special conditions, such as when the simulator compiler or
    elaborator encounters the system task/function name.

3. *Register* the system task or system function name and the associated C language
    routines with the Verilog simulator. This registration tells the simulator about the
    new system task or system function name, and the name of the *calltf routine* asso-
    ciated with that task or system function (along with any other routines).

4. *Compile* the C source files which contain the PLI application routines, and *link*
    the object files into the Verilog simulator.

1.3 User-defined system tasks and system functions

In the Verilog language, a *system task* or a *system function* is a command which is
executed by a Verilog simulator. The name of a system task or a system function
begins with a *dollar sign* (*$*).

A *system task* is used like a programming statement that is executed from a Verilog
process, such as an *initial procedure* or an *always procedure*. For example, to print a
message at every positive edge of a clock, the *$display* system task can be used.

```
always @(posedge clock)
  $display("chip_out = %h", chip_out");
```

A *system function* is used like a programming function which returns a value. A function can be called anywhere that a logic value can be used. For example, to assign a
random number to a vector at every positive edge of a clock, the *$random* system
function can be used.

```
always @(posedge clock)
  vector <= $random();
```

The IEEE 1364 Verilog standard allows system tasks and system functions to be defined in three ways:

- A standard set of built-in system tasks and system functions.

  These are defined as part of the IEEE standard, such as *$display*, *$random*, and *$finish*. All IEEE compliant Verilog simulators will have these standard system tasks and system functions.

- Simulator specific system tasks and system functions.

  These are proprietary commands which are defined as part of a simulator, and may not exist in other simulators. Examples are *$save* to create a simulation check point file or *$db_settrace* to enable debug tracing.

- User-defined system tasks and system functions.

  These are created through the Programming Language Interface. A PLI application developer specifies the name and functionality of the system task/function.

User-defined system task or system function names begin with a dollar sign ($), just as the built-in system tasks and system functions. The user-defined system task/function name is then associated with a user-defined C application. When a Verilog simulator encounters the system task/function name, it will execute the C application associated with the name. This simple association allows Verilog simulation users to extend the capability of a Verilog simulator in any manner desired. Chapter 2 presents the concept of system tasks and system functions in more detail.

1.4 The $hello PLI application example

The well known "Hello world" C program is a quick way to show how PLI applications are created. Though the C program itself is simple, it illustrates one of the most powerful features of the Verilog PLI—the ability for a Verilog designer to extend the Verilog language by having a Verilog simulation dynamically execute a user-defined program while simulation is running.

1.4.1 Step One: defining a $hello system task

The first step in creating a PLI application is create a new system task or system function name. A *$hello* user-defined system task will be created for this PLI application. An example of using *$hello* is:

```
module test;

  initial
    $hello();

endmodule
```

1.4.2 Step Two: writing a calltf routine for $hello

The second step in developing a PLI application is to write a C language routine which will be called when a Verilog simulator executes the *$hello* system task. In this example, the simulator will call a C language routine which prints the message:

```
Hello World!
```

The C language routine which will be called by the simulator is referred to as a *calltf routine*. This routine is a user-defined C function and can be any name. Example 1-1 lists the C source code for the *calltf routine* for *$hello*.

> **CD** The source code for this example is on the CD accompanying this book.
>
> • Application source file: Chapter.01/hello_vpi.c
> • Verilog test bench: Chapter.01/hello_test.v
> • Expected results log: Chapter.01/hello_test.log

Example 1-1: *$hello* — a VPI *calltf routine* for a PLI application

```c
#include <stdlib.h>      /* ANSI C standard library */
#include <stdio.h>       /* ANSI C standard input/output library */
#include "vpi_user.h"    /* IEEE 1364 PLI VPI routine library  */

PLI_INT32 PLIbook_hello_calltf(PLI_BYTE8 *user_data)
{
  vpi_printf("\nHello World!\n\n");
  return(0);
}
```

In the above example:

• The header file *vpi_user.h* contains the VPI library of functions provided by the PLI standard. The library is a collection of C functions which can be used by a PLI application to interact with a Verilog simulator. The *vpi_user.h* header file is part of the IEEE 1364 Verilog standard.

- The **vpi_printf()** function used in the preceding example is very similar to the C printf() function. The difference is that vpi_printf() will write its message to the simulator's output channel and the simulator's output log file, whereas the C printf() writes its message to the operating system's standard out channel.

- The data types **PLI_INT32** and **PLI_BYTE8** are special data types defined in *vpi_user.h*, as a 32-bit C integer and an 8-bit integer, respectively. The Verilog PLI uses these special data types instead of the C language **int** and **char** data types to ensure portability across all operating systems and computer platforms (the ANSI C standard does not guarantee the bit width of int and char data types).

 The fixed width PLI data types were added as part of the IEEE 1364-2001 standard. Prior to that, the PLI libraries used the regular C data types, such as char, int and long, with the non-portable assumption that these data types would be the same size on all operating systems. For example, the IEEE 1364-1995 standard assumed that an int would always be 32 bits, something the ANSI C standard does not guarantee. Existing PLI applications that use the C data types instead of the new PLI data types will still compile and work correctly on most 32-bit operating systems. However, these older applications may need to be updated to the IEEE 1364-2001 PLI data types in order to port correctly to 64-bit operating systems. Section 2.4.4 on page 35 lists the fixed-width types defined in the IEEE 1364-2001 standard.

1.4.3 Step Three: Registering the $hello system task

The third step in creating a new PLI application is to tell the Verilog simulator about the new system task or system function name and the C routines which are associated with the application. This process is referred to a *registering* the PLI application.

The IEEE 1364 Verilog PLI standard provides an ***interface mechanism*** for this process. The interface mechanism defines:

- The type of application, which is a system task or system function.

- The system task or system function name.

- The name of the *calltf routine* and other C routines associated with the system task or system function.

The VPI interface mechanism for PLI applications is part of the IEEE standard, and is implemented in the same way with all IEEE compliant Verilog simulators. The interface mechanism requires two steps:

1. Create a *register function* which specifies the PLI application information in an s_vpi_register_systf structure and calls the vpi_register_systf() VPI routine.

2. List the name of the register function in a C array called *vlog_startup_routines*.

Example 1-2 lists a register function for the *$hello* PLI application. Chapter 2 presents more details of registering PLI applications using the VPI interface mechanism.

Example 1-2: *$hello* — VPI register function for a PLI application

```
void PLIbook_hello_register()
{
  s_vpi_systf_data tf_data;

  tf_data.type        = vpiSysTask;
  tf_data.sysfunctype = 0;
  tf_data.tfname      = "$hello";
  tf_data.calltf      = PLIbook_hello_calltf;
  tf_data.compiletf   = NULL;
  tf_data.sizetf      = NULL;
  tf_data.user_data   = NULL;
  vpi_register_systf(&tf_data);
}
```

1.4.4 Step Four: Compiling and linking the $hello system task

The final step in creating a new PLI application is to compile the C source code containing the application and linking the compiled files to the Verilog simulator. Once the application has been linked to the simulator, the simulator can invoke the *calltf routine* when simulation executes the *$hello* system task name.

Compiling and linking is not part of the IEEE standard. This process is specific to both the simulator as well as the operating system on which simulations are run. Appendix A presents the steps for compiling and linking PLI applications for several major Verilog simulators.

1.4.5 Running simulations with the $hello system task

Once a PLI application has been registered with a Verilog simulator, the new system task or system function can be used in a Verilog model, just as with built-in system tasks and system functions. When the Verilog simulator encounters the new user-defined system task name, it will call the C routine which has been associated with that application.

The following Verilog model can be used to test the *$hello* PLI application:

```
module test;

   initial
     begin
       $hello();
       #10 $stop;
       $finish;
     end
endmodule
```

Example 1-3 shows the results of simulating this Verilog model using the Cadence Verilog-XL simulator.

Example 1-3: $hello — simulation results

```
Hello World!
```

1.5 The $show_value PLI application example

The *$hello* PLI application, though trivial from a C programming aspect, effectively illustrates the powerful capability provided by the Programming Language Interface—the ability to have a Verilog simulator call a user-defined C routine during simulation. The PLI provides far more capabilities, however. The C routine which is called by the simulator can use the VPI library to read information from the simulation data structure, and to dynamically modify logic values and delay values in the data structure.

The system task *$show_value* illustrates using the PLI to allow a C routine to read current logic values within a Verilog simulation. The *$show_value* system task requires one argument, which is the name of a net or reg in the Verilog design. The *calltf routine* will then use the library of VPI routines to read the current logic value of that net or reg, and print its name and logic value to the simulator's output screen. An example of using *$show_value* is:

```
module test;
   reg   a, b, ci;
   wire  sum, co;
```

```
initial
  begin
    ...
    $show_value(sum);
  end
endmodule
```

Two user-defined C routines will be associated with *$show_value*:

- A C routine to verify that *$show_value* has the correct type of argument.

- A C routine to print the name and logic value of the signal.

These programs are presented in the following sections.

1.5.1 Writing a compiletf routine for $show_value

The PLI allows users to provide a C routine to verify that a system task or system function is being used correctly and has the correct types of arguments. This C routine is referred to as a ***compiletf routine***.

The C functions in the VPI library can access the arguments of a system task or system function and determine the Verilog data types of the objects listed in the arguments. This example uses some of these C functions. Only a brief explanation of each function is provided in this chapter. Subsequent chapters will present more details about each function in the VPI library.

The C source code for the *compiletf routine* associated with *$show_value* is listed below. The C functions and constants from the PLI library are shown in bold text.

CD The source code for this example is on the CD accompanying this book.

- Application source file: `Chapter.01/show_value_vpi.c`
- Verilog test bench: `Chapter.01/show_value_test.v`
- Expected results log: `Chapter.01/show_value_test.log`

Example 1-4: *$show_value — a VPI compiletf routine* for a PLI application

```
PLI_INT32 PLIbook_ShowVal_compiletf(PLI_BYTE8 *user_data)
{
  vpiHandle systf_handle, arg_iterator, arg_handle;
  PLI_INT32       arg_type;

  /* obtain a handle to the system task instance */
  systf_handle = vpi_handle(vpiSysTfCall, NULL);
  if (systf_handle == NULL) {
    vpi_printf("ERROR: $show_value failed to obtain systf handle\n");
    vpi_control(vpiFinish,0);  /* abort simulation */
    return(0);
  }

  /* obtain handles to system task arguments */
  arg_iterator = vpi_iterate(vpiArgument, systf_handle);
  if (arg_iterator == NULL) {
    vpi_printf("ERROR: $show_value requires 1 argument\n");
    vpi_control(vpiFinish,0);  /* abort simulation */
    return(0);
  }

  /* check the type of object in system task arguments */
  arg_handle = vpi_scan(arg_iterator);
  arg_type = vpi_get(vpiType, arg_handle);
  if (arg_type != vpiNet && arg_type != vpiReg) {
    vpi_printf("ERROR: $show_value arg must be a net or reg\n");
    vpi_free_object(arg_iterator); /* free iterator memory */
    vpi_control(vpiFinish,0);  /* abort simulation */
    return(0);
  }

  /* check that there are no more system task arguments */
  arg_handle = vpi_scan(arg_iterator);
  if (arg_handle != NULL) {
    vpi_printf("ERROR: $show_value can only have 1 argument\n");
    vpi_free_object(arg_iterator); /* free iterator memory */
    vpi_control(vpiFinish,0);  /* abort simulation */
    return(0);
  }
  return(0);
}
```

Observe the following key points in the above example:

- **vpiHandle** is a special data type defined in the VPI library. This data type is used to store pointers to information about objects in a Verilog simulation data structure.

- The **vpi_handle()** function returns a *handle*, which is a form of pointer, to a specific object in a Verilog simulation data structure. In this example, handles are obtained for the instance of the *$show_value* system task which called the PLI application, and for the object listed as the first argument of the system task.

- The **vpi_iterate()** function obtains an *iterator* for all of a specific type of object. An iterator is essentially a pointer to the next object in a series of objects. The iterator object is stored in a vpiHandle data type. In the above example, an iterator is obtained for all of the arguments to the *$show_value* system task.

- The **vpi_scan()** obtains a handle for the next object that is referenced by an iterator. In the above example, each object returned by vpi_scan() is the next argument of the *$show_value* system task.

- The **vpi_get()** function returns the value of integer properties of a specific object in the simulation data structure. The first input to this function is a constant that defines the property to be obtained. The **vpiType** property used in this example identifies the type of object passed as the argument to *$show_value*. In this example, the test is checking that the argument is a Verilog net or reg data type.

- The **vpi_free_object()** function releases memory allocated by the vpi_iterate() routine. Refer to section 4.5.3 on page 108 in Chapter 4 for more details on when vpi_free_object() needs to be used in a PLI application.

- The **vpi_printf()** function writes a message to the simulator's output channel.

- The **vpi_control()** function allows a PLI application to control certain aspects of a simulation. In this compiletf routine, this function is used to abort simulation if there is a usage error with the *$show_value* system task.

 The vpi_control() routine was added with the IEEE 1364-2001 standard. In the IEEE 1364-1995 standard, there was no VPI routine that would cause a simulator to exit if a PLI application detected an error. Instead, applications needed to use the tf_dofinish() routine from the older TF library.

1.5.2 Writing the calltf routine for $show_value

The *calltf routine* is the C routine which will be executed when simulation encounters the *$show_value* system task during simulation.

The following example lists the C source code for the *calltf routine* associated with *$show_value*. Later chapters define the routines used in this example in full detail. The C functions, constants and data types from the PLI library are shown in bold text.

Example 1-5: *$show_value — a VPI calltf routine* for a PLI application

```
PLI_INT32 PLIbook_ShowVal_calltf(PLI_BYTE8 *user_data)
{
  vpiHandle   systf_handle, arg_iterator, arg_handle, net_handle;
  s_vpi_value current_value;

  /* obtain a handle to the system task instance */
  systf_handle = vpi_handle(vpiSysTfCall, NULL);

  /* obtain handle to system task argument
     compiletf has already verified only 1 arg with correct type */
  arg_iterator = vpi_iterate(vpiArgument, systf_handle);
  net_handle = vpi_scan(arg_iterator);
  vpi_free_object(arg_iterator);  /* free iterator memory */

  /* read current value */
  current_value.format = vpiBinStrVal; /* read value as a string */
  vpi_get_value(net_handle, &current_value);
  vpi_printf("Signal %s ", vpi_get_str(vpiFullName, net_handle));
  vpi_printf("has the value %s\n", current_value.value.str);
  return(0);
}
```

Most of the VPI routines used in the above example were also used in the *compiletf routine*, and are described on the preceding pages. The additional routines used are:

- The **vpi_get_str()** function returns the value of string properties of a specific object in the simulation data structure. The first input to this function is a constant that defines the property to be obtained. The **vpiFullName** property used in this example is the Verilog hierarchical path name of the net listed as an argument to *$show_value*.

- The **vpi_get_value()** function obtains the logic value of a Verilog object. The value is returned into an s_vpi_value structure, which is defined as part of the VPI standard. The vpi_get_value() function allows the value to be obtained in a variety of formats. In this example, the value is obtained as a string, with a binary representation of the value.

1.5.3 Registering the $show_value PLI application

The *register function* is used to inform the Verilog simulator about the PLI application. The information about the application is specified in a s_vpi_register_systf structure.

The following example lists the C source code for the register function for *$show_value*. The next chapter explains the process of registering PLI applications in greater detail.

Example 1-6: *$show_value* — a VPI register function for a PLI application

```
void PLIbook_ShowVal_register()
{
  s_vpi_systf_data tf_data;

  tf_data.type        = vpiSysTask;
  tf_data.sysfunctype = 0;
  tf_data.tfname      = "$show_value";
  tf_data.calltf      = PLIbook_ShowVal_calltf;
  tf_data.compiletf   = PLIbook_ShowVal_compiletf;
  tf_data.sizetf      = NULL;
  tf_data.user_data   = NULL;
  vpi_register_systf(&tf_data);
  return;
}
```

1.5.4 A test case for $show_value

The following Verilog HDL source code is a small test case for the *$show_value*
application. The example lists the values of a simple 1-bit adder after a few input test
values have been applied. Observe how *$show_value* is used as a Verilog program-
ming statement within the test bench.

Example 1-7: *$show_value* — a Verilog HDL test case using a PLI application

```
`timescale 1ns / 1ns
module test;
  reg  a, b, ci, clk;
  wire sum, co;
  addbit i1 (a, b, ci, sum, co);
  initial
    begin
      clk = 0;
      a = 0;
      b = 0;
      ci = 0;
      #10 a = 1;
      #10 b = 1;
      $show_value(sum);
      $show_value(co);
      $show_value(i1.n3);
      #10 $stop;
      $finish;
    end
endmodule
```

```
/*** A gate level 1 bit adder model ***/
'timescale 1ns / 1ns
module addbit (a, b, ci, sum, co);
  input   a, b, ci;
  output sum, co;
  wire   a, b, ci, sum, co,
         n1, n2, n3;
  xor       (n1, a, b);
  xor #2 (sum, n1, ci);
  and       (n2, a, b);
  and       (n3, n1, ci);
  or  #2 (co, n2, n3);
endmodule
```

1.5.5 Output from running the $show_value test case

Following is the output from simulating the test case for the *$show_value* application.

Example 1-8: *$show_value* — sample output

```
Signal test.sum has the value 1
Signal test.co has the value 0
Signal test.i1.n3 has the value 0
```

1.6 Summary

This chapter has presented two short examples which illustrate how PLI applications are created using the VPI routine library in the PLI standard. The *$hello* PLI application showed how a Verilog simulation can execute another C program. The *$show_value* application illustrated how arguments can be passed from a Verilog simulation to a PLI application, and how the application can access information about the arguments.

The principles illustrated by the *$show_value* example can be readily expanded to provide much more powerful and useful capabilities. In Chapter 3, the functionality of *$show_value* will be extended to create an application called *$list_nets*, which automatically finds all nets in a module and prints the values of each net. Another example in the same chapter extends *$show_value* to create *$list_signals*, which automatically finds all nets, regs and variables in a module and prints the values of each one.

The next chapter presents much greater detail on how PLI applications interact with Verilog simulations. Subsequent chapters in Part One of this book will discuss the complete VPI library, and include many examples of how the VPI routines are used to create PLI applications which interact with Verilog simulators.

CHAPTER 2 *Interfacing VPI Applications to Verilog Simulators*

Don't skip this chapter! This chapter defines the terminology used by the Verilog PLI and how PLI applications which use the VPI library are interfaced to Verilog simulators. All remaining chapters in Part One of this book assume the principles covered in this chapter are understood. The general concepts presented are:

- PLI terms as used with the VPI library

- System tasks and system functions

- How VPI routines work

- A complete PLI application example

- Interfacing PLI applications to Verilog simulators

2.1 General PLI terms as used in this book

The Verilog PLI has been in use since 1985, and, over the years, many reference manuals, technical papers and training courses have been written about the PLI. Unfortunately, these documents have not used consistent terms for describing the PLI. This inconsistency can make it difficult to cross reference information about the PLI in different documents.

This book strives to clearly define the PLI terms and consistently use this terminology. The general PLI terms used in this book are.

C program (often abbreviated to *program*)

A complete software program written in the C or C++ programming language. A C program must include a C *main* function.

C function

A function written in the C or C++ programming language. A C function does not include a C *main* function.

Verilog function

A Verilog HDL function, written in the Verilog Hardware Description Language. Verilog functions can only be called from Verilog source code.

User-defined system task or system function (abbreviated to *system task/function*)

A user-defined system task or system function is a construct that is used in Verilog HDL source code. The name of a user-defined system task or system function must begin with a dollar sign (*$*), and it is used in the Verilog language in the same way as a Verilog HDL standard system task or system function. When simulation encounters the user-defined system task or system function, the simulator will execute a *PLI routine* associated with the system task/function.

PLI application (sometimes abbreviated to *application*)

A *user-defined system task or system function* with an associated set of one or more *PLI routines*.

PLI routine

A C function which is part of a *PLI application*. The PLI routine is executed by the simulator when simulation encounters a user-defined system task or system function. The VPI portion of the PLI standard defines several PLI routine types: *calltf routines*, *compiletf routines*, *sizetf routines* and *simulation callback routines*. These terms are defined in more detail later in this chapter.

PLI library

A library of C functions which are defined in the Verilog PLI standard. PLI library functions are called from *PLI routines*, and enable the *PLI routines* to interact with a Verilog simulation. The IEEE 1364 standard provides three PLI libraries, referred to as the *VPI library*, the *TF library* and the *ACC library*.

VPI routines, *TF routines* and *ACC routines*

C functions contained in the *PLI library*. The term *routine* is used for these functions, to avoid confusion with *Verilog functions* and *C functions*.

2.2 System tasks and system functions

The Verilog Hardware Description Language provides constructs called *"system tasks"* and *"system functions"*. A system task or system function is not a hardware modeling construct. It is a command that is executed by a Verilog simulator.

The names of system tasks and system functions always begin with a dollar sign ($). A *system task* is analogous to a subroutine. When the task is called, the simulator branches to a program that executes the functionality of the task. When the task has completed, the simulation returns to the next statement following the task call. A *system function* returns a value. When the function is called, a simulator executes the program associated with the function, which returns a value into the simulation. The return value becomes part of the statement that called the function. The return value of a function may be defined to be an integer, a vector (with a specific bit size) or a floating point number.

2.2.1 Built-in system tasks and system functions

The IEEE 1364 Verilog standard defines a number of system tasks and system functions that are built into all Verilog simulators. Examples of the standard system tasks are *$display*, *$stop*, and *$finish*. Some of the standard system functions are *$time* and *$random*. These system tasks and system functions are part of the Verilog language, and the syntax and usage of these routines are covered in Verilog language books.

The IEEE 1364 standard also allows simulation vendors to add proprietary system tasks and system functions which are specific to a simulator product. For example, the Cadence Verilog-XL simulator adds the command *$db_settrace*, which is a system task that turns on Verilog statement execution tracing to aid in design debugging.

2.2.2 User-defined system tasks and system functions

The Verilog PLI provides a means for Verilog users to add additional system tasks and system functions. When simulation encounters a user-defined system task or system function, the simulator branches to a user-supplied C function, executes that function, and then returns back to executing the Verilog source at the point where the system task/function was called.

User-defined system task and system function names must begin with a dollar sign ($), and may only use the characters that are legal in Verilog names, which are:

    a—z A—Z 1—9 _ $

Following are examples of legal user-defined system task and system function names:

$rand64 $cell_count $GetVector

2.2.3 Overriding built-in system tasks and system functions

A user-defined system task or system function can be given the same name as a system task/function that is built into the simulator. This allows a PLI application developer to override the operation of a built-in system task/function with new functionality. For example, the built-in system function *$random* will return a 32-bit signed random number, with the random sequence defined by the simulator. If the verification of a design requires a special random number sequence, a random number generator could be written in the C language and then associated with a user-defined system function with the same *$random* name. The simulator would then call the user PLI application instead of the built-in random number generator.

2.2.4 System tasks are used as procedural programming statements

System tasks are procedural programming statements in the Verilog HDL. This means a system task may only be called from a Verilog **initial** procedure, an **always** procedure, a Verilog HDL **task** or a Verilog HDL **function**. A system task may not be called outside of a procedure, such as from a continuous assignment statement.

The following example calls a user-defined system task from an `always` procedure:

```
always @(posedge clock)
   $read_test_vector("vectors.pat", input_vector);
```

2.2.5 System functions are used as expressions

System functions are expressions in the Verilog HDL. This means a system function may be called anywhere a logic value may be used. The return value of the system function is considered the result of the expression. System functions may be called from a Verilog **initial** procedure, an **always** procedure, a Verilog HDL **task**, a Verilog HDL **function**, an **assign** continuous assignment statement, or as an operand in a compound expression.

The following examples call a user-defined system function several different ways:

```
always @(posedge clock)
  if (chip_out !== $pow(base,exponent))
    ...

initial
  $monitor("output = %f", $pow(base,exponent));

assign temp = i + $pow(base,exponent);
```

2.2.6 System task/function arguments

A system task or system function may have any number of arguments, including none. The user-supplied PLI application can read the values of the arguments of a system task/function. If the argument is a Verilog reg or variable (integer, real and time), then the PLI application can also write values into the task/function argument.

System task/function arguments are typically numbered from 1 to N, with the left-most argument being argument number 1 (this numbering scheme is only a convention when working with VPI routines, but it is part of the PLI standard when using the older TF/ACC routines). For example:

```
module top (...);
  ...
  reg [15:0] in1;
  ...
  my_chip u1 (in1, out1);              argument 1 is a module
  ...                                  instance name
  initial
    $cell_count(u1);
  ...                                  argument 1 is a string
  always @(posedge clock)
    $read_vector_file("vectors.pat", in1);
  ...                                  argument 2 is a
endmodule                             reg variable
```

2.3 Instantiated Verilog designs

In a Verilog HDL module, a reference to another module is referred to as a *module instance*. In the following Verilog source code, module **top** contains two *instances* of module **bottom.**

```
module top;
  reg   [7:0] in1;
  wire  [7:0] out1;
  bottom b1 (in1[7:4], out1[7:4]);
  bottom b2 (in2[3:0], out2[3:0]);
endmodule

module bottom (in, out);
  ...
endmodule
```

The Verilog PLI requires an instantiated Verilog data structure. This means that a Verilog simulator has created a data structure which contains the complete hierarchy tree that has been represented by one module containing instances of other modules. There is no limit to the number of levels of hierarchy in the Verilog language.

The PLI does not work directly with Verilog HDL source code—The PLI is simply a procedural interface provided by a simulator so that PLI applications can access the data structure of a simulation.

Multiple instances of a system task or system function

The system task or system function which invokes a PLI application can be instantiated any number of times in the Verilog HDL source code. Consider the following Verilog HDL source code fragment:

```
module top;
  ...
  middle m1 (...);              //module instance
  middle m2 (...);              //module instance

  initial
    $my_app_1(in1, out1);       //system task instance

  always @(posedge clock)
    $my_app_1(in2, out2);       //system task instance

endmodule

module middle (...);
  ...
  bottom b1 (...);              //module instance
  bottom b2 (...);              //module instance

endmodule
```

```
module bottom (...);
  ...
  initial
    $my_app_2(in3, out3);      //system task instance

  always @(posedge clock)
    $my_app_2(in4, out4);      //system task instance

endmodule
```

In the preceding example, four different conditions are shown that can exist in a Verilog simulation.

• A single instance of a system task that is invoked one time

In module **top**, the first *instance* of *$my_app_1* is in a Verilog initial procedure, which will be called once at the beginning of a simulation.

• A single instance of a system task that is invoked many times

In module **top**, the second *instance* of *$my_app_1* is in a Verilog always procedure, which will be called every positive edge of the clock signal.

• Multiple instances of a system task that are each invoked one time

In module **bottom**, the first *instance* of *$my_app_2* is in a Verilog initial procedure, which will be called once at the beginning of a simulation.

• Multiple instances of a system task that are each invoked many times

In module **bottom**, the second *instance* of *$my_app_2* is in a Verilog always procedure, which will be called every positive edge of the clock signal.

Note that in the preceding Verilog source code example, module bottom is instantiated two times in module middle. The two instances of bottom, with two instances of *$my_app_2* inside bottom, creates four unique instances of *$my_app_2*. Module middle is instantiated twice in module top, resulting in eight unique instances of *$my_app_2*.

Thus, in the instantiated simulation data structure for the preceding Verilog source code, there are *two instances* of the *$my_app_1* system task, and *eight instances* of the *$my_app_2* system task. Each of these system task instances will have unique logic values for the inputs and outputs. One of the requirements of a PLI application is to allow for multiple unique instances of the application. This is an important consideration when an application allocates static variables or allocates system memory. Section 4.7 on page 122 in Chapter 4 discusses these issues in more depth.

2.4 How PLI applications work

The PLI standard allows a Verilog user to create a user-defined system task/function name and to associate one or more user-defined C routines with the system task/function name. When a Verilog simulator encounters the user-defined system task/function, it will execute the user-defined C routine associated with the name.

```
always @(posedge clock)
    result <= $pow(x,y);  ──────────▶ User-defined PLI routine
                                     ┌──────────────────────────┐
                                     │                          │
                                     │      calculate  x^y      │
                                     │                          │
                                     └──────────────────────────┘
```

2.4.1 The types of PLI routines

The VPI portion of PLI standard defines several types of PLI routines which can be associated with a system task or system function. The type of the routine determines **when** the simulator will execute the routine. Some types of routines are run-time routines, which are invoked during simulation, and some types are elaboration or linking time routines, which are invoked prior to simulation. The types of PLI routines are:

- *calltf routines*
- *compiletf routines*
- *sizetf routines*
- *simulation callback routines*

The purpose of these routines is defined in the following sections of this chapter.

2.4.2 Associating routine types with system task/functions

A system task/function can have multiple PLI routines associated with it, as long as each routine is a different type. A *$read_vector_file* application, for example, might have three routines associated with it—one which is called at elaboration/linking time to perform syntax checking, one which is called when simulation first starts running to open the test vector file, and one which is called during simulation at every positive edge of clock to read vectors from the file.

Up to 3 different C routines may be associated with a system task/function in the VPI portion of the PLI standard: the *calltf routine*, *compiletf routine* and *sizetf routine*. The *simulation callback routine* is not directly associated with a system task/function name. Instead, the *simulation callback routine* is called for various types of activity that can occur during a simulation, such as logic value changes and start or completion of simulation.

2.4.3 C versus C++

Most Verilog simulators expect that PLI applications are written in ANSI C. The PLI libraries are compliant with the ANSI C standard, but also include the appropriate prototypes and other information required for C++. This allows PLI applications to be developed in either C or C++. However, many Verilog simulators were written in the C language, and may or may not support linking C++ applications to the simulator. PLI application developers who wish to work with C++ should first check the limitations of the simulator product to which the applications will be linked. For maximum portability, PLI applications should be written using ANSI C.

2.4.4 Special PLI data types for portability

The VPI library defines several fixed-width data types, which are used by the routines in the VPI library. This is because the ANSI C standard only guarantees the minimum width of C data types, which allows the width to be different on different computer architectures and operating systems. For example, a C **int** could be 16 bits wide, 32 bits wide or 64 bits wide, depending on what computer the application was compiled for. To provide portability across all operating systems, the VPI library defines special data types, such as **PLI_INT32**, which is guaranteed to be a 32-bit signed integer. These data type definitions are made in the vpi_user.h header file, as follows:

Table 2-1: PLI fixed-width data types

PLI_INT32	32-bit signed integer
PLI_UINT32	32-bit unsigned integer
PLI_INT16	16-bit signed integer
PLI_UINT16	16-bit unsigned integer
PLI_BYTE8	8-bit signed integer
PLI_UBYTE8	8-bit unsigned integer

A PLI application should use the special PLI data types so that the application will work correctly on both 32-bit and 64-bit operating systems.

TIP

The vpi_user.h header file can be modified by simulator vendors to define the fixed width data types for a specific computer platform, operating system or C compiler. The default definitions, as defined in the IEEE 1364-2001 standard, are:

```
typedef int            PLI_INT32;
typedef unsigned int   PLI_UINT32;
typedef short          PLI_INT16;
typedef unsigned short PLI_UINT16;
typedef char           PLI_BYTE8;
typedef unsigned char  PLI_UBYTE8;
```

 The fixed width PLI data types were added as part of the IEEE 1364-2001 standard. Prior to that, the PLI libraries used the regular C data types, such as char, int and long, with the non-portable assumption that these data types would be the same size on all operating systems. For example, the IEEE 1364-1995 standard assumed that an int would always be 32 bits, something the ANSI C standard does not guarantee. Existing PLI applications that use the C data types instead of the new PLI data types will still compile and work correctly with the default definitions in the standard vpi_user.h header file. However, these older applications may need to be updated to the PLI data types in order to port correctly to new 64-bit operating systems.

2.5 calltf routines

The ***calltf routine*** is executed when simulation is running. For the *$pow* example that follows, at every positive edge of clock the *calltf routine* associated with *$pow* will be executed by the simulator. The routine is user-defined, and, for this example, will calculate a 32-bit number representing x to the power of y.

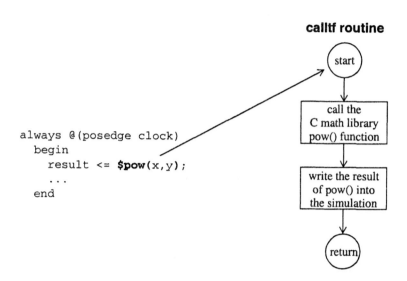

2.6 compiletf routines

The **compiletf routine** is called by the simulator before simulation starts running—in other words, before simulation time 0. The routine may be called by the simulator's compiler or elaborator, when the simulator loads and prepares its simulation data structure. The purpose of *compiletf routine* is to verify that a system task/function is being used correctly (e.g.: to check that the call to the PLI application has the correct number of arguments, and that the arguments are the correct Verilog data types). For example, if the *$pow* system function were passed a module name as its second argument instead of a value, then the C math library would probably get an error when trying to use the name as an exponent value. Bad data can result in errors ranging from incorrect results to program crashes. Therefore, it is important to verify that each usage of a system task/function has valid arguments.

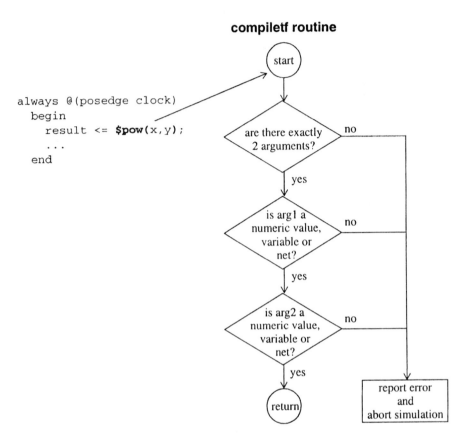

compiletf routine

```
always @(posedge clock)
   begin
      result <= $pow(x,y);
      ...
   end
```

TIP Use a *compiletf routine* to improve the performance of PLI programs. Since this routine is only called one time prior to simulation time 0, the code to check the correctness of arguments is only executed once. The *calltf routine*, which may be invoked millions of times during a simulation, does not need to repeat the syntax checking, and therefore can execute more efficiently.

NOTE The *compiletf routine* will be called one time for each instance of a user-defined system task/function. If a design used the *$pow* user-defined system function in three different places, the *compiletf routine* for *$pow* would be called three times. If a simulator allows system tasks and system functions to be invoked from the simulator's debug command line, then the *compiletf routine* will be called prior to execution of the *calltf routine* for each interactive usage.

Limitations on compiletf callbacks

The intent of the *compiletf routine* is to verify the correctness of the arguments of an instance of a system task/function. The IEEE 1364 standard does not impose explicit restrictions on what VPI routines will, or will not work in a *compiletf routine*. There are, however, considerations which should be observed. Since the *compiletf routine* is called prior to simulation time 0, Verilog variables have not yet been assigned any values, and nets which are assigned values as soon as simulation starts may not yet reflect those values.

 The compiletf routine should only be used for syntax checking! Do not use this routine to perform run-time duties such as allocating memory or opening files. The PLI standard does ensure that activity performed by *compiletf routines* will remain in effect at simulation time. A *simulation callback routine* should be used instead of the *compiletf routine* to perform run-time work at the very beginning of a simulation.

To ensure that a PLI application will be portable, only the following activities should be performed in a *compiletf routine*:

- Accessing the arguments of an instance of a system task/function. Handles for the arguments can be obtained, and the properties of the arguments can be accessed to verify correctness. The values of literal numbers can be read, but variables and nets will not have been initialized. Note that Verilog `parameter` constants can be redefined for each instance of a Verilog module. The IEEE 1364 standard does not state whether the *compiletf routine* will be called before or after parameter redefinitions have occurred. Reading parameter constant values from a *compiletf routine* may not yield the same results on every simulator.

- Using VPI routines such as `vpi_printf()`, which do not access objects in simulation. Attempting to obtain handles and access properties beyond the system task arguments may not work the same in all simulators. Attempting to write logic or delay values onto objects may result in PLI application errors.

- Registering *simulation callback routines* using `vpi_register_cb()` for the end of elaboration/linking or the start of simulation. Using *simulation callback routines* is presented in Chapter 6).

2.7 sizetf routines

The *sizetf routine* is only used with system functions that are registered with the sys-functype as **vpiSizedFunc** or **vpiSizedSignedFunc**. These types indicate that the system function returns scalar or vector values (system functions can also return integer and real number values). Because these function types return a user-specified number of bits, the simulator compiler or elaborator may need to know how many bits to expect from the return, in order to correctly compile the statement from which the system function is called. A *sizetf routine* is called one time, before simulation time 0.

The *sizetf routine* returns to the simulator how many bits wide the return value of system function will be.

 Even if a system function is used multiple times in the Verilog source code, the *sizetf routine* is only called once. The return value from the first call is used to determine the return size for all instances of the system function.

In the *$pow* example, the *calltf routine* will return a 32-bit value. Therefore, the *sizetf routine* associated with *$pow* needs to return a value of 32 to the simulator.

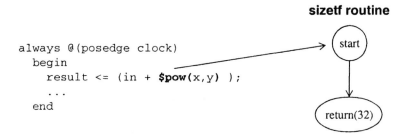

In this example, the return value from *$pow* is added to the signal, in. The Verilog standard requires that the width of an addition operation be the width of all operands in the expression containing the operation. Therefore, the compiler or elaborator must know the width of the return from *$pow* in order to build the simulation data structure for the addition operation.

Limitations on sizetf callbacks

The intent of the *sizetf routine* is to notify the simulator compiler or elaborator of the return size for system functions. The simulator may invoke the *sizetf routine* very early in the elaboration/linking phase of a Verilog design, and, at this early stage, the Verilog hierarchy may not have been generated. In addition, the *sizetf routine* is only called one time for a system function name, and the return size applied to all instances of the system function. For these reasons, only standard C language statements and functions should be used in a *sizetf routine*. An error may result if any VPI routines are called from a *sizetf routine*. Any memory or static variables allocated by a *sizetf routine* may not remain in effect for when simulation starts running.

2.8 VPI Simulation callback routines

The VPI provides a means for PLI applications to be called for specific simulation events. The VPI portion of the PLI standard refers to these types of routines as *simulation callback routines*. The IEEE 1364 standard defines a number of simulation events for which a PLI routine can be called. The full list of callback reasons is presented in Chapter 6. Some examples of simulation related callbacks are:

- The beginning of Verilog simulation (just before the start of simulation time 0).

- Entering debug mode (such as when the *$stop* built-in system task is executed).

- End of simulation (such as when the *$finish* built-in system task is executed).

- Change of value of a signal.

- Execution of a Verilog procedural statement.

A common usage of *simulation callback routines* is to perform tasks at the very beginning and the very end of a simulation. For example, a PLI application to read test vectors might need to open a test vector file at the start of simulation, and close the file at the end of simulation.

```
always @(posedge clock)
    $read_test_vector("vectors.pat", in1);
```

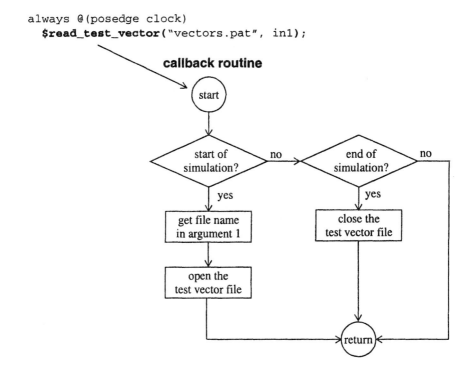

2.9 PLI routine inputs and outputs

All types of PLI routines are C functions, and a Verilog simulator will utilize the inputs and return values of the functions when the simulator calls the functions. The PLI standard defines what the inputs should be, and how the simulator should use the return value.

The input that is passed to the *compiletf routine*, *calltf routine* and *sizetf routine* is a pointer, which points to the user_data value that was specified when the system task/ function was registered. Refer to section 2.12 on page 51 for more details on defining and using the user_data value.

The input that is passed to the *simulation callback routine* is a pointer to a structure containing information about the simulation callback. Refer to Chapter 6 for more details on using *simulation callback routines*.

The *compiletf routine*, *calltf routine*, *sizetf routine* and *simulation callback routine* are expected to be integer functions in the PLI standard. However, the only return value which is used by the simulator is the return of the *sizetf routine*, which represents the bit-width of the system function return value. The return value from the *compiletf routine*, *calltf routine* and *simulation callback routine* are not used, and are ignored by the simulator.

2.10 A complete system function example — $pow

The following example illustrates all the parts of a complete system function, with the user-defined name of **$pow**.

- The system function will return a 32-bit value.
- The system function requires two arguments, a base value and an exponent value. Both arguments must be numeric integer values.

To implement the *$pow* functionality, four user-defined PLI routines are used:

- A *sizetf routine* to establish the return size of *$pow*.
- A VPI *compiletf routine* to verify that the *$pow* arguments are valid values.
- A *calltf routine* to calculate the base to the power of the exponent each time *$pow* is executed by the simulator.
- A VPI *simulation callback routine* to print a message when simulation firsts starts running (immediately prior to simulation time 0).

The *$pow* system function does not really need the VPI *simulation callback routine*. This example PLI application includes the routine, in order to illustrate its usage.

TIP The *compiletf routine, sizetf routine, calltf routine*, and *simulation callback routine* are C functions. These functions may be located in separate files, or they can all be in the same file. Typically, smaller PLI applications place all the routines in a single file, while larger, more complex applications might break them into multiple files.

TIP The *compiletf routine, sizetf routine, calltf routine*, and *simulation callback routine* names can be any legal C function name. However, a typical Verilog HDL design may include several PLI applications. Using unique function name conventions can avoid possible name conflicts when multiple PLI applications are linked together.

For this chapter, the *$pow* application has been simplified to only work with integer numbers. In practice, this application would probably accept both integer and floating point numbers for inputs, and return a floating point number for the result. Example 5-8 on page 179 in Chapter 5 presents another version of *$pow* that works with floating point numbers.

CD The source code for this example is on the CD accompanying this book.

- Application source file: `Chapter.02/pow_vpi.c`
- Verilog test bench: `Chapter.02/pow_test.v`
- Expected results log: `Chapter.02/pow_test.log`

Example 2-1: *$pow* — a system function using VPI routines

```
#include <stdlib.h>    /* ANSI C standard library */
#include <stdio.h>     /* ANSI C standard input/output library */
#include <stdarg.h>    /* ANSI C standard arguments library */
#include "vpi_user.h"  /* IEEE 1364 PLI VPI routine library  */

/* prototypes of PLI application routine names */
PLI_INT32 PLIbook_PowSizetf(PLI_BYTE8 *user_data),
          PLIbook_PowCalltf(PLI_BYTE8 *user_data),
          PLIbook_PowCompiletf(PLI_BYTE8 *user_data),
          PLIbook_PowStartOfSim(s_cb_data *callback_data);

/******************************************************************
 * $pow Registration Data
 * (add this function name to the vlog_startup_routines array)
 ******************************************************************/
void PLIbook_pow_register()
{
  s_vpi_systf_data tf_data;
  s_cb_data        cb_data_s;
  vpiHandle        callback_handle;
```

```
  tf_data.type          = vpiSysFunc;
  tf_data.sysfunctype   = vpiSysFuncSized;
  tf_data.tfname        = "$pow";
  tf_data.calltf        = PLIbook_PowCalltf;
  tf_data.compiletf     = PLIbook_PowCompiletf;
  tf_data.sizetf        = PLIbook_PowSizetf;
  tf_data.user_data     = NULL;
  vpi_register_systf(&tf_data);

  cb_data_s.reason      = cbStartOfSimulation;
  cb_data_s.cb_rtn      = PLIbook_PowStartOfSim;
  cb_data_s.obj         = NULL;
  cb_data_s.time        = NULL;
  cb_data_s.value       = NULL;
  cb_data_s.user_data   = NULL;
  callback_handle = vpi_register_cb(&cb_data_s);
  vpi_free_object(callback_handle); /* don't need callback handle */
}

/*********************************************************************
 * Sizetf application
 *********************************************************************/
PLI_INT32 PLIbook_PowSizetf(PLI_BYTE8 *user_data)
{
  return(32);    /* $pow returns 32-bit values */
}

/*********************************************************************
 * compiletf application to verify valid systf args.
 *********************************************************************/
PLI_INT32 PLIbook_PowCompiletf(PLI_BYTE8 *user_data)
{
  vpiHandle systf_handle, arg_itr, arg_handle;
  PLI_INT32 tfarg_type;
  int       err_flag = 0;

  do { /* group all tests, so can break out of group on error */
    systf_handle = vpi_handle(vpiSysTfCall, NULL);
    arg_itr = vpi_iterate(vpiArgument, systf_handle);
    if (arg_itr == NULL) {
      vpi_printf("ERROR: $pow requires 2 arguments; has none\n");
      err_flag = 1;
      break;
    }
    arg_handle = vpi_scan(arg_itr);
    tfarg_type = vpi_get(vpiType, arg_handle);
    if ( (tfarg_type != vpiReg) &&
         (tfarg_type != vpiIntegerVar) &&
         (tfarg_type != vpiConstant)   ) {
      vpi_printf("ERROR: $pow arg1 must be number, variable or net\n");
      err_flag = 1;
      break;
    }
```

```
    arg_handle = vpi_scan(arg_itr);
    if (arg_handle == NULL) {
      vpi_printf("ERROR: $pow requires 2nd argument\n");
      err_flag = 1;
      break;
    }
    tfarg_type = vpi_get(vpiType, arg_handle);
    if ( (tfarg_type != vpiReg) &&
         (tfarg_type != vpiIntegerVar) &&
         (tfarg_type != vpiConstant)   ) {
      vpi_printf("ERROR: $pow arg2 must be number, variable or net\n");
      err_flag = 1;
      break;
    }
    if (vpi_scan(arg_itr) != NULL) {
      vpi_printf("ERROR: $pow requires 2 arguments; has too many\n");
      vpi_free_object(arg_itr);
      err_flag = 1;
      break;
    }
  } while (0 == 1); /* end of test group; only executed once */

  if (err_flag) {
    vpi_control(vpiFinish, 1);  /* abort simulation */
  }
  return(0);
}

/******************************************************************
 * calltf to calculate base to power of exponent and return result.
 ******************************************************************/
#include <math.h>
PLI_INT32 PLIbook_PowCalltf(PLI_BYTE8 *user_data)
{
  s_vpi_value value_s;
  vpiHandle   systf_handle, arg_itr, arg_handle;
  PLI_INT32   base, exp;
  double      result;

  systf_handle = vpi_handle(vpiSysTfCall, NULL);
  arg_itr = vpi_iterate(vpiArgument, systf_handle);
  if (arg_itr == NULL) {
    vpi_printf("ERROR: $pow failed to obtain systf arg handles\n");
    return(0);
  }

  /* read base from systf arg 1 (compiletf has already verified) */
  arg_handle = vpi_scan(arg_itr);
  value_s.format = vpiIntVal;
  vpi_get_value(arg_handle, &value_s);
  base = value_s.value.integer;
```

```
  /* read exponent from systf arg 2 (compiletf has already verified) */
  arg_handle = vpi_scan(arg_itr);
  vpi_free_object(arg_itr); /* not calling scan until returns null */
  vpi_get_value(arg_handle, &value_s);
  exp = value_s.value.integer;

  /* calculate result of base to power of exponent */
  result = pow( (double)base, (double)exp );

  /* write result to simulation as return value $pow */
  value_s.value.integer = (PLI_INT32)result;
  vpi_put_value(systf_handle, &value_s, NULL, vpiNoDelay);
  return(0);
}

/*****************************************************************
 * Start-of-simulation application
 *****************************************************************/
PLI_INT32 PLIbook_PowStartOfSim(s_cb_data *callback_data)
{
  vpi_printf("\n$pow PLI application is being used.\n\n");
  return(0);
}
```

Simulation results for $pow

Following is the output from running a simulating which uses the *$pow* application.

Example 2-2: *$pow* — sample simulation results

```
$pow PLI application is being used.

$pow(2,3) returns        8
$pow(a,b) returns        1  (a=1 b=0)
```

2.11 Interfacing PLI applications to Verilog simulators

As the previous sections in this chapter have shown, a PLI application may comprise:

- A system task or system function name
- A *calltf routine*

- A *compiletf routine*
- A *sizetf routine*
- Any number of *simulation callback routines*

After these items have been defined, the PLI application must be interfaced with a Verilog simulator. The PLI standard provides an ***interface mechanism*** to make the associations between the system task/function name and the application routines. There are two generations of the interface mechanism, one that was created for the older TF and ACC libraries, and a newer mechanism created for the VPI library. This chapter presents the VPI interface mechanism. The older TF/ACC interface mechanism is presented in Part Two of this book, in Chapter 9.

The VPI interface mechanism involves three basic steps:

1. Create a register function, which associates the system task/function name with the application routines.

2. Notify the Verilog simulator about the registration function.

3. Link the applications into a Verilog simulator, so the simulator can call the appropriate routine when the system task/function name is encountered.

 The PLI standard does not provide any guidelines on how PLI applications should be linked into a Verilog simulator. There are many different C compilers and operating systems available to Verilog users, and each compiler and operating system has unique methods for compiling and linking programs. Since the commands for compiling and linking are not part of the IEEE 1364 standard, this process is not covered in this chapter. Appendix A presents the steps involved to compile and link PLI applications for several Verilog simulators.

The VPI Interface Mechanism is defined as part of the IEEE 1364 standard, providing a consistent method for all Verilog simulators to use. The VPI interface is used to specify:

1. A system task/function ***name***.

2. The application ***type***, which is a *task, sized function, integer function, time function* or *real function*.

3. Pointers to the C functions for a *calltf routine, compiletf routine* and *sizetf routine*, if the routines exist (it is not required—and often not necessary—to provide each class of routine).

4. A ***user_data*** pointer value, which the simulator will pass to the *calltf routine, compiletf routine* and *sizetf routine* each time they are called.

The process of specifying the PLI application information is referred to as *registering* the application. To register a PLI application, the information about the application is specified in an **s_vpi_systf_data** structure. This structure is defined as part of the VPI standard, in the PLI *vpi_user.h* file. The definition is:

```
typedef struct t_vpi_systf_data {
  PLI_INT32   type;                  /* vpiSysTask, vpiSysFunc */
  PLI_INT32   sysfunctype;           /* vpiIntFunc, vpiRealFunc,
                                        vpiTimeFunc, vpiSizedFunc,
                                        vpiSizedSignedFunc */
  PLI_BYTE8  *tfname;                /* quoted task/function name */
  PLI_INT32 (*calltf)(PLI_BYTE8 *);   /* name of C func */
  PLI_INT32 (*compiletf)(PLI_BYTE8 *); /* name of C func */
  PLI_INT32 (*sizetf)(PLI_BYTE8 *);   /* name of C func */
  PLI_BYTE8  *user_data;             /* returned with callback */
} s_vpi_systf_data, *p_vpi_systf_data;
```

Table 48 explains the fields of the s_vpi_systf_data structure:

Table 2-2: VPI interface mechanism s_vpi_systf_data structure fields

s_vpi_systf_data	Definition
type	Defines the type of a PLI application as being either a system task or a system function. This field must be set to the C constant: **vpiSysTask** or **vpiSysFunc**.
sysfunctype	Defines the return type of a system function. This field is only used if the type field is vpiSysFunc, in which case the sysfunctype must be set to the C constant: **vpiIntFunc**, **vpiRealFunc**, **vpiTimeFunc**, **vpiSizedFunc** or **vpiSizedSignedFunc**. Set to 0 if unused.
tfname	Specifies the name of the system task/function as a string.
calltf	Specifies a pointer to the C function that will be called by a simulator for the application's *calltf routine*. Set to NULL if there is no calltf routine.
compiletf	Specifies a pointer to the C function that will be called by a simulator for the application's *compiletf routine*. Set to NULL if there is no compiletf routine.
sizetf	Specifies a pointer to the C function that will be called by a simulator for the application's *sizetf routine*. Set to NULL if there is no sizetf routine.
user_data	Specifies a pointer to application-allocated information—the value of the pointer will be passed to the PLI application routines each time a routine is called. Set to NULL if there is no user_data value.

The 1364-1995 Verilog standard used different constants to represent system function types:

- **vpiIntFunc** was **vpiSysFuncInt**
- **vpiTimeFunc** was **vpiSysFuncTime**
- **vpiRealFunc** was **vpiSysFuncReal**
- **vpiSizedFunc** was **vpiSysFuncSized**

The constant names from the 1364-1995 standard are aliased to the new constant names, to provide backward compatibility.

The **vpiSizedSignedFunc** is new in the IEEE 1364-2001 Verilog standard. This configuration allows a system function to return a signed value of any vector size. This enhancement to the VPI matches the signed arithmetic enhancements in the Verilog HDL for 1364-2001, which allow a net or reg vector of any width to be declared as signed.

2.11.1 The steps required to register a PLI application

To register a system task or system function using the VPI interface mechanism involves the following steps:

1. Create a C function to register the system task/function. The C function name is user-defined and can be any legal C name.

2. Allocate an s_vpi_systf_data C structure.

3. Fill in the fields of the structure with the information about the system task or system function.

4. Register the system task/function by calling the VPI routine **vpi_register_systf()**.

5. Notify the simulator about the name of the register function.

The following example registers the *$pow* application presented in this chapter.

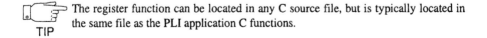 The register function can be located in any C source file, but is typically located in the same file as the PLI application C functions.

TIP

Example 2-3: *$pow* — a VPI register function for a system function

```
/* prototypes of PLI application routine names */
PLI_INT32 PLIbook_PowSizetf(),
          PLIbook_PowCalltf(),
          PLIbook_PowCompiletf(),

void PLIbook_pow_register()
{
  s_vpi_systf_data tf_data;
  s_cb_data    cb_data_s;
  s_vpi_time   time_s;

  tf_data.type        = vpiSysFunc;
  tf_data.sysfunctype = vpiSizedFunc;
  tf_data.tfname      = "$pow";
  tf_data.calltf      = PLIbook_PowCalltf;
  tf_data.compiletf   = PLIbook_PowCompiletf;
  tf_data.sizetf      = PLIbook_PowSizetf;
  tf_data.user_data   = NULL;
  vpi_register_systf(&tf_data);
}
```

2.11.2 Notifying Verilog simulators about PLI applications

Once the register function has been defined, a Verilog simulator must be notified of the name of the register function, so that the simulator can call the functions and register the PLI applications. The IEEE standard provides some leeway on how simulators should be notified. Some common methods are:

- Invocation options which specify the names of the register functions.

- Invocation options which specify a file containing the names of the register functions.

- A special array, called ***vlog_startup_routines***, which contains the names of the register functions.

An example of using a *vlog_startup_routines* array is listed below, with entries for two register functions.

Example 2-4: a *vlog_startup_routines* array with 2 register functions

```
/* prototypes of the PLI application routines */
extern void PLIbook_pow_register(), PLIbook_ShowVal_register();

void (*vlog_startup_routines[])() =
```

```
{
    /*** add user entries here ***/
  PLIbook_pow_register,
  PLIbook_ShowVal_register,
  0 /*** final entry must be 0 ***/
};
```

 Do not place the vlog_startup_routines array in the same C source file as the PLI
TIP ***application!*** In a typical design environment, PLI applications will come from
several sources, such as internally developed applications and 3rd party applications.
If the *vlog_startup_routines* array and a PLI application and are in the same file, then
the source code for the application must be available whenever another PLI
application needs to be added to the start-up array.

Limitations on register functions

The register functions are executed by the simulator before simulation time 0, possibly before the creation of the simulation data structure is complete. This limits what operations can be performed within register functions. The VPI routines which can be used in a register function are:

- **vpi_register_systf()**
- **vpi_register_cb()** (this routine is discussed in Chapter 6, section 6.6.3).
- VPI routines which do not access objects in simulation, such as **vpi_printf()**. It is an error to attempt obtain a handle for an object or any object properties.

2.12 Using the VPI user_data field

When a system task or system function is registered, a **user_data** value can be specified. The user_data is a pointer, which can point to a block of application-allocated data. This user_data value is passed to the *calltf routine, compiletf routine* and *sizetf routine* as a C function input each time the simulator calls one of these routines.

In the following example, two different system tasks, *$get_vector_bin* and *$get_vector_hex*, are associated with the same *calltf routine*. The registration function for these system tasks is:

```
void PLIbook_test_user_data_register()
{
  s_vpi_systf_data tf_data;
  char *id1 = malloc(sizeof(int)); /* allocate storage */
  char *id2 = malloc(sizeof(int)); /* allocate storage */
  *id1 = 1;
  *id2 = 2;

  tf_data.type        = vpiSysFunc;
  tf_data.sysfunctype = vpiSysFuncSized;
  tf_data.tfname      = "$get_vector_bin";
  tf_data.calltf      = PLIbook_GetVectorCalltf;
  tf_data.compiletf   = NULL;
  tf_data.sizetf      = NULL;
  tf_data.user_data   = (PLI_BYTE8 *)id1;
  vpi_register_systf(&tf_data);

  tf_data.type        = vpiSysFunc;
  tf_data.sysfunctype = vpiSysFuncSized;
  tf_data.tfname      = "$get_vector_hex";
  tf_data.calltf      = PLIbook_GetVectorCalltf;
  tf_data.compiletf   = NULL;
  tf_data.sizetf      = NULL;
  tf_data.user_data   = (PLI_BYTE8 *)id2;
  vpi_register_systf(&tf_data);
}
```

Both system tasks invoke the same *calltf routine*, but the user_data value for the two system task names is a pointer to a different block of application-allocated storage. Therefore, the *calltf routine* can check the user_data value to determine which system task name was used to call the routine.

The user_data value is passed to the callback routine as a C function input of type PLI_BYTE8 *. The following example illustrates reading the user_data value for the two system tasks registered in the previous example:

```
PLI_INT32 ReadVectorCalltf(PLI_BYTE8 *user_data)
{
  vpi_printf("\nIn calltf,\tuser_data = %d\n", *user_data);
  if (*user_data == 1) {
    vpi_printf("calltf was invoked by $get_vector_bin()\n");
    /* read test vectors as binary values */
  }
  else if (*user_data == 2) {
    vpi_printf("calltf was invoked by $get_vector_hex()\n");
    /* read test vectors as hex values */
  }
  return(0);
}
```

Application-allocated storage in a register function is not instance specific. The registration functions are only executed once per simulation. Therefore, the user_data value listed in the register routine will be common to all instances of the registered system task. If, for example, the *$get_vector_hex* system task were used in two places in the Verilog design, each task will be passed the same user_data value. Chapter 4, section 4.7 on page 122 discusses how to store data that is unique to each instance of a system task/function.

2.13 Compiling and linking PLI applications

After the C source files for the PLI applications have been defined, they must be compiled and linked into a Verilog simulator. This allows the simulator to call the appropriate PLI routine when the PLI application system task or system function is encountered by the simulator. The compiling and linking process is not part of the IEEE standard for the PLI. This process is defined by the simulator vendor, and is specific to both the simulator and the operating system on which the application is compiled. Appendix A presents the instructions for compiling and linking PLI applications for several of the Verilog simulators that were available at the time this book was written.

2.14 Summary

This chapter has shown the major parts of PLI applications. For PLI applications which are developed using the VPI library, the main parts of an application are: the name of a system task or system function, a *compiletf routine*, a *calltf routine*, a *sizetf routine*, and *simulation callback routines*. A system function called *$pow*, which calculates x^y and returns a 32-bit result, was used in this chapter to illustrate the major parts of a PLI application. The VPI interface mechanism is used to interface a PLI application to a Verilog simulator. The interface mechanism involves registering a PLI application using an application-defined register function, and notifying the simulator about the register function by editing a *vlog_startup_routines* array provided by the simulator. The next four chapters present how to use each of the routines in the VPI library.

CHAPTER 3 *How to Use the VPI Routines*

*T*his chapter introduces the VPI portion of the PLI standard, and shows how to use the VPI routines to access information within a simulation data structure. Two complete PLI applications, *$show_all_nets* and *$show_all_signals*, will be created in this chapter to illustrate how the VPI routines work. The remaining chapters in this part of the book then build on the principles presented in this chapter by explaining the VPI library in much more detail.

The concepts presented in this chapter are:

- An overview of how VPI routines work

- Advantages of the VPI library

- Creating a complete PLI application using the VPI library

- Obtaining handles to Verilog HDL objects

- Accessing properties of Verilog HDL objects

- Reading values of Verilog HDL objects

3.1 Specification of $show_all_nets and $show_all_signals

To show how the VPI routines are used, two PLI applications will be created. The first example presented is an application called *$show_all_nets*. The usage of this application is:

```
$show_all_nets(<module_instance_name>);
```

This PLI application will:

1. Access the first argument of the system task, which is the name of a module instance.

2. Print the hierarchical path and name of that module.

3. Print the current simulation time.

4. Search for all net data type signals in the module.

5. Print the current logic value of each net.

This chapter will first illustrate a *compiletf routine* for *$show_all_nets*, which verifies that the argument provided as an input is a valid module instance name. Then a *calltf routine* will be created to perform the functionality of the system task.

The second example is a PLI application called ***$show_all_signals***. This application prints the current value of all net, reg and variable data types in a module. The usage of this application is:

```
$show_all_signals(<module_instance_name>);
```

To illustrate some additional ways to use the VPI routines, two enhancements to the *$show_all_signals* example will be presented. These are:

*   Use no argument or a null argument to *$show_all_signals* to represent the module instance containing the *$show_all_signals* system task.

*   Allow multiple arguments to *$show_all_signals*, so the values of signals in several modules can be printed with one call to *$show_all_signals*.

3.2 The VPI routine library

"VPI" stands for "Verilog Procedural Interface". The VPI routines are the third of the three generations of the Verilog PLI routines (the TF routines were the first generation, and the ACC routines were the second). The primary purpose of the VPI routines is to provide a PLI application access to the internal data structures of a simulation. The VPI interface provides a consistent layer between a user's PLI application and the underlying data structures of a simulation. By using this procedural interface layer, the PLI application does not need to know the specifics about how the simulator stores its data, and the same application will work with many different simulators.

The VPI routines treat Verilog HDL constructs as *objects*, and several of the VPI routines provide ways to locate specific objects within a simulation data structure. Other VPI routines can then read and modify information about the object.

The VPI library can be divided into five basic groups of routines:

- A *handle* routine obtains a handle for one specific Verilog HDL object.

- An *iterate* routine and a *scan* routine obtain handles for all of a specific type of Verilog object.

- *get* routines access information about an object.

- *set* routines modify information about an object.

- A few *miscellaneous* routines perform a variety of operations.

The library of VPI routines is defined in a C header file called **vpi_user.h**, which is part of the IEEE 1364 standard. This header file also defines a number of C constants and C structures used by the VPI routines. All PLI applications which use VPI routines must include the vpi_user.h file.

The VPI library is designed to work with the standard ANSI C libraries, such as **stdlib.h**, **stdio.h** and **stdarg.h**. An example of including the header files for these libraries is:

```
#include <stdlib.h>    /* ANSI C standard library */
#include <stdio.h>     /* ANSI C standard I/O library */
#include <stdarg.h>    /* ANSI C standard args library */
#include "vpi_user.h"  /* IEEE 1364 PLI VPI library */
```

3.3 Advantages of the VPI library

The VPI routines provide direct access to virtually everything within a Verilog simulation data structure. This direct and all-inclusive access allows a PLI application to fully analyze and interact with a Verilog simulation in any way desired.

The VPI library is a concise set of 37 routines. These routines have a very simple and consistent syntax, and are easy to learn and use. These 37 VPI routines were designed to completely replace the older TF and ACC generations of the PLI standard, which together total more than 200 routines. The older PLI generations had evolved over many years without the guidance of a standards body such as the IEEE. The TF and ACC libraries are bloated, full of redundancies, and very inconsistent in their syntax and semantics. The concise and consistent VPI library is much easier to work with.

As much as possible, the VPI library has been designed that so it can be very efficiently implemented by simulators. Potentially, this can improve the run-time performance of PLI applications compared to applications written with the older ACC generation of the standard.

 The VPI library was designed to replace the TF and ACC libraries with a more concise, more robust, and more versatile procedural interface. The IEEE 1364 standard includes the TF and ACC libraries, in order to provide backward compatibility for older PLI applications. The official policy of the IEEE 1364 standards committee is that, as enhancements are added to the Verilog language, only the VPI library will be expanded to support the new features. The TF and ACC libraries will not be enhanced in future versions of the IEEE 1364 standard.

Some disadvantages of VPI routines

The TF and ACC libraries are supported by virtually every major Verilog simulator, past and present. At the time this book was written, the VPI library was only supported by a few of the leading simulators. The more widespread support of TF and ACC routines makes a PLI application portable to many more simulators and engineering environments. This disadvantage of the VPI routines will disappear as more simulators companies implement the full IEEE PLI standard.

The TF library only provides very limited access to the internals of a simulation data structure. Some simulators, such as the Synopsys VCS simulator and the Cadence NC-Verilog simulator, take advantage of this limited access to highly optimize their data structures when only the TF library is used. This restricted access and additional optimization may provide the best run-time performance of all the PLI libraries, but the restricted access also limits the capabilities of a PLI application.

Many new users of the PLI find that it is sometimes much easier and faster to develop PLI applications using the TF and ACC libraries as compared to the VPI library. The TF and ACC libraries are much larger, and often have built-in routines which take care of a lot of the work that a PLI application needs to accomplish. In the smaller VPI library, the PLI application developer must code much of the corresponding functionality by hand. However, the larger TF and ACC libraries allow, and even encourage, PLI application developers to write poorly structured C code, which can impact simulation performance and becomes difficult to maintain. The VPI library requires a more disciplined, efficient coding style.

3.4 Verilog HDL objects

The VPI routines treat Verilog HDL constructs as **objects**, and several of the VPI routines provide ways to locate any specific object or type of object within a simulation data structure. Other VPI routines can then read and modify information about each object. The simple Verilog HDL example which follows has several objects which can be accessed by the library of VPI routines.

```
module test;
  reg   [1:0] test_in;
  wire  [1:0] test_out;
  buf2 u1 (test_in, test_out);
  initial
    begin
      test_in = 3;
      #50 $display("in=%d, out=5d", test_in, test_out);
    end
endmodule

module buf2 (in, out);
  input   [1:0] in;
  output  [1:0] out;
  wire    [1:0] in, out;
  buf #5 n0 (out[0], in[0]);
  buf #7 n1 (out[1], in[1]);
endmodule
```

In this Verilog HDL example, the objects that a PLI application can access include:

- A top-level module, with the definition name *"test"*. Within this module are:

  - A `reg` signal, with a vector size of 2 and the name *"test_in"*. The signal will have a logic value which can be read and modified by the PLI application.

  - A `wire` net, with a vector size of 2 and the name *"test_out"*. The net reflects a resolved logic value which can be read by the PLI application.

  - A module instance, with the definition name *"buf2"* and the instance name *"u1"*. Within this module are:

    - Two ports, with the names *"in"* and *"out"*. Each port has a vector size and direction.

    - Two `wire` nets, with vector sizes and names. The nets reflect a resolved logic value which can be read by the PLI application.

    - Two primitive instances, with the definition name *"buf"* and the instance names *"n0"* and *"n1"*. Each primitive has a delay value which can be read and modified by the PLI.

- Terminals on each primitive instance, with *"out[0]"* (a bit-select of *"out"*) and *"in[0]"* connected to one instance, and *"out[1]"* and *"in[1]"* connected to the other instance.

- An initial procedure. Within this procedure are:

  - A begin—end statement group.

  - An assignment statement, with expressions on the right-hand side and left-hand side of the assignment.

  - A time control ('#').

  - An instance of a system task and its arguments.

In addition to the Verilog HDL objects, there are simulation objects which do not exist in the Verilog language, but which can be accessed by the PLI. Examples of simulation objects are the simulation event queue, the current simulation time, and propagation delays between two modules (referred to as an inter-connect delay).

3.4.1 The vpiHandle data type

The VPI routines use a special data type, called a ***handle***, to access Verilog HDL and simulation objects. A handle is not a pointer to the actual object, it is a pointer to information about the object. The declaration type for variables to store a handle is **vpiHandle**. The vpiHandle data type is defined in the VPI library (the vpi_user.h file). An example declaration for two handle variables is:

```
vpiHandle  primitive_handle, net_handle;
```

There are several VPI routines that locate an object within a simulation data structure and return the handle for the object. Other VPI routines are used to access information about the object, using the object's handle as a reference point. The information that can be accessed depends on the type of the object, but might include the object's name and current logic value.

 Do not share handles between VPI routines and ACC routines! The ACC routines in the PLI standard also use the concept of a handle for referencing Verilog objects. The IEEE 1364 standard does not guarantee that a handle which is obtained with the ACC library will be the same as a handle which is obtained with the VPI library.

3.4.2 Object relationships

Each object in a Verilog design is related to other objects. In the small example on page 59, module *"buf2"* contains instance *"n0"* of the buffer primitive, and instance *"n0"* of the buffer has two terminals, with signals connected to the terminals.

The VPI standard documents three types of object relationships:

- *One-to-one relationships* occur when an object is related to only one other object. In the example on page 59, the buffer instance *"n0"* is contained in module *"buf2"*, so there is a one-to-one relationship from the instance to the module.

- *One-to-many relationships* occur when an object is related to several other objects of a certain type. In the preceding Verilog HDL example, the buffer instance *"n0"* has multiple terminals, so there is a one-to-many relationship from the primitive to its terminals.

- *Many-to-one relationships* occur when many objects are related to a single other object of a certain type. For example, an input port of a module might be driven by several sources, so there are many drivers for the one port.

All of the possible relationships that an object might have are documented in the VPI standard. This documentation is made in the form of *object diagrams*, which use enclosures to represent objects and arrows to represent the type of relationship. The full details of these object diagrams is not presented in this chapter, which is oriented towards getting started with using the VPI library. These details are presented in the next chapter. Figure 3-1 shows how the VPI diagrams document Verilog object relationships.

Figure 3-1: VPI object relationship for a module to a net (partial diagram)

In this example:

- The double arrow from a module to a net indicates a *one-to-many relationship*. A module can contain any number of nets.

- The single arrow from a net to a module indicates a *many-to-one relationship*. A specific net can only occur within one module.

- The names **vpiModule** and **vpiNet** are referred to as *tags*. These tags are constants which represent the type of object in the relationship. The constants are defined in the VPI library, and the VPI routines will use these tag constants to traverse from one object to another, such as from a net object to its parent module.

3.5 Obtaining object handles

The usage of the *$show_all_nets* PLI application to be created in this chapter is:

```
$show_all_nets(<module_instance_name>);
```

The *$show_all_nets* PLI application will need to obtain a handle for a module instance that is named in the first system task/function argument. The application will then need to obtain handles for all the nets in that module. The VPI library provides special routines that return handles for objects.

There are two important terms used with the VPI routines that obtain object handles:

- **target objects** are the type of objects for which the VPI routine will obtain handles.
- A **reference object** is where the VPI routine will search for the target objects. For example, to find all nets within a module, the reference object is the module.

3.5.1 Obtaining a handle for a one-to-one relationship

vpi_handle() obtains a handle for a target object with a one-to-one relationship from the reference object. The syntax for this routine is:

vpiHandle **vpi_handle (**
   PLI_INT32   **type,**   /* constant representing an object type */
   vpiHandle   **reference)** /* handle for an object */

The PLI application for *$show_all_nets* will need to read the module instance name listed as the argument of the system task. In order to access the argument, the PLI application must first obtain a handle for the instance of the system task which called the application. The object diagram for a system task call shows the following one-to-one relationship:

Figure 3-2: VPI object diagram for a system task call instance (partial diagram)

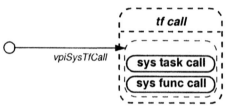

Note: the complete object diagram for a system task call is listed in Appendix D, on page 721.

The small circle in a object diagram indicates that a **NULL** is to be used as the reference object. Therefore, the PLI applications can obtain a handle for the system task call using the following C code.

```
vpiHandle systf_handle;
systf_handle = vpi_handle(vpiSysTfCall, NULL);
```

 The **NULL** (all upper case letters) that is used by VPI routines is defined in the *stdlib.h* ANSI C library file.

3.5.2 Obtaining a handle for a one-to-many relationship

The routines **vpi_iterate()** and **vpi_scan()** are used to obtain handles for all objects when there is a one-to-many relationship.

vpiHandle **vpi_iterate (**
    PLI_INT32   **type,**   /* constant representing an object type */
    vpiHandle   **reference)**   /* handle for an object */

The vpi_iterate() routine returns an ***iterator object***. The routine uses the type constant to determine the type of object for which handles are to be obtained, and the reference_handle to determine the source point of the one-to-many relationship. The ***iterator object*** returned from vpi_iterate() represents the first of the many target objects in the relationship. As each target object is accessed, the iterator is updated automatically to reference the next target object. Conceptually, the many target objects can be thought of as a list of object handles, and the iterator as a pointer to the next object in the list. Note that the usage of a list is purely conceptual—the VPI standard does not require that a simulator create and store lists of target objects. Since the VPI uses the more abstract iterator object to reference each target object, a simulator can maintain its internal storage in any form.

A one-to-many relationship indicates a relationship from a reference object to any number of target objects, including none. For example, a module might not have any nets declared within it, a single net declared, or multiple nets. If there are no target objects in a one-to-many relationship, then vpi_iterate() returns a **NULL** as the iterator value.

vpiHandle **vpi_scan (**
    vpiHandle   **iterator)**   /* handle for an iterator object */

The vpi_scan() routine is provided a single input, the iterator object which was returned from vpi_iterate(). The vpi_scan() routine returns the handle for the next target object which the iterator references. When are there are no more target objects, the next call to `vpi_scan()` will return NULL. In order to access all of the objects in a one-to-many relationship, `vpi_scan()` must be called multiple times, until the return value is NULL.

The following C code fragment uses `vpi_iterate()` and `vpi_scan()` to obtain handles for all nets within a module:

```
vpiHandle module_handle, net_iterator, net_handle;

/* assume a module handle has already been obtained */

net_iterator = vpi_iterate(vpiNet, module_handle);
if (net_iterator == NULL)
  vpi_printf("  No nets found in this module\n");
else {
  while ( (net_handle = vpi_scan(net_iterator)) != NULL ) {
    ... /* code to access information about the net object */
  }
}
```

3.5.3 Comparing VPI handles

A VPI handle is an abstraction used to reference an object within simulation. All VPI routines use this abstraction to access information about the object. This layer of abstraction allows PLI applications to be portable to any number of Verilog simulators, because the abstract handle is a layer between the PLI application and the internal data structures of the simulator. When a VPI application obtains a handle to an object, that handle is not a pointer to the actual object, but rather a pointer to information about the object. When two handles are obtained for the same object, a simulator may allocate a new block of information for each handle, or it may use the same block of information. This level of implementation detail is part of the layer that isolates a user's PLI application from the internal simulation data structure; it is not part of the IEEE standard.

Since the handle for an object is an abstraction, *it is not possible to determine if two handles reference the same object using the C '==' operator*. The handle values may not be equivalent, even if they reference the same object. Instead, the VPI library provides a special routine to see if two handles reference the same object.

PLI_INT32 **vpi_compare_objects (**
    vpiHandle   **object1,**   /* handle for an object */
    vpiHandle   **object2)**   /* handle for an object */

`vpi_compare_objects()` returns a **1** (for true) if two handles reference the same object, and a **0** (for false) if they do not.

3.6 Accessing the arguments of a system task/function

For *$show_all_nets*, the PLI application will need to access the first argument of the system task. which should be a module instance name. For example:

```
$show_all_nets(top.i1);
```

A system task can have any number of arguments. As shown in the object diagram in Figure 3-3, there is a one-to-many relationship, indicated by the double arrow, between a system task/function call and the arguments of the system task/function.

Figure 3-3: VPI object relationship for system task arguments (partial diagram)

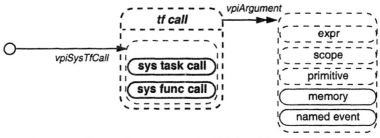

Note: the complete object diagram for a system task call is listed in Appendix D, on page 721

The *$show_all_nets* application can obtain a handle for the module instance that is named in the first system task argument using `vpi_iterate()` and `vpi_scan()`, as shown in the following C code fragment:

```
vpiHandle systf_handle, module_handle,
          net_iterator, net_handle;

systf_handle  = vpi_handle(vpiSysTfCall, NULL);
arg_iterator  = vpi_iterate(vpiArgument, systf_handle);
module_handle = vpi_scan(arg_iterator);
if (module_handle !== NULL)
   vpi_free_object(arg_iterator);   /* free iterator memory */
```

This memory required to store the iterator object is automatically allocated by the simulator when `vpi_iterate()` is called, and is automatically freed when `vpi_scan()` returns NULL after all target objects have been accessed. In the preceding example, however, `vpi_scan()` was only called one time. If the return from `vpi_scan()` is not NULL, then a handle to the object listed as the first task/function argument was obtained. The non-NULL value means the memory for the iterator object has not yet been automatically freed. Since the *$show_all_nets* application has obtained a handle for the first system task argument, the application no longer needs the iterator and should therefore notify the simulator to release the memory for the iterator. The application could continue to call vpi_scan() until a NULL is returned, but the VPI library also provides a special routine to release the iterator object memory when an iterator is no longer needed.

PLI_INT32 **vpi_free_object (**
   vpiHandle   **handle)**   /* handle for an object */

The vpi_free_object() routine is used to release memory which the simulator has allocated for an object. Refer to section 4.5.3 on page 108 in Chapter 4 for more details on when to use vpi_free_object().

3.7 Printing messages from VPI applications

The VPI library uses a special routine for printing text messages.

PLI_INT32 **vpi_printf (**
   PLI_BYTE8 *** format,**   /* character string containing a formatted message */
        ...)   /* arguments to the formatted message string */

vpi_printf() is used to print messages from PLI applications. The syntax for this routine is essentially the same as the C printf() routine, but there is one very important difference: vpi_printf() will print the message to the simulation output channel, and to the simulation output log file. By using vpi_printf(), a PLI application can generate messages which are part of the simulation, without having to determine where the simulator directs its output. vpi_printf() returns the number of characters printed, or EOF if an error occurred.

 The C printf() routine will print its message to the operating system's standard output channel. If the simulator uses some other output channel, such as a graphical window, then printing a message using the C printf() routine might not be seen by the simulator user. Nor will the message be recorded in any output files generated by the simulator.

3.8 Accessing object properties

Every Verilog object has one or more properties which can be accessed by a PLI application. Some properties are the name of a module or net and the logic value of a net. Most properties will be either an integer value or a string value. The VPI identifies these properties using a property constant.

Two VPI routines are provided to read these types of properties.

PLI_INT32 **vpi_get (**
    PLI_INT32   **property,**   /* constant representing an object's property */
    vpiHandle   **object)**   /* handle for an object */

Returns the value associated with integer and boolean properties of an object. Boolean properties return 1 for true and 0 for false.

PLI_BYTE8 ***vpi_get_str (**
    PLI_INT32   **property,**   /* constant representing an object property */
    vpiHandle   **object)**   /* handle for an object */

Returns a pointer to a string containing the value associated with string properties of an object.

3.8.1 Object type properties

Every Verilog object has an integer *type* property, which is accessed using:

```
PLI_INT32 obj_type;
obj_type = vpi_get(vpiType, <object_handle>)
```

The type property identifies what kind of Verilog object is referenced by a VPI handle. This type property is represented by an integer constant, which is defined in the vpi_user.h file. Some example type constants are:

- **vpiModule** — the object handle is referencing a Verilog module instance

- **vpiPrimitive** — the object handle is referencing a Verilog primitive instance

- **vpiNet** — the object handle is referencing a Verilog net data type

- **vpiReg** — the object handle is referencing a Verilog reg data type

The type property can be used many different ways. One common usage is to verify that a handle which was obtained references the type of object expected. For example, the *$show_all_nets* application requires the first task/function argument be a module instance. The following code fragment uses this type property to verify that the argument provided to *$show_all_nets* is correct:

```
vpiHandle systf_handle, arg_iterator, arg_handle;
PLI_INT32 tfarg_type;

systf_handle = vpi_handle(vpiSysTfCall, NULL);
arg_iterator = vpi_iterate(vpiArgument, systf_handle);
arg_handle   = vpi_scan(arg_iterator);
tfarg_type   = vpi_get(vpiType, arg_handle);
if (tfarg_type != vpiModule) {
   /* report error that argument is not correct */
```

TIP

The vpiType property is both an integer and a string property. Using vpi_get(vpiType, <object_handle>) will return the integer value of the type constant. Using vpi_get_str(vpiType, <object_handle>) will return a pointer to a string containing the name of the type constant.

3.8.2 Object name properties

Many Verilog objects have one or more ***name*** properties.

- The property represented by the property constant **vpiName** is the local name of an object. For objects such as nets and variables, the local name is the *declaration name* of the object. For a module or primitive, the local name is the *instance name* within the module in which the module or primitive is used.

- The property represented by the property constant **vpiFullName** is the full hierarchical path name of an object.

- The property represented by the property constant **vpiDefName** is the definition name of a module or primitive.

The following Verilog HDL source code fragment illustrates the difference between the vpiName, vpiFullName and vpiDefName properties.

```
module test;
   wire  a, b, ci, sum, co;                    local name: "u1"
                                                full name: "test.u1"
   addbit u1 (a, b, ci, sum, co);              definition name: "addbit"
endmodule

module addbit (a, b, ci, sum, co);
   input  a, b, ci;                            local name: "sum"
   output sum, co;                             full name: "test.u1.sum"

   wire  a, b, ci, sum, co;                    local name: "g1"
                                               full name: "test.u1.g1"
   xor     g1 (n1, a, b);                      definition name: "xor"
   xor #2  g2 (sum, n1, ci);
   and     g3 (n2, a, b);
   and     g4 (n3, n1, ci);
   or  #2  g5 (co, n2, n3);
endmodule
```

3.9 Reading the logic values of Verilog objects

The VPI library provides a routine to read the logic value of any Verilog object which can contain a value, such as a net or variable.

void **vpi_get_value (**
    vpiHandle   **object,**   /* handle for an object */
    p_vpi_value   **value)**   /* pointer to application-allocated s_vpi_value structure */

The Verilog language uses 4-state logic values, comprising logic 0, 1, Z and X. The vpi_get_value() routine automatically converts Verilog 4-state logic into various C data types for use in PLI applications. A simple way to represent 4-state logic in C is to use character strings, and this is the method used in the *$show_all_nets* application. Chapter 5 presents reading and writing Verilog logic values in more detail.

TIP   Using C strings to represent 4-state logic is a simple method for reading and printing a Verilog logic value. However, the automatic conversion from Verilog values to C strings can be very expensive for the run-time performance of a PLI application. If a PLI application will access a large number of values, or if the application will be called many times during a simulation, it is better to use a more efficient format for reading values. Chapter 5 presents all the formats for reading logic values that are available using VPI routines, and discusses performance considerations.

The PLI application must allocate an s_vpi_value structure prior to calling vpi_get_value(). The definition of the structure is contained in the vpi_user.h header file. The PLI application does not define the structure. The application only allocates memory for the structure. The structure definition is:

```
typedef struct t_vpi_value {
  PLI_INT32     format;  /* vpiBinStrVal,     vpiOctStrVal,
                            vpiDecStrVal,     vpiHexStrVal,
                            vpiScalarVal,     vpiIntVal,
                            vpiRealVal,       vpiStringVal,
                            vpiVectorVal,     vpiTimeVal,
                            vpiStrengthVal,  vpiSuppressVal,
                            vpiObjTypeVal */
  union {
    PLI_BYTE8 *str;      /* if any string format */
    PLI_INT32 scalar;    /* if vpiScalarVal: one of vpi0, vpi1,
                            vpiX, vpiZ, vpiH, vpiL, vpiDontCare */
    PLI_INT32 integer;   /* if vpiIntVal format */
    double    real;      /* if vpiRealVal format */
    struct t_vpi_time         *time;    /* if vpiTimeVal */
    struct t_vpi_vecval       *vector;  /* if vpiVectorVal */
    struct t_vpi_strengthval *strength; /* if vpiStrengthVal */
    PLI_BYTE8 *misc;                     /* not used */
  } value;
} s_vpi_value, *p_vpi_value;
```

The s_vpi_value structure contains two primary fields:

- The **format** field controls how the Verilog logic value should be represented in C. The format is a VPI constant. For example, a format of **vpiBinStrVal** indicates the logic value should be represented using the characters ('0', '1', 'z', and 'x'). A **vpiHexStrVal** format indicates the value should be represented using the characters ('0' through 'F', 'z', and 'x'). Several other formats are available, which are discussed in Chapter 5.

- The **value** field receives the logic value. This field is a union of C data types, and the format constant determines which field within this union that will be used. For formats which receive the logic value as a string, the value.str field will contain a pointer to the string.

The following example retrieves the logic value of a net as a C string:

```
vpiHandle   net_handle;
s_vpi_value current_value;

current_value.format = vpiBinStrVal; /* read as a string */
vpi_get_value(net_handle, &current_value);
```

```
vpi_printf(" net %s  value is  %s (binary)\n",
           vpi_get_str(vpiName, net_handle),
           current_value.value.str);
```

 vpi_get_value() automatically allocates storage for the string which contains the logic value. This storage is temporary, and will automatically be freed when another call is made to vpi_get_value() or when the PLI application exits.

3.10 Reading the current simulation time

In addition to printing the logic values of nets, the *$show_all_nets* application will print the current simulation time for when the application is called.

The Verilog language uses the `timescale compiler directive to establish the time units of a Verilog module, and different modules can represent time in different units. Verilog simulators use an internal simulation time unit, and scale the delays of a module to the internal time unit. The VPI library allows the current simulation time to be represented in either the internal simulation time units or in the time units of any module in the design.

void **vpi_get_time (**
   vpiHandle   **object,**   /* handle for an object, or **NULL** */
   p_vpi_time   **time)**   /* pointer to application-allocated s_vpi_time structure */

The **<object_handle>** is a handle for any object in a design. When simulation time is retrieved in the time scale of a module, the module that is used will be the one containing the object.

The **<time_structure_pointer>** is a pointer to an **s_vpi_time** structure to receive the current simulation time.

The PLI application must allocate an s_vpi_time structure prior to calling vpi_get_time(). The definition of the structure is contained in the vpi_user.h header file. The PLI application does not define the structure. The application only allocates memory for the structure. The structure definition is:

```
typedef struct t_vpi_time {
  PLI_INT32 type;        /* vpiScaledRealTime or vpiSimTime */
  PLI_UINT32 high;       /* when using vpiSimTime */
  PLI_UINT32 low;        /* when using vpiSimTime */
  double     real;       /* when using vpiScaledRealTime */
} s_vpi_time, *p_vpi_time;
```

- The **type** field controls how the Verilog simulation time will be received. The format is set using a constant which is defined in the VPI library.

  - A format of **vpiScaledRealTime** indicates the simulation time will be retrieved as a floating point number and that the time will be scaled to the time units of the module containing the object specified in the call to vpi_get_time(). The **real** field receives the simulation time value when the format field is vpiScaledRealTime.

  - A format of **vpiSimTime** indicates the simulation time should be retrieved as a 64-bit integer and that the time will be in the internal simulation time units. Since not all computer platforms and operating systems have a 64-bit integer data type, the 64-bit value is split into two 32-bit unsigned C integers, as shown in the following diagram:

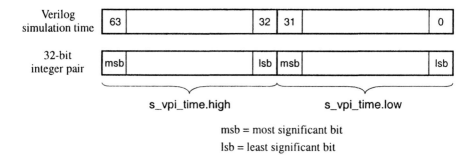

msb = most significant bit
lsb = least significant bit

The *$show_all_nets* application retrieves and prints the current simulation time using vpi_get_time() as follows:

```
vpiHandle    module_handle;
s_vpi_time   current_time;

current_time.type = vpiScaledRealTime;
vpi_get_time(module_handle, &current_time);
vpi_printf("\nAt time %2.2f, nets in module %s (%s):\n",
           current_time.real,
           vpi_get_str(vpiFullName, module_handle),
           vpi_get_str(vpiDefName,  module_handle));
```

3.11 Controlling simulation from PLI applications

There are occasions when a PLI application needs to control what a Verilog simulator is doing. In the *$show_all_nets* application, a *compiletf routine* will be provided to perform syntax checking. If a serious error is detected, such as the argument provided to *$show_all_nets* is not a module instance name, then the PLI application needs to abort simulation execution. That is, to treat the error as a fatal error. The routine vpi_control() is used to abort simulation.

PLI_INT32 **vpi_control (**
  PLI_INT32   **operation,**   /* constant representing the operation to perform */
             ...)   /* variable number of arguments, as required by the
                                operation */

The vpi_control() routine allows a PLI application to control certain aspects of simulation. Returns **1** if successful and **0** if an error occurred. Several operation constants are defined in the IEEE 1364 standard (simulators may add additional flags specific to that product):

- **vpiStop** causes the *$stop()* built-in Verilog system task to be executed upon return of the PLI application. Requires one additional argument of type PLI_INT32, which is the same as the diagnostic message level argument passed to *$stop()*.

- **vpiFinish** causes the *$finish()* built-in Verilog system task to be executed upon return of the PLI application. Requires one additional argument of type PLI_INT32, which is the same as the diagnostic message level argument passed to *$finish()*.

- **vpiReset** causes the *$reset()* built-in Verilog system task to be executed upon return of the PLI application Requires three additional arguments of type PLI_INT32: **stop_value, reset_value** and **diagnostic_level**, which are the same values passed to the *$reset()* system task.

- **vpiSetInteractiveScope** causes a simulator's interactive debug scope to be immediately changed to a new scope. Requires one additional argument of type vpi-Handle, which is a handle to an object in the scope class (see Appendix D, page 727).

 The **vpi_control()** routine was introduced with the IEEE 1364-2001 Verilog standard. The IEEE 1364-1995 standard did not provide the ability for VPI routines to control simulation from a PLI application. Therefore, it was necessary to include the TF routine library, and use the TF simulation control routines such as **tf_error()** and **tf_dofinish()**. Refer to Chapter 10, section 10.4 for a discussion of these routines.

3.12 A complete PLI application using VPI routines

Example 3-1 lists a complete PLI application for the *$show_all_nets* application. This
application includes three C functions:

- A *registration function* to register the PLI application.

- A *compiletf routine* to verify the argument provided to *$show_all_nets*.

- A *calltf routine* to access all nets in a module and print the name and current logic
  value of these nets.

The definition and purpose of the *registration function, compiletf routine* and *calltf
routine* of these PLI routines were presented in Chapter 2.

A short Verilog HDL test case for *$show_all_nets* applications and the simulation
results follow the listing of the PLI application C code.

3.12.1 PLI application source code for $show_all_nets

> **CD** The source code for this example is on the CD accompanying this book.
>
> - Application source file: `Chapter.03/show_all_nets_vpi.c`
> - Verilog test bench: `Chapter.03/show_all_nets_test.v`
> - Expected results log: `Chapter.03/show_all_nets_test.log`

Example 3-1: *$show_all_nets* — using VPI routines to access simulation objects

```
#include <stdlib.h>    /* ANSI C standard library */
#include <stdio.h>     /* ANSI C standard input/output library */
#include <stdarg.h>    /* ANSI C standard arguments library */
#include "vpi_user.h"  /* IEEE 1364 PLI VPI routine library  */

/* prototypes of the PLI application routines */
PLI_INT32 PLIbook_ShowNets_compiletf(PLI_BYTE8 *user_data),
          PLIbook_ShowNets_calltf(PLI_BYTE8 *user_data);

/**********************************************************************
 * $show_all_nets Registration Data
 **********************************************************************/
void PLIbook_ShowNets_register()
{
  s_vpi_systf_data tf_data;

  tf_data.type       = vpiSysTask;
  tf_data.sysfunctype = 0;
```

```
  tf_data.tfname      = "$show_all_nets";
  tf_data.calltf      = PLIbook_ShowNets_calltf;
  tf_data.compiletf   = PLIbook_ShowNets_compiletf;
  tf_data.sizetf      = NULL;
  tf_data.user_data   = NULL;
  vpi_register_systf(&tf_data);
  return;
}

/*********************************************************************
 * compiletf routine
 *********************************************************************/
PLI_INT32 PLIbook_ShowNets_compiletf(PLI_BYTE8 *user_data)
{
  vpiHandle systf_handle, arg_iterator, arg_handle;
  PLI_INT32 tfarg_type;
  int       err_flag = 0;

  /* obtain a handle to the system task instance */
  systf_handle = vpi_handle(vpiSysTfCall, NULL);

  /* obtain handles to system task arguments */
  arg_iterator = vpi_iterate(vpiArgument, systf_handle);
  if (arg_iterator == NULL) {
    vpi_printf("ERROR: $show_all_nets requires 1 argument\n");
    err_flag = 1;
  }
  else {
  /* check the type of object in system task arguments */
  arg_handle = vpi_scan(arg_iterator);
  tfarg_type = vpi_get(vpiType, arg_handle);
  if (tfarg_type != vpiModule) {
    vpi_printf("ERROR: $show_all_nets arg must be module instance\n");
    vpi_free_object(arg_iterator); /* free iterator memory */
    err_flag = 1;
  }
  else {
  /* check that there is only 1 system task argument */
  arg_handle = vpi_scan(arg_iterator);
  if (arg_handle != NULL) {
    vpi_printf("ERROR: $show_all_nets can only have 1 argument\n");
    vpi_free_object(arg_iterator); /* free iterator memory */
    err_flag = 1;
  } } } /* end of if-else-if-else-if sequence */

  if (err_flag) {
    vpi_control(vpiFinish, 1);  /* abort simulation */
  }
  return(0);
}
```

```
/*************************************************************************
 * calltf routine
 *************************************************************************/
PLI_INT32 PLIbook_ShowNets_calltf(PLI_BYTE8 *user_data)
{
  vpiHandle   systf_handle, arg_iterator, module_handle,
              net_iterator, net_handle;
  s_vpi_time  current_time;
  s_vpi_value current_value;

  /* obtain a handle to the system task instance */
  systf_handle = vpi_handle(vpiSysTfCall, NULL);

  /* obtain handle to system task argument */
  /* compiletf has already verified only 1 arg with correct type */
  arg_iterator = vpi_iterate(vpiArgument, systf_handle);
  module_handle = vpi_scan(arg_iterator);
  vpi_free_object(arg_iterator);   /* free iterator memory */

  /* read current simulation time */
  current_time.type = vpiScaledRealTime;
  vpi_get_time(systf_handle, &current_time);

  vpi_printf("\nAt time %2.2f, ", current_time.real);
  vpi_printf("nets in module %s ",
              vpi_get_str(vpiFullName, module_handle));
  vpi_printf("(%s):\n", vpi_get_str(vpiDefName,  module_handle));

  /* obtain handles to nets in module and read current value */
  net_iterator = vpi_iterate(vpiNet, module_handle);
  if (net_iterator == NULL)
    vpi_printf("  no nets found in this module\n");
  else {
    current_value.format = vpiBinStrVal; /* read values as a string */
    while ( (net_handle = vpi_scan(net_iterator)) != NULL ) {
      vpi_get_value(net_handle, &current_value);
      vpi_printf("  net %-10s  value is  %s (binary)\n",
                  vpi_get_str(vpiName, net_handle),
                  current_value.value.str);
    }
  }
  return(0);
}
```

3.12.2 A test bench for $show_all_nets

Example 3-2 lists a simple Verilog HDL design to test *$show_all_nets*. Figure 3-3, which follows, shows the output of running a simulation with this test bench.

Example 3-2: *$show_all_nets* — a Verilog HDL test using a PLI application

```
'timescale 1ns / 1ns
module top;
  reg  [2:0] test;
  tri  [1:0] results;

  addbit i1 (test[0], test[1], test[2], results[0], results[1]);

  initial
    begin
      test = 3'b000;
      #10 test = 3'b011;

      #10 $show_all_nets(top);
      #10 $show_all_nets(i1);

      #10 $stop;
      #10 $finish;
    end
endmodule

/*** A gate level 1 bit adder model ***/
'timescale 1ns / 1ns
module addbit (a, b, ci, sum, co);
  input  a, b, ci;
  output sum, co;

  wire  a, b, ci, sum, co,
        n1, n2, n3;

  xor     (n1, a, b);
  xor #2 (sum, n1, ci);
  and     (n2, a, b);
  and     (n3, n1, ci);
  or  #2 (co, n2, n3);

endmodule
```

3.12.3 Simulation results for $show_all_nets

Example 3-3: *$show_all_nets* — simulation results using a PLI application

```
At time 20.00, nets in module top (top):
  net results       value is  10 (binary)

At time 30.00, nets in module top.i1 (addbit):
  net a             value is   1 (binary)
  net b             value is   1 (binary)
  net ci            value is   0 (binary)
  net sum           value is   0 (binary)
  net co            value is   1 (binary)
  net n1            value is   0 (binary)
  net n2            value is   1 (binary)
  net n3            value is   0 (binary)
```

3.13 Obtaining handles for reg and variable data types

The Verilog HDL defines two general data type groups, **nets** and **variables**. The variable data type group includes the Verilog keywords **reg**, **integer**, **time** and **real**. The PLI treats the reg data type as a unique object, and groups the integer, time and real data types into an object class called **variables**.

Only minor changes are needed to enhance the *$show_all_nets* application so that it can display all the signals of all data types within a module. All that is required is to add additional vpi_iterate() and vpi_scan() statements to access the other signal data types.

The vpi_get_value() is used to read the values of any Verilog data type. The format field in the s_vpi_value structure establishes the C language data type to be used to represent the value. This gives the PLI application developer complete control over how values are represented in the application.

For the *$show_all_signals* application illustrated in this chapter, the following formats will be used:

- For Verilog net data types, values will be represented as a C string, using a binary format.

- For Verilog reg data types, values will be represented as a C string, using a binary format.

- For Verilog integer data types, values will be represented as a 32-bit C integer.

- For Verilog `real` data types, values will be represented as a C `double`.

- For Verilog `time` data types, values will be represented as a pair of 32-bit `unsigned` integers.

The conversion of the Verilog `time` data type into a pair of 32-bit unsigned integers is required because, in the Verilog language, the `time` data type stores a 64-bit unsigned value, but the ANSI C standard does not provide a data type in C that is guaranteed to be a 64-bit wide unsigned integer. Within the VPI, a C structure is used to store the upper portion of the 64 bits in a 32-bit `PLI_UINT32`, and the lower 32 bits in another unsigned `PLI_UINT32`. These two 32-bit C integers can then be concatenated together to display the full 64-bit time value.

The structure to store a time value is `s_vpi_time`, and is defined in the vpi_user.h VPI library file, as follows:

```
typedef struct t_vpi_time {
  PLI_INT32   type;        /* not used by vpi_get_value() */
  PLI_UINT32  high;        /* upper 32-bits of time value */
  PLI_UINT32  low;         /* lower 32-bits of time value */
  double      real;        /* not used by vpi_get_value() */
} s_vpi_time, *p_vpi_time;
```

Note that this same structure definition is used by several VPI routines. In the context of reading a Verilog time variable, the `type` and `real` fields of this structure are not used and can be ignored. The value of the verilog time variable will be stored in the structure, as shown in the following illustration:

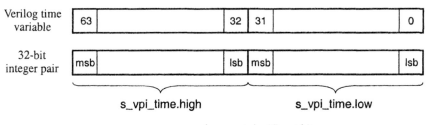

msb = most significant bit
lsb = least significant bit

TIP The C language can be used to perform math operations on the high and low order words that make up the 64-bit time value, but it is important that carries and borrows between the words are handled properly. As a convenience, the TF library of the PLI standard contains several routines to perform math operations on the pair of 32-bit C integer variables used to represent a 64-bit Verilog time value. Refer to section 10.10 on page 335 in Chapter 10 for a list of these TF utility routines.

The following C code fragment illustrates the steps required to read the value of a Verilog time variable and print the value in hexadecimal.

```
vpiHandle    signal_handle;
s_vpi_value current_value;

current_value.format = vpiTimeVal;
vpi_get_value(signal_handle, &current_value);
vpi_printf(" time     %-10s value is  %x%x\n",
              vpi_get_str(vpiName, signal_handle),
              current_value.value.time->high,
              current_value.value.time->low);
```

 vpi_get_value() automatically allocates an s_vpi_time structure when it is called using a vpiTimeVal format. The memory for this structure is temporary, and will automatically be freed when another call is made to vpi_get_value() or when the PLI application exits.

3.13.1 A complete PLI application for $show_all_signals

Example 3-4 lists the complete C code for the *$show_all_signals* PLI application. This application obtains handles for all signals in a module, including the data types of net, reg, integer, time and real.

 Using structured programming techniques makes it easier to maintain or enhance the functionality of a PLI application. This example uses a more structured
TIP   programming style, by creating a separate C function, called *PrintSignalValues()*, to print the current logic value. By moving the printing statements into a separate C function, the *calltf routine* is kept shorter and easier to read, plus there is less duplication of code.

CD   The source code for this example is on the CD accompanying this book.

- Application source file: Chapter.03/show_all_signals_1_vpi.c
- Verilog test bench: Chapter.03/show_all_signals_1_test.v
- Expected results log: Chapter.03/show_all_signals_1_test.log

---

Example 3-4: *$show_all_signals*, version 1 — printing values of all Verilog types

```
#include <stdlib.h>    /* ANSI C standard library */
#include <stdio.h>     /* ANSI C standard input/output library */
#include <stdarg.h>    /* ANSI C standard arguments library */
#include "vpi_user.h"  /* IEEE 1364 PLI VPI routine library  */

/* prototypes of the PLI application routines */
PLI_INT32 PLIbook_ShowSignals_compiletf(PLI_BYTE8 *user_data),
          PLIbook_ShowSignals_calltf(PLI_BYTE8 *user_data);
void      PLIbook_PrintSignalValues(vpiHandle signal_iterator);

/**********************************************************************
 * $show_all_signals Registration Data
 **********************************************************************/
void PLIbook_ShowSignals_register()
{
  s_vpi_systf_data tf_data;

  tf_data.type        = vpiSysTask;
  tf_data.sysfunctype = 0;
  tf_data.tfname      = "$show_all_signals";
  tf_data.calltf      = PLIbook_ShowSignals_calltf;
  tf_data.compiletf   = PLIbook_ShowSignals_compiletf;
  tf_data.sizetf      = NULL;
  tf_data.user_data   = NULL;
  vpi_register_systf(&tf_data);
  return;
}

/**********************************************************************
 * compiletf routine
 **********************************************************************/
PLI_INT32 PLIbook_ShowSignals_compiletf(PLI_BYTE8 *user_data)
{
  vpiHandle systf_handle, arg_iterator, arg_handle;
  PLI_INT32 tfarg_type;
  int       err_flag = 0;

  /* obtain a handle to the system task instance */
  systf_handle = vpi_handle(vpiSysTfCall, NULL);

  /* obtain handles to system task arguments */
  arg_iterator = vpi_iterate(vpiArgument, systf_handle);
  if (arg_iterator == NULL) {
    vpi_printf("ERROR: $show_all_signals requires 1 argument\n");
    err_flag = 1;
  }
  else {
  /* check the type of object in system task arguments */
  arg_handle = vpi_scan(arg_iterator);
  tfarg_type = vpi_get(vpiType, arg_handle);
  if (tfarg_type != vpiModule) {
```

```
    vpi_printf("ERROR: $show_all_signals arg 1");
    vpi_printf(" must be a module instance\n");
    vpi_free_object(arg_iterator); /* free iterator memory */
    err_flag = 1;
  }
  else {
  /* check that there is only 1 system task argument */
  arg_handle = vpi_scan(arg_iterator);
  if (arg_handle != NULL) {
    vpi_printf("ERROR: $show_all_signals can only have 1 argument\n");
    vpi_free_object(arg_iterator); /* free iterator memory */
    err_flag = 1;
  } } } /* end of if-else-if-else-if sequence */
  if (err_flag) {
    vpi_control(vpiFinish, 1);  /* abort simulation */
  }
  return(0);
}

/*****************************************************************************
 * calltf routine
 *****************************************************************************/
PLI_INT32 PLIbook_ShowSignals_calltf(PLI_BYTE8 *user_data)
{
  vpiHandle   systf_handle, arg_iterator, module_handle,
              signal_iterator;
  PLI_INT32   format;
  s_vpi_time  current_time;

  /* obtain a handle to the system task instance */
  systf_handle = vpi_handle(vpiSysTfCall, NULL);

  /* obtain handle to system task argument
     compiletf has already verified only 1 arg with correct type */
  arg_iterator = vpi_iterate(vpiArgument, systf_handle);
  module_handle = vpi_scan(arg_iterator);
  vpi_free_object(arg_iterator);  /* free iterator memory */

  /* read current simulation time */
  current_time.type = vpiScaledRealTime;
  vpi_get_time(systf_handle, &current_time);

  vpi_printf("\nAt time %2.2f, ", current_time.real);
  vpi_printf("signals in module %s ",
              vpi_get_str(vpiFullName, module_handle));
  vpi_printf("(%s):\n", vpi_get_str(vpiDefName, module_handle));

  /* obtain handles to nets in module and read current value */
  signal_iterator = vpi_iterate(vpiNet, module_handle);
  if (signal_iterator != NULL)
    PLIbook_PrintSignalValues(signal_iterator);

  /* obtain handles to regs in module and read current value */
  signal_iterator = vpi_iterate(vpiReg, module_handle);
  if (signal_iterator != NULL)
```

```
    PLIbook_PrintSignalValues(signal_iterator);

  /* obtain handles to variables in module and read current value */
  signal_iterator = vpi_iterate(vpiVariables, module_handle);
  if (signal_iterator != NULL)
    PLIbook_PrintSignalValues(signal_iterator);

  vpi_printf("\n"); /* add some white space to output */
  return(0);
}

void PLIbook_PrintSignalValues(vpiHandle signal_iterator)
{
  vpiHandle    signal_handle;
  PLI_INT32    signal_type;
  s_vpi_value current_value;

  while ( (signal_handle = vpi_scan(signal_iterator)) != NULL ) {
    signal_type = vpi_get(vpiType, signal_handle);
    switch (signal_type) {
      case vpiNet:
        current_value.format = vpiBinStrVal;
        vpi_get_value(signal_handle, &current_value);
        vpi_printf(" net      %-10s  value is  %s (binary)\n",
                   vpi_get_str(vpiName, signal_handle),
                   current_value.value.str);
      break;

      case vpiReg:
        current_value.format = vpiBinStrVal;
        vpi_get_value(signal_handle, &current_value);
        vpi_printf(" reg      %-10s  value is  %s (binary)\n",
                   vpi_get_str(vpiName, signal_handle),
                   current_value.value.str);
      break;

      case vpiIntegerVar:
        current_value.format = vpiIntVal;
        vpi_get_value(signal_handle, &current_value);
        vpi_printf(" integer %-10s  value is  %d (decimal)\n",
                   vpi_get_str(vpiName, signal_handle),
                   current_value.value.integer);
      break;

      case vpiRealVar:
        current_value.format = vpiRealVal;
        vpi_get_value(signal_handle, &current_value);
        vpi_printf(" real     %-10s  value is  %0.2f\n",
                   vpi_get_str(vpiName, signal_handle),
                   current_value.value.real);
      break;

      case vpiTimeVar:
        current_value.format = vpiTimeVal;
        vpi_get_value(signal_handle, &current_value);
        vpi_printf(" time     %-10s  value is  %x%x\n",
```

```
                    vpi_get_str(vpiName, signal_handle),
                    current_value.value.time->high,
                    current_value.value.time->low);
        break;
      }
  }
  return;
}
```

Example 3-5 lists Verilog source code for testing *$show_all_signals*. This test is similar to the test for *$show_all_nets*, with the difference being the addition of more variables in the test bench level and changing the lower level adder model from a gate level model to an RTL model. Example 3-6, which follows, shows the simulation results from running a simulation with *$show_all_signals*.

Example 3-5: *$show_all_signals* — Verilog HDL test for the PLI application

```
`timescale 1ns / 1ns
module top;
  tri   [1:0] results;
  integer    test;
  real       foo;
  time       bar;

  addbit i1 (test[0], test[1], test[2], results[0], results[1]);

  initial
    begin
      test = 3'b000;
      foo = 3.14;
      bar = 0;
      bar[63:60] = 4'hF;
      bar[35:32] = 4'hA;
      bar[31:28] = 4'hC;
      bar[03:00] = 4'hE;
      #10 test = 3'b011;

      #10 $show_all_signals(top);
      #10 $show_all_signals(i1);

      #10 $stop;
      #10 $finish;
    end
endmodule

/*** An RTL level 1 bit adder model ***/
`timescale 1ns / 1ns
module addbit (a, b, ci, sum, co);
  input  a, b, ci;
  output sum, co;
  wire   a, b, ci;
```

```
reg     sum, co;

always @(a or b or ci)
   {co, sum} = a + b + ci;

endmodule
```

Example 3-6: *$show_all_signals* — simulation results using the PLI application

```
At time 20.00, signals in module top (top):
   net      results      value is  10 (binary)
   integer test          value is  3 (decimal)
   real     foo          value is  3.14
   time     bar          value is  f000000ac000000e

At time 30.00, signals in module top.i1 (addbit):
   net      a            value is  1 (binary)
   net      b            value is  1 (binary)
   net      ci           value is  0 (binary)
   reg      sum          value is  0 (binary)
   reg      co           value is  1 (binary)
```

3.14 Obtaining a handle to the current hierarchy scope

The Verilog language allows a system task or system function to be invoked from any hierarchy scope. A *scope* in the Verilog HDL is a level of design hierarchy and can be represented by several constructs:

- Module instances

- Named statement groups

- Verilog HDL tasks

- Verilog HDL function

The following example calls the *$show_all_signals* from a named statement group:

```
module top;
   ...
   always @(posedge clock)
      begin: local
        integer i;
        reg      local_bus;
        ...
        $show_all_signals;
```

```
      end
endmodule
```

A useful enhancement to the *$show_all_signals* example is to allow either no system task argument or a null system task argument to represent the hierarchy scope which called the *$show_all_signals* system task. The difference between no argument and a null argument is shown in the following two examples.

No system task/function arguments:

```
$show_all_signals;
```

A null system task/function argument:

```
$show_all_signals();
```

The following Verilog source code shows the enhanced usage possibilities for the *$show_all_signals* example:

```
module top;
   ...
   addbit i1 (a, b, ci, sum, co);    // instance of an adder
   ...
   always @(sum or co)
     $show_all_signals;        // list signals in this module
   always @(posedge clock)
     begin: local
       integer i;
       reg     local_bus;
       ...
       $show_all_signals;      // list signals in this block
     end
endmodule

module addbit (a, b, ci, sum, co);
   ...
   always @(sum or co)
     $show_all_signals();      // list signals in this instance
endmodule
```

In the above example, the *$show_all_signals* applications should search for signal names in the local hierarchy scope, which in one call is a module, and in another call is a named statement group. The name of any type of hierarchy scope can be passed to the system task as a task/function argument, but in this example, *$show_all_signals* is being called with no arguments. To obtain the local hierarchy scope without being passed the scope name requires two steps:

Figure 3-4: VPI object diagram for a system task/function call (partial diagram)

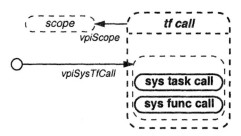

Note: the complete object diagram for a system task call is listed in Appendix D, on page 721.

1. **vpi_handle(vpiSysTfCall, NULL)** returns a handle to the system task/function which called the PLI application.

2. **vpi_handle(vpiScope, systf_handle)** returns a handle to the scope containing the system task handle.

Example 3-7 contains the complete listing of the enhanced *$show_all_signals*, with the ability to use either no arguments or null arguments to represent the local design hierarchy scope. The *compiletf routine* is also enhanced from the previous examples to allow any valid Verilog scope to be used as an argument to *$show_all_signals*.

CD The source code for this example is on the CD accompanying this book.

- Application source file: Chapter.03/show_all_signals_2_vpi.c
- Verilog test bench: Chapter.03/show_all_signals_2_test.v
- Expected results log: Chapter.03/show_all_signals_2_test.log

---

Example 3-7: *$show_all_signals*, version 2 — obtaining the local scope handle

```
#include <stdlib.h>    /* ANSI C standard library */
#include <stdio.h>     /* ANSI C standard input/output library */
#include <stdarg.h>    /* ANSI C standard arguments library */
#include "vpi_user.h"  /* IEEE 1364 PLI VPI routine library */

/* prototypes of the PLI application routines */
PLI_INT32 PLIbook_ShowSignals_compiletf(PLI_BYTE8 *user_data),
          PLIbook_ShowSignals_calltf(PLI_BYTE8 *user_data);
void      PLIbook_PrintSignalValues(vpiHandle signal_iterator);
```

```
/************************************************************************
 * $show_all_signals Registration Data
 ************************************************************************/
void PLIbook_ShowSignals_register()
{
  s_vpi_systf_data tf_data;

  tf_data.type        = vpiSysTask;
  tf_data.sysfunctype = 0;
  tf_data.tfname      = "$show_all_signals";
  tf_data.calltf      = PLIbook_ShowSignals_calltf;
  tf_data.compiletf   = PLIbook_ShowSignals_compiletf;
  tf_data.sizetf      = NULL;
  tf_data.user_data   = NULL;
  vpi_register_systf(&tf_data);
  return;
}

/************************************************************************
 * compiletf routine
 ************************************************************************/
PLI_INT32 PLIbook_ShowSignals_compiletf(PLI_BYTE8 *user_data)
{
  vpiHandle systf_handle, arg_iterator, arg_handle;
  PLI_INT32 tfarg_type;
  int       err_flag = 0;

  /* obtain a handle to the system task instance */
  systf_handle = vpi_handle(vpiSysTfCall, NULL);

  /* obtain handles to system task arguments */
  arg_iterator = vpi_iterate(vpiArgument, systf_handle);
  if (arg_iterator == NULL) {
    return(0); /* no arguments OK; skip remaining checks */
  }

  /* check the type of object in system task arguments */
  arg_handle = vpi_scan(arg_iterator);
  tfarg_type = vpi_get(vpiType, arg_handle);
  switch (tfarg_type) {
    case vpiModule:
    case vpiTask:
    case vpiFunction:
    case vpiNamedBegin:
    case vpiNamedFork:
      break; /* arg is a scope instance; continue to next check */
    case vpiOperation:
      if (vpi_get(vpiOpType, arg_handle) == vpiNullOp)
        break; /* null argument OK; continue to next check */
    default:
      /* wrong type specified for an argument */
      vpi_printf("ERROR: $show_all_signals arg 1");
      vpi_printf(" must be a scope instance or null\n");
```

```
      vpi_free_object(arg_iterator); /* free iterator memory */
      err_flag = 1;
   }
   if (err_flag == 0) {
   /* check that there is only 1 system task argument */
   arg_handle = vpi_scan(arg_iterator);
   if (arg_handle != NULL) {
     vpi_printf("ERROR: $show_all_signals can only have 1 argument\n");
     vpi_free_object(arg_iterator); /* free iterator memory */
     err_flag = 1;
   } } /* end of tests */
   if (err_flag) {
     vpi_control(vpiFinish, 1);  /* abort simulation */
   }
   return(0);
}

/**********************************************************************
 * calltf routine
 **********************************************************************/
PLI_INT32 PLIbook_ShowSignals_calltf(PLI_BYTE8 *user_data)
{
  vpiHandle    systf_handle, arg_iterator, scope_handle,
               signal_iterator;
  PLI_INT32    format;
  s_vpi_time   current_time;

  /* obtain a handle to the system task instance */
  systf_handle = vpi_handle(vpiSysTfCall, NULL);

  /* obtain handle to system task argument */
  arg_iterator = vpi_iterate(vpiArgument, systf_handle);
  if (arg_iterator == NULL) {
    /* no arguments -- use scope that called this application */
    scope_handle = vpi_handle(vpiScope, systf_handle);
  }
  else {
    /* compiletf has already verified arg is scope instance or null */
    scope_handle = vpi_scan(arg_iterator);
    vpi_free_object(arg_iterator);  /* free iterator memory */
    if (vpi_get(vpiType, scope_handle) != vpiModule)
      /* arg isn't a module instance; assume it is null */
    scope_handle = vpi_handle(vpiScope, systf_handle);
  }

  /* read current simulation time */
  current_time.type = vpiScaledRealTime;
  vpi_get_time(systf_handle, &current_time);

  vpi_printf("\nAt time %2.2f, signals in scope %s:\n",
             current_time.real,
             vpi_get_str(vpiFullName, scope_handle));

  /* obtain handles to nets in module and read current value */
  /* nets can only exist if scope is a module */
```

```c
    if (vpi_get(vpiType, scope_handle) == vpiModule) {
      signal_iterator = vpi_iterate(vpiNet, scope_handle);
      if (signal_iterator != NULL)
        PLIbook_PrintSignalValues(signal_iterator);
    }

    /* obtain handles to regs in scope and read current value */
    signal_iterator = vpi_iterate(vpiReg, scope_handle);
    if (signal_iterator != NULL)
      PLIbook_PrintSignalValues(signal_iterator);

    /* obtain handles to variables in scope and read current value */
    signal_iterator = vpi_iterate(vpiVariables, scope_handle);
    if (signal_iterator != NULL)
      PLIbook_PrintSignalValues(signal_iterator);
    vpi_printf("\n"); /* add some white space to output */
    return(0);
  }

  void PLIbook_PrintSignalValues(vpiHandle signal_iterator)
  {
    vpiHandle   signal_handle;
    int         signal_type;
    s_vpi_value current_value;

    while ( (signal_handle = vpi_scan(signal_iterator)) != NULL ) {
      signal_type = vpi_get(vpiType, signal_handle);
      switch (signal_type) {
        case vpiNet:
          current_value.format = vpiBinStrVal;
          vpi_get_value(signal_handle, &current_value);
          vpi_printf("  net     %-10s  value is  %s (binary)\n",
                     vpi_get_str(vpiName, signal_handle),
                     current_value.value.str);
        break;

        case vpiReg:
          current_value.format = vpiBinStrVal;
          vpi_get_value(signal_handle, &current_value);
          vpi_printf("  reg     %-10s  value is  %s (binary)\n",
                     vpi_get_str(vpiName, signal_handle),
                     current_value.value.str);
        break;

        case vpiIntegerVar:
          current_value.format = vpiIntVal;
          vpi_get_value(signal_handle, &current_value);
          vpi_printf("  integer %-10s  value is  %d (decimal)\n",
                     vpi_get_str(vpiName, signal_handle),
                     current_value.value.integer);
        break;

        case vpiRealVar:
          current_value.format = vpiRealVal;
          vpi_get_value(signal_handle, &current_value);
          vpi_printf("  real    %-10s  value is  %0.2f\n",
```

```
                        vpi_get_str(vpiName, signal_handle),
                        current_value.value.real);
        break;

      case vpiTimeVar:
        current_value.format = vpiTimeVal;
        vpi_get_value(signal_handle, &current_value);
        vpi_printf(" time     %-10s   value is   %x%x\n",
                        vpi_get_str(vpiName, signal_handle),
                        current_value.value.time->high,
                        current_value.value.time->low);
        break;
      }
  }
  return;
}
```

3.15 Obtaining handles to multiple task/function arguments

Another useful enhancement to the *$show_all_signals* application is to allow multiple hierarchy scopes to be specified at the same time. For example:

```
$show_all_signals(i1, ,top.local);
```

In the above example, there are three system task/function arguments (the second argument being null, indicating the local hierarchy scope). Only a very minor change is required in the *$show_all_signals* example to support any number of system task arguments. The relationship of a task call to its arguments is a one-to-many relationship. The previous example applications have been accessing the first argument by using vpi_iterate() to obtain an iterator for all arguments, and then calling vpi_scan() just one time in order to access only the first argument. The only change that is needed is to place the call to vpi_scan() in loop to access all of the task arguments instead of just the first argument.

Example 3-8 illustrates using a C while loop to access each system task/function argument. In this example another level of structured programming has been added, in order to keep the *calltf routine* short and easy to maintain. The *calltf routine* contains the loop to access each task argument. Within the loop, a *GetAllSignals* function is called. And, within that function, a *PrintSignalValues* function is called. Structured programming is an important technique to use in PLI applications.

> **CD** The source code for this example is on the CD accompanying this book.
>
> • Application source file: `Chapter.03/show_all_signals_3_vpi.c`
> • Verilog test bench: `Chapter.03/show_all_signals_3_test.v`
> • Expected results log: `Chapter.03/show_all_signals_3_test.log`

Example 3-8: *$show_all_signals*, version 3 — obtaining handles for multiple tfargs

```
#include <stdlib.h>     /* ANSI C standard library */
#include <stdio.h>      /* ANSI C standard input/output library */
#include <stdarg.h>     /* ANSI C standard arguments library */
#include "vpi_user.h"   /* IEEE 1364 PLI VPI routine library  */

/* prototypes of the PLI application routines */
PLI_INT32 PLIbook_ShowSignals_compiletf(PLI_BYTE8 *user_data),
          PLIbook_ShowSignals_calltf(PLI_BYTE8 *user_data);
void      PLIbook_GetAllSignals(vpiHandle scope_handle,
                                p_vpi_time current_time),
          PLIbook_PrintSignalValues(vpiHandle signal_iterator);

/**********************************************************************
 * $show_all_signals Registration Data
 **********************************************************************/
void PLIbook_ShowSignals_register()
{
  s_vpi_systf_data tf_data;

  tf_data.type       = vpiSysTask;
  tf_data.sysfunctype = 0;
  tf_data.tfname     = "$show_all_signals";
  tf_data.calltf     = PLIbook_ShowSignals_calltf;
  tf_data.compiletf  = PLIbook_ShowSignals_compiletf;
  tf_data.sizetf     = NULL;
  tf_data.user_data  = NULL;
  vpi_register_systf(&tf_data);
  return;
}

/**********************************************************************
 * compiletf routine
 **********************************************************************/
PLI_INT32 PLIbook_ShowSignals_compiletf(PLI_BYTE8 *user_data)
{
  vpiHandle systf_handle, arg_iterator, arg_handle;
  PLI_INT32 tfarg_type;
  int       err_flag = 0, tfarg_num = 0;

  /* obtain a handle to the system task instance */
  systf_handle = vpi_handle(vpiSysTfCall, NULL);
```

```
  /* obtain handles to system task arguments */
  arg_iterator = vpi_iterate(vpiArgument, systf_handle);
  if (arg_iterator == NULL) {
    return(0); /* no arguments OK; skip remaining checks */
  }

 /* check each argument */
  while ( (arg_handle = vpi_scan(arg_iterator)) != NULL ) {
    tfarg_num++;

    /* check the type of object in system task arguments */
    tfarg_type = vpi_get(vpiType, arg_handle);
    switch (tfarg_type) {
      case vpiModule:
      case vpiTask:
      case vpiFunction:
      case vpiNamedBegin:
      case vpiNamedFork:
        break; /* arg is a scope instance; continue to next check */
      case vpiOperation:
        if (vpi_get(vpiOpType, arg_handle) == vpiNullOp) {
          break; /* null argument OK; continue to next check */
        }
      default:
        /* wrong type specified for an argument */
        vpi_printf("ERROR: $show_all_signals arg %d", tfarg_num);
        vpi_printf(" must be a scope instance or null\n");
        vpi_free_object(arg_iterator); /* free iterator memory */
        err_flag = 1;
    }
  } /* end of tests */
  if (err_flag) {
    vpi_control(vpiFinish, 1);  /* abort simulation */
  }
  return(0);
}


/*********************************************************************
 * calltf routine
 *********************************************************************/
PLI_INT32 PLIbook_ShowSignals_calltf(PLI_BYTE8 *user_data)
{

  vpiHandle    systf_handle, arg_iterator, scope_handle;
  PLI_INT32    format;
  s_vpi_time   current_time;

  /* obtain a handle to the system task instance */
  systf_handle = vpi_handle(vpiSysTfCall, NULL);

  /* read current simulation time */
  current_time.type = vpiScaledRealTime;
  vpi_get_time(systf_handle, &current_time);
```

```
  /* obtain handle to system task argument */
  arg_iterator = vpi_iterate(vpiArgument, systf_handle);
  if (arg_iterator == NULL) {
    /* no arguments -- use scope that called this application */
    scope_handle = vpi_handle(vpiScope, systf_handle);
    PLIbook_GetAllSignals(scope_handle, &current_time);
  }
  else {
    /* compiletf has already verified arg is scope instance or null */
    while ( (scope_handle = vpi_scan(arg_iterator)) != NULL ) {
      if (vpi_get(vpiType, scope_handle) != vpiModule) {
        /* arg isn't a module instance; assume it is null */
        scope_handle = vpi_handle(vpiScope, systf_handle);
      }
      PLIbook_GetAllSignals(scope_handle, &current_time);
    }
  }
  return(0);
}

void PLIbook_GetAllSignals(vpiHandle scope_handle, p_vpi_time
current_time)
{
  vpiHandle signal_iterator;

  vpi_printf("\nAt time %2.2f, ", current_time->real);
  vpi_printf("signals in scope %s ",
             vpi_get_str(vpiFullName, scope_handle));
  vpi_printf("(%s):\n", vpi_get_str(vpiDefName, scope_handle));

  /* obtain handles to nets in module and read current value */
  /* nets can only exist if scope is a module */
  if (vpi_get(vpiType, scope_handle) == vpiModule) {
    signal_iterator = vpi_iterate(vpiNet, scope_handle);
    if (signal_iterator != NULL)
      PLIbook_PrintSignalValues(signal_iterator);
  }

  /* obtain handles to regs in scope and read current value */
  signal_iterator = vpi_iterate(vpiReg, scope_handle);
  if (signal_iterator != NULL)
    PLIbook_PrintSignalValues(signal_iterator);

  /* obtain handles to variables in scope and read current value */
  signal_iterator = vpi_iterate(vpiVariables, scope_handle);
  if (signal_iterator != NULL)
    PLIbook_PrintSignalValues(signal_iterator);

  vpi_printf("\n"); /* add some white space to output */
  return;
}
```

```
void PLIbook_PrintSignalValues(vpiHandle signal_iterator)
{
  vpiHandle    signal_handle;
  int          signal_type;
  s_vpi_value current_value;

  while ( (signal_handle = vpi_scan(signal_iterator)) != NULL ) {
    signal_type = vpi_get(vpiType, signal_handle);
    switch (signal_type) {
      case vpiNet:
        current_value.format = vpiBinStrVal;
        vpi_get_value(signal_handle, &current_value);
        vpi_printf(" net     %-10s  value is %s (binary)\n",
                   vpi_get_str(vpiName, signal_handle),
                   current_value.value.str);
        break;

      case vpiReg:
        current_value.format = vpiBinStrVal;
        vpi_get_value(signal_handle, &current_value);
        vpi_printf(" reg     %-10s  value is %s (binary)\n",
                   vpi_get_str(vpiName, signal_handle),
                   current_value.value.str);
        break;

      case vpiIntegerVar:
        current_value.format = vpiIntVal;
        vpi_get_value(signal_handle, &current_value);
        vpi_printf(" integer %-10s  value is %d (decimal)\n",
                   vpi_get_str(vpiName, signal_handle),
                   current_value.value.integer);
        break;

      case vpiRealVar:
        current_value.format = vpiRealVal;
        vpi_get_value(signal_handle, &current_value);
        vpi_printf(" real    %-10s  value is %0.2f\n",
                   vpi_get_str(vpiName, signal_handle),
                   current_value.value.real);
        break;

      case vpiTimeVar:
        current_value.format = vpiTimeVal;
        vpi_get_value(signal_handle, &current_value);
        vpi_printf(" time    %-10s  value is %x%x\n",
                   vpi_get_str(vpiName, signal_handle),
                   current_value.value.time->high,
                   current_value.value.time->low);
        break;
    }
  }
  return;
}
```

3.16 Summary

The VPI routines in the PLI standard provide complete access to the internal data structures of a Verilog simulation. This access is done using an object oriented method, where Verilog HDL constructs within the simulation data structure are treated as objects. The VPI routines use *handles* to reference these objects.

Each Verilog object has relationships to other objects. Each object also has specific properties, such a a name or a logic value. The VPI library uses a concise set of routines to access all types of objects. Predefined VPI constants represent each object type and the properties associated with the object. Object object diagrams show the relationships of one object to other objects and the properties of each object type.

This chapter has focused on how to create PLI applications using the VPI library. The following three chapters will present more detail on the syntax and usage of the 37 VPI routines in the PLI standard. These chapters include several additional examples of PLI applications which use the VPI library.

CHAPTER 4 *Details about the VPI Routine Library*

T he VPI routines are a library of 37 C functions that can interact with Verilog simulators. The previous chapter has introduced how PLI applications are developed using the VPI routines. This chapter presents the VPI library in more detail than the previous chapter, and presents how Verilog objects are documented in the VPI standard. Chapter 5 presents how to read and modify the values of Verilog objects, and Chapter 6 presents how to use VPI routines to synchronize PLI applications with Verilog simulation activity. Appendix D presents the complete syntax of the VPI routine library and the complete set of VPI object diagrams.

The concepts presented in this chapter are:

- PLI application performance considerations
- Reading VPI string properties
- Verilog objects and object relationships
- Accessing system task and system function arguments
- Creating a persistent work area unique to each instance of a system task/function
- Traversing Verilog hierarchy to obtain handles for Verilog HDL objects
- Accessing simulation time and time scale factors
- Miscellaneous VPI routines

4.1 PLI application performance considerations

The run-time performance of a simulator can be impacted in either a positive way or a negative way by PLI applications. Often, a complex algorithm can be represented in the C language, using C language data types, much more efficiently than in the hardware-centric Verilog HDL language. The C language can be used for an abstract representation of a design, when 4-state logic, logic transitions, simulation time, and other details are not required, but which a hardware description language must be able to represent. The abstraction that C offers often makes it possible to greatly increase the run-time performance of a simulation algorithm.

However, a poorly thought out PLI application can actually decrease the run-time performance of a simulation. Each call to a routine in the PLI library will take time to be executed. It is important to architect a PLI application to minimize the number of times VPI routines are used.

The following guidelines can help in planning an efficient PLI application:

- Good C programming practices are essential. C programming style and techniques are not discussed within the scope of this book.

- Consider every call to a VPI routine in the VPI library as expensive, and try to minimize the number of calls.

- VPI routines which obtain object handles using an object's name are less efficient than VPI routines which obtain object handles based on a relationship to another object.

- Routines which convert logic values from a simulator's internal representation to C strings, and vice-versa, are less efficient than using other C data types. Strings are a convenient means of representing 4-state values for printing, but strings should be used prudently.

- When the same object must be accessed many times during a simulation, the handle can be obtained once and saved in an application-allocated storage area. Using a pointer to the storage area, a PLI application has immediate access to the object handle, without having to call a VPI routine to obtain the handle each time it is needed. Section 4.7 on page 122 discusses allocating storage within a PLI application.

- Use the VPI library to access the unique abilities of hardware description languages, such as representing hardware parallelism and hardware propagation times. Simulator vendors have invested a great deal in optimizing a simulator's algorithms, and that optimization should be utilized, rather than duplicated in a PLI application.

When developing a PLI application, one primary consideration should be how often a PLI application will be called during a simulation. It is well worth the effort to optimize the performance of an application that is invoked every clock cycle, but may not be as important for an application that is only invoked once during a simulation.

 The objective of this book is to show how the routines in the VPI library are used. Short examples of using many of these routines are shown in the context of complete PLI applications. In order to meet the book's objectives, the examples presented in this book do not always follow the guidelines of efficient C coding and prudent usage of the VPI routines. It is expected that when parts of these example PLI applications are adapted for other applications, the coding style will also be modified to be more efficient and robust.

4.2 The VPI string buffer

A number of VPI objects have properties which are strings, such as the name of a net. The routine `vpi_get_str()` is used to read the value of string properties. This routine returns a pointer to C a character string. The string itself is stored in a temporary string buffer. The buffer will be either freed by the simulator or and may be overwritten by the next call to `vpi_get_str()`. Therefore, a PLI application should use the string pointer returned by a VPI routine immediately. If a string needs to be preserved, the PLI application should copy the string into its own storage space. Following are two examples of using strings in a PLI application.

Read a string and use it immediately:

```
PLI_BYTE8 *string_p;    /* string pointer only, no storage */
string_p = vpi_get_str(vpiName, net_handle);
vpi_printf("string_p points to %s\n", string_p);
```

Read a string and copy it to application-allocated storage for later use:

```
PLI_BYTE8 *string_p;    /* string pointer only, no storage */
char *string_keep;      /* another string pointer only */
string_p = vpi_get_str(vpiName, net_handle);
string_keep = malloc(strlen((char *)string_p)+1);
strcpy(string, (char *)string_p);  /* save string */
```

 If application-allocated storage is to remain valid from one call to the application to another call, then the PLI application must maintain a pointer to the memory. Section 4.7 discusses allocating memory and maintaining pointers.

4.3 VPI error handling

A well written PLI application will perform error checking on the values returned by VPI routines. If a VPI routine failed to return a valid value, passing the invalid return to another VPI routine may lead to unexpected behavior of the Verilog simulator, including program crashes. As an example, if vpi_handle() could not locate a target object, it will return a NULL instead of a valid handle. If the NULL were then passed as an input to vpi_get_str() as the reference argument, an error of some type will occur.

A NULL return from vpi_handle() is that routine's *exception value*, which indicates that an error occurred. A well written PLI application will check for these exception values and act accordingly. For example:

```
task_handle = vpi_handle(vpiSystfCall, NULL)
if (task_handle == NULL) {
  vpi_printf("ERROR: could not obtain task handle\n");
  return(0);
}
```

Most VPI routines have a specific exception return value when the routine cannot perform the operation requested. Routines which return integer or boolean values will return **0** if an error occurs. Routines which return double-precision values will return **0.0**. Routines which return handles will return NULL. Routines which return string pointer will return NULL (note that NULL is not the same a null string).

For VPI routines which return integer, boolean or double values, the exception value could be a legitimate value. Therefore, it might not be possible to determine if an error occurred based on the exception return value. The VPI library provides a useful routine called **vpi_chk_error()**, which is used to check for errors and to report detailed information about an error. This routine returns **0** if the previous call to a VPI routine was successful, and an error severity level code if the call resulted in an error. This return can be used as a true/false test to determine if an error occurred in the previous VPI routine call when the routine does not have an exception value. vpi_chk_error() is also useful for debugging problems in a PLI application.

PLI_INT32 **vpi_chk_error (**
   p_vpi_error_info **info)**   /* pointer to an application-allocated s_vpi_error_info
                                    structure, or **NULL** */

Every VPI routine except vpi_chk_error() will set or clear an internal VPI error status flag, which is common to all VPI routines. When the flag is set by an error, it will remain set until the next call to a VPI routine changes the flag. vpi_chk_error() only reads the error status flag, and does not modify it.

The input to `vpi_chk_error()` is a pointer to an **s_vpi_error_info** structure. If an error in the previous call to VPI routine occurred, `vpi_chk_error()` will fill in the fields of this structure with the simulator product name and version, and the file name and line number containing the system task/function instance which called the PLI application. Using this information, a PLI application can generate meaningful error messages to aid in debugging the cause of an error. If the details about the error are not needed, a NULL can be specified as the input to `vpi_chk_error()`. The definition of the s_vpi_error_info structure in vpi_user.h is as follows:

```
typedef struct t_vpi_error_info {
  PLI_INT32  state;       /* vpiCompile, vpiPLI, vpiRun */
  PLI_INT32  level;       /* vpiNotice, vpiWarning, vpiError,
                             vpiSystem, vpiInternal */
  PLI_BYTE8 *message;
  PLI_BYTE8 *product;
  PLI_BYTE8 *code;
  PLI_BYTE8 *file;
  PLI_INT32  line;
} s_vpi_error_info, *p_vpi_error_info;
```

The Verilog simulator will fill in the fields of the s_vpi_error_info structure when `vpi_chk_error()` is called. The **state** field represents when the error occurred, using one of the constants **vpiCompile, vpiPLI** or **vpiRun**. The **level** field indicates the severity of the error, using one of the constants **vpiNotice, vpiWarning, vpiError, vpiSystem** or **vpiInternal**. The **message, product** and **code** fields are pointers to strings, but the PLI standard does not define the wording of the strings. A simulator may use these fields for any message, and the message may vary from one simulator product to another. The **file** field is a pointer to a string containing Verilog source code file name which contains the instance of the system task/function that called the PLI application. The **line** field will be filled with the line number of the line which contains the system task/function instance.

 It is possible for a VPI call to not be associated with a Verilog HDL source code line (such as when a VPI simulation callback occurs). In these situations, the file name pointer will be set to NULL. Printing a NULL with the **%s** format will result in an error. It is therefore important that a PLI application verify that the file name pointer is valid before printing the file name.

 The `vpi_chk_error()` routine can provide important information about problems in a PLI application, and is a valuable debug utility. However, excessive use of this routine can degrade the run-time performance of simulation. For best performance, an application should use a VPI routine's exception return value for basic error checking, and only use `vpi_chk_error()` when there is not an exception value or when the additional debug information is required.

In the following example, vpi_chk_error() is used to check for an error after vpi_handle() is called, and, if an error occurred, report a detailed error message. In this example, C conditional compilation is used to only include vpi_chk_error() when debugging an application.

```
#define PLIbookDebug 1    /* set to 0 to omit debug messages */
#if PLIbookDebug
  s_vpi_error_info err;   /* allocate a VPI error structure */
#endif

primitive_handle = vpi_handle(vpiPrimitive, NULL);
#if PLIbookDebug    /* if error, generate verbose debug message */
  if (vpi_chk_error(&err)) {
    vpi_printf("\Run time error in $list_nets PLI application:\n");
    vpi_printf(" Product: %s Code: %s\n", err.product, err.code);
    vpi_printf(" Message: %s\n", err.message);
    if (err.file != NULL)
      vpi_printf(" File: %s Line: %d\n\n", err.file, err.line);
  }
#else  /* if error, generate basic error message */
  if (primitive_handle == NULL)
    vpi_printf("\nERROR: could not obtain primitive handle\n");
#endif
```

4.4 VPI object diagrams

The IEEE 1364 PLI standard includes an *object diagram* for each object which VPI routines can access. These object diagrams document:

- The *properties* of the object. For example, a net object has *name*, *vector size*, and *logic value* properties (as well as several other properties).

- The *relationships* of the object. Relationships indicate how an object is connected to or contained within other objects within a Verilog data structure. For example, a net is contained within a module, and may also be connected to other objects, such as a module port or primitive terminal.

The object diagrams use enclosures and arrows. The type of object is listed within each enclosure, and the relationships to other objects are shown as arrows between the enclosures. The properties of the object are listed below the diagrams. Figure 4-1 shows part of the object diagram for a Verilog module object.

Figure 4-1: Partial VPI object diagram for Verilog modules

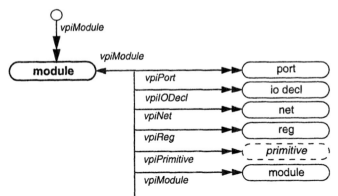

NOTE: Not all objects and properties are listed in this example. Refer to Appendix D.1.4 on page 724 for the complete module diagram.

int / str	vpiType	returns **vpiModule**
bool	vpiTopModule	returns 1 (true) if a module is a top-level module
bool	vpiArray	returns 1 (true) if a module is part of an instance array
int	vpiTimeUnit	returns the module time unit as 2 down to -15, where 2==100 seconds, 1==10s, 0==1s, -1==100ms, -2 ==10ms, -3==1ms, ... -6==1us,... -9==1ns, ... -12==1ps, ... -15==1fs
int	vpiTimePrecision	returns module time precision as 2 to -15, as with time units
str	vpiName	returns the instance name of the module
str	vpiFullName	returns the full hierarchical path name of the module
str	vpiDefName	returns the definition name of the module
str	vpiFile	returns the file name containing the module instance
int	vpiLineNo	returns the file line number of the module instance
str	vpiDefFile	returns the file name where the module is defined
int	vpiDefLineNo	returns the file line number where module is defined

 This book does not use the IEEE 1364 format for object diagrams. The use of enclosures and relationship arrows is similar, but this book adds information which is not shown in the IEEE object diagrams. This book adds:

- The constant names which are used to traverse from one object to another.
- The return values possible for each property constant.

4.4.1 Object diagram symbols

An object's object diagram contains four primary symbols and four font type faces:

- A *solid enclosure*, such as ⬭module⬭ or ⬭port⬭, designates a Verilog object. The name of the object is shown within the enclosure. The font used for the name has significance:

  - A **non-italicized, bold font** designates that this object is being defined in this diagram. In the diagram for module objects, the name, **module**, is in bold.

  - A non-italicized, non-bold font designates that this object is being referenced in this diagram, but is not being defined. The definition will appear in a different diagram. For example, in the diagram for module objects, the name port, is not bolded.

- A *small dotted enclosure*, such as ⌐ *variables* ¬, designates a reference to a named class of Verilog objects. A class of objects is a group of several objects which have something in common. The name of the object class is shown within the enclosure, using an *italicized, non-bold font*. The specific objects within a class are listed in the diagram for the class definition.

- A *large dotted enclosure*, which has small enclosures within it, designates the definition of a class of Verilog objects. The large enclosure contains all of the objects which make up the class. A class of objects may or may not have a name. However, only a named class can be referenced in another diagram. The name of the class is shown at the top of the large enclosure, using an ***italicized, bold font***. Two examples of object class definitions are:

named object class   un-named object class

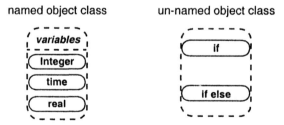

- A *solid circle*, such as ○, designates that a NULL is to be used as the reference object in the VPI routine that traverses to the target object.

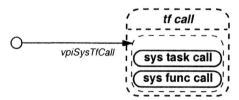

4.4.2 Traversing object relationships

The relationships of one object to another object are shown as arrows in the object diagrams. Each arrow indicates the relationship from a reference object (the originating end of the line) to a target object (the terminating end of the line). The type of arrow at the target object indicates the type of relationship. Most objects can be both a reference object or a target object, depending on which direction the Verilog hierarchy is being traversed.

There are three types of object relationships possible in a Verilog design:

- *One-to-one* relationships are represented by a line which terminates with a single arrow in a object diagram. A one-to-one relationship indicates that a given object is related to, at most, one of another type of object. In the module object diagram shown in Figure 4-1 on page 103, there is a single arrow going from (port) back to (module), which indicates a given port is only contained within one module.

- *One-to-many* relationships are represented by a line which terminates with a double arrow in a object diagram. A one-to-many relationship indicates that a given object is related to any number of another type of object. In the module object diagram shown in Figure 4-1 on page 103, there is a double arrow going from (module) to (port), which indicates there may be any number of ports within a module.

- *many-to-one* relationships are represented by a line which originates with double arrows and terminates with a single arrow. A many-to-one relationship indicates that any number of a given object are related to a single one of another type of object. There is only one type of object in Verilog which can have a many-to-one relationship: an interconnection path between modules (e.g.: a data bus might have many drivers connected to the same load).

In most object relationships, the connecting line both originates and terminates with a single or double arrow. In certain relationships, however, the connecting line has no arrow on the originating end, at the reference object. This indicates that the relationship is one-way. That is, the target object can be accessed from the reference object, but it not possible to get back to the reference object from the target object.

The VPI routines used to traverse from one type of object to another are: vpi_handle(), vpi_iterate(), vpi_scan(), and vpi_handle_multi(). The usage of these routines is presented in the next section.

4.5 Obtaining handles for objects

Several VPI routines provide a means for a PLI application to locate and retrieve handles for Verilog objects. Two of these routines were introduced in the previous chapter on using VPI routines, and are presented again in this section, along with the remaining VPI routines which obtain handles for Verilog objects.

4.5.1 Traversing one-to-one relationships

vpi_handle() returns a handle for a target object which has a one-to-one relationship to a reference object. The routine is passed in two inputs:

vpiHandle **vpi_handle (**
   PLI_INT32   **type,**   /* constant representing an object type */
   vpiHandle   **reference)** /* handle for an object */

In the object diagram for modules shown in Figure 4-1 on page 103, there is a single arrow from (port) to (**module**), indicating a one-to-one relationship. Assuming a PLI application had already obtained a handle for a port, the module which contains the port can be obtained using vpi_handle().

```
vpiHandle  port_handle, module_handle;

/* add code to obtain handle for a port */

module_handle = vpi_handle(vpiModule, port_handle);
```

The constant **vpiModule** used above indicates the type of object to be retrieved by vpi_handle(). The constant name is shown in the object diagram, next to the arrow going to that target object.

4.5.2 Traversing one-to-many relationships

Many objects in Verilog have a one-to-many relationship with other objects. A 2-step process is used to traverse these types of relationships.

1.   Obtain a handle for an *iterator object* for the desired target objects.

2.   Process each object referenced by the iterator object, one-by-one.

vpi_iterate() sets up an iterator object which represents the first of the many target objects, and returns a handle for the iterator. If there are no target objects in the Verilog data structure, then vpi_iterate() will return **NULL**. This routine is passed in two inputs:

vpiHandle **vpi_iterate (**
   PLI_INT32   **type,**   /* constant representing an object type */
   vpiHandle   **reference)**   /* handle for an object */

vpi_scan() returns a handle for each target object referenced by the iterator object. The routine is passed in a single input—the handle for the iterator:

vpiHandle **vpi_scan (**
   vpiHandle   **iterator)**   /* handle for an iterator object */

vpi_scan() returns the next target handle referenced by the iterator each time it is called. Therefore, the routine must be called repeatedly (typically using a loop) in order to access all of the target objects. When vpi_scan() has returned all objects referenced by the iterator, it will return **NULL** the next time it is called. The NULL return can be used as a flag to exit the loop.

In the example object diagram for modules shown in Figure 4-1 on page 103, there is a double arrow from ⟨ **module** ⟩ to ⟨ port ⟩, indicating a one-to-many relationship. Assuming a PLI application had already obtained a handle for a module, the ports of the module can be obtained using vpi_iterate() and vpi_scan().

```
vpiHandle module_handle, port_handle, port_iterator;

/* add code to obtain a handle for a module */

port_iterator = vpi_iterate(vpiPort, module_handle);
while ( (port_handle = vpi_scan(port_iterator)) != NULL) {
   /* code to process the port handle */
}
```

Managing memory for iterator object handles returned by vpi_iterate()

Generally, the memory required for an iterator object is automatically maintained by the simulator. The vpi_iterate() routine will allocate the memory needed for the iterator, and vpi_scan() will automatically free the memory when it returns NULL (which indicates there are no more target objects referenced by the iterator). This automatic memory management relieves the PLI application developer of the need to allocate and de-allocate memory for the iterator object.

 Because the memory for the iterator object is automatically freed, the iterator object set up by vpi_iterate() cannot be reused after vpi_scan() has returned NULL. Attempting to use an iterator handle after the iterator object memory is freed will result in unpredictable behavior in simulation (perhaps a program crash).

4.5.3 When to use vpi_free_object() on iterator handles

Normally, the memory management of iterator objects is automatic, but there is one circumstance which requires special attention. Occasionally, a PLI application might call vpi_iterate() to set up an iterator object, but then the application does not call vpi_scan() at all, or does not call vpi_scan() until the routine returns NULL. In this circumstance, the memory allocated by vpi_iterate() will not be de-allocated automatically. In order to prevent a memory leak in a PLI application, the application must manually free the iterator object memory by calling the routine **vpi_free_object()**. The vpi_free_object() routine returns 1 (for true) if successful, and 0 (for false) if unsuccessful. The syntax for this routine is:

PLI_INT32 **vpi_free_object (**
   vpiHandle   **handle)**   /* handle for an object */

In the following example, vpi_iterate() is used as a simple true/false test to see if a module has any ports. The handles to the ports are not needed, and so vpi_scan() is not called. Therefore, the memory allocated by vpi_iterate() must be explicitly de-allocated using vpi_free_object().

```
int module_has_ports(vpiHandle module_handle)
{
  vpiHandle port_iterator;
  port_iterator = vpi_iterate(vpiPort, module_handle);
  if (port_iterator == NULL)
    return(0);   /* no ports found */
  else
    vpi_free_object(port_iterator); /* release iterator */
  return(1);     /* ports were found */
}
```

 Do not free an iterator object which has already been released! Attempting to do so may result in undefined behavior by a simulator. Once vpi_scan() returns NULL, there is no need to call vpi_free_object() on the iterator, because the simulator has already freed it.

The code fragment that follows calls vpi_iterate() to access the arguments of a system task/function. vpi_scan() is then called twice to obtain handles for the first two arguments. Since vpi_scan() was not called in a loop until it returns NULL, the example performs two important safety checks:

1. Before calling vpi_scan() the second time, the example verifies that the previous call did not return NULL. If the previous return were NULL, then the simulator would have automatically freed the iterator. Any subsequent call to vpi_scan() with that iterator would have resulted in an error and possibly a program crash.

2. After calling. vpi_scan() the second time, the iterator is no longer needed. To
    prevent a memory leak, the PLI application must release the iterator using
    vpi_free_object(). However, before manually releasing the iterator, the
    example verifies that the previous call to vpi_scan() did not return NULL,
    which would mean that the simulator automatically freed the iterator. Calling
    vpi_free_object() with an iterator that was already released would have
    resulted in an error and possibly a program crash.

```
...
arg_iterator = vpi_iterate(vpiArguments, systf_handle);
arg1_handle = vpi_scan(arg_iterator);
if (arg1_handle != NULL) /* verify iterator is still valid */
  arg2_handle = vpi_scan(arg_iterator);
if (arg2_handle != NULL) /* verify iterator is still valid */
  vpi_free_object(arg_iterator);  /* free iterator memory */
...
```

TIP
It is a good practice to always check that the last call to vpi_scan() did not return
NULL before calling vpi_free_object() on an iterator handle. Inadvertently
freeing an iterator that had been automatically released by the simulator can lead to
unpredictable simulation behavior, including program crashes.

4.5.4 Obtaining intermodule path object handles

In the Verilog language, the output port of one module can be connected to one or
more input ports of other modules, using a net data type. The connection from an output to an input is referred to as an ***intermodule path***. In actual hardware, this inter-
connection will have a real delay, but, within the Verilog language, there is no construct to accurately represent that delay. The PLI, however, can add, read and modify
intermodule path delays. A module input port can be driven by any number of module
output ports. An example of multiple inter-connections might be a shared data bus
which can be driven by several components. Because of the possibility of multiple
connections, the VPI library refers to intermodule paths as a many-to-one relationship.

vpi_handle_multi() is used to obtain a handle for an intermodule path. Other VPI
routines can then annotate delays to the path. This routine requires three inputs:

vpiHandle **vpi_handle_multi (**

PLI_INT32	**type,**	/* constant of **vpiInterModPath** */
vpiHandle	**reference1,**	/* handle for an output or inout port */
vpiHandle	**reference2)**	/* handle for an input or inout port */

The constant **vpiInterModPath** is the only type constant supported by
vpi_handle_multi(). The ports specified must be the same vector size, but they
do not need to be in the same level of Verilog hierarchy. If no interconnecting net
exists between the two ports, then vpi_handle_multi() will return NULL. An
example of using this routine is:

```
inter_mod_path_h = vpi_handle_multi(vpiInterModPath,
                                    in_port_handle,
                                    out_port_handle);
if (inter_mod_path_h != NULL)
   /* inter connection path not found -- process an error */
else
   /* read or modify the inter-connect delay values */
```

4.5.5 Obtaining object handles using an object's name

vpi_handle_by_name() obtains a handle for an object using the name of the
object. The handle for any Verilog object with a vpiFullName property in the object
diagrams can be obtained using this routine. The routine requires two inputs:

vpiHandle **vpi_handle_by_name (**
   PLI_BYTE8 *** name,**   /* name of an object */
   vpiHandle   **scope)**   /* handle for a scope object, or **NULL** */

The name provided can be the local name of the object, a relative hierarchical path
name, or a full hierarchical path name. If a name without a full hierarchy path is specified, the routine will only search for the object in the scope specified. If a NULL is
specified for the scope handle, the object will be searched for in the first top level
module found by the routine. If the object cannot be found, then
vpi_handle_by_name() will return NULL.

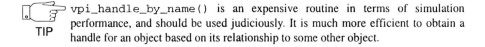

> vpi_handle_by_name() is an expensive routine in terms of simulation
> performance, and should be used judiciously. It is much more efficient to obtain a
> TIP   handle for an object based on its relationship to some other object.

In the following code fragment, the name of a primitive instance and a delay value are
read from a file. The handle for the primitive is obtained using
vpi_handle_by_name(), and the delay is annotated onto the primitive.

```
PLI_BYTE8    prim_name[1025];
vpiHandle    prim_handle;
double       new_delay;
```

```
fscanf(file_p, "%s %f", prim_name, &new_delay);
prim_handle = vpi_handle_by_name(prim_name, NULL);
if (prim_handle != NULL)
  /* add new delay value to the primitive object */
else
  /* error: primitive not found */
```

4.5.6 Obtaining object handles using an object's index number

vpi_handle_by_index() is used to obtain a handle for an object using the object's index position. This routine requires two inputs:

vpiHandle **vpi_handle_by_index (**
    vpiHandle   **parent,**   /* handle for an object with a vpiIndex relationship */
    PLI_INT32   **index)**   /* index number of an object */

The handle for any Verilog object which has a vpiIndex relationship in the object diagrams can be obtained using this routine. These objects are the bits of a vector net or vector reg, and the words of a memory array or a variable array. The bit or word that is represented by an index number is based on the declaration of the vector or array. If an index is out of range, a NULL is returned. Note that vpi_handle_by_index() requires the object have an index *relationship*. Some Verilog objects have an index *property*, which is not the same as an index relationship. Relationship are represented as transition arrows from one object to another in an object diagram.

Assuming a handle for a net vector had already been obtained, the following example would obtain a handle for bit number 2 of the vector:

```
vpiHandle  net_handle, bit_handle;

/* add code to obtain handle for a net vector */

bit_handle  = vpi_handle_by_index(net_handle, 2);
if (bit_handle != NULL)
  /* process the net bit object */
else
  /* error: terminal not found */
```

4.5.7 Obtaining handles for reg, variable and net arrays

IEEE 1364-1995 and earlier generations of Verilog allowed declarations of one-dimensional arrays of reg data types, called *memory arrays*, and one-dimensional arrays of integer and time variables, called *variable arrays*. Since the reg data type allows both a vector size and an array size declaration, some Verilog texts refer to a memory array as two-dimensional; but in actuality, it is a one-dimensional array, where each element in the array can be any vector width. The IEEE 1364-1995 syntax for array declarations is:

```
reg [<msb>:<lsb>] <memory_name> [<first_addr>:<last_addr>];

integer <memory_name> [<first_address>:<last_address>];

time <memory_name> [<first_address>:<last_address>];
```

The IEEE 1364-2001 standard adds several significant enhancements to the Verilog HDL that affect arrays, and how the Verilog PLI accesses arrays:

- Arrays of any data type, including nets, real variables, and named events
- multi-dimensional arrays of all data types
- Bit-selects and part-selects within an array

The following statement in the Verilog HDL declares a 3-dimensional array of 16-bit wide nets (each element in the array is 16 bits wide):

```
wire [15:0] foo [0:127][0:127][0:7]; // 3-D array
```

Examples of a word-select and a bit-select within the preceding array are:

```
vector = foo[25][0][100];    //select full word from array

bit = foo[25][0][100][5];    //select bit 5 from an array word
```

1364-1995 memory arrays

A one-dimensional array of **reg** data types is referred to as a *memory array* in the IEEE Verilog standard. The IEEE 1364-1995 standard provided a special ⊂ **memory** ⊃ object diagram to access memory arrays. The IEEE 1364-2001 standard makes this memory object diagram obsolete, and adds a new ⊂ **reg array** ⊃ diagram, which can access multi-dimensional arrays of **reg** variables. This new reg array object is a super set of the one-dimensional memory arrays. PLI applications should take advantage of the greater flexibility of the reg array object. The older memory object diagram is maintained in the IEEE 1364 standard for backward compatibility. This chapter does not cover the obsolete memory object diagram, however the diagram is included in Appendix D.1.12.

Net, reg and named event arrays

The diagrams for nets, regs and named events are all similar in how these objects and arrays of the objects are represented. A partial diagram for reg and reg array objects is shown here. Refer to Appendix D for the complete diagram on all of these objects.

Figure 4-2: Object diagrams for reg and reg array (partial)

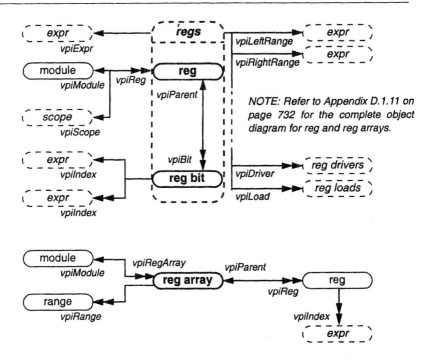

A handle to a multi-dimensional array of nets, regs or named events can be obtained in several ways using the VPI library:

- Pass the name of the array as a task/function argument.

- Iterate on all net, reg or named event array objects, for example using `vpi_iterate(vpiNetArray, module_handle)`.

- Iterate on all memory array objects using `vpi_iterate(vpiMemory, module_or_scope_handle)`; note that this will only obtain one-dimensional reg arrays, and not reg arrays with more than one dimension.

A handle to a word within an array can be obtained:

- As a handle to an expression, when a word-select from an array is used in a Verilog expression.

- With a handle to an array with any number of dimensions, iterate on each word within the array.

- Directly access a specific word by its index in a one-dimensional reg array, using vpi_handle_by_index(memory_handle, index_number).

- Directly access a specific word by its index in a multi-dimensional array of any type, using vpi_handle_by_multi_index(array_handle, array_size, index_array).

vpi_handle_by_multi_index() is used to obtain a handle for a word within an array, or bit-select of a word out of an array object. This routine was added with the IEEE 1364-2001 standard. The routine requires three inputs:

vpiHandle **vpi_handle_by_multi_index (**
 vpiHandle **object,** /* handle for an array object */
 PLI_INT32 **array_size,** /* number or elements in the index array */
 PLI_INT32 *** index_array**) /* pointer to an array of indices */

The order of the indices provided in the index array is the left most select first, progressing to the right most select last. To use vpi_handle_by_multi_index(). The handle for any Verilog object which has a vpiIndex relationship in the object diagrams can be obtained using this routine. These objects are the bits of a vector net or vector reg, and the words of a net array, reg array or variable array. Note that vpi_handle_by_multi_index() requires the object have an index *relationship*. Some Verilog objects have an index *property*. This property is not the same as an index *relationship*. vpi_handle_by_multi_index() is used when there is an index relationship. The bit or word that is represented by an index number is based on the declaration of the vector or array. If an index is out of range, a NULL is returned.

The following statement in the Verilog HDL declares a 3-dimensional array of 16-bit wide nets (each element in the array is 16 bits wide):

```
wire [15:0] foo [0:127][0:127][0:7]; // 3-D array
```

Assuming a handle for the array had already been obtained, the following example would obtain a handle for a specific 16-bit net within the array, specifically foo[25][0][100]:

```
vpiHandle   array_handle, net_handle;
PLI_INT32   indices[10]; /* up to a 10-dimensional array */

/* add code to obtain handle for the array */

indices[0] = 25; indices[1] = 0; indices[2] = 100;
net_handle  = vpi_handle_by_multi_index(array_handle,
                                        3,
                                        indices);
```

Variable arrays

A ⌐ *variables* ¬ object is represented a differently than a reg array, net array or
named event array. A variables object may represent a single variable or it may be an
array of variables (whereas a net, reg or named event object is always a single object,
and never an array).

Figure 4-3: Object diagrams for variables (partial)

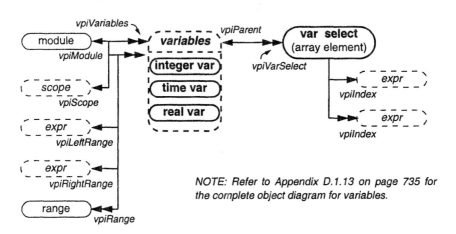

NOTE: Refer to Appendix D.1.13 on page 735 for
the complete object diagram for variables.

A handle for a variable or variable array is obtained the same way, by using
vpi_iterate(vpiVariables, module_or_scope_handle). Once the handle is
obtained, the boolean property **vpiArray** is used to determine if a variable object is
an array.

- If the **vpiArray** property is false, then the variable object is a single variable. The
  vpiSize property will be the number of bits in the variable, and the expressions
  accessed by **vpiLeftRange** and **vpiRightRange** will indicate the most-significant and least-significant bit numbers, respectively.

- If the **vpiArray** property is true, then the variable object is an array of variables. The **vpiSize** property will be the number of elements in the array, and the expressions accessed by **vpiLeftRange** and **vpiRightRange** will indicate the first address and last address of the array, respectively.

Each element in a variable array is a (**var select**). A handle for all var select objects in an array can be obtained using vpi_iterate(vpiVarSelect, array_handle). A var select can also be used in a Verilog expression, and as such is listed in the expression class of objects. A handle for a specific var select can be obtained using vpi_handle_by_index(array_handle, index_number). The expression which can be accessed using **vpiIndex** represents the address of that element within the array, and the **vpiSize** property indicates the vector size of the element. There is no access to the most-significant bit and least-significant bit expressions of a var select object.

4.6 System task and system function objects

The Verilog PLI allows users to create user-defined system tasks and user-defined system functions which can be used in Verilog HDL source code. How user-defined system tasks and system functions are defined was presented in Chapter 2.

4.6.1 System task/function arguments

A user-defined system task or system function can have any number of arguments, including none. The VPI standard does not specify an index number for task/function arguments, but the TF and ACC libraries number the arguments from left to right, starting with 1. For convenience in describing PLI applications, this book uses the same numbering scheme in the VPI chapters. In the following example:

- Task/Function argument number 1 is a string, with the value "vectors.pat".
- Task/Function argument number 2 is a signal, with the name input_bus.

4.6.2 Multiple instances of system tasks and system functions

The Verilog HDL source code can reference the same system task/function any number of times. Section 2.3 on page 31 in Chapter 2 discussed the different ways in which multiple instances of a system task/function can occur in Verilog source code. Each instance of a system task/function is unique, and has unique argument values. However, each instance is associated with the same *calltf routine*. For example:

```
always @(posedge clock)
  $read_test_vector("A.dat", data_bus);

always @(negedge clock)
  $read_test_vector("B.dat", data_bus);
```

In this example, *$read_test_vector* is used two times. The *calltf routine* is a C function, and every instance of *$read_test_vector* will invoke the same C function. Within the simulation data structure, however, each call to the function is unique. At each positive edge of clock the *calltf routine* will be invoked, and when the routine reads the value of its argument 1, it will retrieve the string `"A.dat"`. At the negative edge of clock, the same *calltf routine* will be executed, but when this call to the routine reads the value of its argument 1, it will retrieve the string `"B.dat"`.

 Storage allocated by a PLI application is not unique to each instance! If the application declares static variables or allocates memory, all instances of the application will share those variables or memory.

If a PLI application requires unique storage, such as a storing a different file pointer for each system task instance, then the application must allocate and identify that unique storage. Section 4.7 presents how instance specific storage can be allocated.

4.6.3 Obtaining a handle for a system task/function instance

An instance of a system task or system function is an object, and the VPI routines can obtain a handle for this object. This handle will be unique for each instance of the system task/function. A partial object diagram for a system task/function is shown in Figure 4-4.

Figure 4-4: Object diagram for a task/function call (partial)

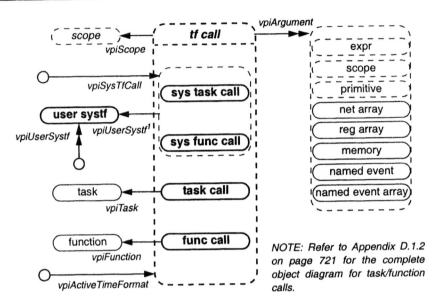

In this diagram, the objects (sys task call) and (sys func call) are part of an object group, called *tf call*. These objects can be accessed from a small circle, using the constant **vpiSysTfCall**. The small circle indicates that a NULL is used as the reference handle for vpi_handle(). Therefore, a handle for the system task/function instance which called a PLI application is obtained using:

```
vpiHandle systf_handle;
systf_handle = vpi_handle(vpiSysTfCall, NULL);
```

4.6.4 Accessing the arguments of system tasks and system functions

Once a handle for an instance of a system task or system function is obtained, the arguments to the system task/function can be accessed by obtaining handles for the arguments. In the Verilog HDL, many different values and data types can be used as a task/function argument—an integer number, a variable name, a net name, a module instance name, or a literal string are just a few of the legal arguments. In the object diagram for *tf call*, one of the types of argument objects is a reference to an *"expression"* class of objects. The expression class is defined in another diagram, and includes several object types which can be used as an argument. Since a system task/ function can have any number of arguments, the line from ⟨ *tf call* ⟩ to ⟨ *expr* ⟩ terminates with a double arrow, indicating a one-to-many relationship.

Examples 4-1 and 4-2, which follow, illustrate using VPI routines to access the arguments of a system task/function. These examples are C functions which are intended to be called from other PLI applications. Example 4-1 is a function called **PLIbook_count_args_vpi()**, which returns the number of arguments in a system task/function instance.

> **CD** The source code for this example is on the CD accompanying this book.
>
> - Application source file: `Chapter.04/count_args_vpi.c`
> - Application test file: `Chapter.04/count_args_vpi_test.c`
> - Verilog test bench: `Chapter.04/count_args_test.v`
> - Expected results log: `Chapter.04/count_args_test.log`

Example 4-1: *PLIbook_count_args_vpi()* — accessing system task/func. arguments

```
int PLIbook_count_args_vpi()
{
  vpiHandle systf_h, arg_itr, arg_h;
  int tfnum = 0;
  s_vpi_error_info err;  /* structure for error handling */

  systf_h = vpi_handle(vpiSysTfCall, NULL);
#if PLIbookDebug /* if error, generate verbose debug message */
  if (vpi_chk_error(&err)) {
    vpi_printf("ERROR: PLIbook_count_args_vpi() could not obtain
               handle to systf call\n");
    vpi_printf("File %s, Line %d: %s\n",
               err.file, err.line, err.message);
  }
#else  /* if error, generate brief error message */
  if (systf_h == NULL)
    vpi_printf("ERROR: PLIbook_count_args_vpi() could not obtain
               handle to systf call\n");
#endif

  arg_itr = vpi_iterate(vpiArgument, systf_h);
#if PLIbookDebug  /* if error, generate verbose debug message */
  if (vpi_chk_error(&err)) {
    vpi_printf("ERROR: PLIbook_count_args_vpi() could not obtain
               iterator to systf args\n");
    vpi_printf("File %s, Line %d: %s\n",
               err.file, err.line, err.message);
  }
#else  /* if error, generate brief error message */
  if (arg_itr == NULL)
    vpi_printf("ERROR: PLIbook_count_args_vpi() could not obtain
               iterator to systf args\n");
#endif
```

```
  while (arg_h = vpi_scan(arg_itr) ) {
    tfnum++;
  }

  return(tfnum);
}
```

TIP The preceding example and the next examples use conditional compilation to enable verbose error messages. These message can provide additional insight for debugging any problems with the applications. The verbose error messages can be omitted for normal usage of the applications.

Example 4-2 lists a C function called **PLIbook_get_arg_handle_vpi()**, which uses the VPI routines to return a handle for a specific system task/function argument. The input to this routine is an index number of the argument, where the left-most argument is index number 1, and the numbers increases from left to right.

CD The source code for this example is on the CD accompanying this book.

- Application source file: `Chapter.04/get_arg_handle_vpi_inefficient.c`
- Application test file: `Chapter.04/get_arg_handle_vpi_test.c`
- Verilog test bench: `Chapter.04/get_arg_handle_test.v`
- Expected results log: `Chapter.04/get_arg_handle_test.log`

Example 4-2: *PLIbook_get_arg_handle_vpi()* — accessing system task/func args

```
vpiHandle PLIbook_get_arg_handle_vpi(int argNum)
{
  vpiHandle systf_h, arg_itr, arg_h;
  int i;
  s_vpi_error_info err;  /* structure for error handling */

  if (argNum < 1) {
    #if PLIbookDebug /* if error, generate verbose debug message */
      vpi_printf("ERROR: PLIbook_get_arg_handle_vpi() arg index of %d is
                  invalid\n", argNum);
    #endif
    return(NULL);
  }

  systf_h = vpi_handle(vpiSysTfCall, NULL);
  #if PLIbookDebug /* if error, generate verbose debug message */
    if (vpi_chk_error(&err)) {
      vpi_printf("ERROR: PLIbook_get_arg_handle_vpi() could not obtain
                  handle to systf call\n");
```

```
        vpi_printf("File %s, Line %d: %s\n",
                    err.file, err.line, err.message);
    }
#else   /* if error, generate brief error message */
    if (systf_h == NULL) {
      vpi_printf("ERROR: PLIbook_get_arg_handle_vpi() could not obtain
                  handle to systf call\n");
      return(NULL);
    }
#endif

    arg_itr = vpi_iterate(vpiArgument, systf_h);
#if PLIbookDebug   /* if error, generate verbose debug message */
    if (vpi_chk_error(&err)) {
      vpi_printf("ERROR: PLIbook_get_arg_handle_vpi() could not obtain
                  iterator to systf args\n");
      vpi_printf("File %s, Line %d: %s\n",
                    err.file, err.line, err.message);
    }
#else   /* if error, generate brief error message */
    if (systf_h == NULL) {
      vpi_printf("ERROR: PLIbook_get_arg_handle_vpi() could not obtain
                  iterator to systf args\n");
      return(NULL);
    }
#endif

    for (i=1; i<=argNum; i++) {
      arg_h = vpi_scan(arg_itr);
      #if PLIbookDebug   /* if error, generate verbose debug message */
        if (vpi_chk_error(&err)) {
          vpi_printf("ERROR: PLIbook_get_arg_handle_vpi() could not obtain
                      handle to systf arg %d\n", i);
          vpi_printf("File %s, Line %d: %s\n",
                      err.file, err.line, err.message);
        }
      #endif
      if (arg_h == NULL) {
        #if PLIbookDebug   /* if error, generate verbose debug message */
          vpi_printf("ERROR: PLIbook_get_arg_handle_vpi() arg index of %d
                      is out-of-range\n",
                      argNum);
        #endif
        return(NULL);
      }
    }
    if (arg_h != NULL)
      vpi_free_object(arg_itr); /* free iterator--didn't scan all args */

    return(arg_h);
}
```

The preceding two examples are not as efficient for simulation run-time performance as they could be. In both examples, the routines `vpi_iterate()` and `vpi_scan()` must be called every time the functions are invoked, in order to access the system task/function arguments. A more efficient method is presented in the following section.

4.7 Storing data for each instance of a system task/function

There are many circumstances where a PLI application may need to preserve information from one call to a PLI application to another call to the same application. The preceding *PLIbook_count_args_vpi()* and *PLIbook_get_arg_handle_vpi()* examples could be made much more efficient if the argument count and argument handles were saved the first time the application were called. Each subsequent call could simply return the saved information without having to call `vpi_iterate()` and `vpi_scan()` to retrieve the information again.

Important! The system task or function that calls a PLI application can occur many times in the Verilog HDL source code. Each instance of the system task/function is unique, and may have different task/function arguments. For example:

```
always @(posedge clock)
  $ALU(a_bus, b_bus, opcode, result_bus, overflow);

always @(negedge clock)
  $ALU(in1, in2, control, out1);
```

In order to store the system task/function arguments, unique storage must be allocated for each instance of a system task or function which uses the *PLIbook_count_args_vpi()* or *PLIbook_get_arg_handle_vpi()* applications. A PLI application is a C function, which is called by the Verilog simulator. Local variables within a C function are automatic, which means any variables (such as an array to store the argument handles) do not remain allocated from one call of the function to another call.

A common solution in the C programming language to preserve data is to declare a static variables, instead of automatic variables. Another method is to use global variables instead of local variables. ***These C programming techniques can cause serious problems in PLI applications!*** The Verilog HDL allows a system task/function to be used multiple times in the Verilog source code, and each module which uses the task/function can be instantiated multiple times. Each occurrence in each instance of a module becomes a unique ***instance*** of the system task/function. However, in the PLI application, the same C function will be called by the simulator for each instance. A static or global variable cannot hold different values for each instance of the system task/function. In the example:

The *$ALU* instance that is called at the positive edge of clock has different arguments than the *$ALU* that is called at the negative edge of clock. Since both instances of *$ALU* will invoke the same C function, static or global variables cannot be used, because the two instances would share the same variables. The last $ALU called would overwrite the information from the other $ALU. *Using static or global variables is a sure way to have problems when there are multiple instances of a system task or system function.* Therefore, a PLI application must allocate storage that is unique to each instance of a system task/function.

The VPI standard provides a special storage location for each instance of a system task or system function. This instance-specific storage is allocated automatically by the simulator, and is available for use by a PLI application whenever needed. Special VPI routines are provided to read and write values in the instance-specific storage area. The storage area is shared by all routines which are associated with the system task/function, which are the *calltf, compiletf* and *sizetf routines*.

PLI_INT32 **vpi_put_userdata (**
   vpiHandle   **tfcall,**   /* handle for a system task or system function call */
   void   *** data)**   /* pointer to application-allocated storage */

void ***vpi_get_userdata (**
   vpiHandle   **tfcall)**   /* handle for a system task or system function call */

vpi_put_userdata() stores a pointer to application-allocated storage into simulator-allocated storage for an instance of a system task or function. The routine returns 1 if successful and 0 if an error occurred. The simulation-allocated storage will persist throughout simulation. vpi_get_userdata() retrieves a pointer to the data that was stored using vpi_put_userdata(). The routine returns NULL if no data has been stored.

4.7.1 Storing a single value in the VPI instance-specific storage area

The instance-specific storage area is defined to be a pointer, which can store a single value. The value to be stored should be cast to a void pointer. An example of storing a single integer value in the work area is:

```
int arg_count;
/* add code to count number of task/function arguments */
vpi_put_userdata(systf_handle, (void *)arg_count);
```

The count stored in the instance-specific storage area will be preserved throughout simulation. Each time the same instance of the system task or function calls a PLI application, it can retrieve the count value, without the need to iterate and scan for all arguments.

4.7.2 Storing multiple values in the VPI instance-specific storage area

Since the instance-specific storage area is a character pointer, it can be used to store a pointer to an application-allocated memory location. Multiple values can be stored by allocating a block of memory and storing a pointer to the memory in the storage area. In the example:

```
always @(posedge clock)
  $ALU(a_bus, b_bus, opcode, result_bus, overflow);

always @(negedge clock)
  $ALU(in1, in2, control, out1);
```

The first time a PLI application for one of the instances of $ALU calls either *PLIbook_count_args_vpi()* or *PLIbook_get_arg_handle_vpi()*, the function can allocate an array to store both the count and all the handles to the task/function arguments. A pointer to the array can be stored in the instance-specific storage area for that system task/function instance.

```
vpiHandle *arg_array; /* array pointer */

/* count number of task/function arguments */

arg_array = (vpiHandle *)malloc(sizeof(vpiHandle)*(args+1));

/* put argument count into arg_array[0] */
/* fill rest of array with argument handles */

vpi_put_userdata(systf_h, (void *)arg_array);
```

Example 4-3 shows a more efficient version of the *PLIbook_count_args_vpi()* and *PLIbook_get_arg_handle_vpi()* applications. The first time either of these functions is called, it will call a sub function to allocate an array, save the argument count and argument handles, and save a pointer to the array in the instance-specific storage area. Since the storage is associated with the system task/function instance, both *PLIbook_count_args_vpi()* and *PLIbook_get_arg_handle_vpi()* will have access to the pointer stored in the storage area.

CD The source code for this example is on the CD accompanying this book.

- Application source file: `Chapter.04/get_arg_handle_vpi_efficient.c`
- Application test file: `Chapter.04/get_arg_handle_vpi_test.c`
- Verilog test bench: `Chapter.04/get_arg_handle_test.v`
- Expected results log: `Chapter.04/get_arg_handle_test.log`

Example 4-3: *PLIbook_get_arg_handle_vpi()* — efficient version

```
/*******************************************************************
 * PLIbook_count_args_vpi() -- Efficient Version
 *******************************************************************/
int PLIbook_count_args_vpi()
{
  vpiHandle systf_h, arg_itr, arg_h;
  int tfnum = 0;
  vpiHandle *arg_array;    /* array pointer to store arg handles */
  #if PLIbookDebug
    s_vpi_error_info err;  /* structure for error handling */
  #endif

  systf_h = vpi_handle(vpiSysTfCall, NULL);
  #if PLIbookDebug /* if error, generate verbose debug message */
    if (vpi_chk_error(&err)) {
      vpi_printf("ERROR: PLIbook_count_args_vpi() could not obtain
                 handle to systf call\n");
      vpi_printf("File %s, Line %d: %s\n",
                 err.file, err.line, err.message);
    }
  #else  /* if error, generate brief error message */
    if (systf_h == NULL)
      vpi_printf("ERROR: PLIbook_count_args_vpi() could not obtain
                 handle to systf call\n");
  #endif

  /* retrieve pointer to array with all argument handles */
  arg_array = (vpiHandle *)vpi_get_userdata(systf_h);
  if (arg_array == NULL) {
    /* array with all argument handles doesn't exist, create it */
    arg_array = create_arg_array(systf_h);
  }

  return((int)arg_array[0]);
}
```

```
/**********************************************************************
 * PLIbook_get_arg_handle_vpi() -- Efficient Version
 **********************************************************************/
vpiHandle PLIbook_get_arg_handle_vpi(int argNum)
{
  vpiHandle  systf_h, arg_h;
  vpiHandle *arg_array;    /* array pointer to store arg handles */
  #if PLIbookDebug
    s_vpi_error_info err;  /* structure for error handling */
  #endif

  if (argNum < 1) {
    #if PLIbookDebug  /* if error, generate verbose debug message */
      vpi_printf("ERROR: PLIbook_get_arg_handle_vpi() arg index of %d is
                 invalid\n",
                 argNum);
    #endif
    return(NULL);
  }

  systf_h = vpi_handle(vpiSysTfCall, NULL);
  #if PLIbookDebug /* if error, generate verbose debug message */
    if (vpi_chk_error(&err)) {
      vpi_printf("ERROR: PLIbook_get_arg_handle_vpi() could not obtain
                 handle to systf call\n");
      vpi_printf("File %s, Line %d: %s\n",
                 err.file, err.line, err.message);
    }
  #else /* if error, generate brief error message */
    if (systf_h == NULL) {
      vpi_printf("ERROR: PLIbook_get_arg_handle_vpi() could not obtain
                 handle to systf call\n");
      return(NULL);
    }
  #endif
  /* retrieve pointer to array with all argument handles */
  arg_array = (vpiHandle *)vpi_get_userdata(systf_h);
  if (arg_array == NULL) {
    /* array with all argument handles doesn't exist, create it */
    arg_array = create_arg_array(systf_h);
  }
  if (argNum > (int)arg_array[0]) {
    #if PLIbookDebug  /* if error, generate verbose debug message */
      vpi_printf("ERROR: PLIbook_get_arg_handle_vpi() arg index of %d is
                 out-of-range\n",
                 argNum);
    #endif
    return(NULL);
  }
  /* get requested tfarg handle from array */
  arg_h = (vpiHandle)arg_array[argNum];
  return(arg_h);
}
```

```
/***********************************************************************
 * Subroutine to allocate an array and store the number of arguments
 * and all argument handles in the array.
 ***********************************************************************/
vpiHandle *create_arg_array(vpiHandle systf_h)
{
  vpiHandle  arg_itr, arg_h;
  vpiHandle *arg_array; /* array pointer to store arg handles */
  int        i, tfnum = 0;
#if PLIbookDebug
    s_vpi_error_info err;   /* structure for error handling */
#endif

  /* allocate array based on the number of task/function arguments */
  arg_itr = vpi_iterate(vpiArgument, systf_h);
  if (arg_itr == NULL) {
    vpi_printf("ERROR: PLIbook_numargs_vpi() could not obtain iterator
               to systf args\n");
    return(NULL);
  }
  while (arg_h = vpi_scan(arg_itr) ) {  /* count number of args */
    tfnum++;
  }
  arg_array = (vpiHandle *)malloc(sizeof(vpiHandle) * (tfnum + 1));

  /* store pointer to array in simulator-allocated user_data storage
     that is unique for each task/func instance */
  vpi_put_userdata(systf_h, (void *)arg_array);

  /* store number of arguments in first address in array */
  arg_array[0] = (vpiHandle)tfnum;

  /* fill the array with handles to each task/function argument */
  arg_itr = vpi_iterate(vpiArgument, systf_h);
#if PLIbookDebug /* if error, generate verbose debug message */
  if (vpi_chk_error(&err)) {
    vpi_printf("ERROR: PLIbook_get_arg_handle_vpi() could not obtain
               iterator to systf args\n");
    vpi_printf("File %s, Line %d: %s\n",
               err.file, err.line, err.message);
  }
#else /* if error, generate brief error message */
  if (systf_h == NULL) {
    vpi_printf("ERROR: PLIbook_get_arg_handle_vpi() could not obtain
               iterator to systf args\n");
    return(NULL);
  }
#endif
  for (i=1; i<=tfnum; i++) {
    arg_h = vpi_scan(arg_itr);
    #if PLIbookDebug /* if error, generate verbose debug message */
      if (vpi_chk_error(&err)) {
        vpi_printf("ERROR: PLIbook_get_arg_handle_vpi() could not obtain
                   handle to systf arg %d\n", i);
```

```
            vpi_printf("File %s, Line %d: %s\n",
                      err.file, err.line, err.message);
      }
   #endif
   arg_array[i] = arg_h;
  }
  if (arg_h != NULL)
    vpi_free_object(arg_itr); /* free iterator--didn't scan all args */

  return(arg_array);
}
```

4.8 Traversing Verilog hierarchy using object relationships

Using the object diagrams, the Verilog design hierarchy can be traversed from one
object to any other object anywhere in a Verilog design. Traversing the design hierarchy often requires following object relationships across several diagrams. For example, suppose a PLI application had obtained a handle for a module, and the application
needs to locate every module output, where the output is also connected to a module
path delay. The following Verilog source code shows the starting and ending objects
for which handles are desired in this example.

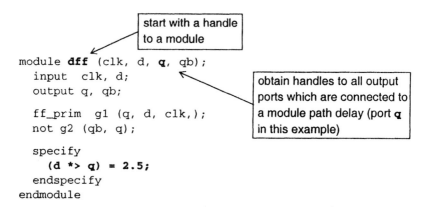

The following steps show how the relationships in the object diagrams can be followed to traverse from the starting point in a Verilog design to the desired destination:

1. The **module** diagram (shown in Figure 4-5, which follows) shows a one-to-many
    connection from ⬭ **module** ⬭ to ⬭ **mod path** ⬭ .

    Using **vpi_iterate(vpiModPath,   module_handle)**, an iterator for all
    module paths in a module can be obtained.

Figure 4-5: Object diagram for Verilog modules (partial)

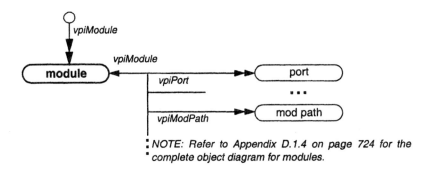

2. The ***mod path*** object diagram (shown in Figure 4-6, on the next page) shows a one-to-many relationship from (**mod path**) to (**path term**). This connection can be traversed with specific constants to obtain handles for the path input terminals, output terminals, or the data terminal. Since this example application is looking for the output ports connected to module paths, the output terminal is the relationship which needs to be followed. Using **vpi_iterate(vpiModPathOut, path_handle)**, an iterator for all path outputs can be obtained.

Figure 4-6: Object diagram for module paths (partial)

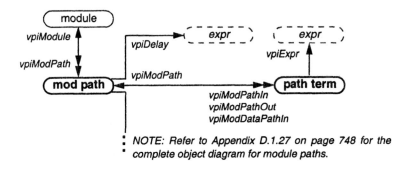

3. The (**path term**) object in the preceding ***mod path*** object diagram shows a one-to-one connection to an ⌐ ⎯ ⎯ _expr_ ⎯ ⎯ ⌐ object group (expr stands for expression). This object relationship is traversed using **vpi_handle(vpiExpr, path_term_handle)**. The expressions object diagram is shown in Figure 4-7.

Figure 4-7: Object diagram for expressions (partial)

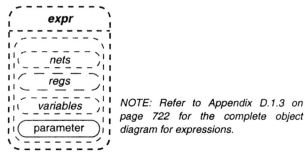

NOTE: Refer to Appendix D.1.3 on page 722 for the complete object diagram for expressions.

The expression class is a general group that contains the types of Verilog HDL objects that can be used in expressions anywhere in the Verilog HDL language. Once a handle to an expression object is obtained, calling vpi_get(vpiType, expression_handle) will show what type of object in the expression class is connected to the module path terminal. The diagram for that object will then show what properties and relationships are available for that connection. Often, however, Verilog HDL syntax will not allow some types of objects in the expression class to be used in a specific context. In this example, the expression handle was obtained from a module path output terminal. Verilog HDL syntax requires that path outputs must be connected to a module port, and that only a net data type can be used for this connection. Therefore, calling vpi_get(vpiType, expression_handle) is not really necessary in this example, as the context limits the expression to only a net or a net bit.

4. The *nets* object diagram (shown in Figure 4-8, below) shows two types of one-to-many connections from ⟨ *nets* ⟩ to ⟨ *ports* ⟩. One connection is accessed using **vpiPorts**, and the other using **vpiPortInst**.

Figure 4-8: Object diagram for nets (partial)

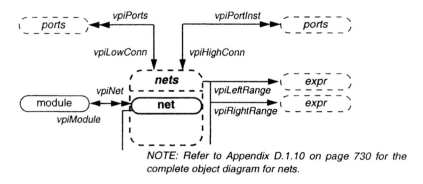

NOTE: Refer to Appendix D.1.10 on page 730 for the complete object diagram for nets.

The connection accessed by vpiPorts is the connection to a port of the module which contains the net. The connection accessed by vpiPortInst is the connection to a port of a module instance within the current module. The following Verilog model illustrates the difference between the two types of objects:

```
module chip (clock, in1, out1, out2);
    ...
    wire in1;
    ...
    dff u1 (clock, in1, out1, out2);
endmodule
```

module ports are accessed using **vpiPorts**

module *instance* ports are accessed using **vpiPortInst**

For the example being illustrated in this section (finding the module ports which are connected to a module path output), the type of port object needed is accessed using **vpi_iterate(vpiPorts, net_handle)**, which will return a list of all ports of the module to which the net is connected.

Example summary

In the example described in this section, four object connections were followed to traverse the Verilog hierarchy from a module object to the output ports connected to a path delay. The connection path is from (module) to (mod path) to (path term) to (nets) (in the expression group) to (ports). The following PLI source code illustrates how these multiple connections through the Verilog hierarchy are traversed.

CD The source code for this example is on the CD accompanying this book.

- Application source file: Chapter.04/list_pathout_ports_vpi.c
- Verilog test bench: Chapter.04/list_pathout_ports_test.v
- Expected results log: Chapter.04/list_pathout_ports_test.log

Example 4-4: *$list_pathout_ports* — traversing Verilog hierarchy

```
PLI_INT32 PLIbook_ListPorts_calltf(PLI_BYTE8 *user_data)
{
  vpiHandle module_handle, systf_h, arg_itr,
            path_itr, path_handle,
            term_itr, term_handle,
            port_itr, port_handle,
            net_handle;
```

```
/* get module handle from first system task argument.  Assume the  */
/* compiletf routine has already verified correct argument type.   */
systf_h = vpi_handle(vpiSysTfCall, NULL);
if (systf_h == NULL) {
  vpi_printf("ERROR: list_pathout_ports could not obtain handle to
             systf call\n");
  return(0);
}
arg_itr = vpi_iterate(vpiArgument, systf_h);
if (systf_h == NULL) {
  vpi_printf("ERROR: list_pathout_ports could not obtain iterator to
             systf args\n");
  return(0);
}
module_handle = vpi_scan(arg_itr);
vpi_free_object(arg_itr); /* free itr since didn't scan until null */
vpi_printf("\nModule %s\n", vpi_get_str(vpiDefName, module_handle));

path_itr = vpi_iterate(vpiModPath, module_handle);
if (path_itr == NULL) {
  vpi_printf("  No module paths found\n");
  return(0);
}
while (path_handle = vpi_scan(path_itr)) {
  term_itr = vpi_iterate(vpiModPathOut, path_handle);
  if (term_itr == NULL) {
    vpi_printf("  No path output terminal found\n");
    break; /* go to next path */
  }
  while (term_handle = vpi_scan(term_itr)) {
    net_handle = vpi_handle(vpiExpr, term_handle);
    port_itr = vpi_iterate(vpiPorts, net_handle);
    if (port_itr == NULL) {
      vpi_printf("  Path output does not connect to a port\n");
      break; /* go to next path output terminal */
    }
    while (port_handle = vpi_scan(port_itr)) {
      vpi_printf("  Port %s is connected to a path delay output\n",
                 vpi_get_str(vpiName, port_handle));
    }
  }
}
return(0);
}
```

4.8.1 Traversing hierarchy across module ports

The Verilog hierarchy connections can be traversed across module boundaries by following the relationships within and without module ports. A module port has two connections, as shown in the following diagram.

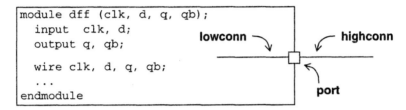

The *lowconn* (hierarchically lower connection) is the signal *inside* the module that is connected to a port. In the object diagrams, the internal connection is shown as an *expression*, because a number of different types of objects can be connected to the port. Within a module, the Verilog HDL syntax restricts the expression connected to a port to the Verilog data types of *nets*, *regs* and *variables* (except real variables). The expression connected to the port can be a scalar signal (1-bit wide), a vector, a bit or part select of a vector, or a concatenation of any valid data types. Any of these expression types can be connected to an output port, but only expressions using the *net* class of data types can be connected to an input or inout port. The Verilog HDL syntax also allows the bits of the *lowconn* connections to have different port directions, referred to as a *mixed I/O* port.

The *highconn* (hierarchically higher connection) is the signal *outside* the module that is connected to a port. External to a module, the expression connected to the port can be a scalar signal, a vector, a bit or part select of a vector, or a concatenation of signals, a constant, a literal value, an operation, or the return of a function call. Any of these expression types can be connected to an input port, but only expressions using the *net* class of data types can be connected to an output or inout port.

The object diagram for `ports` (shown in Figure 4-9) shows a one-to-one connection to either the *highconn* and *lowconn* connection to the port. The connecting object is an `expr` object group, to allow for different types of objects which may be connected to the port. Using the handle for the `expr` object, the type property of the object can be accessed to determine what is connected to the port. The object diagram for that type of object can then be used to continue traversing the Verilog design hierarchy. The object type property is accessed using `vpi_get(vpiType, <expression_handle>)`.

Figure 4-9: Object diagram for ports (partial)

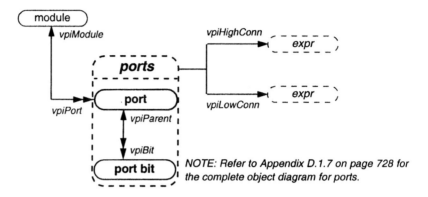

The following example accesses all ports of a module and lists the port name, port size, port direction, and type of object connected as the *lowconn* and *highconn*. The name, size and direction are all properties of the module port.

> **CD** The source code for this example is on the CD accompanying this book.
>
> • Application source file: Chapter.04/port_info_vpi.c
> • Verilog test bench: Chapter.04/port_info_test.v
> • Expected results log: Chapter.04/port_info_test.log

Example 4-5: *$port_info* — traversing hierarchy across module ports

```
PLI_INT32 PLIbook_PortInfo_calltf(PLI_BYTE8 *user_data)
{
  vpiHandle systf_h, arg_itr, mod_h,
            port_itr, port_h, lowconn_h, highconn_h;
  PLI_INT32 lowconn_type, highconn_type;

  /* get module handle from first system task argument.  Assume the  */
  /* compiletf routine has already verified correct argument type.   */
  systf_h = vpi_handle(vpiSysTfCall, NULL);
  if (systf_h == NULL) {
    vpi_printf("ERROR: list_pathout_ports could not obtain handle to
                systf call\n");
    return(0);
  }
  arg_itr = vpi_iterate(vpiArgument, systf_h);
  if (systf_h == NULL) {
    vpi_printf("ERROR: list_pathout_ports could not obtain iterator to
                systf args\n");
    return(0);
  }
```

```
mod_h = vpi_scan(arg_itr);
vpi_free_object(arg_itr); /* free itr since did not scan until null */

vpi_printf("\nModule %s (instance %s)\n",
            vpi_get_str(vpiDefName, mod_h),
            vpi_get_str(vpiFullName, mod_h));

port_itr = vpi_iterate(vpiPort, mod_h);
if (!port_itr) {
  vpi_printf("   No ports found\n");
  return(0);
}
while (port_h = vpi_scan(port_itr)) {
  vpi_printf("  Port name is %s\n", vpi_get_str(vpiName, port_h));
  vpi_printf("    Size is %d\n", vpi_get(vpiSize, port_h));
  switch (vpi_get(vpiDirection, port_h)) {
    case vpiInput:  vpi_printf("    Direction is input\n");  break;
    case vpiOutput: vpi_printf("    Direction is output\n"); break;
    case vpiInout:  vpi_printf("    Direction is inout\n");  break;
  }

  lowconn_h = vpi_handle(vpiLowConn, port_h);
  lowconn_type = vpi_get(vpiType, lowconn_h);
  vpi_printf("    Low conn data type is %s\n",
      vpi_get_str(vpiType, lowconn_h)); /* Verilog-2001 property */

  highconn_h = vpi_handle(vpiHighConn, port_h);
  if (!highconn_h) {
    vpi_printf("    No high conn\n");
    return(0);
  }
  highconn_type = vpi_get(vpiType, highconn_h);
  vpi_printf("    High conn data type is %s\n",
      vpi_get_str(vpiType, highconn_h)); /* Verilog-2001 property */
}
return(0);
}
```

NOTE The preceding example prints the *name* of the type constant which represents the type property of the object connected to a port. The IEEE 1364-2001 Verilog provides access to both the integer value of type constant, using `vpi_get(vpiType, object_handle)`, as well as the name of the type constants, using `vpi_get_str(vpiType, object_handle)`. Accessing the name of the constant was not supported in the IEEE 1364-1995 standard.

4.8.2 Traversing through multiple levels of hierarchy

In some PLI applications, it may be necessary to traverse through multiple levels of design hierarchy. For instance, an application might need to locate every Verilog primitive in an entire design. This requires starting at the very top of the design hierarchy tree, and descending down through every possible branch of the tree, searching for primitive instances. The PLI application will need to traverse an unknown number of hierarchy levels, because the number of branches in the tree, and the depth of the hierarchy are not available.

The most straightforward programming method to traverse an unknown number of hierarchy levels is to use recursive calls to a C function. A C function is passed in a handle for a module, and then searches for module instances within that module. For each module instance found, the function calls itself, passing the module instance handle as an input.

Example 4-6 lists the PLI source code for an application called *$count_all_prims*. The application does not require any inputs. Instead, the application searches for all top level modules, and then traverses all levels from the modules on down and counts the number of primitives found in each module instance as the hierarchy is traversed. A summary report is printed after all hierarchy levels are traversed.

CD The source code for this example is on the CD accompanying this book.

- Application source file: `Chapter.04/count_all_prims_vpi.c`
- Verilog test bench: `Chapter.04/count_all_prims_test.v`
- Expected results log: `Chapter.04/count_all_prims_test.log`

Example 4-6: *$count_all_prims* — traversing multiple levels of hierarchy

```
/****************************************************************
 * calltf routine
 ****************************************************************/
PLI_INT32 PLIbook_CountPrims_calltf(PLI_BYTE8 *user_data)
{
  vpiHandle top_mod_itr, top_mod_h;
  int total_prims = 0;
  top_mod_itr = vpi_iterate(vpiModule, NULL); /*get top modules*/
  while (top_mod_h = vpi_scan(top_mod_itr)) {
    total_prims += PLIbook_find_child_mod(top_mod_h);
  }
  vpi_printf("\nTotal number of primitives is %d\n\n", total_prims);
  return(0);
}
```

```
/******************************************************************
 * Function to look for module instances in local scope.
 * THIS FUNCTION CALLS ITSELF RECURSIVELY.
 ******************************************************************/
int PLIbook_find_child_mod(vpiHandle this_mod_h)
{
  vpiHandle child_mod_itr, child_mod_h;
  int prims_in_child;

  prims_in_child = PLIbook_count_local_prims(this_mod_h);
  child_mod_itr = vpi_iterate(vpiModule, this_mod_h);
  if (child_mod_itr != NULL)
    while (child_mod_h = vpi_scan(child_mod_itr))
      prims_in_child += PLIbook_find_child_mod(child_mod_h);
  return(prims_in_child);
}

/******************************************************************
 * Function to count primitives in local scope.
 ******************************************************************/
int PLIbook_count_local_prims(vpiHandle module_h)
{
  vpiHandle prim_itr, prim_h;
  int prims_in_mod = 0;

  prim_itr = vpi_iterate(vpiPrimitive, module_h);
  if (prim_itr != NULL)
    while (prim_h = vpi_scan(prim_itr))
      prims_in_mod++;
  return (prims_in_mod);
}
```

4.9 Writing messages to files

The VPI library provides routines for writing information to disk files, and special routines for controlling those files. The files can be opened from either the PLI application or from the Verilog HDL before the application is called.

In the IEEE 1364-2001 Verilog standard, the *$fopen* built-in system function supports two types of file pointers:

- A *multi-channel descriptor* (*mcd*) file pointer will have a single bit set to represent an open file. 30 files can be opened at a time, plus one output channel which is reserved by the simulator. The reserved *mcd* is bit 0, which represents the simulator's output channel (typically stdout) and the simulator's output log file, if open. Bit 31 of an *mcd* must always be zero. Since only one bit in an *mcd* is set, multiple *mcd*'s can be ORed together to represent multiple open files.

- A *single-channel descriptor* (*fd*) file pointer will have multiple bits set to represent an open file. 2^{30} application-specified files can be opened at a time with a single-file descriptor. Bit 31 of an *fd* must always be set, to indicate that the descriptor represents a single file. It is illegal to OR *fd*'s together. Single-channel descriptors were added with the IEEE 1364-2001 Verilog standard.

The VPI library provides routines for writing to files that are represented with either mcd's or fd's. However, the VPI library can only open files represented with an *mcd* file pointer.

4.9.1 Opening and closing files

Opening a file from the PLI application is done with the following VPI routine:

PLI_UINT32 **vpi_mcd_open (**
  PLI_BYTE8 ***file_name**) /* name of a file to be opened */

This routine opens a file for writing from a PLI application. The input to the routine is a string containing the file name to be opened. The routine returns an integer, which is a *multi-channel-descriptor* (*mcd*). If the file cannot be opened, a value of **0** is returned for the *mcd*. If the file is already open, the value of the file's current *mcd* or *fd* is returned.

A file that has been opened can be closed with the routine:

PLI_UINT32 **vpi_mcd_close (**
  PLI_UINT32 **mcd_or_fd**) /* multi-channel descriptor or file descriptor */

vpi_mcd_close() can close one or more files that were opened with either vpi_mcd_open() or with *$fopen* in the Verilog HDL. The file pointer can be either an *mcd* of an *fd*. If the file was opened from the Verilog HDL, the *mcd* or *fd* must be past to the PLI application as a system task/function argument. It is illegal to use an *mcd* value which represent simulator's output channel (bit zero set). The routine returns **0** if successful, and the *mcd* or *fd* of files not closed if an error occurs.

4.9.2 Writing to files

The VPI library provides two routines to write to a file that has been opened either from a PLI application or from the Verilog HDL. If the file was opened from the Verilog HDL, the *mcd* or *fd* must be past to the PLI application as a system task/function argument.

PLI_INT32 **vpi_mcd_printf (**
   PLI_INT32   **mcd,**   /* multi-channel descriptor */
   PLI_BYTE8 *** format,**   /* character string with formatted message */
        ...)   /* arguments to the formatted message string */

vpi_mcd_printf() writes a formatted message to one or more open files represented with an *mcd*, using a format string with formatting controls similar to the C
printf() function. The routine cannot write to a file represented by an *fd* file
descriptor (indicated by the most significant bit being set). The routine returns the
number of characters written, or **EOF** if an error occurred. vpi_mcd_printf() uses
a format string with formatting controls similar to the C printf() function. Multiple
mcd values can be ORed together to write to multiple files at the same time.

The IEEE 1364-2001 standard adds another VPI routine for writing to files:

PLI_INT32 **vpi_mcd_vprintf (**
   PLI_INT32   **mcd_or_fd,** /* multi-channel descriptor of open files */
   PLI_BYTE8 *** format,**   /* quoted character string of formatted message */
   va_list   **arg_list)** /* an already started list of variable arguments */

This routine writes a formatted message similar to the C vprintf() routine to one or
more open files represented with an *mcd* or *fd*. It uses a list of variable arguments that
has already been started using the variable argument functions in the C stdarg.h
library. The routine returns the number of characters written, or **EOF** if an error
occurred. Note: the *va_list* data type is defined in the ANSI C *stdarg.h* library, so the
PLI application must include this library.

4.9.3 Other VPI file management routines

PLI_BYTE8 ***vpi_mcd_name (**
   PLI_UINT32   **mcd_or_fd)** /* multi-channel descriptor or file descriptor */

vpi_mcd_name() returns the name of a file represented by an *mcd* or *fd*. An *mcd*
value must represent a single file (e.g.: only 1 bit may be set in the *mcd*).

PLI_INT32 **vpi_mcd_flush (**
   PLI_UINT32   **mcd_or_fd)** /* multi-channel descriptor or file descriptor */

vpi_mcd_close() flushes the output buffer for one or more files that were opened
with either vpi_mcd_open() in a PLI application, or with *$fopen* in the Verilog
HDL. The routine returns 0 if successful, and non-zero if an error occurs.

4.10 Reading and using simulation times

The VPI routines can retrieve the current simulation time and information about the time scaling applied to the model. The VPI routines use two terms when referring to simulation time:

- *Module time units* are the units of time within a specific Verilog module. The `timescale` directive in Verilog indicates what time units are used in the modules which follow the directive. The time scale also indicates a time precision, which is how many decimal points of accuracy are permitted. Time values with more decimal points than the precision are rounded off. Each module in a design can have a different time scale, so one model can specify delays in nanoseconds, with two decimal points of precision, and another model can specify delays in microseconds, with three decimal points.

- *Simulation time units* are the units of time used internally within the simulator. The simulator scales the times in all modules to the simulation time units. Most Verilog simulators will determine the finest precision of all modules in a simulation, and set the simulation time units to that precision.

Within a Verilog HDL model, time can be specified either as an integer or as a floating point real number. Within a Verilog simulation, however, time is represented as a 64 bit unsigned integer. As a Verilog model is compiled or loaded into simulation, time values in the module are rounded off to the specified precision and then scaled to the simulation time units. As an example:

```
'timescale 1ns/10ps
module A;
   . . .
   nand #5.63 n1 (y, a, b);
   . . .
endmodule
'timescale 1us/100ns
module B;
   . . .
   nand #3.581 n1 (y, a, b);
   . . .
endmodule
```

In this example, ten-picosecond units of time is the finest precision used by all modules, and will be set as the simulation time units. The simulator will then scale the 5.63 nanosecond delay in module A to 563 10-picosecond units, and scale the 3.581 microsecond delay in module B to 360,000 10-picosecond units (rounding to the module's time precision occurs before scaling to the simulator's time units).

4.10.1 Reading the current simulation time

A special routine is used to retrieve the current simulation time.

void **vpi_get_time (**

| *vpiHandle* | **object,** | /* handle for an object, or **NULL** */ |
| *p_vpi_time* | **time)** | /* pointer to application-allocated s_vpi_time structure */ |

The vpi_get_time() routine retrieves simulation time and places in an **s_vpi_time** structure. The time can be represented in the simulation time units or a module's time units. An object handle must be specified when time is to retrieved in a module's time units. vpi_get_time() will scale the time value to the time scale of the module containing the object.

The s_vpi_time structure can represent the simulation time as either a pair of 32-bit C integers (to hold the 64-bit Verilog time value), or as a double-precision value. The s_vpi_time structure is defined in the vpi_user.h file, as follows:

```
typedef struct t_vpi_time {
  PLI_INT32   type;    /* vpiScaledRealTime or vpiSimTime */
  PLI_UINT32  high;    /* when using vpiSimTime */
  PLI_UINT32  low;     /* when using vpiSimTime */
  double      real;    /* when using vpiScaledRealTime */
} s_vpi_time, *p_vpi_time;
```

The **type** field in the structure is set by the PLI application to the constant: **vpiSimTime** or **vpiScaledRealTime**. This field determines whether the simulation time should be retrieved as a real number or as integers.

- If the type is set to **vpiScaledRealTime**, then the current simulation time is placed into the **real** field of the structure. The simulation time is returned in the time units of the module containing the reference object specified as an input to vpi_get_time(). If a NULL is specified for the reference object, the simulation time is returned in the simulator's time units.

- If the type is set to **vpiSimTime**, then the current simulation time is placed into the pair of 32-bit C integers in the structure, so that the lower 32 bits of time are in the **low** integer, and the upper 32 bits of time are in the **high** integer. The simulation time is always returned in the simulator's time units. The object handle is not used and can be set to NULL.

When time is retrieved as a pair of 32-bit C integers, the time will be stored in the structure, as show in the following illustration:

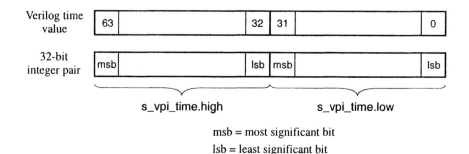

msb = most significant bit
lsb = least significant bit

Using vpiSimTime is better for simulation performance. The Verilog language
standard specifies that time is a 64-bit unsigned integer. Retrieving the simulation
TIP   time as integers is faster than requiring the simulator convert time to a real number.

The following code fragment retrieves the current simulation time, scaled to a module's time scale, using vpi_get_time(). There are three basic steps required to
retrieve the time:

1. Allocate an s_vpi_time structure.

2. Set the type field.

3. Call vpi_get_time() with an object handle and a pointer to the time structure
    as inputs.

```
vpiHandle  module_handle;
s_vpi_time current_time;

/* get handle for the module (or any object in the module */

current_time.type = vpiScaledRealTime;
vpi_get_time(module_handle, &current_time);

vpi_printf("Current time is %f\n", current_time.real);
```

4.10.2 Reading time scale factors

The PLI application can also read the time scale factors of a module. The time units
and time precision are properties of a module, and are accessed using vpi_get() for
the properties of **vpiTimeUnit** and **vpiTimePrecision**. Both properties are integers which represents a unit of time, as shown in table 4-1.

The simulation time unit can also be accessed using the vpiTimeUnit and vpi-TimePrecision properties, and using a NULL for the reference handle.

The time units and time precision are represented as the magnitude of 1 second, which is the exponent of 1 second times 10^n. For example, 1 nanosecond is 1 second times 10^{-9}, so the integer value used to represent nanoseconds is -9.

Table 4-1: Time unit values for vpiTimeUnit and vpiTimePrecision

Time Unit or Time Precision	Unit of Time Represented
2	100 seconds (1×10^2)
1	10 seconds (1×10^1)
0	1 second (1×10^0)
-1	100 milliseconds (1×10^{-1})
-2	10 milliseconds (1×10^{-2})
-3	1 millisecond (1×10^{-3})
-4	100 microseconds (1×10^{-4})
-5	10 microseconds (1×10^{-5})
-6	1 microsecond (1×10^{-6})
-7	100 nanoseconds (1×10^{-7})
-8	10 nanoseconds (1×10^{-8})
-9	1 nanosecond (1×10^{-9})
-10	100 picoseconds (1×10^{-10})
-11	10 picoseconds (1×10^{-11})
-12	1 picosecond (1×10^{-12})
-13	100 femtoseconds (1×10^{-13})
-14	10 femtoseconds (1×10^{-14})
-15	1 femtosecond (1×10^{-15})

4.11 User-defined invocation options

The PLI provides a means for application developers to create user-defined invocation options. This capability makes it possible to configure PLI applications or pass data to an application from the invocation command line of a Verilog simulator.

PLI_INT32 **vpi_get_vlog_info (**
   p_vpi_vlog_info **info)**   /* pointer to an application-allocated s_vpi_vlog_info
                                    structure */

vpi_get_vlog_info() returns information about the Verilog simulator that is running the PLI application. The routine returns 1 (for true) if it was successful, and 0 (for false) if an error occurred in retrieving the simulator information. The information retrieved includes:

- The simulator product name.

- The simulator version.

- The number of invocation options with which simulation was invoked.

- The values of the invocation options.

The input to vpi_get_vlog_info() is a pointer to an **s_vpi_vlog_info** structure which has been allocated by the PLI application. The routine will fill in the fields of the structure. Note that for string values, such as the product name, a pointer to the string is placed into the structure. The string is stored in a temporary buffer which is allocated and maintained by the simulator. The structure is:

```
typedef struct t_vpi_vlog_info {
  PLI_INT32    argc;
  PLI_BYTE8  **argv;
  PLI_BYTE8   *product;
  PLI_BYTE8   *version;
} s_vpi_vlog_info, *p_vpi_vlog_info;
```

The **product** and **version** strings are defined by the simulator, and will be different for each simulator product.

The **argc** and **argv** values are the same command line values defined in the C language. argc is the number of invocation command arguments, and argv is a pointer to an array of strings, where each string is one argument from the command line.

The ability to check the simulator command line options makes it possible to create user-defined invocation options. For example, a PLI application could use user-defined options to:

- Enable debug messages when debugging a PLI application. A verbose mode invocation option could be used to specify that debug messages should be printed.

- Pass file names or other values from the command line to a PLI application.

Parsing the -f command file invocation option

The IEEE 1364 Verilog standard does not specify any invocation options for Verilog simulators. Every Verilog simulator, however, has adopted a small number of de facto standard invocation options. One of these is the **-f** option. This invocation option specifies that the file name which follows the option contains additional command line invocation arguments. When the argv value is -f, the next argv will be a pointer to a NULL terminated array of pointers to strings. Element 0 in the array will contain the name of the file specified with the -f option, and the remaining elements in the array will be invocation commands contained in the file (comments are not included).

For example, assume a command file named *run.f* contained the following:

```
my_chip.v
my_test.v
+my_debug
```

If simulation were invoked with the command:

```
verilog -f run.f -s
```

Then argv would point to the following array of strings:

```
argv[0] -> "verilog"
    [1] -> "-f"
    [2] -------> [0] -> "run.f"
                 [1] -> "my_chip.v"
                 [2] -> "my_test"
                 [3] -> "+my_debug"
                 [5] -> NULL
    [3] -> "-s"
```

Example 4-7 illustrates a PLI application called *$test_invoke_options*. This application uses vpi_get_vlog_info() to test to see if simulation was invoked with any user-specified option on the operating system command line or in a simulation command file, and return a true/false result. An example of using this applications is:

```
if ($test_invoke_options("+verbose"))
  $monitor("data_bus = %b", data_bus);
```

To show how invocation options are retrieved, this example also prints out all invocation options with which a simulation was invoked.

CD The source code for this example is on the CD accompanying this book.

- Application source file: Chapter.04/invoke_options_vpi.c
- Verilog test bench: Chapter.04/invoke_options_test.v
- Expected results log: Chapter.04/invoke_options_vpi.log

Example 4-7: *$test_invoke_options* — testing for user-defined invocation options

```
/***********************************************************************
 * sizetf application
 ***********************************************************************/
PLI_INT32 PLIbook_TestInvokeOptions_sizetf(PLI_BYTE8 *user_data)
{
  return(1);     /* $test_invoke_options returns a 1-bit value */
}

/***********************************************************************
 * calltf routine
 ***********************************************************************/
PLI_INT32 PLIbook_TestInvokeOptions_calltf(PLI_BYTE8 *user_data)
{
  vpiHandle        systf_h, arg_itr, arg_h;
  char             *option_name;
  s_vpi_value      value_s;
  s_vpi_vlog_info  sim_info;
  int              found;

  vpi_get_vlog_info(&sim_info);

  /* get system function arg--compiletf already verified correctness */
  systf_h = vpi_handle(vpiSysTfCall, NULL);
  arg_itr = vpi_iterate(vpiArgument, systf_h);
  arg_h   = vpi_scan(arg_itr);
  vpi_free_object(arg_itr); /* free iterator -- did not scan to null */

  /* read target option name from first tfarg */
  value_s.format = vpiStringVal;
  vpi_get_value(arg_h, &value_s);
  option_name = value_s.value.str;

  /* test for target option and return true/false to system function */
  found = PLIbook_GetOptions((char *)option_name,
                             (int)sim_info.argc,
                             (char **)sim_info.argv );
  value_s.format = vpiIntVal;
  value_s.value.integer = found;
  vpi_put_value(systf_h, &value_s, NULL, vpiNoDelay);
  return(0);
}
```

```
int PLIbook_optfound = 0; /* global variable for option found flag */
int PLIbook_indent = 0;   /* global variable to format text indenting */

int PLIbook_GetOptions(char *option, int argc, char **argv)
{
  int i;
  PLIbook_optfound = 0;
  PLIbook_indent = 0;
  for (i=0; i<argc; i++) {
    #if PLIbook_verbose
      vpi_printf("%s\n", *argv);
    #endif
    if (strcmp(*argv, option) == 0) {
      PLIbook_optfound = 1;
    }
    if (strcmp(*argv, "-f") == 0) {
      argv++;  /* next arg is address to array of strings */
      i++;
      PLIbook_ScanCommandFile(option, (char **)*argv);
    }
    argv++; /* increment to next argument */
  }
  return(PLIbook_optfound);
}

void PLIbook_ScanCommandFile(char *option, char **arg)
{
  int i;

  #if PLIbook_verbose
    PLIbook_indent += 4; /* increase text indentation */
  #endif
  while ( *arg != NULL ) { /* loop until null termination */
    #if PLIbook_verbose
      for (i=0; i<=PLIbook_indent; i++)
        vpi_printf(" ");
      vpi_printf("%s\n", *arg);
    #endif
    if (strcmp(*arg, option) == 0) {
      PLIbook_optfound = 1;
    }
    if (strcmp(*arg, "-f") == 0) {
      arg++;  /* next arg is address to array of strings */
      PLIbook_ScanCommandFile(option, (char **)*arg);
    }
    arg++;
  }
  #if PLIbook_verbose
    PLIbook_indent -= 4; /* decrease text indentation */
  #endif
  return;
}
```

```
/*********************************************************************
 * compiletf routine
 *********************************************************************/
PLI_INT32 PLIbook_TestInvokeOptions_compiletf(PLI_BYTE8 *user_data)
{
  vpiHandle    systf_h, arg_itr, arg_h;
  int          err = 0;

  systf_h = vpi_handle(vpiSysTfCall, NULL);
  arg_itr = vpi_iterate(vpiArgument, systf_h);
  if (arg_itr == NULL) {
    vpi_printf("ERROR: $test_invoke_options requires 1 argument\n");
    vpi_control(vpiFinish, 1);  /* abort simulation */
    return(0);
  }
  arg_h = vpi_scan(arg_itr);  /* get handle for first tfarg */
  if (vpi_get(vpiType, arg_h) != vpiConstant) {
    vpi_printf("$test_invoke_options arg must be a quoted name\n");
    err = 1;
  }
  else if (vpi_get(vpiConstType, arg_h) != vpiStringConst) {
    vpi_printf("$test_invoke_options arg must be a string\n");
    err = 1;
  }
  if (vpi_scan(arg_itr) != NULL) {
    vpi_printf("test_invoke_options requires only 1 argument\n");
    vpi_free_object(arg_itr);
    err = 1;
  }
  if (err)
    vpi_control(vpiFinish, 1);  /* abort simulation */

  return(0);
}
```

4.12 Controlling Simulations

Several of the examples presented thus far in this book have used the VPI routine
vpi_control() to abort simulation elaboration/linking if a fatal error was encountered in a compiletf application.

PLI_INT32 **vpi_control (**

    PLI_INT32   **operation,**   /* constant representing the operation to perform */

               ...)   /* variable number of arguments, as required by the
                            operation */

The routine vpi_control() can be used to control other aspects of simulation as well. The IEEE 1364-2001 standard defines certain standard operation constants, and states that simulators may add additional constants specific to that product. vpi_control() takes a variable number of arguments. The number and type of arguments is determined by the operation constant. The standard operation constants and their effect are:

vpiFinish causes the *$finish()* built-in Verilog system task to be executed upon return of the PLI application. This operation requires one additional argument of type PLI_INT32. The value of the integer is the same as the diagnostic message level argument passed to *$finish()*, as defined in the Verilog HDL, where: 0 prints no message. 1 prints the simulation time *$finish()* was executed and the source file location containing the *$stop()* command. 2 prints the simulation time, file location, and simulation memory usage information.

vpiStop causes the *$stop()* built-in Verilog system task to be executed upon return of the PLI application. This operation requires one additional argument of type PLI_INT32. The value of the integer is the same as the diagnostic message level argument passed to *$stop()*, as defined in the Verilog HDL, where: 0 prints no message. 1 prints the simulation time *$stop()* was executed and the source file location containing the *$stop()* command. 2 prints the simulation time, file location, and simulation memory usage information.

vpiReset causes the *$reset()* built-in Verilog system task to be executed upon return of the PLI application This operation requires three additional arguments of type PLI_INT32: the stop_value, reset_value and diagnostic_level, which are the same values passed to the *$reset()* system task.

vpiSetInteractiveScope causes a simulator's interactive debug scope to be immediately changed to a new scope. The command requires one additional argument of type vpiHandle, which is a handle to an object in the scope class. Refer to the scope object diagram in Annex D, on page 727 for the objects in the scope class.

For all control operations, vpi_control() returns 1 if successful and 0 if an error occurred.

NOTE The vpi_control() routine was added with the IEEE 1364-2001 standard. In the IEEE 1364-1995 standard, there was no VPI routine that would cause a simulator to exit if a PLI application detected an error. Instead, applications needed to use the tf_dofinish() routine from the older TF library.

4.13 Summary

The VPI routines in the PLI standard provide a simple, yet powerful, way to access what is happening in a Verilog simulation. The VPI routines provide this access by treating Verilog HDL constructs as objects, and using *handles* to reference these objects.

The VPI object diagrams serve as the primary documentation on the relationships of one object to another. The diagrams also document every property that is available for each object, such as names and vector sizes. Using the object diagrams, it is possible to traverse from any point in a Verilog design hierarchy to any other point in the hierarchy.

Previous chapters introduced a number of routines from the VPI library. This chapter has presented more details on most of the VPI routines in the library. The next chapter will present the VPI routines which are used to read and modify logic values and delay values. Chapter 6 presents synchronizing PLI applications with activity occurring within a simulation. These chapters include several complete examples of PLI applications which use the VPI library of routines.

CHAPTER 5 *Reading and Modifying Values Using VPI Routines*

T his chapter applies many of the concepts presented in the previous chapter on obtaining handles for objects, and adds the concepts of reading and modifying simulation values using the VPI routines.

The concepts presented in this chapter are:

- Objects which have logic values
- Reading logic values from Verilog simulations
- Writing logic values into Verilog simulations
- Objects which have delay values
- Reading delay values from Verilog simulations
- Writing delay values into Verilog simulations

5.1 Accessing objects which have logic values

The VPI routines can read the values of several different types of objects, and can modify the values of certain object types. In order to read an object's logic value or write a new value into an object, the object must have a value property in the VPI object diagrams. The objects which have this property are:

- Any *net* data type: scalar, vector, part-selects and bit-selects of vectors
- The *reg* data type: scalar, vector, part-selects and bit-selects of vectors
- The *integer* and *time* variable data types, part-selects and bit-selects

- The *real* and *realtime* variable data types
- A *memory* word select, or a part-select or bit-select of a memory word
- An *array* word select, or a part-select or bit-select of an array word
- A *parameter* constant (read only)
- A *specparam* constant, including a *specparam attribute* (read only)
- A literal *integer* value (read only)
- A literal *real* value (read only)
- A literal *string* value (read only)
- A *function call*
- A *system function call*
- A *sequential user-defined primitive*
- A *user-defined primitive table entry* (read only)

To access the value of an object, a handle for the object must be obtained. The routines to obtain handles were presented in the previous chapter.

5.2 Reading object logic values

Using the handle for an object, a PLI application can read the logic value property of the object. A single VPI routine is used to read the logic value of any type of object and any data type of logic value.

5.2.1 Working with a 4-logic value, multiple strength level system

The Verilog HDL supports 4 logic values, **0**, **1**, **z** and **x** and multiple levels of signal strength. There are also two ambiguous logic values, represented by **L** (low) and **H** (high), and many ambiguous strength values. The C programming language does not directly represent the same information, and so the VPI routines provide several ways to automatically translate values between Verilog and C. The translation converts Verilog 4-state logic into the following C types:

- *A 32-bit C integer*: Verilog scalar and vector logic values are converted to a single 32-bit C integer. The Verilog 4-state logic is converted to 2-state logic values of 0 and 1. Logic values of z and x values are converted to 0, and strength levels are ignored.

- *A C double*: Verilog real number values, scalar values and vector values are converted to a C double precision value. The real data type in Verilog is defined to be double-precision floating point, which converts directly to C doubles. Scalar and vector values in Verilog use 4-state logic, which is converted to 2-state logic. Strength levels are ignored. Verilog real numbers are converted to a string representation with a period character for the decimal point.

- *A C string*: Verilog scalar and vector logic values are converted to a C character string. The Verilog 4-state logic is converted to the characters "0", "1", "z" and "x". Logic strength levels are ignored.

- *A C constant*: Verilog scalar values are converted to a C integer constants. The Verilog 4-state logic is converted to the constants **vpi0**, **vpi1**, **vpiZ**, **vpiX**, **vpiH**, **vpiL**. Logic strength levels are ignored. A UDP table entry can also be represented with these constants, plus the constant **vpiDontCare**.

- *A C aval/bval structure*: Verilog scalar and vector logic values are converted to a C structure which encodes each bit of a Verilog 4-state value to a pair of bits in C, referred to as an *aval/bval* pair. An array of aval/bval pairs is used to encode Verilog vectors of any size. The logic strength levels are ignored.

- *A C strength structure*: Verilog scalar logic values, or each bit of Verilog vector values, are converted to a C structure. The logic value is represented as a C constant of **vpi0**, **vpi1**, **vpiZ** or **vpiX**. The logic strengths as two 32-bit C integer values, which represent the strength encoding defined in the Verilog HDL standard.

TIP   The format in which a value is read can have an impact on the run-time performance of a PLI application. The fastest run-time performance will be achieved when a value is retrieved in a format closest to the format in which a value is saved in the simulation structure. For Verilog scalar and vector nets and regs, this format is the aval/bval pair. For Verilog integers, C integers are most efficient, and for Verilog reals, C doubles are most efficient. The least efficient method for run-time performance is to retrieve a logic value as a C string.

5.2.2 The vpi_get_value() routine

The VPI routine which reads logic values is **vpi_get_value()**. The routine automatically converts Verilog logic values to C representations. The syntax for this routine is:

void **vpi_get_value (**
   vpiHandle   **object,**   /* handle for an object */
   p_vpi_value **value)**   /* pointer to application-allocated s_vpi_value structure */

The object's logic value is retrieved into an **s_vpi_value** structure. The PLI application must allocate this structure, and pass a pointer to the structure as an input to vpi_get_value().

Following are two ways of allocating an `s_vpi_value` structure:

To allocate local, automatic storage which will be freed when the PLI applications exits:

```
s_vpi_value  arg_info;
vpi_get_value(tfarg_handle, &arg_info);
```

To allocate persistent storage which can be used by future calls to the PLI application (a pointer to the storage must be maintained by the application):

```
p_vpi_value  arg_info;
arg_info = (p_vpi_value)malloc(sizeof(s_vpi_value));
vpi_get_value(tfarg_handle, arg_info);
```

 An `s_vpi_value` structure can also be statically allocated, which will create storage which remains in effect throughout simulation. However, caution should be observed when using static variables in PLI applications. All instances of a system task or system function will share the same static variable. When different storage is needed for each instance of a system task, it is better to allocate a separate structure for each instance. Section 4.7 on page 122 in Chapter 4 discusses allocation of instance specific storage.

The **s_vpi_value** structure is defined in vpi_user.h, and is listed below.

```
typedef struct t_vpi_value {
  PLI_INT32     format;   /* vpiBinStrVal,    vpiOctStrVal,
                             vpiDecStrVal,    vpiHexStrVal,
                             vpiScalarVal,    vpiIntVal,
                             vpiRealVal,      vpiStringVal,
                             vpiVectorVal,    vpiTimeVal,
                             vpiStrengthVal,  vpiSuppressVal,
                             vpiObjTypeVal */
  union {
    PLI_BYTE8 *str;      /* if any string format */
    PLI_INT32 scalar;    /* if vpiScalarVal: one of vpi0, vpi1,
                            vpiX, vpiZ, vpiH, vpiL, vpiDontCare */
    PLI_INT32 integer;   /* if vpiIntVal format */
    double    real;      /* if vpiRealVal format */
    struct t_vpi_time          *time;    /* if vpiTimeVal */
    struct t_vpi_vecval        *vector;  /* if vpiVectorVal */
    struct t_vpi_strengthval *strength;  /* if vpiStrengthVal */
    PLI_BYTE8 *misc;                      /* not used */
  } value;
} s_vpi_value, *p_vpi_value;
```

This s_vpi_value structure has two fields, **format** and **value**. The format field controls what C language data type should be used to receive the value, and the value field receives the value of the object. The definitions of the format values and the value union fields are listed in the following table.

Table 5-1: The s_vpi_value structure format constants

Format	Definition
vpiBinStrVal	represents a Verilog logic value as a C string, using binary numbers, with 4-state logic
vpiOctStrVal	represents a Verilog logic value as a C string, using octal numbers, with 4-state logic
vpiDecStrVal	represents a Verilog logic value as a C string, using decimal numbers, with 4-state logic
vpiHexStrVal	represents a Verilog logic value as a C string, using hexadecimal numbers, with 4-state logic
vpiScalarVal	represents a Verilog scalar logic value as C constants, which represents Verilog 4-state logic
vpiStrengthVal	represents a Verilog scalar value as a VPI strength structure containing the logic value and Verilog strength levels
vpiVectorVal	represents a Verilog vector value as an array of VPI aval/bval structures, encoded to represent Verilog 4-state logic
vpiIntVal	represents a Verilog logic value as a 32-bit C integer, with 2-state logic
vpiTimeVal	represents a Verilog time value as a VPI time structure
vpiRealVal	represents a Verilog logic value as a C double precision number, with 2-state logic
vpiStringVal	represents a Verilog value as a C string
vpiObjTypeVal	allows simulation to pick the most efficient way to represent a value

The value field of the s_vpi_value structure is a union of C data types. The format field of the structure controls which C data type is used to receive the Verilog value:

- The **scalar** field in the value union is used if the format field is vpiScalarVal. The vpi_get_value() routine will fetch the object's logic value, and return one of the constants: **vpi0**, **vpi1**, **vpiZ**, **vpiX**, **vpiH**, **vpiL**, or **vpiDontCare**. The latter constant is used when reading the values of UDP tables.

- The **integer** field in the value union is used if the format field is vpiIntVal. The `vpi_get_value()` routine will fetch the object's logic value, convert the value into 2-state logic (logic X and Z are converted to 0), and return the value as a 32-bit C integer. The maximum number of bits which can be retrieved using this format is 32 bits. If the number of bits of the Verilog object is greater than 32 bits, the most significant bits (the left-most bits) of the value are truncated. If the number of bits of the Verilog object is less than the size of an integer, the value is retrieved into the right-most bits of the integer, and the left-most bits are filled following the same rules as a Verilog HDL assignment statement (unsigned values are zero extended, signed values are sign extended). Refer to section 5.2.3 on page 157 for more details on reading Verilog integer values.

- The **real** field in the value union is used if the format field is `vpiRealVal`. The `vpi_get_value()` routine will fetch the object's logic value, convert the value into 2-state logic (logic X and Z are converted to 0), and return the value as a C double. Refer to section 5.2.4 on page 158 for more details on reading real values.

- The **str** field in the value union is used if the format field is `vpiBinStrVal`, `vpi-OctStrVal`, `vpiDecStrVal`, `vpiHexStrVal`, or `vpiStringVal`. The `vpi_get_value()` routine will fetch the object's logic value, convert the value into an ASCII string (which can represent all Verilog logic values, including X and Z), and return a pointer to the string. Refer to section 5.2.5 on page 160 for more details on reading Verilog string values as C strings, and to section 5.2.6 on page 161 for more details on reading Verilog logic values as C strings.

- The **vector** field in the value union is used if the format field is `vpiVectorVal`. The `vpi_get_value()` routine will fetch the object's logic value into an array of `s_vpi_vecval` structures. Refer to section 5.2.7 on page 163 for more details on reading Verilog vector values.

- The **strength** field in the value union is used if the format field is `vpiStrength-Val`. The `vpi_get_value()` routine will fetch the object's logic value into an `s_vpi_strengthval` structure. Refer to section 5.2.8 on page 167 for more details on reading Verilog strength values.

- The **time** field in the value union is used if the format field is `vpiTimeVal`. The `vpi_get_value()` routine will fetch the object's logic value into an `s_vpi_time` structure. Refer to section 5.2.9 on page 170 for more details on reading Verilog time variable values.

When reading logic values, the memory required to store the value for any string, time, strength or vector is allocated and maintained by the simulator. The simulator will place a pointer to the simulation-allocated memory in the appropriate field of the `value` union in the `s_vpi_value` structure. *The storage allocated by the simulator is temporary, and will only remain valid until the next call to* ***vpi_get_value()*** *or until the PLI application exits.* If the PLI application needs to preserve the value, the application must allocate memory for its own storage, and copy the values from the simulator's temporary storage.

5.2.3 Reading 2-state logic as 32-bit C integers

The `vpi_get_value()` can be set to read Verilog values into a 32-bit C integer by setting the format flag in the `s_vpi_value` structure to **vpiIntVal**. Verilog 4-state logic will be converted into C 2-state logic by converting logic Z and X to 0. As a standard data type in C, the value can be easily manipulated if needed.

The Verilog value which is read can be any Verilog data type, and it will be converted to a 32-bit C integer. Verilog real values will be rounded to an integer. An integer value in Verilog can be a 1-bit scalar value or a vector of any bit size. The maximum value which can be read is constrained by the size of the 32-bit C integer. If a Verilog vector is wider than the 32-bit C integer, then the left-most bits (the most significant bits) of the Verilog vector are truncated. If the Verilog vector is less than 32-bits it will be left-extended to 32-bits. Unsigned Verilog values will be zero extended. Signed Verilog values will be sign extended.

The steps to read the value of an object as a 32-bit C integer are:

1. Allocate an `s_vpi_value` structure.

2. Set the `format` field in the structure to **vpiIntVal**.

3. Call `vpi_get_value()`, giving a pointer to the `s_vpi_value` structure as an input, along with a handle for the object from which to read the logic value.

4. Read the logic value of the object from the **value.integer** field of the `s_vpi_value` structure.

Example 5-1, on the following page, lists a C function called *PLIbook_get_arg_int_val_vpi()*. This routine reads the logic value of a Verilog system task/function argument as an integer value. The input provided to the routine is the index number of the task/function argument, where the left-most argument is number 1, and the index number increases from left to right. Note that this example calls an example presented in the previous chapter, *PLIbook_get_arg_handle_vpi()*, to obtain a handle for a system task/function argument (example 4-3 on page 125).

CD The source code for this example is on the CD accompanying this book.

- Application source file: `Chapter.05/vpi_get_arg_val_vpi.c`
- Verilog test bench: `Chapter.05/vpi_get_arg_val_test.v`
- Expected results log: `Chapter.05/vpi_get_arg_val_test.log`

Example 5-1: *PLIbook_get_arg_int_val_vpi()* —reading values as 32-bit C integers

```
PLI_INT32 PLIbook_get_arg_int_val_vpi(int argNum)
{
  vpiHandle      arg_h;
  s_vpi_value    argVal;
  s_vpi_error_info err;  /* structure for error handling */

  arg_h = PLIbook_get_arg_handle_vpi(argNum);
  if (arg_h == NULL) {
    vpi_printf("ERROR: PLIbook_get_arg_int_val_vpi() could not obtain
               arg handle\n");
    return(0);
  }
  argVal.format = vpiIntVal;
  vpi_get_value(arg_h, &argVal);
  if (vpi_chk_error(&err)) {
    #if PLIbookDebug  /* if error, generate verbose debug message */
      vpi_printf("ERROR: PLIbook_get_arg_int_val_vpi() could not obtain
                 arg value\n");
      vpi_printf("File %s, Line %d: %s\n", err.file, err.line,
                 err.message);
    #endif
    return(0);
  }
  return(argVal.value.integer);
}
```

5.2.4 Reading 2-state logic as C doubles

The vpi_get_value() reads Verilog values into a C double by setting the format flag in the s_vpi_value structure to **vpiRealVal**. Verilog 4-state logic will be converted into C as 2-state logic by converting logic Z and X to 0. The Verilog value which is read can be any Verilog data type, and it will be converted to a C double. Reading a string value as a double is an error, and will return a value of 0.0.

The steps to read the value of an object as a C double are:

1. Allocate an s_vpi_value structure.

2. Set the format field in the structure to **vpiRealVal**.

3. Call vpi_get_value(), giving a pointer to the s_vpi_value structure as an input, along with a handle for the object from which to read the logic value.

4. Read the logic value of the object from the **value.real** field of the s_vpi_value structure.

Example 5-2 lists a C function called *PLIbook_get_arg_real_val_vpi()*, which reads the logic value of a system task/function argument as a C double. The input provided to the routine is the index number of the task/function argument, where the left-most argument is number 1, and the index number increases from left to right. The example calls an example presented in the previous chapter, *PLIbook_get_arg_handle_vpi()*, to obtain a handle for a system task/function argument (example 4-3 on page 125).

CD The source code for this example is on the CD accompanying this book.

- Application source file: Chapter.05/vpi_get_arg_val_vpi.c
- Verilog test bench: Chapter.05/vpi_get_arg_val_test.v
- Expected results log: Chapter.05/vpi_get_arg_val_test.log

Example 5-2: *PLIbook_get_arg_real_val_vpi()* — reading values as C doubles

```
double PLIbook_get_arg_real_val_vpi(int argNum)
{
  vpiHandle arg_h;
  s_vpi_value argVal;
  s_vpi_error_info err;  /* structure for error handling */

  arg_h = PLIbook_get_arg_handle_vpi(argNum);
  if (arg_h == NULL) {
    vpi_printf("ERROR: PLIbook_get_arg_real_val_vpi() could not obtain
               arg handle\n");
    return(0);
  }
  argVal.format = vpiRealVal;
  vpi_get_value(arg_h, &argVal);
  if (vpi_chk_error(&err)) {
    #if PLIbookDebug  /* if error, generate verbose debug message */
      vpi_printf("ERROR: PLIbook_get_arg_real_val_vpi() could not obtain
                 arg value\n");
      vpi_printf("File %s, Line %d: %s\n", err.file, err.line,
                 err.message);
    #endif
    return(0.0);
  }
  return(argVal.value.real);
}
```

5.2.5 Reading Verilog string values into C strings

The Verilog language stores strings differently than the C language. C stores strings in a character array, the first character in the string stored in the element 0 of the array. and the string is terminated with a \0 character. Strings stored in a Verilog vector are stored with the last character of the string in least-significant byte of the vector, and proceeding to the left-most byte of the vector. If the vector is larger than needed for the string, the left-most bits are zero padded. Verilog strings do not have a \0 null termination as in C. The `vpi_get_value()` routine converts Verilog strings into a C string by setting the format flag in the `s_vpi_value` structure to **vpiStringVal**. A pointer to the string will be placed in the **value.str** field of the `s_vpi_value` structure. If the Verilog object is a vector, each 8 bits of the vector will be considered an ASCII character, beginning with the left-most byte. If the Verilog object is a real (floating point) value, the value is converted to a string representation with a period character to represent the decimal point.

The steps for reading strings is the same as for reading integer and double values, except that the format specified is vpiStringVal.

 The character array to hold the C string is allocated by the simulator, and is only guaranteed to be valid until the next call to vpi_get_value(), or until the PLI application exits. If the string need to be maintained, the PLI application must allocate its own character array, and copy the string from the simulator's storage.

The following example lists ***PLIbook_get_arg_string_val_vpi()***. This example reads the string value of a system task/function argument as a C string.

CD The source code for this example is on the CD accompanying this book.

- Application source file: `Chapter.05/vpi_get_arg_val_vpi.c`
- Verilog test bench: `Chapter.05/vpi_get_arg_val_test.v`
- Expected results log: `Chapter.05/vpi_get_arg_val_test.log`

Example 5-3: *PLIbook_get_arg_string_val_vpi()* — reading Verilog string values

```
PLI_BYTE8 *PLIbook_get_arg_string_val_vpi(int argNum)
{
  vpiHandle arg_h;
  s_vpi_value argVal;
  s_vpi_error_info err;  /* structure for error handling */
  static char null_string[1] = {'\0'};
```

```
arg_h = PLIbook_get_arg_handle_vpi(argNum);
if (arg_h == NULL) {
  vpi_printf("ERROR: PLIbook_get_arg_string_val_vpi() could not obtain
             arg handle\n");
  return(0);
}
argVal.format = vpiStringVal;
vpi_get_value(arg_h, &argVal);
if (vpi_chk_error(&err)) {
  #if PLIbookDebug  /* if error, generate verbose debug message */
    vpi_printf("ERROR: PLIbook_get_arg_string_val_vpi() could not
               obtain arg value\n");
    vpi_printf("File %s, Line %d: %s\n", err.file, err.line,
               err.message);
  #endif
  return(null_string);
}
return(argVal.value.str);
}
```

5.2.6 Reading 4-state logic as C strings

The vpi_get_value() can read any Verilog logic value as a C string by setting the format flag in the s_vpi_value structure to one of the constants: **vpiBinStrVal**, **vpiOctStrVal, vpiDecStrVal** or **vpiHexStrVal**. The easiest method of printing Verilog 4-state logic in the C language is to use C character strings. The logic values Z and X are represented with the characters "z" and "x". As a character string, Verilog vectors of any width can be represented. The values can be represented in binary, octal, decimal or hexadecimal by setting format field to the appropriate constant in the s_vpi_value structure. The C string functions contained in standard C libraries, such as strings.h, can be used to manipulate the string values.

TIP

Representing a Verilog logic value as a C string is a simple way to represent 4-state logic in a printable form in the C language, and a simple way to read Verilog vectors of any bit width. However, the conversion from a Verilog logic value to a C string is very expensive for simulation run-time performance. The best simulation performance will be achieved when Verilog values are read into the C format that most closely resembles how the value is stored in a Verilog simulation.

Example 5-4 finds all the nets in a Verilog module and reads each net's logic value as a binary string value. A module instance name is passed to the routine as the first system task/function argument. This example was presented in Chapter 3, as Example 3-1 on page 74. Only the *calltf* routine is listed in this chapter.

CD The source code for this example is on the CD accompanying this book.

- Application source file: Chapter.03/show_all_nets_vpi.c
- Verilog test bench: Chapter.03/show_all_nets_test.v
- Expected results log: Chapter.03/show_all_nets_test.log

Example 5-4: using vpi_get_value() to read 4-state values as a C string

```
PLI_INT32 PLIbook_ShowNets_calltf(PLI_BYTE8 *user_data)
{
  vpiHandle   systf_handle, arg_iterator, module_handle,
              net_iterator, net_handle;
  s_vpi_time  current_time;
  s_vpi_value current_value;

  /* obtain a handle to the system task instance */
  systf_handle = vpi_handle(vpiSysTfCall, NULL);

  /* obtain handle to system task argument */
  /* compiletf has already verified only 1 arg with correct type */
  arg_iterator = vpi_iterate(vpiArgument, systf_handle);
  module_handle = vpi_scan(arg_iterator);
  vpi_free_object(arg_iterator);  /* free iterator memory */

  /* read current simulation time */
  current_time.type = vpiScaledRealTime;
  vpi_get_time(systf_handle, &current_time);

  vpi_printf("\nAt time %2.2f, ", current_time.real);
  vpi_printf("nets in module %s ",
              vpi_get_str(vpiFullName, module_handle));
  vpi_printf("(%s):\n", vpi_get_str(vpiDefName, module_handle));

  /* obtain handles to nets in module and read current value */
  net_iterator = vpi_iterate(vpiNet, module_handle);
  if (net_iterator == NULL)
    vpi_printf("  no nets found in this module\n");
  else {
    current_value.format = vpiBinStrVal; /* read values as a string */
    while ( (net_handle = vpi_scan(net_iterator)) != NULL ) {
      vpi_get_value(net_handle, &current_value);
      vpi_printf("  net %-10s  value is  %s (binary)\n",
                 vpi_get_str(vpiName, net_handle),
                 current_value.value.str);
    }
  }
  return(0);
}
```

5.2.7 Reading Verilog 4-state logic vectors as encoded aval/bval pairs

The `vpi_get_value()` routine with a **vpiVectorVal** format will retrieve an object's 4-state logic value as an encoded pair of 32-bit C integers. The encoding uses an *aval/bval* pair of 32-bit integers to represent the 4 logic values of Verilog, with each bit of the aval/bval pair representing one bit of a Verilog value. The encoding is:

Table 5-2: aval/bval logic value encoding

aval/bval pair	Verilog logic value represented
0/0	0
1/0	1
0/1	Z
1/1	X

The VPI library defines C integers which are 32-bits wide, and uses the aval/bval pair to encode up to 32 bits of a Verilog vector. By using an array of aval/bval integer pairs, vector lengths of any size may be represented. The representation of a 40-bit vector in Verilog can be visualized as:

For the Verilog declaration: **reg [39:0] data;**

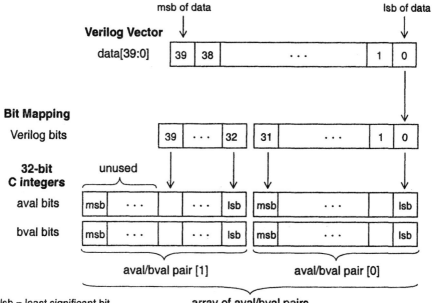

The Verilog language supports any numbering convention for a vector's bit numbers. The least significant bit of the Verilog vector can be the smallest bit number, such as bit 0 (which is referred to as little endian convention). Or, the least significant bit of the Verilog vector can be the largest bit number, such as bit 39 (which is referred to as big endian convention). Verilog does not require there be a bit zero at all. Each of the following examples are valid vector declarations in Verilog:

```
reg [39:0] data;    /* little endian -- LSB is bit 0  */
reg [0:39] data2;   /* big endian    -- LSB is bit 39 */
reg [40:1] data3;   /* little endian -- LSB is bit 1  */
```

The bit numbering used in Verilog does not affect the aval/bval representation of the Verilog vector. In the array of aval/bval pairs, the LSB of the Verilog vector will always be the LSB of the first 32-bit C integer in the array, and the MSB of the Verilog vector will always be the last bit in the array which is used. The following diagram illustrates the aval/bval array for a Verilog vector declared with a big endian convention.

For the Verilog declaration: `reg [1:40] data;`

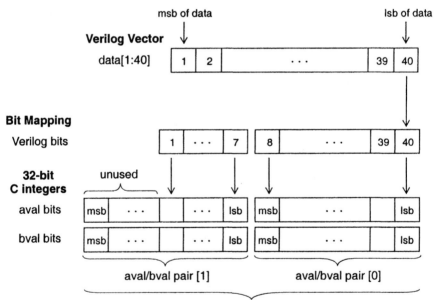

lsb = least significant bit
msb = most significant bit

The aval/bval pair is declared in an **s_vpi_vecval** structure, which is defined in the vpi_user.h VPI library. The structure definition is:

```
typedef struct t_vpi_vecval {
  PLI_INT32 aval, bval;     /* bit encoding: a/b:
                               0/0==0, 1/0==1, 1/1==X, 0/1==Z */
} s_vpi_vecval, *p_vpi_vecval;
```

To read a task/function argument's 4-state logic value using vpi_get_value() involves four basic steps:

1. Allocate an s_vpi_value structure.

2. Set the format field in the structure to vpiVectorVal.

3. Call vpi_get_value(), giving a pointer to the s_vpi_value structure as an input, along with a handle for the object from which to read the logic value.

4. Read the logic value of the object from the value.vector field of the s_vpi_value structure.

The value.vector field is a pointer to an array of s_vpi_vecval structures, with each element in the array representing 32 bits of a Verilog vector. The formula for determining the number of elements is:

```
number_of_array_elements = ((vector_size - 1) / 32 + 1);
```

The vector size is a property of the Verilog vector object, and can be accessed using:

```
vector_size = vpi_get(vpiSize, vector_handle);
```

Once the number of elements are known, the value of each 32-bit group of the Verilog vector can be accessed by reading the aval/bval pair of each s_vpi_vecval structure in the array. Within each aval/bval pair, an individual bit of the Verilog vector can be accessed by masking out all of the other bits in the aval/bval pair.

 NOTE The array of s_vpi_vecval structures is allocated by the simulator, and is only guaranteed to be valid until the next call to vpi_get_value(), or until the PLI application exits. If the values need to be maintained, the PLI application must allocate its own array of structures and copy the values from the simulator's storage.

Example 5-5 illustrates using vpi_get_value() to read the aval/bval encoded 4-state value of a vector that is passed to the PLI application as the first system task/function argument. The value of the vector is then printed one bit at a time. For simplicity, it is assumed that the least-significant bit of the Verilog vector is bit 0.

CD The source code for this example is on the CD accompanying this book.

- Application source file: `Chapter.05/read_vecval_vpi.c`
- Verilog test bench: `Chapter.05/read_vecval_test.v`
- Expected results log: `Chapter.05/read_vecval_test.log`

Example 5-5: *$read_vecval* — reading 4-state vector values as aval/bval pairs

```
PLI_INT32 PLIbook_ReadVecVal_calltf(PLI_BYTE8 *user_data)
{
  vpiHandle   systf_h, arg_itr, arg_h, vector_h;
  s_vpi_value vector_val;       /* structure to receive vector value */
  PLI_INT32   vector_size, array_size;
  int  i, bit_num, avalbit, bvalbit;
  char vlogval;

  /* obtain a handle to the system task instance */
  systf_h = vpi_handle(vpiSysTfCall, NULL);

  /* obtain handle to system task argument
     compiletf has already verified only 1 arg with correct type */
  arg_itr = vpi_iterate(vpiArgument, systf_h);
  vector_h = vpi_scan(arg_itr);
  vpi_free_object(arg_itr);  /* free iterator memory */

  vector_size = vpi_get(vpiSize, vector_h); /* determine number of...*/
  array_size  = ((vector_size-1) / 32 + 1); /* ...elements in array  */

  vector_val.format = vpiVectorVal;         /* set value format field */

  vpi_get_value(vector_h, &vector_val); /* read vector's logic value */

  vpi_printf("\nVector %s encoded value:\n",
             vpi_get_str(vpiName,vector_h));
  for (i=0; i<array_size; i++) {
    /* for simplicity, the following loop assumes the Verilog LSB    */
    /* is bit 0.  More robust code would obtain handles for the      */
    /* vpiRightRange and vpiLeftRange expressions and determine the  */
    /* vector declaration                                            */
    for (bit_num=0; bit_num<=31; bit_num++) {
      avalbit=PLIbook_getbit(vector_val.value.vector[i].aval, bit_num);
      bvalbit=PLIbook_getbit(vector_val.value.vector[i].bval, bit_num);
      vlogval=PLIbook_get_4state_val(avalbit, bvalbit);
      vpi_printf("  bit[%2d]  aval/bval = %d/%d  4-state value = %c\n",
                 (i*32+bit_num), avalbit, bvalbit, vlogval);
      /* quit when reach last bit of Verilog vector */
      if ((i*32+bit_num) == vector_size-1) break;
    }
  }
  return(0);
}
```

```
/*********************************************************************
 * Function to determine if a specific bit is set in a 32-bit word.
 * Sets the least-significant bit of a mask value to 1 and shifts the
 * mask left to the desired bit number.
 *********************************************************************/
int PLIbook_getbit(PLI_INT32 word, int bit_num)
{
  int mask;
  mask = 0x00000001 << bit_num;
  return((word & mask)? 1: 0);   /* 1 == TRUE, 0 == FALSE */
}


/*********************************************************************
 * Function to convert aval/bval encoding to 4-state logic represented
 * as a C character.
 *********************************************************************/
char PLIbook_get_4state_val(int aval, int bval)
{
  if       (!bval && !aval) return('0');
  else if (!bval &&  aval) return('1');
  else if ( bval && !aval) return('z');
  else                      return('x');
}
```

5.2.8 Representing Verilog strength values in C

The vpi_get_value() routine with a **vpiStrengthVal** format can be used to retrieve the 4-state logic value and strength level of a scalar net object, or each bit of a vector net object. The information is retrieved into an array of **s_vpi_strengthval** structures, with one structure for each bit. The first structure in the array will represent the least-significant of the Verilog vector. The memory for the structures is allocated by the simulator, and a pointer to the memory is returned in the value.strength field of the s_vpi_value structure. If the object is an array, the pointer will point to an array of s_vpi_strengthval structures with as many elements as there are bits in the vector.

The **s_vpi_strengthval** structure definition is:

```
typedef struct t_vpi_strengthval {
  PLI_INT32 logic;   /* one of: vpi0, vpi1, vpiX, vpiZ */
  PLI_INT32 s0, s1; /* Logical-OR of the constants:
                vpiSupplyDrive, vpiStrongDrive,
                vpiPullDrive, vpiWeakDrive, vpiLargeCharge,
                vpiMediumCharge, vpiSmallCharge, vpiHiZ */
} s_vpi_strengthval, *p_vpi_strengthval;
```

The logic value of each bit is represented as one of the constants: **vpi0**, **vpi1**, **vpiZ** or **vpiX**.

The strength level is represented as a pair of 32-bit C integers which contain the logic 0 and logic 1 strength components of the net. The Verilog language has 8 named strength levels for logic zero and 8 strength levels for a logic one, plus a number of ambiguous strength levels which are represented by a range of strength levels. Each named strength level is represented by a Verilog keyword, as shown in Table 5-3, which follows:

Table 5-3: Verilog HDL strength levels and keywords

Strength Level	Strength Name	Specification Keyword	
7	Supply Drive	**supply0**	**supply1**
6	Strong Drive	**strong0**	**strong1**
5	Pull Drive	**pull0**	**pull1**
4	Large Capacitance	**large**	
3	Weak Drive	**weak0**	**weak1**
2	Medium Capacitance	**medium**	
1	Small Capacitance	**small**	
0	High Impedance	**highz0**	**highz1**

Within Verilog, the strength of a signal is stored as two 8-bit bytes, as shown in the diagram below:

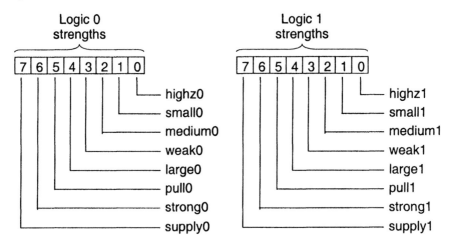

The full details on how the two strength bytes represent Verilog logic strength are defined in the Verilog language, and are outside the scope of this book. Refer to the list of Verilog HDL books on page 7 for suggestions on where to find more information on Verilog HDL strengths.

In the VPI, the strength value returned by vpi_get_value() is represented by a pair of 32-bit C integers with a logical OR of one or more constants. Each constant represents one of the named strength levels. The logical OR of the constants will result in a value from 00 (hex) to FF (hex). The value indicates which bits of the corresponding Verilog strength byte are set (only 8 bits of each 32-bit C integer are used). Note that both integers represent bit 0, the highz bit, as the least significant bit of the 32-bit C integer.

 The s_vpi_strengthval structure is allocated by the simulator, and is only guaranteed to be valid until the next call to vpi_get_value(), or until the PLI application exits. If the values need to be maintained, the PLI application must allocate its own structure, and copy the strength value from the simulator's storage.

The following example illustrates how to read the 4-state logic value and strength value of scalar nets using vpi_get_value(). The example assumes that a scalar Verilog net is passed in as the first system task/function argument.

CD The source code for this example is on the CD accompanying this book.

- Application source file: Chapter.05/read_strengthval_vpi.c
- Verilog test bench: Chapter.05/read_strengthval_test.v
- Expected results log: Chapter.05/read_strengthval_test.log

Example 5-6: *$read_strengthval* — reading logic and strength values

```
PLI_INT32 PLIbook_ReadStrengthVal_calltf(PLI_BYTE8 *user_data)
{
  vpiHandle    systf_h, arg_itr, arg_h, net_h;
  s_vpi_value net_val;              /* structure to receive net value */

  /* obtain a handle to the system task instance */
  systf_h = vpi_handle(vpiSysTfCall, NULL);

  /* obtain handle to system task argument
     compiletf has already verified only 1 arg with correct type */
  arg_itr = vpi_iterate(vpiArgument, systf_h);
  net_h = vpi_scan(arg_itr);
  vpi_free_object(arg_itr);  /* free iterator memory */

  net_val.format = vpiStrengthVal;      /* set value format field */
```

```
   vpi_get_value(net_h, &net_val);        /* read net's strength value */

   vpi_printf("\nNet %s:  ", vpi_get_str(vpiName, net_h));
   vpi_printf("value=%c  strength0=%2x(hex)  strength1=%2x(hex)\n\n",
             PLIbook_DecodeBitValue(net_val.value.strength->logic),
             net_val.value.strength->s0,
             net_val.value.strength->s1);
   return(0);
}


/**********************************************************************
 * Function to convert VPI logic constant to a character
 **********************************************************************/
char PLIbook_DecodeBitValue(int bit_constant)
{
  switch (bit_constant) {
    case vpi0:        return('0'); break;
    case vpi1:        return('1'); break;
    case vpiZ:        return('Z'); break;
    case vpiX:        return('X'); break;
    case vpiL:        return('L'); break;
    case vpiH:        return('H'); break;
    case vpiDontCare: return('?'); break;
    default:          return('U'); /* undefined value passed in */
  }
}
```

5.2.9 Reading Verilog time variables

The Verilog language stores a time variable as a 64-bit unsigned integer. The vpi_get_value() routine, with a **vpiTimeVal** format, can be used to retrieve the value stored in a time variable. The information is retrieved into an **s_vpi_time** structure. The memory for the structure is allocated by the simulator, and a pointer to the memory is returned in the value.time field of the s_vpi_value structure.

The s_vpi_time structure is defined in the vpi_user.h file, as follows:

```
typedef struct t_vpi_time {
  PLI_INT32  type;           /* not used by vpi_get_value() */
  PLI_UINT32 high;           /* upper 32-bits of time value */
  PLI_UINT32 low;            /* lower 32-bits of time value */
  double     real;           /* not used by vpi_get_value() */
} s_vpi_time, *p_vpi_time;
```

When reading the value of a time variable, the 64-bit time value will be retrieved as two unsigned 32-bit C integers, with the upper 32 bits of the variable stored in the **high** field, and the lower 32 bits of the variable stored in the **low** field of the structure. The `type` field and `real` field in the structure are not used when time variables are read.

 The `s_vpi_timeval` structure is allocated by the simulator, and is only guaranteed to be valid until the next call to `vpi_get_value()`, or until the PLI application exits. If the values need to be maintained, the PLI application must allocate its own structure, and copy the time value from the simulator's storage.

The basic steps required to retrieve the value are:

1. Allocate an `s_vpi_value` structure.

2. Set the `format` field to **vpiTimeVal** field.

3. Call `vpi_get_value()`, with an object handle and a pointer to the value structure as inputs.

4. Read the time value from the **value.time->high** and **value.time->low** fields of the time structure.

The calltf application shown in Example 5-7 retrieves the value of a time variable, using `vpi_get_value()`.

> *CD* The source code for this example is on the CD accompanying this book.
>
> • Application source file: `Chapter.05/read_timeval_vpi.c`
> • Verilog test bench: `Chapter.05/read_timeval_test.v`
> • Expected results log: `Chapter.05/read_timeval_test.log`

Example 5-7: *$read_timeval* — reading time variable values

```
PLI_INT32 PLIbook_ReadTimeVal_calltf(PLI_BYTE8 *user_data)
{
  vpiHandle    systf_h, arg_itr, arg_h, timevar_h;
  s_vpi_value timevar_val;    /* structure to receive variable value */

  /* obtain a handle to the system task instance */
  systf_h = vpi_handle(vpiSysTfCall, NULL);

  /* obtain handle to system task argument
     compiletf has already verified only 1 arg with correct type */
  arg_itr = vpi_iterate(vpiArgument, systf_h);
  timevar_h = vpi_scan(arg_itr);
  vpi_free_object(arg_itr);  /* free iterator memory */

  timevar_val.format = vpiTimeVal;          /* set value format field */

  vpi_get_value(timevar_h, &timevar_val);   /* read variable's value */
```

```
vpi_printf("\nTime Variable %s:  ", vpi_get_str(vpiName, timevar_h));
vpi_printf("  hi-word/lo-word = %d/%d\n\n",
            timevar_val.value.time->high,
            timevar_val.value.time->low);
return(0);
}
```

5.2.10 Reading Verilog values without determining the Verilog data type

The vpi_get_value() provides a convenient format mode that makes it possible to read any Verilog data type into C data types, without having to first determine the data type of the Verilog object. When the format field of the s_vpi_value structure is set to **vpiObjTypeVal**, the simulator will retrieve the Verilog data type into the C data type that most closely matches the Verilog type. The simulator automatically allocates whatever storage is needed, and places a pointer to the storage in the appropriate value field of the s_vpi_value structure. The simulator also sets the format field of the s_vpi_value structure with the corresponding format constant to indicate how the retrieved value is stored, overriding the vpiObjTypeVal set before vpi_get_value() was called.

The Verilog data types are converted to C types as follows:

- Verilog integer variables are converted using the **vpiIntVal** format.
- Verilog real variables are converted using the **vpiRealVal** format.
- Verilog strings are converted using the **vpiStringVal** format.
- Verilog time variables are converted using the **vpiRealVal** format.
- Verilog scalar reg or bit select of a vector reg or integer or time variable are converted using the **vpiScalarVal** format.
- Verilog scalar net or bit select of a vector net are converted using the **vpiStrengthVal** format.
- Verilog vectors are converted using the **vpiVecVal** format.

The automatic Verilog type conversion makes it easy for a PLI application to read any Verilog data type, by simply obtaining a handle to an object with a logic value, setting the format to vpiObjTypeVal, calling vpi_get_value(), and then examining to format field of the s_vpi_value structure, to determine how the value was retrieved.

 Any structures to receive the value are allocated by the simulator, and are only guaranteed to be valid until the next call to vpi_get_value(), or until the PLI application exits. If the values need to be maintained, the PLI application must allocate its own structures, and copy the value from the simulator's storage.

5.3 Writing values to Verilog objects

Writing a value to a Verilog object is simply a reverse process from reading a value. A single VPI routine is used to write values onto any type of object.

vpi_put_value() converts a value represented as C data type into Verilog 4-state logic, and writes the value to a Verilog object. The value to be written can be represented a variety of ways in the C language. These representations are the same as was described in section 5.2.1 on page 152 for reading values, which include representing Verilog logic values as a character string, as a scalar value represented using VPI constants, as an integer, as a double, as an array of s_vpi_vecval structures, as an s_vpi_time structure, or as an s_vpi_strengthval structure (if the object is a vector, the value.strength pointer must point to an array of s_vpi_strengthval structures with at least as many elements as there are bits in the vector).

The syntax for vpi_put_value() is:

vpiHandle **vpi_put_value (**

vpiHandle	**object,**	/* handle for an object */
p_vpi_value	**value,**	/* pointer to application-allocated s_vpi_value structure */
p_vpi_time	**time,**	/* pointer to application-allocated s_vpi_time structure */
PLI_INT32	**flag)**	/* constant representing the delay propagation method */

The **s_vpi_value** structure that is pointed to in the value field must be allocated by the PLI application, and the fields within the structure set to the value to be written to the object.

 When writing logic values, the PLI application must allocate and maintain all storage elements required to represent the logic value in C. Pointers to this user-allocated storage will be passed to vpi_put_value(). This is exactly the opposite of vpi_get_value(), where the simulator allocates and maintains any necessary storage.

The third argument to vpi_put_value() is pointer to an **s_vpi_time** structure. This structure is allocated by the PLI application and set to a propagation delay value. The vpi_put_value() routine can write the value to the object immediately or using the simulator's event scheduling mechanism. By using the event scheduler, a value can be written to an object later in the current simulation time step, or scheduled to occur in any future simulation time step.

The last argument to `vpi_put_value()` is a propagation flag, indicating how the propagation delay should be applied, using the simulator's event scheduler. The flag is one of the following constants:

- **vpiNoDelay** indicates no propagation delay is to be used. The object may be a Verilog reg, variable, memory word, variable array word select, sequential UDP or system function. When this flag is used, the propagation delay argument is not used and can be set to NULL.

- **vpiInertialDelay** indicates inertial delay propagation is to be used. Any pending events which are scheduled for the object are cancelled. The object may be a Verilog reg, variable, memory word, variable array word select or sequential UDP.

- **vpiPureTransportDelay** indicates transport delay propagation is to be used. Any pending events which are scheduled for the object remain scheduled (no events are cancelled). The object may be a Verilog reg, variable, memory word, variable array word select or sequential UDP.

- **vpiTransportDelay** indicates a modified transport delay propagation is to be used. Any pending events are cancelled which are scheduled for the object at a later time than this new event. The object must be a Verilog reg, variable, memory word, variable array word select or sequential UDP.

- **vpiForceFlag** indicates the value is to be forced onto the object, overriding any existing values. No other changes can occur on the object except another force until the force is released. No propagation delay is used, and the `time_p` argument may be set to NULL. The object may be any Verilog object which has a value property. Only one force value may exist for an object at a time. Setting a force value on an object will replace any existing force value, regardless of whether the force was set within the PLI or within the Verilog HDL.

- **vpiReleaseFlag** indicates any existing force on the object is to be released.

- **vpiReturnEvent** indicates that `vpi_put_value()` should return a handle for the scheduled event. Using this handle, the event can be removed from the simulator's event queue. This flag needs to be logically OR'ed with any of the flags vpiInertialDelay, vpiTransportDelay or vpiPureTransportDelay. If this flag is not set, `vpi_put_value()` will return NULL. A NULL is also returned if `vpi_put_value()` is called with no delay.

- **vpiCancelEvent** indicates an event should be removed from the simulator's queue. To use this flag, the handle passed to `vpi_put_value()` must be an event object which was returned by a previous call to `vpi_put_value()` which had used the vpiReturnEvent flag. It is not an error to attempt to cancel an event which has already transpired—the call to `vpi_put_value()` simply does nothing. The `value` and `time` arguments to `vpi_put_value()` are not needed for cancelling an event, and can be set to NULL.

To write a value to an object requires the following steps:

1. Obtain a handle for an object—the object must have a value property, in order to write a value to the object, or be a named event (putting a value on a named event will trigger an event; the value structure is not required, and the value_p argument can be NULL).

2. Allocate an **s_vpi_value** structure.

3. Allocate an **s_vpi_time** structure (if a propagation delay will be used).

4. Allocate variables or structures to store the logic value to be written onto the object, using the appropriate C data type for the format of the value.

5. Load the value memory with the value to be written.

6. Set the appropriate field in the **value** union of the s_vpi_value structure to a pointer to the value.

7. Set the **format** field in the s_vpi_value structure to indicate how the logic value is represented in C.

8. Set the **type** field in the s_vpi_time structure to indicate how the delay value is represented. The delay type is one of the following VPI constants:

   • **vpiScaledRealTime** indicates the delay is represented as a C double. The delay value will be scaled to the time units and precision of the module containing the object onto which the value will be written.

   • **vpiSimTime** indicates the delay is represented as a pair of 32-bit C integers, which contain the high order 32 bits and the low order 32 bits of the 64-bit simulation time. The delay value is in the simulator's internal time units, and will *not* be scaled to the object's time scale.

9. Set the delay value in the appropriate field of the s_vpi_time structure.

10. Call vpi_put_value(), with pointers to the s_vpi_value and s_vpi_time structures, and the constant representing how the value is to be scheduled in simulation time.

The following code fragment takes a value represented as a C character string, and writes the value onto the second argument of a system task, using transport delay event scheduling. The procedure for writing a value represented in any of the other C types is very similar to this example, and therefore separate examples of each value representation are not shown. The code fragment shown in this example is adapted from the *$read_stimulus_ba* application, which is presented in full in the next chapter as Example 6-3 on page 229 in Chapter 6.

```
PLI_INT32 PLIbook_ReadNextStim(PLI_BYTE8 *user_data)
{
   ...
   s_vpi_time        time_s;
   s_vpi_value       value_s;

   /* obtain system task handle and Verilog vector handle */
   ...
   /* read next line from the file */
   ...

   time_s.type = vpiScaledRealTime;
   time_s.real = delay;
   value_s.format = vpiBinStrVal;
   value_s.value.str = vector;
   vpi_put_value(vector_h, &value_s, &time_s,
                 vpiTransportDelay);

   return(0);
}
```

No delay versus zero delay

The Verilog PLI standard makes a distinction between putting a value into simulation
with no delay and putting a value into simulation with zero delay.

- The **vpiNoDelay** delay flag indicates that the value will be written into Verilog
  simulation instantly. When the PLI application returns back to simulation, any values written to an object or system function return using these routines will already
  be in effect for Verilog HDL statements to use.

- The **vpiIntertialDelay**, **vpiTransportDelay** and **vpiPureTransport-
  Delay** flags schedule a value to be written into simulation. If a delay of zero is
  specified, the value is scheduled to be written to the object later in the current simulation time step. When the system task returns back to simulation, the scheduled
  value will not yet have taken effect, and other Verilog HDL statements scheduled to
  execute in the same simulation time step may or may not see the new value of the
  object (depending on where the value change which was scheduled by the PLI falls
  in the simulator's event queue in relation to other Verilog HDL events).

The following simple Verilog HDL source code illustrates the potential problem of
putting a value into simulation using a delay of zero.

```
module test;
   reg [7:0] reg1, reg2;
   initial
     begin
       reg1 = 0; reg2 = 0;
       $put_value(reg1, reg2);
       $display("reg1=%d   reg2=%d", reg1, reg2);
       $strobe ("reg1=%d   reg2=%d", reg1, reg2);
       #1 $finish;
     end
endmodule
```

If the *calltf routine* for *$put_value* writes a value to reg1 using **vpiNoDelay**, then when *$put_value* returns to the simulation, and the *$display* statement prints the value of reg1, the <u>new</u> value will be printed.

If, however, the *calltf routine* for *$put_value* writes a value to reg2 using **vpiIner-tialDelay** or **vpiScaledRealTime** with a delay of zero, then, when *$put_value* returns to the simulation, the *$display* statement will probably print the <u>old</u> value of reg2. The old value is printed because the value written by the PLI has been scheduled to take place in the current time step, but it will not yet have taken effect. The *$strobe* statement which follows the *$display* will print the new value of both reg1 and reg2, because the definition of *$strobe* is to print its message at the end of the current simulation time step, after all value changes for that moment in time have taken effect.

5.4 Returning logic values of system functions

The vpi_put_value() routine is also used to write the return value of a system function into simulation. The object handle for where the value is to be written is the system function which called the PLI application.

Rules for returning values to system functions

There are two important restrictions on returning values to a system function:

- A value can only be written to the return of a system function from a *calltf routine*, which is when the system function is active. The *calltf routine* is invoked when the system function is encountered by the simulator when simulation is running. Return values cannot be written from VPI *simulation callback routines*, because the simulation is not executing the statement containing the function at the times these routines are invoked.

- When returning a value to a system function, the **vpiNoDelay** propagation flag must be used. It is illegal to specify a propagation delay when returning a value to a system function. If a delay is specified, the value will not be written and the VPI error flag will be set. Use vpi_chk_error() to determine if the flag is set.

Types of system functions

The VPI allows for several types of system functions. The type of function is established when the system function is registered through a call to vpi_register_systf(). Refer to section 2.11 on page 46 in Chapter 2 for a full description of registering system functions. In brief, the process involves allocating an s_vpi_systf_data structure, setting the type field to **vpiSysFunction**, and setting the sysfunctype field to one of the following VPI constants:

- **vpiIntFunc** indicates the system function will return a Verilog integer value. Verilog integers are 32-bit signed values.

- **vpiRealFunc** indicates the system function will return a Verilog real value. Verilog reals are double precision floating point values.

- **vpiTimeFunc** indicates the system function will return a Verilog time value. Verilog time variables are 64-bit unsigned values.

- **vpiSizedFunc** indicates the system function will return a Verilog scalar or vector value. Verilog scalars are 1-bit wide and vectors can be any width. The function returns unsigned values.

- **vpiSizedSignedFunc** indicates the system function will return a Verilog scalar or vector value. Verilog scalars are 1-bit wide and vectors can be any width. The function returns signed values. This function return type is a new feature in the IEEE 1364-2001 Verilog standard, to support the signed arithmetic extensions in the Verilog HDL, which allow a net or reg of any vector size to be declared as signed.

 The 1364-1995 Verilog standard used different constants for system function types:
- **vpiIntFunc** was **vpiSysFuncInt**
- **vpiTimeFunc** was **vpiSysFuncTime**
- **vpiRealFunc** was **vpiSysFuncReal**
- **vpiSizedFunc** was **vpiSysFuncSized**

The constant names from the 1364-1995 standard are aliased to the new constant names, to provide backward compatibility.

Setting the system function return size

In order for a Verilog simulator to compile Verilog source code, it must determine the return widths of functions. The system function types **vpiIntFunc**, **vpiRealFunc** and vpiTimeFunc have predefined return sizes. The return size for vpiSizedFunc and vpiSizedSignedFunc system functions, however, is defined by the PLI application. The return size is established using the VPI *sizetf routine*. This routine is called by the simulator's compiler or elaborator prior to simulation time 0. The routine uses the C return value to inform the simulator of the vector size that the *calltf routine* will write back to the simulation. The *sizetf routine* is only used if the type of system function is vpiSizedFunc or vpiSizedSignedFunc. If no *sizetf routine* is registered for these system function types, the default return size is 32 bits. The following example illustrates a *sizetf routine* to specify that the *calltf routine* will return a 64-bit vector.

```
PLI_INT32 rand64_sizetf(PLI_BYTE8 *user_data)
{
   return(64);    /* $rand64 returns 64-bit values */
}
```

 If a system function is used multiple times in the Verilog source code, the *sizetf routine* is only called once. The return value from this call is used to determine the return size for all instances of the system function, including any interactive calls to the function.

Example 5-8, shown below, implements a *$realpow* system function. This application returns a double-precision value representing **mⁿ** (m to the power of n). This example is an enhanced version of the *$pow* that was shown as Example 2-1 on page 43 in Chapter 2, which returned a 32-bit sized vector value.

> ***CD*** The source code for this example is on the CD accompanying this book.
>
> • Application source file: Chapter.05/realpower_vpi.c
> • Verilog test bench: Chapter.05/realpower_test.v
> • Expected results log: Chapter.05/realpower_test.log

Example 5-8: *$realpow* — returning values of system functions

```
#include <math.h>       /* ANSI C standard input/output library */
PLI_INT32 PLIbook_RealPow_calltf(PLI_BYTE8 *user_data)
{
```

```
s_vpi_value value_s;
vpiHandle   systf_handle, arg_itr, arg_handle;
double      base, exp, result;

value_s.format = vpiRealVal;

/* obtain handle to system task arguments;
   compiletf has already verified only 2 args with correct types */
systf_handle = vpi_handle(vpiSysTfCall, NULL);
arg_itr = vpi_iterate(vpiArgument, systf_handle);

/* read base value of system function arg 1 */
arg_handle = vpi_scan(arg_itr);
vpi_get_value(arg_handle, &value_s);
base = value_s.value.real;

/* read base value of system function arg 2 */
arg_handle = vpi_scan(arg_itr);
vpi_get_value(arg_handle, &value_s);
exp = value_s.value.real;
vpi_free_object(arg_itr); /* free iterator--did not scan till null */

/* calculate result of base to power of exponent */
result = pow(base,exp );

/* write result to simulation as return value $realpow_func */
value_s.value.real = result;
vpi_put_value(systf_handle, &value_s, NULL, vpiNoDelay);

return(0);
}
```

5.5 Reading and writing automatic variables

Verilog HDL tasks and functions have been static in the IEEE 1364-1995 Verilog standard, and all previous generations of Verilog,. A static task or function uses one copy of internal storage for all calls to the task or function. Static tasks cannot be invoked in parallel. That is, a second call to a task cannot be made while a previous call is still executing. The second call will overwrite the storage of the first call. A static function cannot be called recursively, because each call will overwrite the values of its parent.

IEEE 1364-2001 adds automatic tasks and functions. An automatic task is re-entrant; it can be called any number of times, and each call allocates unique storage. Automatic functions can be called recursively; each call is stacked, and has its own unique local storage. Automatic tasks and functions are declared with the automatic keyword, which is a new reserved word in the IEEE 1364-2001 standard.

The VPI supports automatic tasks and functions through a special **frame** object diagram: A frame represents a currently active call to a Verilog task or Verilog function. There is at most one active frame at a time. The parent of a frame may be the task or function call, or another frame, which is not currently active. Any local storage within a frame can be access by iterating on the method vpiAutomatic.

Figure 5-1: Object diagram for frames (partial)

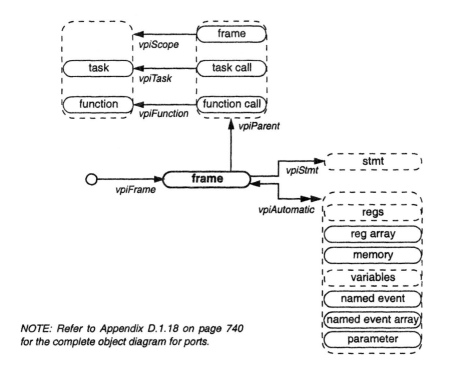

NOTE: Refer to Appendix D.1.18 on page 740 for the complete object diagram for ports.

The value of an automatic object is read or written the same as any other object.

 The storage within a frame is temporary. A handle to that storage will not be valid once that call to the task or function has exited. The boolean property vpiValid can be used to determine if a frame or any objects accessed using vpiAutomatic are still active. It is important to verify that an object is still valid before reading or writing values.

5.6 Reading IEEE 1364-2001 Verilog attributes

Often a PLI application needs information about a model which is not part of the model functionality. An interface to an analog simulator, for example, may need to know resistance and capacitance information about an object, which is not part of the normal Verilog HDL digital model.

The IEEE 1364-2001 Verilog standard adds a special attribute construct which allows non-Verilog information to be associated with specific objects. An attribute can also be associated with a module definition. Attributes associated with modules apply to all objects within that module. An attribute is defined with the (* and *) tokens. Attributes can have integer, real or string values. Handles to the attributes associated with an object are obtained using vpi_iterate(vpiAttribute, object_handle).

The value of an attribute is read using vpi_get_value() the same as other logic values.

5.7 Reading IEEE 1364-1995 specparam constant attributes

The IEEE 1364-1995 Verilog standard and earlier generations of the Verilog language created pseudo-attributes using Verilog parameter constants or specparam constants. As an example, a standard cell delay calculator might need the rise, slope and load factors of the cells used in a design. This information that is needed by the PLI application could be stored within the Verilog model as constants. The PLI application can then obtain a handle for the constant and read its value.

The older ACC generation of the PLI standard supports a special usage of the Verilog parameter or specparam constant, called an *attribute constant*. This special usage requires adding a dollar sign ($) to the end of the constant name. An attribute constant can be associated with all objects within a module, or with specific objects in a module.

- A general attribute constant is a parameter or specparam with a name which ends with a dollar sign.

- An object-specific attribute constant is a parameter or specparam with a base name which ends with a dollar sign, followed by the name of some object in the module.

The following example illustrates three specparam attribute constants:

```
module AN2 (o, a, b); // 2-input AND gate standard cell
  output o;
  input   a, b;

  and (o, a, b);

  specify
    specparam BaseDelay$  = 2.2; //general attribute
    specparam InputLoad$a = 0.2; //object-specific attribute
    specparam InputLoad$b = 0.3; //object-specific attribute
  endspecify
endmodule
```

A special set of routines in the ACC portion of the PLI standard can read the value of the attribute constant using the constant name and a handle for the object associated with that attribute constant. These ACC routines by-pass the need to first obtain a handle for the parameter or specparam. The ACC routines will search first for an object-specific attribute constant, then a general attribute constant, and finally return a default value.

The VPI library does not have a direct counterpart to the ACC attribute constant routines. However, it is easy to implement the same functionality. Example 5-9 lists a C function to fetch the value of a parameter or specparam attribute constant. The name of the attribute constant and a handle for an object are passed into the function. The attribute value is returned as a C double. If no attribute can be found, a default value is returned. The steps performed by this example are:

1. Access the name of the object

2. Concatenate the name of the object to the end of the attribute name.

3. Use vpi_handle_by_name() to try to obtain a handle for the constant with the attribute name and object name (the object-specific attribute constant).

4. If an object-specific attribute constant is not found, use vpi_handle_by_name() to try to obtain a handle for the constant with just the constant name (the general attribute constant).

5. If a general attribute constant is not found, then return the default value. The default value is passed into the function as its third input.

CD The source code for this example is on the CD accompanying this book.

- Application source file: Chapter.05/read_attribute_const_vpi.c
- Verilog test bench: Chapter.05/read_attribute_const_test.v
- Expected results log: Chapter.05/read_attribute_const_test.log

Example 5-9: using vpi_get_value() to read attribute constant values

```
PLI_INT32 PLIbook_ReadAttributeConst_calltf(PLI_BYTE8 *user_data)
{
  vpiHandle   systf_h, mod_h, arg_itr, arg2_h, port_itr, port_h;
  double      attribute_value;
  PLI_BYTE8  *attribute_name;
  s_vpi_value attribute_name_s;

  /* obtain handle to system task arguments;
     compiletf has already verified only 2 args with correct types */
  systf_h = vpi_handle(vpiSysTfCall, NULL);
  arg_itr = vpi_iterate(vpiArgument, systf_h);
  mod_h = vpi_scan(arg_itr);

  /* read base value of system function arg 2 */
  arg2_h = vpi_scan(arg_itr);
  vpi_free_object(arg_itr); /* free iterator--did not scan till null */
  attribute_name_s.format = vpiStringVal;
  vpi_get_value(arg2_h, &attribute_name_s);
  attribute_name = attribute_name_s.value.str;

  vpi_printf("\nModule %s:\n", vpi_get_str(vpiDefName, mod_h));
  port_itr = vpi_iterate(vpiPort, mod_h);
  while ( (port_h = vpi_scan(port_itr)) != NULL) {
    attribute_value = PLIbook_GetAttribute(port_h, attribute_name, 9.9);
    vpi_printf("  Port name = %s, attribute %s for this port = %2.2f\n",
               vpi_get_str(vpiName, port_h),
               attribute_name, attribute_value);
  }
  return(0);
}

/*******************************************************************
 * Function to return a specparam attribute value.
 *******************************************************************/
double PLIbook_GetAttribute(vpiHandle obj_h, PLI_BYTE8 *attribute,
                            double default_value)
{
  vpiHandle module_h, param_h;
  char *object_name;
  char param_name[1024];  /* character string to hold attribute name */
  s_vpi_value param_val;  /* structure to receive attribute value */

  param_val.format = vpiRealVal;
  module_h = vpi_handle(vpiModule, obj_h); /* get parent module */

  /* build specparam name out of object name and attribute name */
  object_name = vpi_get_str(vpiName, obj_h);
  strcpy(param_name, attribute);
  strcat(param_name, object_name);

  /* try to get a handle to the object specific attribute */
  param_h = vpi_handle_by_name(param_name, module_h);
  if (!vpi_chk_error(NULL)) {
```

```
    vpi_get_value(param_h, &param_val);
    if (!vpi_chk_error(NULL))
      return(param_val.value.real);  /* found specific attribute */
  }
  /* try to get a handle to a general attribute */
  strcpy(param_name, attribute);
  param_h = vpi_handle_by_name(param_name, module_h);
  if (!vpi_chk_error(NULL)) {
    vpi_get_value(param_h, &param_val);
    if (!vpi_chk_error(NULL))
      return(param_val.value.real);  /* found general attribute */
  }

  /* failed to find object-specific or general attribute specparam */
  return(default_value);
}
```

5.8 Accessing Verilog net, reg, memory and arrays

The VPI routines can be used to both read and modify the contents of Verilog arrays.

The IEEE 1364-1995 standard supported two types of arrays that have logic values:

- Memory arrays, which are one dimensional arrays of the Verilog reg data type.
- Variable arrays of Verilog integer and time data types.

The IEEE 1364-2001 standard adds:

- Arrays of Verilog net and real data types.
- multi-dimensional arrays of net, reg, and variable data types.

The logic value of any element in a memory or variable array can be read or modified by obtaining a handle for the array element and using vpi_get_value() or vpi_put_value(). Chapter 4, section 4.5.7 discusses how to obtain handles for arrays and words within the arrays. The entire contents of an array can be accessed by obtaining a handle for the array, and using vpi_iterate() to obtain handles for all of the elements in the array. It is not possible to access all of the values of an array at once using the VPI library; only the values of each individual object within an array can be accessed.

 The TF library also provides limited—and crude—access to the values within a memory array. This capability is presented in Chapter 11, section 11.8 on page 378.

5.9 Reading and modifying delay values

There are several types of Verilog objects which have delays values. The VPI routines can both read and modify the delays of these objects.

5.9.1 Verilog objects which have delay values

Several types of constructs in the Verilog language can have delay values. How delays are represented for these different types of objects is part of the Verilog HDL standard, and is outside the scope of this book. The objects which can have delays are:

- *Primitive instances* can have delays specified for 1, 2 or 3 output transitions.
- *Module paths* can have delays specified for 1, 2, 3, 6 or 12 output transitions.
- *Module input ports* can have delays specified for 1, 2, or 3 output transitions. There is no construct in the Verilog language to represent Module Input Port Delays (MIPD's); only the PLI can add or modify delays on input ports.
- *Module interconnect paths* can have delays specified for 1, 2, or 3 output transitions. An interconnect path is the connection from the output of one module to the input of another module. There is no construct in the Verilog language to represent interconnect path delays; only the PLI can add or modify delays on input ports.
- *Timing constraint checks* can have delays specified for 1 limit for each constraint in the check. Timing constraints are represented in Verilog using *$setup*, *$hold*, etc. The IEEE 1364-2001 Verilog standard adds new timing checks and new arguments to some existing timing checks. The VPI timing check object diagram (Appendix D, section D.1.30 on page 750) was enhanced for IEEE 1364-2001 to support these new features.
- *Continuous assignments* can have delays specified for 1, 2, or 3 output transitions.
- *Procedural time controls (#)* can have 1 delay specified. This delay represents the time before a statement is executed, rather than an output transition delay.

In addition to multiple delay transitions, each delay value can be a set of delays representing a minimum, typical and maximum delay range. Just a few of the ways propagation delays can be represented for a Verilog bufif1 tri-state buffer are:

A tri-state buffer with no delays:

```
bufuf1 g1 (...);
```

A tri-state buffer with delay of 5 for rising, falling, and turn-off transitions:

```
bufuf1 #5 g2 (...);
```

A tri-state buffer with separate delays for rising, falling, and turn-off transitions:

```
bufif1 #(3, 4, 5) g3(...);
```

A tri-state buffer with separate minimum:typical:maximum delay sets for rising, falling and turn-off transitions:

```
bufif1 #(2:3:4, 3:4:5, 5:6:7) g4 (...);
```

5.9.2 The VPI internal delay terminal

The Verilog HDL provides a means to model pin-to-pin path delays, which are internal to a module, and represent the propagation delay from a module input port to a module output port. This path can span any amount of complex logic within the module. While this is a real path, there may not be an actual, single object in the Verilog model to represent the path. Therefore, the Verilog PLI needs to create an internal place holder to store The Verilog PLI can specify delays on module input ports, and on the interconnections between modules (from the output of one module to the input of another.

To represent these types of delays a simulator may—but is not required to—add an internal delay objects. The IEEE 1364-2001 standard adds a special delay device object diagram, which allows a VPI application to obtain handles and information about the delay. Refer to D.1.29 on page 749 in Appendix D for the delay device object diagram.

5.9.3 Reading an object's delay values

vpi_get_delays() reads the delay values of any type of object which has a delay property. The routine has two inputs, a handle for an object, and a pointer to an **s_vpi_delay** structure.

void **vpi_get_delays (**
  | *vpiHandle* | **object,** | /* handle for an object */ |
  | *p_vpi_delay* | **delay)** | /* pointer to application-allocated s_vpi_delay structure */ |

The **s_vpi_delay** structure is defined in vpi_user.h, as follows.

```
typedef struct t_vpi_delay {
  struct t_vpi_time  *da; /* pointer to application allocated
                             array of delay values */
  PLI_INT32 no_of_delays; /* number of delay transitions */
  PLI_INT32 time_type;    /* vpiScaledRealTime, vpiSimTime */
  PLI_INT32 mtm_flag;     /* set to 0 (false) to retrieve a
                             single delay value for each
                             transition; set to 1 (true) to
                             retrieve min:typ:max delay sets
                             for each transition */
  PLI_INT32 append_flag;  /* not used by vpi_get_delays() */
  PLI_INT32 pulsere_flag; /* set to 0 (false) to retrieve only
                             the delays of the object; set to
                             1 (true) to read both delays and
                             pulse control values */
} s_vpi_delay, *p_vpi_delay;
```

There are six fields in the s_vpi_delay structure, which must be set by the PLI application before calling vpi_get_delays().

- The **da** field of the structure must be set to an array of s_vpi_time structures. The delay values retrieved from the object will be loaded into this array.

 The PLI application must allocate the array of s_vpi_time structures, and the array must have at least as many elements as the number of delay values which are to be read. Refer to table 5-4 for how the number of elements in the array is determined.

- The **no_of_delays** field of the structure is set by the PLI application to the number of delay *transitions* which are to be retrieved from the object. Legal values are 1, 2, 3, 6 and 12. vpi_get_value() will the number of transitions specified, regardless of the number of transitions represented in the Verilog HDL source code. The delays in the source code are mapped to the number of transitions requested.

- The **time_type** field of the structure is set by the PLI application to one of the constants **vpiScaledRealTime** or **vpiSimTime**. The type indicates how the simulator should retrieve and store the delay values of the object. For vpiScaled-RealTime, the delays are retrieved as double precision values and stored in the real field of the s_vpi_time structures pointed to by the da field. The delay values will be scaled to the time scale of the module containing the object from which the values are read. For vpiSimTime, the delays are retrieved as a pair of integer values, and stored in the high and low fields of the s_vpi_time structures pointed to by the da field. The delays are represented in the simulator's time units.

- The **mtm_flag** field is set by the PLI application to **0** (false) or **1** (true). False indicates that a single delay (the typical delay value) should be retrieved for each transition. True indicates that a minimum, typical, maximum delay set should be retrieved for each transition. Note that most Verilog simulators only store a single delay value for each transition, regardless of what is represented in the original Verilog source code. By default, these simulators store the typical delay value for each transition, but through invocation options or compiler directives, may store the minimum delay value or the maximum delay value for each transition. If the mtm_flag is set to true, and the simulator has only stored a single value for each transition, that one value will be returned for all three delays of the minimum, typical, maximum set.

- The **append_flag** field is not used by vpi_get_delays().

- The **pulsere_flag** field is set by the PLI application to **0** (false) or **1** (true). False indicates that a single delay or delay set (depending on the mtm_flag) should be retrieved for each transition. True indicates that a pulse control set of values should be retrieved for each delay. The pulse control set comprises of the delay value, a pulse *reject limit*, and a *pulse error* limit. If the mtm_flag is also true, then a pulse control set will be retrieved for each delay in the minimum:typical:maximum set.

Before calling vpi_get_delays(), the PLI application must allocate an array of s_vpi_time structures to receive the delay values. The total number of elements in the array must be at least as large as the number of delay values to be retrieved. This total is controlled by the settings of **no_of_delays**, **mtm_flag**, and **pulsre_flag**. The number of required elements in the array can vary from 1 element (1 delay transition, typical delay mode, and no pulse control values) to 108 elements (12 delay transitions, minimum:typical:maximum delays, and pulse control values). Table 5-4 shows how to determine the total number of delays which will be retrieved.

Table 5-4: Number of elements and order of the delay array

Flag Values	Number of Array Elements	Order of retrieved delays		
		array element	**object delay**	
mtm_flag = 0 pulsere_flag = 0	no_of_delays * 1	[0] receives [1] receives ...	1st delay 2nd delay ...	
mtm_flag = 1 pulsere_flag = 0	no_of_delays * 3	[0] receives [1] [2] [3] receives [4] [5] ...	1st delay 2nd delay ...	min value typ value max value min value typ value max value ...

Table 5-4: Number of elements and order of the delay array (continued)

Flag Values	Number of Array Elements	Order of retrieved delays		
		array element	object delay	
mtm_flag = 0 pulsere_flag = 1	no_of_delays * 3	[0] receives 1st delay	delay value	
		[1]	reject limit	
		[2]	error limit	
		[3] receives 2nd delay	delay value	
		[4]	reject limit	
		[5]	error limit	
		... ...	...	
mtm_flag = 1 pulsere_flag = 1	no_of_delays * 9	[0] receives 1st delay	min value	
		[1]	min reject limit	
		[2]	min error limit	
		[3]	typ value	
		[4]	typ reject limit	
		[5]	typ error limit	
		[6]	max value	
		[7]	max reject limit	
		[8]	max error limit	
		[9] receives 2nd delay	min value	
		[10]	min reject limit	
		[11]	min error limit	
		[12]	typ value	
		[13]	typ reject limit	
		[14]	typ error limit	
		[15]	max value	
		[16]	max reject limit	
		[17]	max error limit	
		... ...	...	

To read the delay values of an object requires the following basic steps:

1. Obtain a handle for an object with a delay property.

2. Allocate memory for an s_vpi_delay structure.

3. Allocate an array of s_vpi_time structures.

4. Set the appropriate fields in the s_vpi_delay structure.

5. Call vpi_get_delays() with a handle for the object and a pointer to the s_vpi_delay structure.

Example 5-10 lists a C function which reads the minimum, typical, maximum rise and fall delays of a module path. Note that module paths do not have a name property in the Verilog object diagrams. In order to print something meaningful, this example creates a module path name by retrieving the names of the nets connected to the first input of the path and the first output of the path, and concatenating these names together with a '$' in between the two names.

CD The source code for this example is on the CD accompanying this book.

- Application source file: `Chapter.05/read_delays_vpi.c`
- Verilog test bench: `Chapter.05/read_delays_test.v`
- Expected results log: `Chapter.05/read_delays_test.log`

Example 5-10: *$read_delays* — reading module path delays

```
PLI_INT32 PLIbook_ReadDelays_calltf(PLI_BYTE8 *user_data)
{
  vpiHandle  systf_h, arg_itr, mod_h, path_itr, path_h;

  /* obtain a handle to the system task instance */
  systf_h = vpi_handle(vpiSysTfCall, NULL);

  /* obtain handle to system task argument; */
  /* compiletf has already verified only 1 arg with correct type */
  arg_itr = vpi_iterate(vpiArgument, systf_h);
  mod_h = vpi_scan(arg_itr);
  vpi_free_object(arg_itr);  /* free iterator--did not scan to null */

  vpi_printf("\nModule %s paths:\n", vpi_get_str(vpiDefName, mod_h));
  path_itr = vpi_iterate(vpiModPath, mod_h);
  if (path_itr != NULL)
    while ((path_h = vpi_scan(path_itr)) != NULL) {
      PLIbook_PrintDelays(path_h);
    }
  else
    vpi_printf("  No path delays found.\n");
  return(0);
}

void PLIbook_PrintDelays(vpiHandle modpath_h)
{
  char *path_name;              /* pointer to path name string */
  s_vpi_delay delay_struct;     /* structure to setup delays   */
  s_vpi_time  delay_array[6];   /* structure to receive delays */

  delay_struct.da           = delay_array;
  delay_struct.no_of_delays = 2;
  delay_struct.time_type    = vpiScaledRealTime;
  delay_struct.mtm_flag     = 1;
  delay_struct.append_flag  = 0;
  delay_struct.pulsere_flag = 0;

  vpi_get_delays(modpath_h, &delay_struct);
  path_name = PLIbook_BuildPathName(modpath_h);
```

```c
  vpi_printf("Delays for %s = (%.2f:%.2f:%.2f, %.2f:%.2f:%.2f)\n",
    path_name,
    delay_array[0].real, delay_array[1].real, delay_array[2].real,
    delay_array[3].real, delay_array[4].real, delay_array[5].real);
}

char *PLIbook_BuildPathName(vpiHandle modpath_h)
{
  vpiHandle term_itr, term_h, net_h;
  static char path_name[2050]; /* character array to hold path name */
  char *term_name;

  path_name[0] = '\0';    /* clear the path name string */

  term_itr = vpi_iterate(vpiModPathIn, modpath_h);
  if (term_itr == NULL)
    return("UNKNOWN PATH NAME");
  term_h = vpi_scan(term_itr);
  net_h = vpi_handle(vpiExpr, term_h);
  if (net_h == NULL)
    return("UNKNOWN PATH NAME");
  term_name = vpi_get_str(vpiName, net_h);
  strcat(path_name, term_name);
  vpi_free_object(term_itr); /* free iterator--did not scan to null */

  strcat(path_name, "$");

  term_itr = vpi_iterate(vpiModPathOut, modpath_h);
  if (term_itr == NULL)
    return("UNKNOWN PATH NAME");
  term_h = vpi_scan(term_itr);
  net_h = vpi_handle(vpiExpr, term_h);
  if (net_h == NULL)
    return("UNKNOWN PATH NAME");
  term_name = vpi_get_str(vpiName, net_h);
  strcat(path_name, term_name);
  vpi_free_object(term_itr); /* free iterator--did not scan to null */

  return(path_name);
}
```

5.9.4 Writing delay values into an object

vpi_put_delays() writes delays values into an object. This routine requires two inputs, a handle for an object which has a delay property, and a pointer to an s_vpi_delay structure. The syntax of this routine is:

void **vpi_put_delays (**

vpiHandle	**object,**	/* handle for an object */
p_vpi_delay	**delay)**	/* pointer to an application-allocated s_vpi_delay structure containing delay information */

The process of writing delays into an object is nearly the reverse of reading delays, with the exception that the PLI application fills in the array of s_vpi_time structures with the values to be written into the object. The steps are:

1. Obtain a handle for an object with a delay property

2. Allocate memory for an s_vpi_delay structure.

3. Allocate an array of s_vpi_time structures.

4. Fill in the appropriate fields of the s_vpi_time structures with the delay values to be written onto the object. The delay values can be represented as double precision values or as pairs of 32-bit C integers. Which way delays are represented is controlled by the setting of the type field in the s_vpi_delay structure. The type also determines whether or not the delay values will be scaled to the object's time scale. A type of vpiScaledReadTime will scale the delay values to the time scale of the module which contains the object, and vpiSimTime will not scale the delay values (the delays are in the internal simulation time units).

5. Set the appropriate fields in the s_vpi_delay structure. The settings for these fields are nearly the same as with vpi_get_delays(), but with one additional flag. The **append_flag** is set to a **0** (false) or **1** (true). A false indicates that vpi_put_delays() will replace any existing delays for the object with the delays being written. A true indicates vpi_put_delays() will add the delays being written to any existing delays for the object.

6. Call vpi_put_delays() with a handle for the object and a pointer to the s_vpi_delay structure.

Example 5-11 shows a useful PLI application called *$mipd_delays*. The Verilog HDL does not have a construct to represent module input port delays—these delays may only be represented through the PLI. *$mipd_delays* provides a means for a Verilog model to add delays to the input port of a module.

The usage of this application is:

```
$mipd_delays(<port_name>, <t_rise>, <t_fall>, <t_toZ>)
```

Example:

```
$mipd_delays(in1, 2.7, 2.2, 1.0);
```

> **CD** The source code for this example is on the CD accompanying this book.
>
> • Application source file: Chapter.05/set_mipd_delays_vpi.c
> • Verilog test bench: Chapter.05/set_mipd_delays_test.v
> • Expected results log: Chapter.05/set_mipd_delays_test.log

Example 5-11: *$mipd_delays* — using vpi_put_delays()

```
PLI_INT32 PLIbook_SetMipd_calltf(PLI_BYTE8 *user_data) {
  vpiHandle tfarg_itr, tfarg_h, port_itr, port_h;
  int i;
  s_vpi_delay delay_struct;      /* structure to delay setup */
  s_vpi_time  delay_array[3];    /* structure to hold delays */
  s_vpi_value tfarg_value;       /* structure to hold tfarg values */

  delay_struct.da            = delay_array;
  delay_struct.no_of_delays  = 3;
  delay_struct.time_type     = vpiScaledRealTime;
  delay_struct.mtm_flag      = 0;
  delay_struct.append_flag   = 0;
  delay_struct.pulsere_flag  = 0;
  tfarg_value.format         = vpiRealVal;

  /* obtain handle to system task arguments;
     compiletf has already verified only 4 args with correct types */
  tfarg_itr = vpi_iterate(vpiArgument, vpi_handle(vpiSysTfCall, NULL));
  tfarg_h = vpi_scan(tfarg_itr); /* read 1st tfarg */
  port_itr = vpi_iterate(vpiPorts, tfarg_h);
  port_h = vpi_scan(port_itr);
  if (port_h != NULL)
    vpi_free_object(port_itr); /* free iterator-did not scan to null */
  for (i=0; i<=2; i++) {
    vpi_get_value(vpi_scan(tfarg_itr), &tfarg_value); /* read tfargs */
    delay_array[i].real = tfarg_value.value.real;
  }
  vpi_free_object(tfarg_itr); /* free iterator--did not scan to null */

  vpi_put_delays(port_h, &delay_struct);
}
```

5.10 Summary

This chapter has presented the VPI routines which read and modify logic values and delay values in Verilog simulations. The VPI routines provide a means for PLI applications to access the logic and delay values of any object within simulation. This interface layer makes it possible for a PLI application to be portable to any IEEE 1364 compliant simulator, without concern for how the each simulator stores this complex information. Several short PLI applications illustrate the concepts presented.

The next chapter presents another powerful aspect of the VPI routines, the ability to synchronize PLI application activity with simulation activity.

CHAPTER 6 *Synchronizing to Simulations Using VPI Callbacks*

*T*his chapter presents how to use a powerful feature of the VPI library—the ability to schedule with a Verilog simulator to invoke a PLI application for specific simulation events and simulation times. This ability to use the simulator's event scheduling mechanism makes it possible to synchronize PLI applications with activities that occur during a simulation.

The concepts presented in this chapter are:

- Sharing data between the routines of a PLI application
- Registering callbacks for when simulation actions occur
- Registering callbacks for specific simulation times
- Registering callbacks for when logic value changes occur
- Registering callbacks for when procedural statements are executed

6.1 PLI application callbacks

The Programming Language Interface provides a means whereby Verilog simulator users can write C language functions which are called by the Verilog simulator. The VPI standard refers to these calls as *callback routines* to the PLI application. The VPI provides two types of callback routines:

- *System task/function callback routines* — are routines that executed when the simulator encounters a user-defined system task or system function. The system task/function callback routines are *calltf routines*, *compiletf routines* and *sizetf routines*.

- *Simulation callback routines* — are routines that executed when specific types of events occur, generally during simulation, such as time advancing to a specified time step or a logic value change.

The two types of callback routines need to be *registered* with the simulator. The VPI library provides routines for performing this registration, as well as a routine to remove a requested callback which has not yet transpired. There are two registries for the two types of routines. Chapter 2 presented how system task/function callback routines are registered. This chapter presents how to register and use the *simulation callback routines*.

6.2 Sharing information between callback routines

An example test vector reader called *$read_test_vector* is used in this chapter. The arguments for this application are a file name and a Verilog reg vector. An example of using *$read_test_vector* is:

```
always @(posedge clock)
    $read_test_vector("A.dat", data_bus);
```

For this example, a *simulation callback* routine will be used to open the file specified in the first argument. This *simulation callback* will be registered to occur at the very start of simulation. In this way, the file can be opened for reading just one time, instead of each time the *calltf routine* is invoked. (this is not the only way this particular example could be implemented; using a simulation callback at the start of simulation is used in this example to illustrate that type of callback routine).

The *simulation callback* at the start of simulation will need to save the pointer to the file that was opened. As the simulation runs, each time the *calltf routine* is invoked it will need to access that file pointer. Therefore, the *simulation callback* routine and the *calltf routine* will need to share the file pointer information. Each of these routines is a C function. In the C programming language, a common programming method to share information between two functions is to use a global variable. ***Don't use global variables to share information between VPI callback routines!***

The Verilog HDL permits multiple instances of system tasks and functions. Each instance can have unique argument values. For this *$read_test_vector* example, the Verilog source code might contain:

```
always @(posedge clock)
    $read_test_vector("A.dat", data_bus);
```

```
always @(negedge clock)
  $read_test_vector("B.dat", data_bus);
```

In these two instances, the *$read_test_vector* instance that is called at the positive edge of clock needs to save a file pointer to *"A.dat"*, and the *$read_test_vector* that is called at the negative edge of clock needs to save a file pointer to *"B.dat"*. Since both instances of *$read_test_vector* will invoke the same C function to open the file, a global variable cannot be used, because the two instances would share the same single copy of the variable. The variable can only hold a single file pointer, and there is no simple way of knowing which instance of *$read_test_vector* was the last one to store a value in the variable.

Using global variables is a sure way to have problems when there are multiple instances of a system task or system function. A PLI application must have storage that is unique to each instance of a system task/function. The VPI library provides an instance-specific storage area for this very purpose. This storage area, commonly referred to as the system task/function *work area*, will allow a different file pointer to be stored for each instance of the *$read_test_vector* system task. Section 4.7 on page 122 in Chapter 4 presented details on the VPI instance-specific storage area.

Storing multiple values in the instance-specific storage area

Since the VPI instance-specific work area is a pointer, the storage area can be used to store a pointer to a memory location. Multiple values can be stored in the work area by allocating a block of memory and storing a pointer to the memory in the work area.

Often, each call to a PLI application will need to access the same objects. In the *$read_test_vector* example, the vector value read from a file will be applied to the second task argument. The *calltf routine* could go through the steps of obtaining a handle for the system task, traversing to the second argument and obtaining a handle for that argument. But, the *calltf routine* for *$read_test_vector* will be called every clock cycle in a simulation, possibly millions of times. The additional steps of obtaining the same handle every time the *calltf routine* is invoked is not very efficient for simulation run-time performance.

 Use the instance specific storage areas to improve the performance of PLI applications by reducing the number of times simulation hierarchy needs to be traversed to obtain object handles.
TIP

The instance-specific storage area can be used to improve the performance of a PLI application. All handles which will be used in every call to the application can be retrieved one time and saved in the work area. Future calls to the PLI application can retrieve the handles from the work area much more quickly than repeatedly traversing the Verilog hierarchy to obtain the handles. This means, however, that there are now two things to store in this single instance-specific storage area, the file pointer and the handle to the second task argument. The PLI application fragment shown below defines a structure called PLIbook_MyAppData. This structure can store the two values needed.

At the start of simulation, a *simulation callback* routine that is registered for *$read_test_vector* is used to perform three operations:

- Allocate memory for one PLIbook_Data structure.

- Fill the fields in the structure.

- Store a pointer to the allocated memory in the instance-specific storage area.

Creating and registering *simulation callback routines* for the start-of-simulation is presented later in section 6.6 on page 206. The complete *$read_test_vector* is presented in that section, as Example 6-1 on page 213.

```
/****************************************************************************
 * Define storage structure for file pointer and vector handle.
 ****************************************************************************/
typedef struct PLIbook_Data {
  FILE *file_ptr;      /* test vector file pointer */
  vpiHandle  obj_h;  /* pointer to store handle for a Verilog object */
} PLIbook_Data_s, *PLIbook_Data_p;

/****************************************************************************
 * StartOfSim callback (partial code)
 *   Opens the test vector file and saves the pointer and the handle
 *   for the 2nd system task argument in the work area storage.
 ****************************************************************************/
PLI_INT32 PLIbook_StartOfSim(p_cb_data cb_data)
{
  FILE         *vector_file;
  vpiHandle    systf_h, arg_itr, arg1_h, arg2_h;

  PLIbook_Data_p vector_data;  /* pointer to a ReadVecData structure */

  vector_data = (PLIbook_Data_p)malloc(sizeof(PLIbook_Data_s));

  /* get argument handles (compiletf already verified only 2 args) */
  /* read file name from first tfarg and open file */

  /* store file pointer and tfarg2_h in work area for this instance */
  vector_data->file_ptr = vector_file;
  vector_data->obj_h = arg2_h;
  vpi_put_userdata(systf_h, (void *)vector_data);

  return(0);
}
```

```
/*******************************************************************
 * calltf routine (partial code)
 *******************************************************************/
PLI_INT32 PLIbook_ReadVector_calltf(PLI_BYTE8 *user_data)
{
  FILE           *vector_file;
  vpiHandle       systf_h, arg2_h;
  PLIbook_Data_p vector_data;  /* pointer to a ReadVecData structure */

  systf_h = vpi_handle(vpiSysTfCall, NULL);

  /* get ReadVecData pointer from work area for this task instance */
  vector_data = (PLIbook_Data_p)vpi_get_userdata(systf_h);
  vector_file = vector_data->file_ptr;
  arg2_h      = vector_data->obj_h;

  /* read next line from the file */
  /* write the vector to the second system task argument */
  return(0);
}
```

6.3 Registering simulation callback routines

The vpi_register_cb() routine registers a callback to a *simulation callback routine* for various simulation activity that can occur during simulation. These callbacks can occur any time during a simulation, and are not related the execution of a user-defined system task/function. The types of activity which can occur during simulation that can result in a PLI application being called are divided into three major categories:

- **Simulation actions**, such as the start of simulation or the end of simulation.

- **Simulation time activity**, such as the end of the current simulation time step or the advancement to a specific simulation time.

- **Simulation events**, such as a logic value change or procedural statement execution.

The syntax of the vpi_register_cb() routine is:

vpiHandle **vpi_register_cb (**
    p_cb_data   **data)**   /* pointer to an application-allocated s_cb_data structure
                                 containing callback information */

A callback to a PLI routine for any of these activities is not automatic. Some other PLI routine, such as a *calltf routine*, must register a callback request with the simulator for when the desired activity occurs. To register a callback, the following steps are required:

1. Allocate an **s_cb_data** structure.

2. Fill in the fields of the structure.

3. Call **vpi_register_cb()** with a pointer to the structure as an input.

The s_cb_data structure is defined in vpi_user.h, as follows:

```
typedef struct t_cb_data {
  PLI_INT32    reason;        /* callback reason */
  PLI_INT32    (*cb_rtn)(struct t_cb_data *); /* routine name */
  vpiHandle    obj;           /* trigger object */
  p_vpi_time   *time;         /* callback time */
  p_vpi_value  *value;        /* trigger object value */
  PLI_INT32    index;         /* index of memory word or var
                                 select that changed value */

  PLI_BYTE8    *user_data;
} s_cb_data, *p_cb_data;
```

The fields within the structure are:

*   **reason** — an integer constant which represents what simulation activity will cause the callback to occur. The constants are defined in the vpi_user.h. The names and descriptions of these reason constants are presented on the following pages, in sections 6.6 through 6.8.

*   **cb_rtn** — the name of the PLI routine which should be called when the specified simulation activity occurs.

*   **obj** — a handle for an object. Not all callback reasons require an object handle. If a callback is requested for logic value changes on a specific net, for example, then a handle for that net must be provided in this field. The descriptions of each callback reason on the following pages indicate if an object handle is required for that callback.

*   **time** — a pointer to an s_vpi_time structure. If a callback is requested at some specific simulation time, the time structure needs to be filled with time value of when the callback should occur. The descriptions of each callback reason on the following pages indicate if a time structure is required for that callback.

*   **value** — a pointer to an s_vpi_value structure. If a callback is requested for logic value changes on a specific object, then when the change occurs, the simulator will automatically pass the new logic value to the callback routine. When the callback is registered, the format field in the value structure determines how the simulator will return the value of the object. The descriptions of each callback reason on the following pages indicate if a value structure is required for that callback.

*   **index** — not used when a callback is registered.

- **user_data** — a user data value. The value placed in user_data will be passed to the routine that is called when the specified simulation activity occurs. If this field is not needed by the routine, it should be set to NULL.

Several examples of registering simulation activity callbacks are presented in the following sections of this chapter.

6.3.1 C function inputs to a simulation callback routine

When the simulator calls the registered routine, a pointer to an **s_cb_data** structure is passed as an input to the routine. Within this structure is information about the callback.

- For *simulation action callbacks* and *simulation feature callbacks*, the structure will contain the user_data value specified when the callback was registered.

- For *simulation time callbacks*, the structure will contain the current simulation time and the user_data value.

- For *simulation event callbacks*, the structure will contain the current simulation time, the logic value of the object on which the event occurred, the index and object handle, if they pertain to the type of event, and the user_data value.

 The structure passed into the *simulation callback routine* is allocated and maintained by the simulator, and will be freed after the *simulation callback routine* exits.

6.3.2 Using the user_data value

The user data value is not required for any simulation activity callback, but can be a very useful way to pass information to a callback. The user_data is a pointer, which can be used to store a single value or to point to a block of data. The user_data value is passed to the *simulation callback routine* each time the simulator calls the routine.

 If a pointer to a block of data is specified in the user_data field, the PLI application must maintain this memory. The PLI application should allocate memory to store the data, and free the memory when it is no longer needed. Local variables in the function that registers the callback should not be placed in the user_data field, as local variables are automatically freed and will not be available when the callback occurs.

6.3.3 Accessing system task/function arguments in a simulation callback

The *$read_test_vector* example used in this chapter illustrates another important aspect of simulation callbacks. The usage of this example is:

```
always @(posedge clock)
  $read_test_vector("A.dat", data_bus);
```

A simulation callback routine is not directly associated with an instance of a system task or system function, and therefore cannot directly obtain a handle for a specific task/function instance. In many PLI applications, however, the simulation callback may have been registered from a *compiletf routine* or *calltf routine* for a specific task/function instance, and the simulation callback may need to access information about that instance, such as the system task/function arguments.

For the *$read_test_vector* example, a *simulation callback* routine will be registered for the start of simulation to open the file named in the first task/function argument. In order to access this argument, the simulation callback routine will need to obtain a handle for the instance of the system task containing the file name. Only *sizetf routines*, *compiletf routines* and *calltf routines* are directly associated with each instance of a system task or function, and can therefore obtain a handle for that instance. In the *$read_test_vector* example, the *compiletf routine* registers the simulation callback for the start of simulation. The *compiletf routine* can obtain the handle for the instance of the system task that invoked the compiletf routine, and pass the handle for that instance to the simulation callback through the **user_data** field as the callback is registered. The *simulation callback routine* at start-of-simulation can then retrieve the system task/function handle from the user_data, and access the arguments of that instance of the system task/function.

The partial code for *$read_test_vector* presented on page 200 illustrates using user_data to pass a system task/function instance handle to the simulation callback routine. The complete *$read_test_vector* is presented in Example 6-1 on page 213.

6.4 Removing scheduled callbacks

A callback which has been requested, but which has not yet transpired, is referred to as a ***scheduled callback***. A scheduled callback can be removed at any time using **vpi_remove_cb()**. The syntax for this routine is:

PLI_INT32 **vpi_remove_cb (**
   vpiHandle   **cb_object)**   /* handle for a callback object */

The `vpi_remove_cb()` routine removes callbacks to PLI applications which were registered with `vpi_register_cb()`. The routine returns 1 (for true) if successful, and 0 (for false) if an error occurred. The callback handle is no longer valid after the callback is removed. It is not an error to attempt to remove a callback that has already transpired. If `vpi_remove_cb()` is called to remove a callback which has already transpired, the removal request is ignored.

When a callback is registered, `vpi_register_cb()` returns a handle for the scheduled callback. If a PLI application might need to remove a callback at a future time, the application should save the scheduled callback handle. The handle cam be placed in application-allocated storage, and a pointer to the storage placed in the instance-specific storage area (using `vpi_put_userdata()`) or in the `user_data` field for a simulation callback.

6.5 Avoiding memory leaks with simulation callbacks

When a simulation callback is registered, the handle for the callback will persist the remainder of simulation, unless the callback is removed. The only usage of the callback handle is for removing the callback using `vpi_remove_cb()`. If the PLI application has no need to remove the callback, then the handle for the callback should be freed immediately after the callback is registered. The recommended method of registering most simulation callbacks is:

```
callback_handle  = vpi_register_cb(&cb_data_s);
vpi_free_object(callback_handle); /* free callback handle */
```

Freeing the callback handle immediately after registering the callback is registered prevents a memory leak in simulation. For example, if a new simulation callback were registered every positive edge of a design clock, and the design were simulated millions of clock cycles, then millions of handles would be created. If the handles were not released, a serious memory usage problem would occur.

 NOTE Do not call `vpi_free_object()` on a callback handle if the callback has been removed using `vpi_remove_cb()`. The latter routine automatically releases the callback handle, making the handle invalid.

6.6 Simulation action-related callbacks

Simulation callbacks to a PLI application can be scheduled for when specific actions occur during a simulation. These callbacks are registered using constants to represent the reason for the callback, as described in Table 6-1, below.

One-time callbacks and repeating callbacks

Some types of simulation callbacks will only occur one-time during simulation, such as a callback registered for the start of simulation. Other callbacks, such as for whenever simulation enters interactive mode, can occur any number of times during simulation. In the table below, footnote tokens indicate if the callback occurs one-time (represented by †), or will repeat (represented by ‡). When a repeatable simulation callback is no longer needed by a PLI application, it can be removed using vpi_remove_cb().

Table 6-1: VPI simulation action-related callback constants

Constant	Definition
cbEndOfCompile†	calls a PLI application at the end of simulation data structure elaboration or build (immediately before time 0)
cbStartOfSimulation†	calls a PLI application at the very start of simulation (beginning of the time 0 simulation cycle)
cbEndOfSimulation†	calls a PLI application at the end of simulation (e.g.: the *$finish* system task was executed)
cbStartOfSave†	calls a PLI application when a simulation checkpoint is started (e.g.: a *$save* command is executed)
cbEndOfSave†	calls a PLI application when a simulation checkpoint is completed
cbStartOfRestart†	calls a PLI application when a restart from a simulation checkpoint is started (e.g.: a *$restart* command is executed)
cbEndOfRestart†	calls a PLI application when a restart from a simulation checkpoint is completed
cbEnterInteractive‡	calls a PLI application when simulation enters interactive debug mode (e.g.: a *$stop* system task executed)
cbExitInteractive‡	calls a PLI application when simulation exits interactive mode
cbInteractiveScopeChange‡	calls a PLI application when a simulation command to change interactive scope is executed

Table 6-1: VPI simulation action-related callback constants (continued)

Constant	Definition
cbError[‡]	calls a PLI application if a run-time error occurred while executing the Verilog HDL portion of the simulation
cbPLIError[‡]	calls a PLI application if a run-time error occurred while executing a PLI application
cbTchkViolation[‡]	calls a PLI application if a Verilog HDL timing check violation occurred
cbSignal[‡]	calls a PLI application if an operating system signal occurred
cbStartOfReset[‡]	calls a PLI application when a restart from a simulation reset is started (e.g.: a *$reset* command is executed)
cbEndOfReset[‡]	calls a PLI application when a restart from a simulation checkpoint is completed
cbUnresolvedSystf[‡]	calls a PLI application when a system task/function is encountered that is not pre-defined in the simulator or is not registered using vpi_register_systf()
	simulators may add any number of product-specific callback constants to support features of that product

[†] Indicates the callback is a one-time event. The simulator will automatically remove the call-back after it occurs. *Note*: the callback handle will still be valid, and must be freed by the PLI application using vpi_free_object() to avoid a memory leak (see Section 6.5).

[‡] Indicates the callback can occur multiple times. The PLI application must unregister the callback using vpi_remove_cb() when no longer needed. *Note*: vpi_remove_cb() also frees the callback handle (see Section 6.4).

6.6.1 Required settings to register simulation action-related callbacks

To register a callback for a specific simulation action or feature, the following fields in the **s_cb_data** structure must be set:

- **reason** must be set to one of the simulation action or feature constants.

- **cb_rtn** must specify the name of the PLI application which should be called when the action occurs.

- **user_data** (optional) can be set to point to a user data value, if needed. If this field is not used, it should be set to NULL.

The remaining fields are not required for simulator action-related callbacks. These unused fields should be set to NULL.

A simulator might not implement all of the simulation actions listed in table 6-1 on the preceding pages. For example, some simulators do not have *$save* and *$restart* check pointing capabilities. It is not an error to register a callback for an action-related that is not implemented in a simulator. The constants for these callback reasons are defined in the IEEE 1364 standard, which means the PLI application will compile and link correctly with any IEEE 1364 compliant simulator. If the simulator has not implemented the action, the callback will simply never occur.

More caution is needed for product-specific feature callbacks. If a simulator vendor adds a reason constant to a simulator, that constant may not exist, or might be defined with a different value in another simulator. If a PLI application uses product-specific callback reasons, then it is recommended the application be written in such a way that those callbacks can be omitted when the application is compiled for other simulators (perhaps using #if conditional compilation directives).

6.6.2 When a simulation action-related callback occurs

When a simulator action-related callback occurs, the simulator will allocate an **s_cb_data** structure and pass a pointer to the structure as an input to the *callback routine* which is called. For action-related callbacks, the only field in the s_cb_data structure that is used is the **user_data** field. This field will contain whatever user_data value was specified when the callback was registered. In the example which follows, the user_data field will contain a handle for the specific instance of the system task which requested the simulation action callback.

6.6.3 How to register a simulation callback before time zero

Most simulation callbacks will be registered from another VPI routine that has been called during simulation, such as a *calltf routine*. In order to register a simulation callback for cbEndOfCompile or cbStartOfSimulation, however, the call to vpi_register_cb() must occur prior to simulation time zero. There are two places where a simulation time callback can be registered prior to time zero, each with unique and important capabilities:

- A C function with the call to vpi_register_cb() can be listed in the *vlog_startup_routines* array. The functions listed in this array are executed by the simulator prior to the end of elaboration/linking. ***The functions listed in the start up array are only executed one time.*** This means the end-of-compile or start-of-simulation callback will only occur one time, regardless of how many instances of a system task/function occur in the Verilog hierarchy. Also note that the functions called from the start up array are not associated with any specific instance of a system task or system function, and therefore cannot obtain a handle for a specific system task/function.

- The call to vpi_register_cb() can be listed in a ***compiletf routine***. The *compiletf routine* is invoked automatically as the simulation data structure is being built, so both end-of-compile and start-of-simulation callbacks can be registered from the *compiletf* routine. ***The compiletf routine is invoked for each instance of a system task/function.*** This means the end-of-compile or start-of-simulation callback may be executed multiple times, once for each instance of the system task/function. Also the *compiletf routine* is associated with a task/function instance, and can access information about the task/function to pass on to the simulation callback, such as the handle to that instance of a system task/function.

6.6.4 Limitations on callbacks registered before time zero

The cbEndOfCompile and cbStartOfSimulation callbacks have certain limitations which must be observed. These callbacks occur after the simulation data structure has been fully created, but before any simulation values have been propagated. This means all reg and variables will be uninitialized, and nets may not yet reflect the values of their drivers. Reading logic values prior to time zero is not an error, but probably has little meaning. Other than reading logic values, there are no other restrictions on what can be done at these callback times. Some activity that is common and appropriate for these callback reasons are:

- Traverse the design hierarchy to collect handles or other information that will be needed by the *calltf routine* or other callbacks as simulation time advances.
- Allocate memory for use during simulation.
- Open files that will be needed throughout simulation.
- Other activity that only needs to be performed one time during simulation.

6.6.5 Simulation save and restart checkpoint files

Some Verilog simulators provide the means to save the simulation state, and restart simulation from the saved state. Save and restart are sometimes referred to as checkpointing. A *$save* built-in system task typically creates the checkpoint file, and a *$restart* built-in system task or an invocation option resumes simulation from the saved state in the checkpoint file.

Since a PLI application may allocate its own storage, it is important that the PLI application save and restore its storage as part of the simulation save/restart checkpoint file. An event-related *simulation callback routine* can be registered for the reasons **cbStartOfSave** or **cbEndOfSave** when the simulator executes a save command, and **cbStartOfRestart** or **cbEndOfRestart** when the simulator restarts from a saved file.

Saving data

When the simulation callback routine is called at the start or end of a save checkpoint, the routine can have the simulator append any number of blocks of data to the checkpoint file. The routine to append to a checkpoint file is vpi_put_data(). This routine can only be called from a simulation callback routine that was registered for the reasons cbStartOfSave or cbEndOfSave.

PLI_INT32 **vpi_put_data (**
    PLI_INT32   **id,**   /* a save/restart ID */
    PLI_BYTE8 *** data_addr,**   /* address of application-allocated storage */
    PLI_INT32   **byte_count)** /* number of bytes to be added to the simulator's save/
                                       restart file; must be greater than 0 */

vpi_put_data() adds a specified number of bytes of data, beginning at a specified location, to the simulator's save/restart checkpoint file. vpi_put_data() returns the number of bytes written into the save/restart checkpoint file, or 0 if an error occurred.

Since vpi_put_data() can be called any number of times, the routine also adds an ID value to each block of data saved. The ID is obtained as follows:

```
id = vpi_get(vpiSaveRestartID, NULL);
```

When the saved information is retrieved, the ID value must be specified, and the data for that ID will be retrieved. This means the ID must be passed to the simulation callback that will be invoked for cbStartOfRestart or cbEndOfRestart. The recommended way to pass the ID value is for the simulation callback that saves the data to also register the simulation callback for a restart. The ID can then be passed as the user_data value when the restart callback is registered. The ID value allows any number of PLI applications to append data to the simulation checkpoint file. Each application can retrieve its saved data by using its ID value.

The same ID can be used for multiple calls to vpi_put_data(). Data for that ID is retrieved in the same order in which the data was added. This allows a PLI application to save several pieces of information, and retrieve the information as one block of data, or as separate blocks of data.

The name of the checkpoint file can be accessed using vpi_get_str(vpiSaveRestartLocation, NULL).

 Handles and pointers that are saved in a save/restart checkpoint file may not be valid after a restart occurs. The PLI application should regenerate handles and pointers after a restart.

Retrieving saved data

When a simulation callback routine is called for the reasons cbStartOfRestart or cbEndOfRestart, the routine can retrieve data that was previously written to a save/restart checkpoint file. The routine to retrieve checkpoint data is vpi_get_data(). This routine can only be called from a simulation callback routine that was registered for the reasons cbStartOfRestart or cbEndOfRestart.

PLI_INT32 **vpi_get_data (**
    PLI_INT32   **id,**   /* a save/restart ID */
    PLI_BYTE8 *** data_addr,**   /* address of application allocated storage */
    PLI_INT32   **byte_count)** /* number of bytes to be retrieved from the simulator's
                                  save/restart file; must be greater than 0 */

This routine retrieves the number of bytes requested from the saved data for the ID specified. The information is placed into the location pointed to **data_addr**. The routine returns the actual number of bytes retrieved, or 0 if an error occurred. *The memory pointed to by data_addr must first be allocated by the PLI application before calling vpi_get_data().*

The ID value can only be created during a save operation. The value of the ID can be passed to the simulation callback for a restart through the user_data field of the callback. For a given ID value, any number of calls can be made to vpi_get_data(). Each call will begin retrieving data where the previous call for that ID left off. Data will always be retrieved in the same order in which it was written for that ID. The number of bytes retrieved does not need to match the number of bytes written. This allows a great deal of flexibility on how information can be written into a checkpoint file and then retrieved. Some possibilities are:

- The data to be written into the save/restart checkpoint file can be saved is in a structure. The C sizeof() function can be used to determine the number of bytes in the structure, and therefore the number of bytes to be retrieved.

- The number of bytes written into the save/restart checkpoint file can be passed to the simulation routine for the restart, using the user_data field of the callback. This requires allocating a structure to store both the byte count and the ID, as both values must be passed as the user_data value.

- When data is written, the first byte, or group of bytes, can store the total number of bytes saved. vpi_get_data() can then be called once, to retrieve the count. Using that count, vpi_get_data() can be called a second time to retrieve the remainder of the saved information.

- vpi_put_data() for a given ID can be called multiple times, to save several blocks of data in the save/restart checkpoint file. The total number of calls, and the byte count of each call, could be passed to the simulation through the user_data, or saved at the beginning of the data written into the checkpoint file.

When simulation is restarted, any pointers which were saved may no longer be valid. A file pointer, for example, can no longer be used, because the file may have been closed between the save and restart operations. Any application-allocated memory may have been freed or relocated, making pointers to the memory invalid. Any handles that had been obtained prior to the restart may also be invalid. The simulation callback routine for a restart should re-open files, re-allocate memory, and re-obtain any necessary handles.

 The vpi_put_data() and vpi_get_data() routines are enhancements to the VPI library which were added as part of the IEEE 1364-2001 Verilog standard.

6.6.6 Callbacks for operating system signals

vpi_register_cb() can be used to set up an operating system signal handler, by setting the reason field to **cbSignal** and setting the index field to one of the legal signals specified by the operating system. When this signal occurs, the simulator will trap the signal, proceed to a safe point (if possible), and then call the callback routine.

 The vpiSignal simulation callback was added as part of the IEEE 1364-2001 Verilog standard.

6.6.7 An example of using simulation action-related callbacks

Example 6-1 lists the complete *$read_test_vector* PLI application, which reads test vectors from a file and passes the vector value to simulation. This example utilizes a number of concepts that have been presented in this chapter and in previous chapters. Some of the key VPI concepts used are:

- Using multiple C functions to create a complete PLI application, including a *calltf routine*, a *compiletf routine*, and a *callback routine* for a start-of-simulation action.
  - Using of the *compiletf routine* to a register simulation callback for each instance of *$read_test_vector*.
  - Using *simulation action callbacks*; specifically a callback for the reason **cbStartOfSimulation**.
- Using an instance-specific storage area to preserve data over simulation time.

The application reads vectors from a file one line at a time. Therefore *$read_test_vector* should be invoked in a loop that will read vectors until the end of simulation or the end of the vector file. For example:

```
always @(posedge clock)
  $read_test_vector("<file_name>", <reg_vector>);
```

> **CD** The source code for this example is on the CD accompanying this book.
>
> • Application source file: Chapter.06/read_test_vector_vpi.c
> • Verilog test bench: Chapter.06/read_test_vector_test.v
> • Expected results log: Chapter.06/read_test_vector_test.log

Example 6-1: *$read_test_vector* — using VPI simulation action callbacks

```
/******************************************************************
 * Define storage structure for file pointer and vector handle.
 ******************************************************************/
typedef struct PLIbook_Data {
  FILE *file_ptr;     /* test vector file pointer */
  vpiHandle  obj_h;   /* pointer to store handle for a Verilog object */
} PLIbook_Data_s, *PLIbook_Data_p;

/******************************************************************
 * VPI Registration Data
 ******************************************************************/
void PLIbook_ReadVector_register()
{
  s_vpi_systf_data tf_data;

  tf_data.type        = vpiSysTask;
  tf_data.sysfunctype = 0;
  tf_data.tfname      = "$read_test_vector";
  tf_data.calltf      = PLIbook_ReadVector_calltf;
  tf_data.compiletf   = PLIbook_ReadVector_compiletf;
  tf_data.sizetf      = NULL;
  tf_data.user_data   = NULL;
  vpi_register_systf(&tf_data);
}

/******************************************************************
 * compiletf routine
 ******************************************************************/
PLI_INT32 PLIbook_ReadVector_compiletf(PLI_BYTE8 *user_data)
{
  s_cb_data   cb_data_s;
  vpiHandle   systf_h, arg_itr, arg_h, cb_h;
  PLI_INT32   tfarg_type;
  int         err = 0;
  char        *file_name;
```

```
  systf_h = vpi_handle(vpiSysTfCall, NULL);
  arg_itr = vpi_iterate(vpiArgument, systf_h);
  if (arg_itr == NULL) {
    vpi_printf("ERROR: $read_test_vector requires 2 arguments\n");
    vpi_control(vpiFinish, 1);  /* abort simulation */
    return(0);
  }
  arg_h = vpi_scan(arg_itr);  /* get handle for first tfarg */
  if (vpi_get(vpiType, arg_h) != vpiConstant) {
    vpi_printf("$read_test_vector arg 1 must be a quoted file name\n");
    err = 1;
  }
  else if (vpi_get(vpiConstType, arg_h) != vpiStringConst) {
    vpi_printf("$read_test_vector arg 1 must be a string\n");
    err = 1;
  }
  arg_h = vpi_scan(arg_itr);  /* get handle for second tfarg */
  tfarg_type = vpi_get(vpiType, arg_h);
  if ( (tfarg_type != vpiReg) &&
       (tfarg_type != vpiIntegerVar) &&
       (tfarg_type != vpiTimeVar) ) {
    vpi_printf("$read_test_vector arg 2 must be a register type\n");
    err = 1;
  }
  if (vpi_scan(arg_itr) != NULL) {
    vpi_printf("read_test_vector requires only 2 arguments\n");
    vpi_free_object(arg_itr);
    err = 1;
  }
  if (err) {
    vpi_control(vpiFinish, 1);  /* abort simulation */
    return(0);
  }

  /* No syntax errors, setup a callback for start of simulation */
  cb_data_s.reason    = cbStartOfSimulation;
  cb_data_s.cb_rtn    = PLIbook_StartOfSim;
  cb_data_s.obj       = NULL;
  cb_data_s.time      = NULL;
  cb_data_s.value     = NULL;
  /* use user_data to pass systf_h to simulation callback so that the
     callback can access the system task arguments */
  cb_data_s.user_data = (PLI_BYTE8 *)systf_h;
  cb_h = vpi_register_cb(&cb_data_s);
  vpi_free_object(cb_h);  /* free callback handle -- don't need it */

  return(0);  /* no syntax errors detected */
}


/*************************************************************************
 * StartOfSim callback -- opens the test vector file and saves the
 * pointer and the system task handle in the work area storage.
 *************************************************************************/
```

```
PLI_INT32 PLIbook_StartOfSim(p_cb_data cb_data)
{
   s_vpi_value  argVal;
   char         *file_name;
   FILE         *vector_file;
   vpiHandle    systf_h, arg_itr, arg1_h, arg2_h;

   PLIbook_Data_p vector_data;   /* pointer to a ReadVecData structure */

   vector_data = (PLIbook_Data_p)malloc(sizeof(PLIbook_Data_s));

   /* retrieve system task handle from user_data */
   systf_h = (vpiHandle)cb_data->user_data;

   /* get argument handles (compiletf already verified only 2 args) */
   arg_itr = vpi_iterate(vpiArgument, systf_h);
   arg1_h  = vpi_scan(arg_itr);
   arg2_h  = vpi_scan(arg_itr);
   vpi_free_object(arg_itr); /* free iterator -- did not scan to null */

   /* read file name from first tfarg */
   argVal.format = vpiStringVal;
   vpi_get_value(arg1_h, &argVal);
   if (vpi_chk_error(NULL)) {
     vpi_printf("ERROR: $read_test_vector could not get file name\n");
     vpi_control(vpiFinish, 1);  /* abort simulation */
     return(0);
   }
   file_name = argVal.value.str;
   if ( !(vector_file = fopen(file_name, "r")) ) {
     vpi_printf("$read_test_vector could not open file %s", file_name);
     vpi_control(vpiFinish, 1);  /* abort simulation */
     return(0);
   }

   /* store file pointer and tfarg2_h in work area for this instance */
   vector_data->file_ptr = vector_file;
   vector_data->obj_h = arg2_h;
   vpi_put_userdata(systf_h, (void *)vector_data);

   return(0);
}

/*****************************************************************
 * calltf routine
 *****************************************************************/
PLI_INT32 PLIbook_ReadVector_calltf(PLI_BYTE8 *user_data)
{
   s_cb_data    data_s;
   s_vpi_time   time_s;
   s_vpi_value  value_s;
   FILE         *vector_file;
   vpiHandle    systf_h, arg2_h;
```

```
PLIbook_Data_p vector_data;   /* pointer to a ReadVecData structure */

char             vector[1024]; /* fixed vector size, could use malloc*/

systf_h = vpi_handle(vpiSysTfCall, NULL);

/* get ReadVecData pointer from work area for this task instance   */
/* the data in the work area was loaded at the start of simulation */
vector_data = (PLIbook_Data_p)vpi_get_userdata(systf_h);
vector_file = vector_data->file_ptr;
arg2_h      = vector_data->obj_h;

/* read next line from the file */
if ( (fscanf(vector_file,"%s\n", vector)) == EOF) {
  vpi_printf("$read_test_vector reached End-Of-File\n");
  fclose(vector_data->file_ptr);
  vpi_control(vpiFinish, 1);  /* abort simulation */
  return(0);
}

/* write the vector to the second system task argument */
value_s.format = vpiBinStrVal;
value_s.value.str = vector;
vpi_put_value(arg2_h, &value_s, NULL, vpiNoDelay);

return(0);
}
```

6.7 Simulation time-related callbacks

Simulation callbacks can be registered to occur at specific times during a simulation. These types of callbacks utilize the event scheduling mechanism of the simulator to schedule the simulation callback at a specific point in simulation time. All time-related simulation callbacks are one-time events—once the callback occurs, it will not be repeated. The reason constants for simulation time callbacks are listed in table 6-2.

Table 6-2: VPI simulation time callback constants

Constant	Definition
cbReadWriteSynch[†]	calls a PLI application after the execution of all known events in the specified time step. The PLI can schedule additional events for the same simulation time
cbReadOnlySynch[†]	calls a PLI application after the execution of all events in the specified time step. The PLI *cannot* schedule additional events for the same simulation time. However, another cbReadOnlySynch callback can be scheduled.
cbNextSimTime[†]	calls a PLI application before the execution of any events in the next simulation time step which has simulation events scheduled
cbAtStartOfSimTime[†]	calls a PLI application at a specific simulation time (absolute to time 0), before the execution of any simulation events in the time step
cbAfterDelay[†]	calls a PLI application after a specified amount of time (relative to the current time), before execution of any simulation events in that time step
	simulators may add any number of product-specific callback constants to support features of that product

† Indicates the callback is a one-time event. The simulator will automatically remove the callback after it occurs. *Note*: the callback handle will still be valid, and must be freed by the PLI application using vpi_free_object() to avoid a memory leak (see Section 6.5).

Required settings to register simulation time-related callbacks

To register a simulation time-related callback, set following fields in the **s_cb_data** structure:

- **reason** must be set to one of the simulation time constants.

- **cb_rtn** must specify the name of the PLI routine which should be called when the specified time activity occurs.

- **time** must specify a pointer to an **s_vpi_time** structure. The **type** field controls how time is specified for when the callback is to occur, and the way that time is passed to the callback routine when the callback occurs. The vpiSuppressTime constant is not allowed when registering simulation time callbacks.

- **obj** must specify a handle for an object if the type field in the time structure is vpiScaledRealTime. The simulation time scale for the module which contains the object will be used to scale the time value. If the time type is vpiSimTime, the obj field is not used, and should be set to NULL.

- **user_data** (optional) can be set to a user data value, if needed, or NULL.

The value and index fields are not required for simulator time callbacks, and should be set to NULL.

When a simulation time-related callback occurs

When a simulator time-related callback occurs, the simulator will allocate an **s_cb_data** structure and pass a pointer to the structure as an input to the *simulation callback routine* which is called. The fields in the s_cb_data structure which the simulator will fill are:

- The **user_data** field of the structure will contain what ever user_data value was specified when the callback was registered.

- The **time** field will contain a pointer to a time structure. Within the time structure, will be the simulation time in which the callback occurred (the current simulation time). The time value will be stored in the format that was specified in the time type field when the callback was registered—vpiScaledRealTime will store time values in the real field, and vpiSimTime will store time values in the high and low fields.

6.7.1 Simulation callbacks at the end of a time step

A PLI application can schedule simulation time callbacks to a *simulation callback routine* at the end of the current simulation time step or at the end of a future time step. This capability allows PLI applications to synchronize themselves with the activity in simulation.

A **cbReadWriteSynch** callback will call a PLI application after the execution of all known events of specific types in the specified time step. The PLI can schedule additional events in that simulation time step.

A **cbReadOnlySynch** callback will call a PLI application after the execution of all events in the specified time step. The PLI *cannot* schedule additional events in that simulation time step. However, a PLI application is permitted to schedule another cbReadOnlySynch callback.

For both of these callbacks, the time step in which the callback occurs is set using the s_vpi_time structure when the callback is registered. When the type field in the time structure is vpiSimTime or vpiScaledRealTime, the callback will occur at the time specified in the high and low fields or the real field of the time structure, using a relative delay from the time the callback was registered.

There are many reasons a PLI application might need to synchronize activity to the end of a simulation time step. One situation might be to communicate all logic value changes in simulation for that moment in simulation time to a C language model, and pass any C model value changes back to the Verilog simulation. This type of callback would require a read-write synchronization, which would use a simulation time call-back scheduled with a cbReadWriteSynch reason.

Another example of where synchronization to the end of a simulation time step is shown in the following example:

```
always @(a or b)              always @(a or b)
    $my_strobe(sum);             sum = a + b;
```

parallel (concurrent) activity

The *calltf routine* for *$my_strobe* will be called every time **a** or **b** changes, at which time the routine might need to read the value of argument 1, which is sum. At the same moment in simulation time, however, sum is also scheduled to change value in simulation. This is a classic race condition in hardware simulation that is caused by the concurrent activity of reading and writing values of the same object in the same simulation time step. The Verilog standard states that most types of concurrent activity can be processed in any order by the simulation, which means the *calltf routine* can be executed either before or after the sum signal changes at the positive edge of the clock. The outcome of this race condition is not predictable.

The race condition in the above example can be resolved by synchronizing the *$my_strobe* application to the very end of the simulation time step in which the change on **a** or **b** occurs. The *$my_strobe* would require a read-only synchronization, which is a *simulation time-related callback* scheduled with the cbReadOnlySynch reason. When the simulation callback routine is called at the end of the time step, all statements for that simulation time will have been executed, and the values of the arguments to the system task or system function will be at their most current value. For the *$my_strobe* example, when the *calltf routine* is invoked when **a** changes, it can schedule a *simulation callback* at the end of that simulation time step in read-only mode. The simulation callback routine can then read the value of sum, and be assured that it has the most current value for that moment in time.

The VPI provides two methods of synchronizing within a simulation time step:

cbReadWriteSynch schedules a simulation time-related callback after certain types of events in the specified simulation time. The simulation callback routine will be in a *read-write* mode, meaning the application can both read the values of Verilog objects, as well as modify the values of objects in the current simulation time (using vpi_put_value() with no delay or zero delay). The application can also schedule values to be written in future simulation times.

 Changing the value of an object may cause additional events in the current simulation time. The new events will be processed, but the cbReadWriteSynch simulation callbacks which have already been executed for that time step will *not* be re-invoked. Only new events will be processed.

cbReadOnlySynch schedules a simulation time-related callback at the end of a specified simulation time in a *read-only* mode. In this mode, the simulation callback routine is only permitted to read values. The routine is not allowed to modify values in the current time. Values can scheduled to be written at a future time using vpi_put_value() with a delay.

Two fundamental differences should be noted between VPI routines and TF routines regarding synchronization to the end of a time step. First, the equivalent TF routines (tf_synchronize() and tf_rosynchronize()) can only schedule a callback to a *misctf routine* at the current simulation time—they cannot schedule synchronization callbacks at a future time. Second, in a read-only synchronization callback using TF routines, it is illegal to write values into simulation at any time, including future times. The VPI *simulation callback routines* provide more capability and flexibility for synchronizing PLI applications to simulation time.

The IEEE 1364 Verilog standard contains a generalized description of an event scheduling algorithm for Verilog simulators. This algorithm is complex, and beyond the scope of this book. The following paragraphs present a simplified version of the IEEE algorithm, in order to show the differences between **cbReadWriteSynch** and **cbReadOnlySynch** simulation callbacks. In essence, there are four distinct regions of events within a simulation time step in Verilog, which are referred to a *slots* in the following figure. Within each slot, certain types of simulation events are scheduled to be executed. *The Verilog standard allows simulators to optimize many of the events within a slot in any order, and to interleave the types of events within each slot in any order.*

Figure 6-1: Organization of events in a Verilog simulation time step

Current Simulation Time:

    Slot 1: Active Events

events within
a slot may be
interleaved
in any order!

        Evaluate right-hand side of nonblocking assignments

        Evaluate right-hand-side & change left-hand-side of blocking assignments

        Evaluate right-hand-side & change left-hand-side of continuous assignments

        Evaluate changes on primitive inputs, & schedule changes to outputs

        Print outputs from scheduled $display and $write tasks

        Call PLI calltf routines for scheduled system tasks and system functions

    Slot 2: Nonblocking Assignment Update Events

        Change left-hand side of nonblocking assignments

    Slot 3: Read-write Synchronization

        Call registered `cbReadWriteSynch` *simulation callbacks*

    Slot 4: Read-only Synchronization

        Print outputs from scheduled $strobe and $monitor tasks

        Call registered `cbReadOnlySynch` *simulation callbacks*

Next Simulation Time:

    . . .

NOTE *Some simulators reverse the order of nonblocking assignment update events and read-write synchronization!* The IEEE 1364 description of event scheduling is much more complex than illustrated above, and allows simulators latitude on the ordering of some types of events.

The IEEE 1364 Verilog standard is purposely written in such a way as to allow simulators a certain degree of latitude on how to implement their internal event scheduling algorithms. This allows simulator vendors the flexibility needed to create competitive products. The latitude is also necessary to allow simulators the freedom to order events differently as required for the behavior of different types of hardware. CMOS hardware does not behave exactly the same as ECL hardware, for example. The flexibility provided in the Verilog standard is one of the strengths of Verilog. But, for PLI application developers, the latitude in event scheduling permitted by the Verilog standard can be a source of frustration. It means that when synchronizing PLI applications within a simulation time step, the `cbReadWriteSynch` event may occur in a different order in different simulators. There is no simple fix for this; it is just the nature of the Verilog language. Note, however, that the `cbReadOnlySynch` does not have this ambiguity. The read-only synchronization events will occur after all other types of simulation events within a time step, in all Verilog simulators.

In the preceding illustration, a simulation will first execute all scheduled events in slot 1, the active events. Processing an active event may cause another active event to be scheduled. Consider the following Verilog HDL code:

```
always @(a or b)
  begin
    sum = a + b;
    $my_strobe(sum, zero_flag);
  end

assign zero_flag = ~|sum;
```

In this example, when either a or b changes, the statements `sum = a + b;` and `$my_strobe(sum, zero_flag);` are scheduled as active events. Because these events are within a sequential `begin`—`end` statement group, the Verilog standard guarantees that these active events will be executed in the order they are listed. However, when `sum` changes, the statement `assign zero_flag = ~|sum;` is also scheduled as an active event. The `assign` statement is a concurrent (parallel) process, and simulators are allowed to interleave these concurrent events between sequential events. That is, the new active event to assign `zero_flag` can be scheduled to occur either before or after the active event to execute the *calltf* routine for *$my_strobe*. For simulation results, the affects of the event ordering should not matter—everything appears to have happened at the same time in simulation time.

This latitude to interleave active events as provided by the IEEE standard may affect the results of PLI *calltf routines*. In one simulator, the *calltf routine* might be executed before the concurrent continuous assignment statement has been executed, and therefore will see the value of `zero_flag` before it changes. In another simulator, however, the assignment to `zero_flag` might be interleaved before the *calltf routine* is executed, and therefore the *calltf routine* will see value of `zero_flag` after the assignment is executed. This situation is very easy to rectify. Instead of immediately reading the value of `zero_flag`, the *calltf routine* can register a simulation callback using the reason **cbReadWriteSynch**. This callback will be scheduled in the read-write synchronization event list, after all active events have been processed.

Simulation will not proceed to slot 2 until it has executed all scheduled active events within slot 1, which includes executing the *calltf* routine for *$my_strobe* in the preceding example. Simulators then move the events in slot 2 to the active event list, and begin processing those events. As these new active events are processed, they may cause additional active events, nonblocking assignment update events or read-write synchronization events to be scheduled.

Once all the events that were moved from slot 2 to the active event list have been processed, simulators will move the events in slot 3 to the active event list, and begin processing those. Once again, these new active events may cause new active events, nonblocking assignment update events or read-write synchronization events to be scheduled.

Most Verilog simulators will execute all the originally scheduled events in slots 2 and 3, by moving those events to the active event slot. The simulator will then return to slot 2, and move any new events that have been scheduled to the active event list and process those new events. Then the simulator will move any new slot 3 events to the active list and process those. This loop through slots 1, 2 and 3 will continue until all three slots are completely empty.

It is critical to note that when simulation returns to a slot that was previously processed, only *new* events will be processed—events which have already been executed have been removed from that slot's event list, and will not be executed a second time. Assuming there are no zero-delay infinite loops in the Verilog code (which is possible), simulation will eventually complete all events in slots 1, 2 and 3, and then proceed to slot 4.

In the read-write synchronization slot, which may be either slot 2 or 3, any registered simulation time-related callbacks with **cbReadWriteSynch** will be invoked. In this mode, the PLI is allowed to schedule new events in the current simulation time step. These new events will be scheduled as active events in slot 1. The simulation callback routine can also register another callback for cbReadWriteSynch. The new callback will be scheduled as a new read-write synchronization event, which will not be executed until after the simulator revisits the read-write synchronization slot. Any scheduled active events and nonblocking assignment update events will be executed before the read-write synchronization slot is revisited.

Once all events in slots 1, 2 and 3 have been processed, simulation will proceed to slot 4, where any registered simulation time-related callbacks with **cbReadOnlySynch** will be invoked. In the read-only synchronize mode, the PLI is allowed to read information from the simulation, but the PLI may *not* schedule new events in the current simulation time step. Slot 4 represents the true end of the current simulation time step.

Read-write synchronization ambiguity

 A critical latitude permitted in the IEEE 1364 Verilog standard is that slot 2 and slot 3 in the illustration shown in Figure 6-1 can be reversed. This ambiguity means that in some simulators, a read-write synchronization will occur after nonblocking assignments have been updated. In other Verilog simulators, a read-write synchronization will occur before nonblocking assignments are updated.

Slot 2 in the illustration shown in Figure 6-1 on page 221 may be either the nonblocking assignment update events or the read-write synchronization events. If it is the latter, then the PLI application will be able to see the results of all active events, but the left-hand side of the nonblocking assignment update events will not yet have been processed. Therefore the PLI application will not see the results of those events. The following Verilog code illustrates this potential problem:

```
always @(posedge clock)
  begin
    sum <= a + b;        //non-blocking assignment to sum
    $my_strobe(sum);
  end
```

In this example, using a read-write synchronization will not resolve the ambiguity possible between different simulators. A simulator might process the read-write synchronize events before processing the nonblocking assignment update events, in which case a PLI application called as a read-write synchronize event will see the value of sum before it changes. A coding style that can be used to reduce this ambiguity is for the first call to a *simulation callback routine* for **cbReadWriteSynch** can register a second callback for **cbReadWriteSynch**. This second callback will occur after all scheduled nonblocking assignment update events have been processed. However, there is no way to ensure that new nonblocking assignment update events have not been scheduled and not yet processed. The only sure way to ensure that a PLI application has been called after all nonblocking assignment update events have been processed is to register a simulation callback for **cbReadOnlySynch**.

 There has been discussion of adding two additional synchronization callback reasons to provide PLI applications greater control over when read-write synchronization callbacks occur. **cbBeforeNBA** synchronization would occur after all scheduled active events have been processed, but before any scheduled nonblocking assignment updates occur. **cbAfterNBA** synchronization callbacks would occur after all scheduled nonblocking assignment updates have been processed. Note that these proposed synchronization callbacks and constant names are part of the IEEE standard, and may not be implemented in all Verilog simulators. Simulators that do implement these additional synchronization callbacks can still be IEEE 1364 compliant, however, because the Verilog standard allows simulators to add simulator-specific callback reasons.

Example 6-2, below, illustrates using a simulation time callback registered with cbReadOnlySynch to implement *$my_strobe*. In this example, the *calltf routine* for *$my_strobe* will be executed as an active event. However, at the same moment in simulation time the signal sum may also be scheduled to change, possibly as an active event (by a blocking assignment) or a nonblocking assignment update event (by a nonblocking assignment). Therefore, the *calltf routine* does not read the value of sum. Instead, the *calltf routine* schedules a *read-only synchronize* callback to a *simulation callback routine* at the end of the current simulation time. The value of sum is then read from the *simulation callback routine*.

> **CD** The source code for this example is on the CD accompanying this book.
>
> - Application source file: Chapter.06/my_strobe_vpi.c
> - Verilog test bench: Chapter.06/my_strobe_test.v
> - Expected results log: Chapter.06/my_strobe_test.log

Example 6-2: *$my_strobe* — simulation callbacks at the end of the current time

```
/*****************************************************************
 * calltf routine
 *****************************************************************/
PLI_INT32 PLIbook_MyStrobe_calltf(PLI_BYTE8 *user_data)
{
  vpiHandle   systf_h, arg_itr, arg_h, cb_h;
  s_vpi_time  time_s;
  s_cb_data   cb_data_s;

  /* obtain a handle to the system task instance */
  systf_h = vpi_handle(vpiSysTfCall, NULL);

  /* obtain handle to system task argument; */
  /* compiletf has already verified only 1 arg with correct type */
  arg_itr = vpi_iterate(vpiArgument, systf_h);
  arg_h = vpi_scan(arg_itr);
  vpi_free_object(arg_itr);  /* free iterator--did not scan to null */

  /* setup end-of-time step callback */
  time_s.low  = 0;
  time_s.high = 0;
  time_s.type = vpiSimTime;
  cb_data_s.reason    = cbReadOnlySynch;
  cb_data_s.cb_rtn    = PLIbook_EndOfTimeStep_callback;
  cb_data_s.obj       = arg_h;
  cb_data_s.time      = &time_s;
  cb_data_s.value     = NULL;                  /* not used */
  cb_data_s.index     = 0;                     /* not used */
  cb_data_s.user_data = (PLI_BYTE8 *)arg_h;
  cb_h = vpi_register_cb(&cb_data_s);
  vpi_free_object(cb_h); /* don't need callback handle */

  return(0);
}


/*****************************************************************
 * Value change callback application
 *****************************************************************/
PLI_INT32 PLIbook_EndOfTimeStep_callback(p_cb_data cb_data_p)
{
  s_vpi_time   time_s;
  s_vpi_value  value_s;
  vpiHandle    arg_h;
```

```
    arg_h = (vpiHandle)cb_data_p->user_data;
    value_s.format = vpiBinStrVal;
    vpi_get_value(arg_h, &value_s);
    vpi_printf("$my_strobe: At %d: \t %s = %s\n",
               cb_data_p->time->low,
               vpi_get_str(vpiFullName, arg_h),
               value_s.value.str);
    return(0);
}
```

6.7.2 Simulation time-related callbacks at a future simulation time

The VPI provides three ways in which a PLI application can schedule simulation callbacks at a future simulation time.

cbAtStartOfSimTime schedules a simulation time callback at a specific simulation time, using the absolute time in which the callback should occur. Absolute time is always relative to time 0, and is not affected by the current simulation time. The callback will occur at the beginning of the specified time step, before any other simulation events are processed in that time step. The time step in which the callback is to occur is set using the s_vpi_time structure when the callback is registered.

cbAfterDelay schedules a simulation time callback at a future simulation time, using a relative time from the current simulation time. The callback will occur at the beginning of the specified time step, before any other simulation events are processed in that time step. The time step in which the callback is to occur is set using the s_vpi_time structure when the callback is registered.

cbNextSimTime schedules a simulation time callback at a future simulation time, which is determined by whenever the simulator has scheduled its next event to be executed. The callback will occur at the beginning of the specified time step, before any other simulation events are processed in that time step. If the current simulation time is 100, and the next scheduled event is time 125, then the simulation callback will be scheduled before any other simulation events are processed in time 125.

cbReadWriteSynch and **cbReadOnlySynch**, which were discussed in the previous section, can also be used schedule a simulation time callback synchronized within a future time step. The time step in which the callback is to occur is set using the s_vpi_time structure when the callback is registered.

When the `type` field in the time structure is `vpiSimTime` or `vpiScaledRealTime`, the callback will occur at the time specified in the `high` and `low` fields or the `real` field of the time structure. The `vpiSuppressTime` time type is not allowed with simulation time related callbacks.

The `vpi_remove_cb()` routine can be used to remove a scheduled callback which has not transpired. In order to remove a callback, the handle for the callback must have been preserved by the PLI application. The callback handle is returned by `vpi_register_cb()` when the callback is registered.

An example of scheduling simulation callbacks at future times

Example 6-3 illustrates using a simulation callback for a `cbAfterDelay` reason. The example implements a system task called *$read_stimulus_ba*, which uses a *callback routine* to read one line at a time from a stimulus vector file. Each line contains a simulation time and a test vector, as follows:

```
time        vector
----    ----------------
  10    11111111xxxxxxxx
  17    00000000zzzzzzzz
  30    1000000011011101
```

The *callback routine* for *$read_stimulus_ba* will need to process the following loop as long as there are more test vectors in the file:

1. Read a time and test vector from the file.

2. Schedule test vector to be applied to simulation at the desired time.

3. Schedule the *simulation callback routine* to called again once the test vector has been applied.

The usage for this example is:

```
initial
  $read_stimulus_<base><delay_type>("file", verilog_reg);
```

where:

```
<base>        is b or h (for binary or hex vectors)
<delay_type>  is a or r for absolute or relative times
"file"        is the name of the file to be read, in quotes
verilog_reg   is a verilog register data type
```

For example:

```
initial
    $read_stimulus_ba("readstim.pat", input_vector);
```

Notice that *$read_stimulus_ba* is called from a Verilog `initial` procedure, which means the *calltf routine* for *$read_stimulus_ba* will only be invoked one time throughout simulation. The *calltf routine* is used to schedule an immediate callback to a *simulation callback routine*. The *simulation callback routine* will then read the first line from the file, and schedule the test vector to be applied at the simulation time specified by the delay (using `vpi_put_value()`). After scheduling the vector to be applied, the *simulation callback routine* will schedule another callback to itself, this time using the same delay with which the test vector was scheduled to be applied. At that future simulation time, the *simulation callback routine* is invoked again, and the cycle is repeated. When the last line of the file is read, the *simulation callback routine* will cause the simulation to exit.

Example 6-3 on page 229 presents the *$read_stimulus_ba* PLI application. This PLI application utilizes several types of VPI callbacks:

- A *compiletf system task callback* is used to perform syntax checking and to schedule a simulation callback for the start of simulation.

- A *calltf system task callback* is used to open the stimulus file and invoke the first call to a simulation callback which reads one line at a time from the file.

- A *simulation action callback* scheduled for the start of simulation is used to allocate an instance-specific block of persistent memory to be used by the other *simulation callback routines*. A pointer to the memory is saved in an instance-specific storage area.

- A *simulation time callback* scheduled for a future time is used to retrieve a stimulus pattern and a simulation time from the file, and schedule the simulator to apply the test pattern and the designated simulation time.

- A *simulation time callback* scheduled for the end of the current time step is used to terminate simulation when *$read_stimulus_ba* has reached the end of the stimulus pattern file. Instead of terminating simulation immediately on end of file, a read-only synchronous callback is scheduled to allow any other activity in the current simulation time step to complete first.

This example also illustrates other important concepts which have been presented in this and earlier chapters on the VPI library. Some of these concepts are:

- Using invocation option to configure a PLI application. In this example, invoking simulation with **+readstim_debug** will enable additional error checking and output messages as the lines area read from the stimulus pattern file.

- Using instance-specific work areas to store information that needs to be shared by different *simulation callback routines*.

- Using the system task/function user_data to allow different system task names to call the same PLI application *simulation callback routines*. In this example, the stimulus values contained in the stimulus file can be specified as either hex patterns or binary patterns, and the simulation times can be specified as relative delays or as absolute time from time 0. This flexibility is accomplished by registering several versions of the *$read_stimulus_ba* system task name, all of which call the same PLI application, but with different user-data values:

  - *$read_stimulus_ba* has a user-data of *"ba"* —binary vectors, absolute delays

  - *$read_stimulus_br* has a user-data of *"br"* —binary vectors, relative delays

  - *$read_stimulus_ha* has a user-data of *"ha"* —hex vectors, absolute delays

  - *$read_stimulus_hr* has a user-data of *"hr"* —hex vectors, relative delays

As with many of the examples in this book, the purpose of the example is to show how specific PLI routines are used. In order to make the code functionality more obvious, the example does not emphasize efficient C programming techniques. The format of the stimulus file has also been kept simple.

CD The source code for this example is on the CD accompanying this book.

- Application source file: `Chapter.06/PLIbook_read_stimulus_vpi.c`
- Verilog test bench: `Chapter.06/PLIbook_read_stimulus_test.v`
- Expected results log: `Chapter.06/PLIbook_read_stimulus_test.log`

Example 6-3: *$read_stimulus_ba* — scheduling callbacks at a future time

```
/*****************************************************************
 * Define storage structure for file pointer and vector handle.
 *****************************************************************/
typedef struct ReadStimData {
  FILE *file_ptr;      /* test vector file pointer */
  vpiHandle  obj_h;    /* pointer to store handle for a Verilog object */
  int mode;            /* 0 & 1 = binary values, 2 & 3 = hex values   */
                       /* 0 & 2 = absolute time, 1 & 3 = relative time */
  int debug;           /* print debug messages if true */
} s_ReadStimData, *p_ReadStimData;

/*****************************************************************
 * VPI Registration Data
 *****************************************************************/
void PLIbook_ReadStim_register()
{
```

```
  s_vpi_systf_data tf_data;
  tf_data.type        = vpiSysTask;
  tf_data.sysfunctype = 0;
  tf_data.calltf      = PLIbook_ReadStim_calltf;
  tf_data.compiletf   = PLIbook_ReadStim_compiletf;
  tf_data.sizetf      = NULL;

  tf_data.tfname      = "$read_stimulus_ba"; /* binary, absolute time*/
  tf_data.user_data   = "ba";
  vpi_register_systf(&tf_data);

  tf_data.tfname      = "$read_stimulus_br"; /* binary, relative time*/
  tf_data.user_data   = "br";
  vpi_register_systf(&tf_data);

  tf_data.tfname      = "$read_stimulus_ha"; /* hex, absolute time */
  tf_data.user_data   = "ha";
  vpi_register_systf(&tf_data);

  tf_data.tfname      = "$read_stimulus_hr"; /* hex, relative time */
  tf_data.user_data   = "hr";
  vpi_register_systf(&tf_data);
}

/*********************************************************************
 * compiletf routine
 *********************************************************************/
PLI_INT32 PLIbook_ReadStim_compiletf(PLI_BYTE8 *user_data)
{
  s_cb_data    cb_data_s;
  vpiHandle    systf_h, arg_itr, arg_h, cb_h;
  PLI_INT32    tfarg_type;
  int          err = 0;
  char         *file_name;

  systf_h = vpi_handle(vpiSysTfCall, NULL);
  arg_itr = vpi_iterate(vpiArgument, systf_h);
  if (arg_itr == NULL) {
    vpi_printf("ERROR: $read_stimulus_?? requires 2 arguments\n");
    vpi_control(vpiFinish, 1);  /* abort simulation */
    return(0);
  }
  arg_h = vpi_scan(arg_itr);  /* get handle for first tfarg */
  if (vpi_get(vpiType, arg_h) != vpiConstant) {
    vpi_printf("$read_stimulus_?? arg 1 must be a quoted file name\n");
    err = 1;
  }
  else if (vpi_get(vpiConstType, arg_h) != vpiStringConst) {
    vpi_printf("$read_stimulus_?? arg 1 must be a string\n");
    err = 1;
  }
  arg_h = vpi_scan(arg_itr);  /* get handle for second tfarg */
```

```
    tfarg_type = vpi_get(vpiType, arg_h);
    if ( (tfarg_type != vpiReg) &&
         (tfarg_type != vpiIntegerVar) &&
         (tfarg_type != vpiTimeVar) ) {
      vpi_printf("$read_stimulus_?? arg 2 must be a register type\n");
      err = 1;
    }
    if (vpi_scan(arg_itr) != NULL) {
      vpi_printf("read_stimulus_?? requires only 2 arguments\n");
      vpi_free_object(arg_itr);
      err = 1;
    }
    if (err) {
      vpi_control(vpiFinish, 1);  /* abort simulation */
      return(0);
    }

    /* setup a callback for start of simulation */
    cb_data_s.reason    = cbStartOfSimulation;
    cb_data_s.cb_rtn    = PLIbook_StartOfSim;
    cb_data_s.obj       = NULL;
    cb_data_s.time      = NULL;
    cb_data_s.value     = NULL;
    cb_data_s.user_data = (PLI_BYTE8 *)systf_h; /* pass systf_h */
    cb_h = vpi_register_cb(&cb_data_s);
    vpi_free_object(cb_h); /* don't need callback handle */

    return(0);  /* no syntax errors detected */
}

/**********************************************************************
 * calltf routine -- registers an immediate callback to the
 * ReadNextStim application.
 *********************************************************************/
PLI_INT32 PLIbook_ReadStim_calltf(PLI_BYTE8 *user_data)
{
    s_cb_data        data_s;
    s_vpi_time       time_s;
    vpiHandle        systf_h, cb_h;
    p_ReadStimData   StimData; /* pointer to a ReadStimData structure */

    /* get ReadStimData pointer from work area for this task instance */
    systf_h  = vpi_handle(vpiSysTfCall, NULL);
    StimData = (p_ReadStimData)vpi_get_userdata(systf_h);

    /* look at user data to set the stimulus mode flag */
    if      (strcmp(user_data, "ba") == 0) StimData->mode = 0;
    else if (strcmp(user_data, "br") == 0) StimData->mode = 1;
    else if (strcmp(user_data, "ha") == 0) StimData->mode = 2;
    else                                   StimData->mode = 3;

    /* setup immediate callback to ReadNextStim routine */
    systf_h      = vpi_handle(vpiSysTfCall, NULL);
    time_s.type  = vpiSimTime;
    time_s.low   = 0;
```

```
  time_s.high      = 0;
  data_s.reason    = cbReadWriteSynch;
  data_s.cb_rtn    = PLIbook_ReadNextStim;
  data_s.obj       = NULL;
  data_s.time      = &time_s;
  data_s.value     = NULL;
  data_s.user_data = (PLI_BYTE8 *)systf_h;
  cb_h = vpi_register_cb(&data_s);
  vpi_free_object(cb_h); /* don't need callback handle */

  return(0);
}

/**********************************************************************
 * ReadNextStim callback -- Reads a time and vector from a file.
 * Schedules the vector to be applied at the specified time.
 * Schedules a callback to self that same time (to read next line).
 **********************************************************************/
PLI_INT32 PLIbook_ReadNextStim(p_cb_data cb_data)
{
  char vector[1024]; /* fixed max. size, should use malloc instead */
  int             delay;
  vpiHandle       systf_h, cb_h;
  s_cb_data       data_s;
  s_vpi_time      time_s;
  s_vpi_value     value_s;
  p_ReadStimData  StimData; /* pointer to a ReadStimData structure */

  /* retrieve system task handle from user_data */
  systf_h = (vpiHandle)cb_data->user_data;

  /* get ReadStimData pointer from work area for this task instance */
  StimData = (p_ReadStimData)vpi_get_userdata(systf_h);

  /* read next line from the file */
  if ( (fscanf(StimData->file_ptr,"%d %s\n", &delay, vector)) == EOF) {
    /* At EOF, schedule ReadStimEnd callback at end of this time */
    time_s.type      = vpiSimTime;
    time_s.low       = 0;
    time_s.high      = 0;
    data_s.reason    = cbReadOnlySynch;
    data_s.cb_rtn    = PLIbook_ReadStimEnd;
    data_s.obj       = NULL;
    data_s.time      = &time_s;
    data_s.value     = NULL;
    data_s.user_data = (PLI_BYTE8 *)StimData->file_ptr;
    cb_h = vpi_register_cb(&data_s);
    vpi_free_object(cb_h); /* don't need callback handle */
    return(0);
  }

  if (StimData->debug) {
    vpi_printf("Values read from file: delay=%d vector=%s\n",
               delay, vector);
  }
```

```
  /* convert absolute delay from file to relative delay if needed */
  time_s.type = vpiScaledRealTime;
  if (StimData->mode == 0 || StimData->mode == 2) {
    vpi_get_time(cb_data->obj, &time_s);
    time_s.real = ((double)delay - time_s.real);
  }
  else
    time_s.real = (double)delay;

  /* schedule the vector to be applied after the delay period */
  if (StimData->mode == 0 || StimData->mode == 1)
    value_s.format = vpiBinStrVal;
  else
    value_s.format = vpiHexStrVal;
  value_s.value.str = vector;
  vpi_put_value(StimData->obj_h, &value_s, &time_s, vpiTransportDelay);

  /* schedule callback to this routine when time to read next vector */
  data_s.reason     = cbAfterDelay;
  data_s.cb_rtn     = PLIbook_ReadNextStim;
  data_s.obj        = systf_h; /* object required for scaled delays */
  data_s.time       = &time_s;
  data_s.value      = NULL;
  data_s.user_data  = (PLI_BYTE8 *)systf_h;
  cb_h = vpi_register_cb(&data_s);
  if (vpi_chk_error(NULL))
    vpi_printf("An error occurred registering ReadNextStim callback\n");
  else
    vpi_free_object(cb_h); /* don't need callback handle */
  return(0);
}

/*****************************************************************************
 * StartOfSim callback -- opens the test vector file and saves the
 * file pointer and other info in an instance-specific work area.
 *****************************************************************************/
PLI_INT32 PLIbook_StartOfSim(p_cb_data cb_data)
{
  char             *file_name;
  FILE             *vector_file;
  vpiHandle        systf_h, tfarg_itr, tfarg1_h, tfarg2_h;
  s_cb_data        data_s;
  s_vpi_time       time_s;
  s_vpi_value      argVal;
  s_vpi_vlog_info  options_s;
  p_ReadStimData   StimData;  /* pointer to a ReadStimData structure */
  int              i, debug;

  /* retrieve system task handle from user_data */
  systf_h = (vpiHandle)cb_data->user_data;

  /* get tfarg handles (compiletf already verified args are correct) */
  tfarg_itr = vpi_iterate(vpiArgument, systf_h);
```

```
  tfarg1_h = vpi_scan(tfarg_itr);
  tfarg2_h = vpi_scan(tfarg_itr);
  vpi_free_object(tfarg_itr); /* free iterator--did not scan to null */

  /* read file name from first tfarg */
  argVal.format = vpiStringVal;
  vpi_get_value(tfarg1_h, &argVal);
  if (vpi_chk_error(NULL)) {
    vpi_printf("ERROR: $read_stimulus_?? could not get file name\n");
    return(0);
  }
  file_name = argVal.value.str;
  if ( !(vector_file = fopen(file_name,"r")) ) {
    vpi_printf("$read_stimulus_?? could not open file %s\n",file_name);
    vpi_control(vpiFinish, 1);  /* abort simulation */
    return(0);
  }

  /* check for +readstim_debug invocation option */
  debug = 0;  /* assume not invoked with debug flag */
  vpi_get_vlog_info(&options_s);
  for (i=1; i<options_s.argc; i++) {
    if (strcmp(options_s.argv[i], "+readstim_debug") == 0) {
      debug = 1; /* invocation option found */
      break;
    }
  }

  /* allocate memory to store information about this instance */
  StimData = (p_ReadStimData)malloc(sizeof(s_ReadStimData));
  StimData->file_ptr = vector_file;
  StimData->obj_h = tfarg2_h;
  StimData->debug = debug;
  vpi_put_userdata(systf_h, (PLI_BYTE8 *)StimData);

  return(0);
}

/************************************************************************
 * End-Of-Simulation callback -- close file and exit simulation.
 ************************************************************************/
int PLIbook_ReadStimEnd(p_cb_data cb_data_p)
{
  vpi_printf("$read_stimulus_?? reached End-Of-File.\n");
  fclose((FILE *)cb_data_p->user_data);
  vpi_control(vpiFinish, 1);  /* abort simulation */
  return(0);
}
```

6.8 Simulation event-related callbacks

Simulation callbacks can be registered for when specific events transpire during simulation. All event-related callbacks can occur more than one-time. The callback will automatically repeated.

Table 6-3: VPI simulation event callback constants

Constant	Definition
cbValueChange[‡]	calls a PLI application after a logic value change or strength value change on an expression or terminal
cbStmt[‡]	calls a PLI application before execution of a procedural statement
cbAssign[‡]	calls a PLI application after a procedural assign has been executed on a simple expression
cbDeassign[‡]	calls a PLI application after a procedural de-assign has been executed on a simple expression
cbForce[‡]	calls a PLI application after a force has occurred on a simple expression; it is illegal to place a cbForce callback on a variable bit select
cbRelease[‡]	calls a PLI application after a release has occurred on a simple expression; it is illegal to place a cbRelease callback on a variable bit select
cbDisable[‡]	calls a PLI application after a procedural disable has been executed on a block of code containing a system task or system functions; the callback object must be a handle to a system task call, a system function call, a named begin statement, a named fork statement, a Verilog task or a Verilog function
	simulators may add any number of product-specific callback constants to support features of that product

‡ Indicates the callback can occur multiple times. The PLI application must unregister the callback using vpi_remove_cb() when no longer needed. *Note*: vpi_remove_cb() also frees the callback handle (see Section 6.4).

Required settings to register simulation event-related callbacks

To register a simulation event callback, the following fields in the **s_cb_data** structure must be set:

- **reason** must be set to one of the simulation event constants.

- **cb_rtn** must specify the name of the PLI routine which should be called when the specified event activity occurs.

- **obj** must specify a handle for an object for most simulation event callback reasons. The object must be appropriate for the callback reason. If the reason is cbForce or cbRelease, and the object field is NULL, the simulation callback will be called for every force or release.

- **time** must specify a pointer to an **s_vpi_time** structure. The time structure **type** field in the time structure can be set to **vpiScaledRealTime** or **vpiSimTime**. The time type determines how the simulation time will be passed to the *simulation callback routine* when a callback occurs. If the simulation time is not needed by the *simulation callback routine*, the type field can be set to **vpiSuppressTime**.

- **value** must specify a pointer to an **s_vpi_value** structure for all event reasons except cbStmt. The **format** field in the value structure must be set to a value format constant to indicate how the logic value of the object will be returned to the *simulation callback routine*. The format constants are listed in table 6-4, which follows. Refer to section 5.2 on page 152 in Chapter 5 for a full description of each value format and how the Verilog logic value is represented. If the new logic value of the object is not needed by the *simulation callback routine*, the format field can be set to **vpiSuppressVal**. For cbStmt simulation event callbacks, the value field is not used, and should be set to NULL.

- **user_data** (optional) can be set to a user data value, if needed. If this field is not used, it should be set to NULL.

Table 6-4: The s_vpi_value structure format constants

Format	Definition
vpiBinStrVal	represents a Verilog logic value as a C string, using binary numbers, with 4-state logic
vpiOctStrVal	represents a Verilog logic value as a C string, using octal numbers, with 4-state logic
vpiDecStrVal	represents a Verilog logic value as a C string, using decimal numbers, with 4-state logic
vpiHexStrVal	represents a Verilog logic value as a C string, using hexadecimal numbers, with 4-state logic
vpiScalarVal	represents a Verilog scalar logic value as C constants, which represents Verilog 4-state logic
vpiStrengthVal	represents a Verilog scalar value as a VPI strength structure containing the logic value and Verilog strength levels

Table 6-4: The s_vpi_value structure format constants (continued)

Format	Definition
vpiVectorVal	represents a Verilog vector value as an array of VPI aval/bval structures, encoded to represent Verilog 4-state logic
vpiIntVal	represents a Verilog logic value as a 32-bit C integer, with 2-state logic
vpiTimeVal	represents a Verilog time value as a VPI time structure
vpiRealVal	represents a Verilog logic value as a C double precision number, with 2-state logic
vpiStringVal	represents a Verilog string value as a C string
vpiObjTypeVal	allows simulation to pick the best way to represent a value
vpiSuppressVal	no logic value is retrieved when a simulation callback occurs

Objects on which a **cbValueChange** callback reason can be registered include: nets, regs, variables, memories, memory word selects, variable arrays, variable array selects, module ports, and primitive terminals. A value change is any event in Verilog simulation that results in either the logic value or strength value of the object to change. If a cbValueChange callback is registered and the value format is set to vpiStrengthVal then the callback will occur whenever the object changes strength, including strength changes that do not result in a value change.

When a simulation event-related callback occurs

When a simulator event callback occurs, the simulator will allocate an **s_cb_data** structure, and pass a pointer to the structure as an input to the PLI routine which is called. The fields in the s_cb_data structure which are filled in by the simulator are:

- The **user_data** field of the structure will contain the user_data value which was specified when the callback was registered.

- The **time** field will contain a pointer to a time structure with the simulation time in which the callback occurred (the current simulation time). The time will be stored in the appropriate time value field, based on the time type used when the callback was registered.

- The **value** field will contain a pointer to a value structure, which will contain the new logic value of the specified object, in the format specified when the callback was registered. For **cbForce** and **cbAssign** callbacks, the value in the value structure will show the resultant value of the left-hand side of the statement. For **cbRelease** and **cbDeassign** statements, the value will contain the value of the object after the release has occurred. The value field is not used for **cbStmt** callbacks.

- For **cbValueChange** callback reasons, the **obj** field will be a handle for the object which changed (the same object for which the callback was registered). For **cbForce**, **cbRelease**, **cbAssign** and **cbDeassign** callback reasons, the **obj** field will be a handle for the force, release, assign or deassign statement which was executed on the specified object.

- If the object for which the callback was registered is an element within an array, then the **index** field will contain the index into the array.

All memory used by the s_cb_data structure that is passed to the *simulation callback routine* is allocated and maintained by the simulator. This is temporary storage, and the pointers to the time and value structures will not remain valid after the *simulation callback routine* exits. If the PLI application needs to preserve the time or logic values, then the application must allocate its own memory and copy the information.

An example of using simulation event callbacks for logic value changes

The following example implements an application called *$my_monitor*. This application is passed a module instance name as an input. The application locates all nets within the module, and registers for **cbValueChange** callbacks for each net. When the callback occurs, the application registers a simulation time callback to end of the current time step, and prints the new value of the net, along with the net's name and the current simulation time.

The usage of *$my_monitor* is:

```
initial
  $my_monitor(<module_instance_name>);
```

CD The source code for this example is on the CD accompanying this book.

- Application source file: Chapter.06/my_monitor_vpi.c
- Verilog test bench: Chapter.06/my_monitor_test.v
- Expected results log: Chapter.06/my_monitor_test.log

Example 6-4: *$my_monitor* — scheduling simulation callbacks at a future time

```
/*********************************************************************
 * calltf routine
 *********************************************************************/
PLI_INT32 PLIbook_MyMonitor_calltf(PLI_BYTE8 *user_data)
{
  vpiHandle     systf_h, arg_itr, mod_h, net_itr, net_h, cb_h;
  s_vpi_time    time_s;
  s_vpi_value   value_s;
  s_cb_data     cb_data_s;
  PLI_BYTE8     *net_name_temp, *net_name_keep;

  /* setup value change callback options */
  time_s.type      = vpiScaledRealTime;
  value_s.format   = vpiBinStrVal;

  cb_data_s.reason = cbValueChange;
  cb_data_s.cb_rtn = PLIbook_MyMonitor_callback;
  cb_data_s.time   = &time_s;
  cb_data_s.value  = &value_s;

  /* obtain a handle to the system task instance */
  systf_h = vpi_handle(vpiSysTfCall, NULL);

  /* obtain handle to system task argument */
  /* compiletf has already verified only 1 arg with correct type */
  arg_itr = vpi_iterate(vpiArgument, systf_h);
  mod_h = vpi_scan(arg_itr);
  vpi_free_object(arg_itr);  /* free iterator--did not scan to null */

  /* add value change callback for each net in module named in tfarg */
  vpi_printf("\nAdding monitors to all nets in module %s:\n\n",
             vpi_get_str(vpiDefName, mod_h));

  net_itr = vpi_iterate(vpiNet, mod_h);
  while ((net_h = vpi_scan(net_itr)) != NULL) {
    net_name_temp = vpi_get_str(vpiFullName, net_h);
    net_name_keep = malloc(strlen((char *)net_name_temp)+1);
    strcpy((char *)net_name_keep, (char *)net_name_temp);
    cb_data_s.obj = net_h;
    cb_data_s.user_data = net_name_keep;
    cb_h = vpi_register_cb(&cb_data_s);
    vpi_free_object(cb_h); /* don't need callback handle */
  }
  return(0);
}
```

```
/**********************************************************************
 * Value change callback application
 **********************************************************************/
PLI_INT32 PLIbook_MyMonitor_callback(p_cb_data cb_data_p)
{
  vpi_printf("At time %0.2f:\t %s = %s\n",
             cb_data_p->time->real,
             cb_data_p->user_data,
             cb_data_p->value->value.str);
  return(0);
}
```

6.9 Summary

This chapter has presented one of the more powerful aspects of the VPI routines and the Programming Language Interface—the ability to synchronize PLI application calls with different types of simulation activity. The examples presented in this chapter have shown ways to synchronize PLI applications with the start of simulation, with the end of a simulation time step, with future simulation times, and with logic value changes. Another important topic presented in this chapter was how to allocate and store instance-specific data, and share this data between the different routines that make up a PLI application. A critical consideration in sharing and preserving data is that the data must be *instance specific*. This means unique storage for the data must be allocated for each instance of a system task or system function.

The next chapter presents how the PLI can be used to interface C language models with Verilog simulation. This interface will utilize many of the concepts which were presented in this chapter.

CHAPTER 7 *Interfacing to C Models Using VPI Routines*

*I*nterfacing C language models to Verilog simulations is a common and powerful application of the Programming Language Interface. The VPI *simulation callback routines* presented in the previous chapter make it easy to create this interface, and to synchronize activity with logic value changes and with simulation time. This chapter shows several ways in which a C model can be interfaced to a Verilog simulation using the VPI routine library (Chapter 13 presents using the TF library for interfacing to C models, and Chapter 18 shows how to use the ACC library to accomplish this same task).

The concepts presented in this chapter are:

- Representing hardware models in C
- Verilog HDL shell modules
- Combinational logic interfaces to C models
- Sequential logic interfaces to C models
- Synchronizing with the end of a simulation time step
- Synchronizing with a future simulation time step
- Multiple instances of a C model
- Creating instance specific storage within C models
- Representing propagation delays in C models

TIP

One reason for representing hardware models in the C language is to achieve faster simulation performance. The C programming language allows a very abstract, algorithmic representation of hardware functionality, without representing detailed timing, multi-state logic, hardware concurrency and other hardware specific details offered by the Verilog language.

The PLI can be a means to access the efficiency of a highly abstract C model. However, a poorly written PLI application can become a bottleneck that offsets much of the efficiency gains. Care must be taken to write PLI applications that execute as efficiently as possible. Some guidelines that can help maximize the efficiency and run-time performance of PLI applications are:

- Good C programming practices are essential. General C programming style and techniques are not discussed within the scope of this book.

- Consider every call to a VPI routine in the VPI library as expensive, and try to minimize the number of calls.

- Routines which convert logic values from a simulator's internal representation to C strings, and vice-versa, are very expensive in terms of performance. Best efficiency is attained when the value representation in C is as similar as possible to the value representation in Verilog.

- Use the Verilog language to model the things hardware description languages do well, such as representing hardware parallelism and hardware propagation times. Simulator vendors have invested a great deal in optimizing a simulator's algorithms, and that optimization should be utilized.

The objective of this book is to show several ways in which the VPI library can be used to interface to C models. Short examples are presented that are written in an easy to follow C coding style. In order to meet the book's objectives, the examples do not always follow the guidelines of efficient C coding and prudent usage of the PLI routines. It is expected that if these example PLI applications are adapted for other applications, the coding style will be modified to be more efficient and robust.

7.1 How to interface C models with Verilog simulations

The power and flexibility of the C programming language and the Verilog PLI provide a wide variety of methods that can be used to interface a Verilog simulation with a C language model. All methods, however, have three essential concepts in common:

- Value changes which occur in the Verilog simulator must be passed to the C model.

- Value changes within the C model must be passed to the Verilog simulation.

- Simulated time in both the Verilog simulation and the C model must remain synchronized.

This chapter will present one of the more common methods of interfacing a Verilog simulation with a C model. The method presented is by no means the only way this interface can be accomplished, and may not always be the most efficient method. However, the method presented does have many advantages, including simplicity to implement, portability to many types of Verilog simulators, and the ability to use the C model any number of times and at any level in the hierarchy of a Verilog design.

The fundamental steps that are presented in this chapter are:

1. Create the C language model as an independent block of pure C code that does not use the PLI routines in any way. The C model will have inputs and outputs, but it will not know where the inputs come from or where the outputs go to. The C code to implement the model might be in the form of a C function with no main function, or it might be a complete C program with its own main function.

2. Create a Verilog HDL *shell module*, also called a *wrapper module* or *bus functional module* (BFM), to represent the inputs and outputs of the C language model. This module will be written completely in the Verilog language, but will not contain any functionality. To represent the functionality of the model, the shell module will call a PLI application.

3. Create a PLI application to serve as an interface between the C model and the Verilog shell module. The PLI application is a communication channel, which:

   • Uses the PLI routines to retrieve data from the Verilog HDL shell module and pass the data to the C model via standard C programming.

   • Uses standard C programming to receive data from the C model, and passes the data to the Verilog shell module via PLI routines.

The following diagram shows how the blocks which are created in these three steps interact with each other.

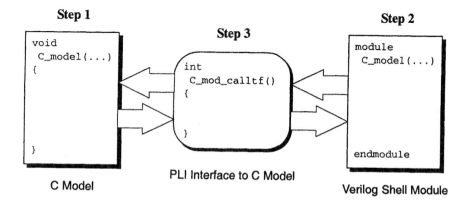

This chapter presents steps 2 and 3 of this interface method in detail. Step 1 is to model some desired functionality or algorithm in the C language. This step is pure C programming, which does not directly involve the Verilog language or the Verilog PLI. This chapter does not cover how to implement ideas in the C language—the focus is on how to interface that implementation with a Verilog simulation. To maintain this focus, the C model example presented in this chapter will be a practical example, but relatively simple to model in C. The C model example used illustrates all of the important concepts of integrating C models into a Verilog simulation.

7.2 Creating the C language model

A hardware model can be represented in the C programming language in two basic forms, either as a C function or as an independent C program.

7.2.1 Using functions to represent the C model

When the C model is represented as a C function, that function can be linked into the Verilog simulator, together with the PLI application that serves as the interface to the model. The PLI application can then call the function when needed, passing inputs to the function, and receiving outputs from the function. One advantage of representing a C model as a function is the simplicity of passing values to and from the model. Another advantage is ease of porting to different operating systems, since the C model is called directly from the PLI application as a C function. A disadvantage of using a function to represent the C model is that the C model may need to contain additional code to allow a Verilog design to instantiate the C model multiple times. The model needs to specifically create unique storage for each instance.

7.2.2 Using independent programs to represent the C model

When the C model is represented as an independent program, which means it has its own C *main* function, then the Verilog simulation and the C model can be run as parallel processes on the same or on different computers. The PLI application which serves as an interface between the simulation and the model will need to create and maintain some type of communication channel between the two programs. This communication can be accomplished several ways, such as using the exec command in the C standard library. On Unix operating systems, the fork or vfork commands with either Unix pipes or Unix sockets is an efficient method to communicate with the C model program. On PC systems running a DOS or windows operating system, the spawn command can be used to invoke the C model program and establish two-way communications between the PLI application and the C model process.

One of the advantages of representing the C model as an independent model is the ability to have parallel processes running on the same computer or separate computers. Another advantage is that when a Verilog design instantiates multiple instances of the C model, each instance will be a separate process with its own memory storage. The major disadvantage of independent programs when compared to using a C function to represent the C model is that the PLI interface to invoke and communicate with the separate process is more complex, and might be operating system dependent.

David Roberts, of Cadence Design Systems, who reviewed many of the chapters of this book, has provided a full example of representing a C model as a separate C program. This example is included with the CD that accompanies this book.

7.3 A C model example

The C model used for different PLI interfaces shown in this chapter is a scientific Arithmetic Logic Unit, which utilizes the C math library. The C model is represented as a C function, which will be called from the PLI interface mechanism. This model is written entirely with the standard C library routines and C data types, without reference to any PLI routines or PLI data types. This same example is also used in other chapters, to show how a PLI interface to C models can be created using the TF and ACC libraries of the PLI.

The inputs and outputs of the scientific ALU C model are shown below, and Table 7-1 shows the operations which the ALU performs.

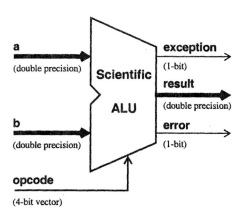

exception is set to 1 whenever an operation results in a value which is out of range of the double-precision result.

error is set to 1 whenever an input to an operation is out of range for the operation.

Table 7-1: Scientific ALU C model operations

Opcode	C Math Library Operation
0	pow(a,b) — returns a to the power of b
1	sqrt(a) — returns the square root of a
2	exp(a) — returns the natural exponent of a
3	ldexp(a,b) — returns a * (2 to the power of b)
4	fabs(a) — returns the absolute of a
5	fmod(a,b) — returns the floating remainder of a / b
6	ceil(a) — returns smallest whole number not less than a
7	floor(a) — returns largest whole number not more than a
8	log(a) — returns the natural log of a
9	log10(a) — returns the base 10 log of a
A	sin(a) — returns the sine of a
B	cos(a) — returns the cosine of a
C	tan(a) — returns the tangent of a
D	asin(a) — returns the arcsine of a
E	acos(a) — returns the arccosine of a
F	atan(a) — returns the arctangent of a

The source code for the scientific ALU is listed in Example 7-1. This version of the ALU uses combinational logic (the outputs change whenever an input changes).

CD The source code for this example is on the CD accompanying this book.

- Application source file: Chapter.07/sci_alu_combinational_vpi.c
- Verilog shell module: Chapter.07/sci_alu_combinational_shell.v
- Verilog test bench: Chapter.07/sci_alu_combinational_test.v
- Expected results log: Chapter.07/sci_alu_combinational_test.log

Example 7-1: scientific ALU C model — combinational logic version

```
#include <stdlib.h>
#include <math.h>
#include <errno.h>

void PLIbook_ScientificALU_C_model(
        double *result,   /* output from ALU */
        int    *excep,    /* output; set if result is out of range */
        int    *err,      /* output; set if input is out of range */
        double a,         /* input */
        double b,         /* input */
        int    opcode)    /* input */
{
  switch (opcode) {
    case 0x0: *result = pow   (a, b);      break;
    case 0x1: *result = sqrt  (a);         break;
    case 0x2: *result = exp   (a);         break;
    case 0x3: *result = ldexp (a, (int)b); break;
    case 0x4: *result = fabs  (a);         break;
    case 0x5: *result = fmod  (a, b);      break;
    case 0x6: *result = ceil  (a);         break;
    case 0x7: *result = floor (a);         break;
    case 0x8: *result = log   (a);         break;
    case 0x9: *result = log10 (a);         break;
    case 0xA: *result = sin   (a);         break;
    case 0xB: *result = cos   (a);         break;
    case 0xC: *result = tan   (a);         break;
    case 0xD: *result = asin  (a);         break;
    case 0xE: *result = acos  (a);         break;
    case 0xF: *result = atan  (a);         break;
  }
  *excep = (errno == ERANGE);   /* result of math func. out-of-range */
  #ifdef WIN32   /* for Microsoft Windows compatibility */
   *err   = (_isnan(*result) || /* result is not-a-number, or */
             errno == EDOM);    /* arg to math func. is out-of-range */
  #else
   *err   = (isnan(*result) ||  /* result is not-a-number, or */
             errno == EDOM);    /* arg to math func. is out-of-range */
  #endif
  if (*err) *result = 0.0;      /* set result to 0 if error occurred */
  errno = 0;                    /* clear the error flag */
  return;
}
```

7.4 Creating a Verilog shell module

A *shell module* allows a Verilog design to reference a C model using standard Verilog HDL syntax. The shell module is a Verilog module which has the same input and output ports as the C model, but the module has no functionality modeled within. To represent the module's functionality, the shell module invokes a PLI application, which in turn invokes the C model. A shell module is sometimes referred to as a *wrapper module*, because the module is wrapped around the call to a PLI application.

The shell module for a combinational logic version of the scientific ALU is listed in Example 7-2.

Example 7-2: Verilog shell module for the scientific ALU C model

```
`timescale 1ns / 1ns
module scientific_alu(a_in, b_in, opcode,
                      result_out, exception, error);
  input  [63:0] a_in, b_in;
  input  [3:0] opcode;
  output [63:0] result_out;
  output        exception, error;

  reg           exception, error;
  real          a, b, result; // real variables used in this module

  // convert real numbers to/from 64-bit vector port connections
  assign result_out = $realtobits(result);
  always @(a_in) a = $bitstoreal(a_in);
  always @(b_in) b = $bitstoreal(b_in);

  //call the PLI application which interfaces to the C model
  initial
    $scientific_alu(a, b, opcode, result, exception, error);

endmodule
```

In this scientific ALU example, the primary inputs and outputs of the model are double-precision floating point values, represented as Verilog `real` data types. The Verilog language does not permit real numbers to be connected to module ports. However, the language provides built-in system functions which convert real numbers to 64-bit vectors, and vice-versa, so the real values can be passed through a module port connection. These built-in system functions are *$realtobits()* and *$bitstoreal()*.

The Verilog shell module that represents the C model can be instantiated in a design in the same way as any other Verilog module. For example:

```
module chip (...)
   ...
   scientific_alu u1 (a, b, opcode, result, excep, err);
   ...
endmodule
```

Creating a shell module to represent the C model is not mandatory—the PLI application could be called directly from any place in a Verilog design. However, there are important advantages to using a shell module to represent the C model:

- The shell module provides a simple method to encapsulate the C model.

- The shell module can be instantiated anywhere in a Verilog design hierarchy.

- The shell module can be instantiated any number of times in a Verilog design.

- The shell module can add Verilog HDL delays to the C model, which can accurately represent rise and fall delay delays, state-dependent delays, and timing constraints such as setup times.

- Delays within a shell module can be annotated using delay calculators or SDF files to provide additional delay accuracy for each instance of the shell module.

Section 7.10 later in this chapter discusses how delays can be represented in the Verilog shell module.

7.5 Creating a combinational logic interface to a C model

In a combinational logic model, the outputs of the model continuously reflect the input values of the model. The inputs are asynchronous—when any input changes value, the model outputs are re-evaluated to reflect the input change.

The VPI library can access the arguments of a system task. In the discussion of the Verilog shell module in the previous section of this chapter, it was suggested that a system task be created to represent the C model interface, and that this task list all of the C model inputs and outputs as arguments. This gives a PLI application easy access to these inputs and outputs using the VPI library.

A *simulation callback routine* provides an simple method of creating a combinational logic interface to a C model. A callback can be used to schedule a *simulation callback routine* whenever an argument of a system task changes value. The routine can then read the input values from the system task arguments, and pass the input values to the C model. The outputs of the C model are then passed back to the Verilog simulation by writing the results onto the system task arguments in the Verilog shell module. Chapter 6 presented how to register callbacks to a *simulation callback routine* for logic value changes.

The basic steps involved with using a *simulation callback routine* to implement a combinational logic interface are:

1. Create a PLI application system task to represent the interface between the Verilog shell module and the C model. The system task is invoked from the shell module, and all of the C model inputs and outputs are listed as arguments to the system task. For example:

    ```
    initial
       $scientific_alu(a, b, opcode, result, exception, error);
    ```

    Note that, in this example, the system task is called from a Verilog **initial** procedure, which means the system task will only be invoked one time for each instance of the shell module.

2. In the *calltf routine* associated with the system task, register logic value change callbacks for a *simulation callback routine*. A callback is registered for each system task argument which represents an input to the C model, and each callback invokes the same *simulation callback routine*.

3. In the *simulation callback routine*, which is called whenever a system task argument changes value, read the values of all inputs and pass the values to the C model. The output values of the C model are returned to the same *simulation callback routine*, which then writes the values to the system task arguments that represent the outputs of the C model.

Obtaining object handles within simulation callback routines

 The VPI *simulation callback routine* is not directly associated with the system task which represents the C model interface. This means the *callback routine* cannot obtain handles to the system task arguments unless the callback is passed a handle to the system task or is passed the argument handles.

The *simulation callback routine* needs the handles for the system task arguments which represent the C model, in order to read the input values and write the C model output values. Since the *simulation callback routine* is not associated with the system task, access to the task arguments must be passed to the routine. This information can be passed in at least two ways:

- When the *calltf routine* registers the callbacks, the *calltf routine* can obtain a handle for the system task instance, and pass the handle to the *simulation callback routine* through the user_data field. The *simulation callback routine* can then use the system task handle to obtain the handles for the task arguments.

- When the *calltf routine* registers callbacks, the *calltf routine* can allocate persistent storage, obtain handles for all task arguments, store the handles in the storage block, and pass a pointer to the storage to the *simulation callback routine* through the user_data field. The *simulation callback routine* can then obtain the argument handles from the storage block.

Each of these methods has advantages. Saving just the handle for the system task instance is simpler, and makes the instance handle available for other uses, such as scheduling a callback to the *simulation callback routine* for that instance of the system task. However, each time a value change simulation callback occurs, the *simulation callback routine* must call vpi_iterate() and vpi_scan() to obtain handles for each of the *$scientific_alu* arguments. Saving all handles in a block of memory means the handles do not need to be obtained each time a value change occurs. Example 7-3, which follows, illustrates passing the system task instance handle. Example 7-4 on page 257 illustrates allocating a block of memory to store the handles for all system task arguments, and passing a pointer to the memory to the consumer routine.

The following example implements a combinational logic interface for the scientific ALU C model.

CD The source code for this example is on the CD accompanying this book.

- Application source file: `Chapter.07/sci_alu_combinational_vpi.c`
- Verilog shell module: `Chapter.07/sci_alu_combinational_shell.v`
- Verilog test bench: `Chapter.07/sci_alu_combinational_test.v`
- Expected results log: `Chapter.07/sci_alu_combinational_test.log`

Example 7-3: combinational logic C model interface using VPI routines

```
/*********************************************************************
 * Value change simulation callback routine: Serves as an interface
 * between Verilog simulation and the C model.  Called whenever the
 * C model inputs change value, passes the values to the C model, and
 * puts the C model outputs into simulation.
 *********************************************************************/
PLI_INT32 PLIbook_ScientificALU_interface(p_cb_data cb_data)
{
  double      a, b, result;
  int         opcode, excep, err;
  s_vpi_value value_s;
  vpiHandle   instance_h, arg_itr;
  vpiHandle   a_h, b_h, opcode_h, result_h, excep_h, err_h;

  /* Retrieve handle to $scientific_alu instance from user_data */
  instance_h = (vpiHandle)cb_data->user_data;

  /* obtain handles to system task arguments */
  /* compiletf has already verified arguments are correct */
  arg_itr  = vpi_iterate(vpiArgument, instance_h);
  a_h      = vpi_scan(arg_itr); /* 1st arg is a input */
  b_h      = vpi_scan(arg_itr); /* 2nd arg is b input */
  opcode_h = vpi_scan(arg_itr); /* 3rd arg is opcode input */
  result_h = vpi_scan(arg_itr); /* 4th arg is result output */
  excep_h  = vpi_scan(arg_itr); /* 5th arg is excep output */
  err_h    = vpi_scan(arg_itr); /* 6th arg is error output */
  vpi_free_object(arg_itr);  /* free iterator--did not scan to null */

  /* Read current values of C model inputs from Verilog simulation */
  value_s.format = vpiRealVal;
  vpi_get_value(a_h, &value_s);
  a = value_s.value.real;

  vpi_get_value(b_h, &value_s);
  b = value_s.value.real;

  value_s.format = vpiIntVal;
  vpi_get_value(opcode_h, &value_s);
  opcode = (int)value_s.value.integer;
```

```
/****** Call the C model ******/
PLIbook_ScientificALU_C_model(&result, &excep, &err, a, b, opcode);

/* Write the C model outputs onto the Verilog signals */
value_s.format = vpiRealVal;
value_s.value.real = result;
vpi_put_value(result_h, &value_s, NULL, vpiNoDelay);

value_s.format = vpiIntVal;
value_s.value.integer = (PLI_INT32)excep;
vpi_put_value(excep_h, &value_s, NULL, vpiNoDelay);

value_s.value.integer = (PLI_INT32)err;
vpi_put_value(err_h, &value_s, NULL, vpiNoDelay);

return(0);
}

/**********************************************************************
 * calltf routine: Registers a callback to the C model interface
 * whenever any input to the C model changes value
 **********************************************************************/
PLI_INT32 PLIbook_ScientificALU_calltf(PLI_BYTE8 *user_data)
{
  vpiHandle    instance_h, arg_itr, a_h, b_h, opcode_h, cb_h;
  s_vpi_value  value_s;
  s_vpi_time   time_s;
  s_cb_data    cb_data_s;

  /* obtain a handle to the system task instance */
  instance_h = vpi_handle(vpiSysTfCall, NULL);

  /* obtain handles to system task arguments for model inputs */
  /* compiletf has already verified arguments are correct */
  arg_itr = vpi_iterate(vpiArgument, instance_h);
  a_h      = vpi_scan(arg_itr); /* 1st arg is a input */
  b_h      = vpi_scan(arg_itr); /* 2nd arg is b input */
  opcode_h = vpi_scan(arg_itr); /* 3rd arg is opcode input */
  vpi_free_object(arg_itr);   /* free iterator--did not scan to null */

  /* setup value change callback options */
  time_s.type      = vpiSuppressTime;
  cb_data_s.reason = cbValueChange;
  cb_data_s.cb_rtn = PLIbook_ScientificALU_interface;
  cb_data_s.time   = &time_s;
  cb_data_s.value  = &value_s;

  /* add value change callbacks to all signals which are inputs to  */
  /* pass handle to $scientific_alu instance as user_data value */
  cb_data_s.user_data = (PLI_BYTE8 *)instance_h;
  value_s.format = vpiRealVal;
  cb_data_s.obj = a_h;
```

```
  cb_h = vpi_register_cb(&cb_data_s);
  vpi_free_object(cb_h); /* don't need callback handle */

  cb_data_s.obj = b_h;
  cb_h = vpi_register_cb(&cb_data_s);
  vpi_free_object(cb_h); /* don't need callback handle */

  value_s.format = vpiIntVal;
  cb_data_s.obj = opcode_h;
  cb_h = vpi_register_cb(&cb_data_s);
  vpi_free_object(cb_h); /* don't need callback handle */

  return(0);
}

/***********************************************************************
 * compiletf routine: Verifies that $scientific_alu() is used correctly
 *    Note: For simplicity, only limited data types are allowed for
 *    task arguments.  Could add checks to allow other data types.
 ***********************************************************************/
PLI_INT32 PLIbook_ScientificALU_compiletf(PLI_BYTE8 *user_data)
{
  vpiHandle systf_h, arg_itr, arg_h;
  int       err = 0;

  systf_h = vpi_handle(vpiSysTfCall, NULL);
  arg_itr = vpi_iterate(vpiArgument, systf_h);
  if (arg_itr == NULL) {
    vpi_printf("ERROR: $scientific_alu requires 6 arguments\n");
    vpi_control(vpiFinish, 1);  /* abort simulation */
    return(0);
  }

  arg_h = vpi_scan(arg_itr); /* 1st arg is a input */
  if (vpi_get(vpiType, arg_h) != vpiRealVar) {
    vpi_printf("$scientific_alu arg 1 (a) must be a real variable\n");
    err = 1;
  }

  arg_h = vpi_scan(arg_itr); /* 2nd arg is b input */
  if (vpi_get(vpiType, arg_h) != vpiRealVar) {
    vpi_printf("$scientific_alu arg 2 (b) must be a real variable\n");
    err = 1;
  }

  arg_h = vpi_scan(arg_itr); /* 3rd arg is opcode input */
  if (vpi_get(vpiType, arg_h) != vpiNet) {
    vpi_printf("$scientific_alu arg 3 (opcode) must be a net\n");
    err = 1;
  }
  else if (vpi_get(vpiSize, arg_h) != 4) {
    vpi_printf("$scientific_alu arg 3 (opcode) must be 4-bit vector\n");
    err = 1;
```

```
    }

    arg_h = vpi_scan(arg_itr); /* 4th arg is result output */
    if (vpi_get(vpiType, arg_h) != vpiRealVar) {
      vpi_printf("$scientific_alu arg 4 (result) must be a real var.\n");
      err = 1;
    }

    arg_h = vpi_scan(arg_itr); /* 5th arg is exception output */
    if (vpi_get(vpiType, arg_h) != vpiReg) {
      vpi_printf("$scientific_alu arg 5 (exception) must be a reg\n");
      err = 1;
    }
    else if (vpi_get(vpiSize, arg_h) != 1) {
      vpi_printf("$scientific_alu arg 5 (exception) must be scalar\n");
      err = 1;
    }

    arg_h = vpi_scan(arg_itr); /* 6th arg is error output */
    if (vpi_get(vpiType, arg_h) != vpiReg) {
      vpi_printf("$scientific_alu arg 6 (error) must be a reg\n");
      err = 1;
    }
    else if (vpi_get(vpiSize, arg_h) != 1) {
      vpi_printf("$scientific_alu arg 6 (error) must be scalar\n");
      err = 1;
    }

    if (vpi_scan(arg_itr) != NULL) { /* should not be any more args */
      vpi_printf("ERROR: $scientific_alu requires only 6 arguments\n");
      vpi_free_object(arg_itr);
      err = 1;
    }

    if (err) {
      vpi_control(vpiFinish, 1);  /* abort simulation */
      return(0);
    }
    return(0);  /* no syntax errors detected */
}
```

7.6 Creating a sequential logic interface to a C model

In a sequential logic model, the outputs of the model change synchronously with an input strobe, such as a positive edge of a clock. There may also be one or more asynchronous inputs, such as a reset signal.

The VPI *simulation callback routine* provides a straightforward way to model a sequential logic interface to a C model. Callbacks are only registered for a logic value change on the clock input and any asynchronous inputs of the C model. Only when those inputs change, are the new input values passed to the C model. The outputs of the C model are then passed back to the Verilog simulation by writing the results onto the system task arguments in the Verilog shell module which represents the outputs of the C model.

The basic steps involved with using the Value Change Link to implement a synchronous sequential logic interface are very similar to implementing a combinational logic interface. The one difference is that simulation callbacks are only registered for specific C model inputs instead of all inputs.

Regardless of whether the interface is combinational or sequential logic, the *simulation callback routine* needs the handles for the system task arguments which represents the C model, in order to read the inputs values and write the C model output values. Access to the task arguments must be passed to the *simulation callback routine* through the callback's `user_data` field.

Example 7-4, which follows, implements a sequential logic interface for the scientific ALU C model, where all inputs are synchronized to value changes of a clock input.

Note that this example also improves on the efficiency of the previous example, Example 7-3, in another way. This example uses the *calltf routine* to obtain the handles to all of the inputs to the $scientific_alu instance, and saves the handle in a structure allocated by the *calltf routine*. A pointer to this saved instance is then passed to the *simulation callback* routine that serves as the interface to the C model. This is much more efficient than passing the instance handle, and having the simulation callback routine iterate and scan for all argument handles each time the routine is called.

CD The source code for this example is on the CD accompanying this book.

- Application source file: `Chapter.07/sci_alu_sequential_vpi.c`
- Verilog shell module: `Chapter.07/sci_alu_sequential_shell.v`
- Verilog test bench: `Chapter.07/sci_alu_sequential_test.v`
- Expected results log: `Chapter.07/sci_alu_sequential_test.log`

Example 7-4: sequential logic C model interface using VPI routines

```
/*********************************************************************
 * Definition for a structure to hold the data to be passed from
 * calltf routine to the ALU interface.
 *********************************************************************/
typedef struct PLIbook_ScientificALU_data {
  vpiHandle  clock_h, a_h, b_h, opcode_h, result_h, excep_h, err_h;
} PLIbook_ALU_data_s, *PLIbook_ALU_data_p;

/*********************************************************************
 * Value change simulation callback routine: Serves as an interface
 * between Verilog simulation and the C model.  Called whenever the
 * C model inputs change value, passes the values to the C model, and
 * puts the C model outputs into simulation.
 *********************************************************************/
PLI_INT32 PLIbook_ScientificALU_interface(p_cb_data cb_data)
{
  double      a, b, result;
  int         clock, opcode, excep, err;
  s_vpi_value value_s;

  PLIbook_ALU_data_p  ALUdata;

  /* Retrieve pointer to ALU data structure from callback user_data. */
  /* The structure contains the handles for the $scientific_alu args */
  ALUdata = (PLIbook_ALU_data_p)cb_data->user_data;

  /* Read current values of C model inputs from Verilog simulation */
  value_s.format = vpiIntVal;
  vpi_get_value(ALUdata->clock_h, &value_s);
  clock = (int)value_s.value.integer;
  if (clock != 1) /* abort if not a positive edge of the clock input */
    return(0);

  value_s.format = vpiRealVal;
  vpi_get_value(ALUdata->a_h, &value_s);
  a = value_s.value.real;

  vpi_get_value(ALUdata->b_h, &value_s);
  b = value_s.value.real;

  value_s.format = vpiIntVal;
  vpi_get_value(ALUdata->opcode_h, &value_s);
  opcode = (int)value_s.value.integer;

  /****** Call the C model ******/
  PLIbook_ScientificALU_C_model(&result, &excep, &err, a, b, opcode);

  /* Write the C model outputs onto the Verilog signals */
  value_s.format = vpiRealVal;
  value_s.value.real = result;
  vpi_put_value(ALUdata->result_h, &value_s, NULL, vpiNoDelay);

  value_s.format = vpiIntVal;
  value_s.value.integer = (PLI_INT32)excep;
  vpi_put_value(ALUdata->excep_h, &value_s, NULL, vpiNoDelay);
```

```
    value_s.value.integer = (PLI_INT32)err;
    vpi_put_value(ALUdata->err_h, &value_s, NULL, vpiNoDelay);

    return(0);
}

/**********************************************************************
 * calltf routine: Registers a callback to the C model interface
 * whenever any input to the C model changes value
 **********************************************************************/
PLI_INT32 PLIbook_ScientificALU_calltf(PLI_BYTE8 *user_data)
{
    vpiHandle    instance_h, arg_itr, cb_h;
    s_vpi_value  value_s;
    s_vpi_time   time_s;
    s_cb_data    cb_data_s;

    PLIbook_ALU_data_p ALUdata;

    /* allocate storage to hold $scientific_alu argument handles */
    ALUdata = (PLIbook_ALU_data_p)malloc(sizeof(PLIbook_ALU_data_s));

    /* obtain a handle to the system task instance */
    instance_h = vpi_handle(vpiSysTfCall, NULL);

    /* obtain handles to system task arguments */
    /* compiletf has already verified arguments are correct */
    arg_itr = vpi_iterate(vpiArgument, instance_h);
    ALUdata->clock_h  = vpi_scan(arg_itr); /* 1st arg is clock input */
    ALUdata->a_h      = vpi_scan(arg_itr); /* 2nd arg is a input */
    ALUdata->b_h      = vpi_scan(arg_itr); /* 3rd arg is b input */
    ALUdata->opcode_h = vpi_scan(arg_itr); /* 4th arg is opcode input */
    ALUdata->result_h = vpi_scan(arg_itr); /* 5th arg is result output */
    ALUdata->excep_h  = vpi_scan(arg_itr); /* 6th arg is excep output */
    ALUdata->err_h    = vpi_scan(arg_itr); /* 7th arg is error output */
    vpi_free_object(arg_itr);  /* free iterator--did not scan to null */

    /* setup value change callback options */
    time_s.type      = vpiSuppressTime;
    cb_data_s.reason = cbValueChange;
    cb_data_s.cb_rtn = PLIbook_ScientificALU_interface;
    cb_data_s.time   = &time_s;
    cb_data_s.value  = &value_s;

    /* add value change callbacks to clock input to the C model, */
    /* pass pointer to storage for handles as user_data value */
    value_s.format      = vpiSuppressVal;
    cb_data_s.user_data = (PLI_BYTE8 *)ALUdata;
    cb_data_s.obj       = ALUdata->clock_h;
    cb_h = vpi_register_cb(&cb_data_s);
    vpi_free_object(cb_h); /* don't need callback handle */

    return(0);
}
```

7.7 Synchronizing with the end of a simulation time step

Within a simulation, it is possible for several signals to change at the same moment of simulation time. In Verilog simulators, the *simulation callback routine* may be called in the middle of a simulation time step for asynchronous value changes, before all input value changes have occurred for that time step.

With a combinational logic interface, the *simulation callback routine* will be called for every input change. In the combinational logic interface presented in Example 7-3, the C model is called each time the *simulation callback routine* is called for an input value change. This is the correct functionality for combinational logic—at the completion of a simulation time step, the outputs from the C model represent the most current input values. However, by synchronizing the call to the C model with the end of the simulation time step in which changes occur, the multiple calls to the C model within a time step could be optimized to a single call.

With a sequential logic interface synchronized to a clock, when the *simulation callback routine* is called at a clock change, other input changes at that moment in simulation time may or may not have occurred. It may be desirable to ensure that the C model is not called until all inputs have their most current value for the time step in which the clock changes. By using another *simulation callback routine*, both combinational logic and sequential logic C model interfaces can be synchronized to the end of a current simulation time step. This is done by using the asynchronous value change callback to the *simulation callback routine* to schedule a synchronous callback to a different *simulation callback routine* for end of the current time step. Chapter 6 presented how to register callbacks at the end of the current simulation time step, and this chapter uses those concepts.

Example 7-5 modifies the combinational logic interface presented in Example 7-3. This modified version schedules a simulation callback at the end of a simulation time step in which an input changed value. This example uses a flag to indicate when a synchronous callback for the end of the current time step has already been scheduled for the current simulation time. This flag is used in the C model interface to prevent more than one synchronous callback to the *simulation callback routine* being requested in the same time step. The flag is stored in the same block of memory that was allocated to store the handles for the *$scientific_alu* arguments. Since this memory block is unique for each instance of a system task, each instance of the C model will have a unique flag.

 The IEEE 1364 Verilog standard allows PLI read-write synchronization events to be processed either before or after nonblocking assignments. This ambiguity can lead to different results from different simulators. The reason the Verilog standard allows this latitude in event scheduling is discussed in Chapter 6, on page 221.

CD The source code for this example is on the CD accompanying this book.

- Application source file: `Chapter.07/sci_alu_synchronized_vpi.c`
- Verilog shell module: `Chapter.07/sci_alu_synchronized_shell.v`
- Verilog test bench: `Chapter.07/sci_alu_synchronized_test.v`
- Expected results log: `Chapter.07/sci_alu_synchronized_test.log`

Example 7-5: C model interface synchronized to the end of a time step

```c
/*************************************************************************
 * Definition for a structure to hold the data to be passed from
 * calltf routine to the ALU interface.
 *************************************************************************/
typedef struct PLIbook_ScientificALU_data {
  vpiHandle  clock_h, a_h, b_h, opcode_h, result_h, excep_h, err_h;
  short int  sync_flag;
} PLIbook_ALU_data_s, *PLIbook_ALU_data_p;

/*************************************************************************
 * Value change simulation callback routine: Schedules a read-write
 * synchronize simulation callback at the end of the current time step.
 *************************************************************************/
PLI_INT32 PLIbook_ScientificALU_interface(p_cb_data cb_data)
{
  vpiHandle          cb_h;
  s_cb_data          cb_data_s;
  s_vpi_time         time_s;
  PLIbook_ALU_data_p ALUdata;
  /* Retrieve pointer to ALU data structure from user_data field.   */
  /* The structure contains a flag indicating if a synchronize      */
  /* callback has already been scheduled (the sync_flag is set by    */
  /* this routine, and cleared by the read-write synch callback.     */
  ALUdata = (PLIbook_ALU_data_p)cb_data->user_data;
  if (!ALUdata->sync_flag) {
    /* Schedule a synchronize simulation callback for this instance */
    ALUdata->sync_flag  = 1; /* set sync_flag */
    time_s.type         = vpiSimTime;
    time_s.high         = 0;
    time_s.low          = 0;
    cb_data_s.reason    = cbReadWriteSynch;
    cb_data_s.user_data = (PLI_BYTE8 *)ALUdata;
    cb_data_s.cb_rtn    = PLIbook_EndOfTimeStep_callback;
    cb_data_s.obj       = NULL;
    cb_data_s.time      = &time_s;
    cb_data_s.value     = NULL;
    cb_h = vpi_register_cb(&cb_data_s);
    vpi_free_object(cb_h); /* don't need callback handle */
  }
  return(0);
}
```

```c
/**********************************************************************
 * Read-write synchronize simulation callback routine: Serves as an
 * interface between Verilog simulation and the C model.  Passes the
 * values to the C model, and puts the C model outputs into simulation.
 **********************************************************************/
PLI_INT32 PLIbook_EndOfTimeStep_callback(p_cb_data cb_data)
{
  double      a, b, result;
  int         opcode, excep, err;
  s_vpi_value value_s;

  PLIbook_ALU_data_p  ALUdata;

  /* Retrieve pointer to ALU data structure from callback user_data. */
  /* The structure contains the handles for the $scientific_alu args */
  ALUdata = (PLIbook_ALU_data_p)cb_data->user_data;

  /* Set the sync_flag to 0 to indicate that this callback has been */
  /* processed */
  ALUdata->sync_flag = 0;

  /* Read current values of C model inputs from Verilog simulation */
  value_s.format = vpiRealVal;
  vpi_get_value(ALUdata->a_h, &value_s);
  a = value_s.value.real;

  vpi_get_value(ALUdata->b_h, &value_s);
  b = value_s.value.real;

  value_s.format = vpiIntVal;
  vpi_get_value(ALUdata->opcode_h, &value_s);
  opcode = (int)value_s.value.integer;

  /****** Call the C model  ******/
  PLIbook_ScientificALU_C_model(&result, &excep, &err, a, b, opcode);

  /* Write the C model outputs onto the Verilog signals */
  value_s.format = vpiRealVal;
  value_s.value.real = result;
  vpi_put_value(ALUdata->result_h, &value_s, NULL, vpiNoDelay);

  value_s.format = vpiIntVal;
  value_s.value.integer = (PLI_INT32)excep;
  vpi_put_value(ALUdata->excep_h, &value_s, NULL, vpiNoDelay);

  value_s.value.integer = (PLI_INT32)err;
  vpi_put_value(ALUdata->err_h, &value_s, NULL, vpiNoDelay);

  return(0);
}
/**********************************************************************
 * calltf routine: Registers a callback to the C model interface
 * whenever any input to the C model changes value
 **********************************************************************/
PLI_INT32 PLIbook_ScientificALU_calltf(PLI_BYTE8 *user_data)
{
```

```
vpiHandle      instance_h, arg_itr, cb_h;
s_vpi_value    value_s;
s_vpi_time     time_s;
s_cb_data      cb_data_s;

PLIbook_ALU_data_p  ALUdata;

/* allocate storage to hold $scientific_alu argument handles */
ALUdata = (PLIbook_ALU_data_p)malloc(sizeof(PLIbook_ALU_data_s));

/* obtain a handle to the system task instance */
instance_h = vpi_handle(vpiSysTfCall, NULL);

/* obtain handles to system task arguments */
/* compiletf has already verified arguments are correct */
arg_itr = vpi_iterate(vpiArgument, instance_h);
ALUdata->a_h      = vpi_scan(arg_itr); /* 1st arg is a input */
ALUdata->b_h      = vpi_scan(arg_itr); /* 2nd arg is b input */
ALUdata->opcode_h = vpi_scan(arg_itr); /* 3rd arg is opcode input */
ALUdata->result_h = vpi_scan(arg_itr); /* 4th arg is result output */
ALUdata->excep_h  = vpi_scan(arg_itr); /* 5th arg is excep output */
ALUdata->err_h    = vpi_scan(arg_itr); /* 6th arg is error output */
vpi_free_object(arg_itr);  /* free iterator--did not scan to null */

/* setup value change callback options */
time_s.type       = vpiSuppressTime;
cb_data_s.reason  = cbValueChange;
cb_data_s.cb_rtn  = PLIbook_ScientificALU_interface;
cb_data_s.time    = &time_s;
cb_data_s.value   = &value_s;

/* add value change callbacks to all signals which are inputs to  */
/* pass pointer to storage for handles as user_data value */
cb_data_s.user_data = (PLI_BYTE8 *)ALUdata;
value_s.format = vpiRealVal;
cb_data_s.obj = ALUdata->a_h;
cb_h = vpi_register_cb(&cb_data_s);
vpi_free_object(cb_h); /* don't need callback handle */

cb_data_s.obj = ALUdata->b_h;
cb_h = vpi_register_cb(&cb_data_s);
vpi_free_object(cb_h); /* don't need callback handle */

value_s.format = vpiIntVal;
cb_data_s.obj = ALUdata->opcode_h;
cb_h = vpi_register_cb(&cb_data_s);
vpi_free_object(cb_h); /* don't need callback handle */

/* clear the callback sync_flag to indicate that no read-write */
/* synchronize callbacks have been processed */
ALUdata->sync_flag = 0;

return(0);
}
```

7.8 Synchronizing with a future simulation time step

In certain C model applications, it may be necessary to synchronize C model activity with future simulation activity. The VPI routines can also be used to schedule callbacks to a *simulation callback routine* at a specific amount of time in the future, relative to the current simulation time.

The VPI routines can also determine the future simulation time in which the next simulation event is scheduled to occur. This provides a way for a PLI application to synchronize activity for when the Verilog simulator is processing simulation events.

Using VPI routines to schedule future simulation callbacks is discussed in Chapter 6.

7.9 Allocating storage within a C model

Special attention and care must be taken when a C model uses static variables or allocates memory. The Verilog language can instantiate a model any number of times. Each instance of the Verilog shell module creates a unique instance of the system task which invokes the PLI interface to the C model. Therefore, the *calltf routine* which is invoked by a system task instance will be unique. Any memory which is allocated by the *calltf routine* will also be unique for each instance of the system task. However, the *simulation callback routine* used to communicate with the C model is <u>not</u> instance specific. Any memory allocated in the *simulation callback* routine will be shared by all instances of the C model.

When a C model is represented as an independent program, multiple instances of the model are not a problem, as each instance will invoke a new process with unique storage for each process. When the C model is represented as a C function, however, multiple instances of the model will share the same function. The model must allow for the possibility of multiple instances, and provide unique storage for each instance.

Example 7-6 presents a latched version of the scientific ALU, which can store the result of a previous operation indefinitely, and Example 7-7 presents a combinational logic interface to the model. This example allocates storage within the C model, which is unique storage for each version of the C model. The storage for the C model output values is allocated as part of the same instance-specific storage that holds the handles for the $scientific_alu arguments. The storage is allocated by the *calltf routine* for each system task instance. A pointer to the storage is passed to the *simulation callback routine* as the user_data value, which then passes the pointer to the C model as an input to the model function.

> **CD** The source code for this example is on the CD accompanying this book.
>
> • Application source file: `Chapter.07/sci_alu_latched_vpi.c`
> • Verilog shell module: `Chapter.07/sci_alu_latched_shell.v`
> • Verilog test bench: `Chapter.07/sci_alu_latched_test.v`
> • Expected results log: `Chapter.07/sci_alu_latched_test.log`

Example 7-6: scientific ALU C model with latched outputs

```c
#include <stdlib.h>
#include <math.h>
#include <errno.h>

/*******************************************************************
 * Structure definition to store output values when the ALU is latched.
 * The structure is allocated in the calltf routine for each instance
 * of the C model
 *******************************************************************/
typedef struct PLIbook_ScientificALU_outputs {
  double result;     /* stored result of previous operation */
  int     excep;
  int     err;
} PLIbook_ALU_outputs_s, *PLIbook_ALU_outputs_p;

void PLIbook_ScientificALU_C_model(
        double *result,   /* output from ALU */
        int     *excep,   /* output; set if result is out of range */
        int     *err,     /* output; set if input is out of range */
        double  a,        /* input */
        double  b,        /* input */
        int     opcode,   /* input */
        int     enable,   /* input; 0 = latched */
        PLIbook_ALU_outputs_p  LatchedOutputs) /* input */
{
  if (enable) { /* ALU is not latched, calculate outputs and store */
    switch (opcode) {
        case 0x0: LatchedOutputs->result = pow   (a, b);      break;
        case 0x1: LatchedOutputs->result = sqrt  (a);         break;
        case 0x2: LatchedOutputs->result = exp   (a);         break;
        case 0x3: LatchedOutputs->result = ldexp (a, (int)b); break;
        case 0x4: LatchedOutputs->result = fabs  (a);         break;
        case 0x5: LatchedOutputs->result = fmod  (a, b);      break;
        case 0x6: LatchedOutputs->result = ceil  (a);         break;
        case 0x7: LatchedOutputs->result = floor (a);         break;
        case 0x8: LatchedOutputs->result = log   (a);         break;
        case 0x9: LatchedOutputs->result = log10 (a);         break;
        case 0xA: LatchedOutputs->result = sin   (a);         break;
        case 0xB: LatchedOutputs->result = cos   (a);         break;
        case 0xC: LatchedOutputs->result = tan   (a);         break;
        case 0xD: LatchedOutputs->result = asin  (a);         break;
        case 0xE: LatchedOutputs->result = acos  (a);         break;
```

```
            case 0xF: LatchedOutputs->result = atan    (a);              break;
      }
      LatchedOutputs->excep = (errno == ERANGE);/* result out-of-range */
      #ifdef WIN32  /* for Microsoft Windows compatibility */
       LatchedOutputs->err  = (_isnan(*result) ||  /* not-a-number, or */
                                errno == EDOM);     /* arg out-of-range */
      #else
       LatchedOutputs->err  = (isnan(*result) ||    /* not-a-number, or */
                                errno == EDOM);     /* arg out-of-range */
      #endif
      if (LatchedOutputs->err) LatchedOutputs->result = 0.0;
      errno = 0;                                   /* clear error flag */
    }

    /* return the values stored in the C model */
    *result = LatchedOutputs->result;
    *err    = LatchedOutputs->err;
    *excep  = LatchedOutputs->excep;

    return;
}
```

Example 7-7: combinational logic interface to the latched scientific ALU C model

```
/*****************************************************************
 * Definition for a structure to hold the data to be passed from
 * calltf routine to the ALU interface. Also allocates a structure
 * to store the latched output values of the ALU. This storage is
 * allocated in the calltf routine for each instance of the C model.
 *****************************************************************/
typedef struct PLIbook_ScientificALU_data {
  vpiHandle  enable_h, a_h, b_h, opcode_h, result_h, excep_h, err_h;
  PLIbook_ALU_outputs_s LatchedOutputs;  /* storage for outputs */
} PLIbook_ALU_data_s, *PLIbook_ALU_data_p;

/*****************************************************************
 * Value change simulation callback routine: Serves as an interface
 * between Verilog simulation and the C model. Called whenever the
 * C model inputs change value, passes the values to the C model, and
 * puts the C model outputs into simulation.
 *****************************************************************/
PLI_INT32 PLIbook_ScientificALU_interface(p_cb_data cb_data)
{
  double      a, b, result;
  int         opcode, excep, err, enable;
  s_vpi_value value_s;

  PLIbook_ALU_data_p  ALUdata;

  /* Retrieve pointer to ALU data structure from callback user_data. */
  /* The structure contains the handles for the $scientific_alu args */
  ALUdata = (PLIbook_ALU_data_p)cb_data->user_data;
```

```
/* Read current values of C model inputs from Verilog simulation */
value_s.format = vpiRealVal;
vpi_get_value(ALUdata->a_h, &value_s);
a = value_s.value.real;

vpi_get_value(ALUdata->b_h, &value_s);
b = value_s.value.real;

value_s.format = vpiIntVal;
vpi_get_value(ALUdata->opcode_h, &value_s);
opcode = (int)value_s.value.integer;

vpi_get_value(ALUdata->enable_h, &value_s);
enable = (int)value_s.value.integer;

/****** Call the C model ******/
PLIbook_ScientificALU_C_model(&result, &excep, &err, a, b, opcode,
                              enable, &ALUdata->LatchedOutputs);

/* Write the C model outputs onto the Verilog signals */
value_s.format = vpiRealVal;
value_s.value.real = result;
vpi_put_value(ALUdata->result_h, &value_s, NULL, vpiNoDelay);

value_s.format = vpiIntVal;
value_s.value.integer = (PLI_INT32)excep;
vpi_put_value(ALUdata->excep_h, &value_s, NULL, vpiNoDelay);

value_s.value.integer = (PLI_INT32)err;
vpi_put_value(ALUdata->err_h, &value_s, NULL, vpiNoDelay);

return(0);
}


/***********************************************************************
 * calltf routine: Registers a callback to the C model interface
 * whenever any input to the C model changes value. Memory to store
 * the $scientific_alu argument handles and the latched output values
 * is allocated here, because the calltf routine is called for each
 * instance of $scientific_alu.
 ***********************************************************************/
PLI_INT32 PLIbook_ScientificALU_calltf(PLI_BYTE8 *user_data)
{
  vpiHandle    instance_h, arg_itr, cb_h;
  s_vpi_value  value_s;
  s_vpi_time   time_s;
  s_cb_data    cb_data_s;

  PLIbook_ALU_data_p  ALUdata;

  /* allocate storage to hold $scientific_alu argument handles */
  ALUdata = (PLIbook_ALU_data_p)malloc(sizeof(PLIbook_ALU_data_s));

  /* obtain a handle to the system task instance */
  instance_h = vpi_handle(vpiSysTfCall, NULL);

  /* obtain handles to system task arguments */
  /* compiletf has already verified arguments are correct */
```

```
arg_itr = vpi_iterate(vpiArgument, instance_h);
ALUdata->enable_h = vpi_scan(arg_itr); /* 1st arg is enable input */
ALUdata->a_h     = vpi_scan(arg_itr); /* 2nd arg is a input */
ALUdata->b_h     = vpi_scan(arg_itr); /* 3rd arg is b input */
ALUdata->opcode_h = vpi_scan(arg_itr); /* 4th arg'is opcode input */
ALUdata->result_h = vpi_scan(arg_itr); /* 5th arg is result output */
ALUdata->excep_h = vpi_scan(arg_itr); /* 6th arg is excep output */
ALUdata->err_h   = vpi_scan(arg_itr); /* 7th arg is error output */
vpi_free_object(arg_itr);  /* free iterator--did not scan to null */

/* setup value change callback options */
time_s.type      = vpiSuppressTime;
cb_data_s.reason = cbValueChange;
cb_data_s.cb_rtn = PLIbook_ScientificALU_interface;
cb_data_s.time   = &time_s;
cb_data_s.value  = &value_s;

/* add value change callbacks to all signals which are inputs to  */
/* pass pointer to storage for handles as user_data value */
cb_data_s.user_data = (PLI_BYTE8 *)ALUdata;
value_s.format = vpiRealVal;
cb_data_s.obj = ALUdata->a_h;
cb_h = vpi_register_cb(&cb_data_s);
vpi_free_object(cb_h); /* don't need callback handle */

cb_data_s.obj = ALUdata->b_h;
cb_h = vpi_register_cb(&cb_data_s);
vpi_free_object(cb_h); /* don't need callback handle */

value_s.format = vpiIntVal;
cb_data_s.obj = ALUdata->opcode_h;
cb_h = vpi_register_cb(&cb_data_s);
vpi_free_object(cb_h); /* don't need callback handle */

cb_data_s.obj = ALUdata->enable_h;
cb_h = vpi_register_cb(&cb_data_s);
vpi_free_object(cb_h); /* don't need callback handle */

return(0);
}
```

7.10 Representing propagation delays in a C model

Propagation delays from an input change to an output change in a C model can be represented in two ways:

- Using delays in the PLI interface.
- Using delays in the Verilog shell module.

Delays in the PLI interface are represented by specifying a delay value with the
vpi_put_value() routine, which writes values onto the system task arguments.
Either inertial or transport event propagation can be used, depending on the requirements of the C model. However, using vpi_put_value() does not offer a great deal
of flexibility on creating delays which are different for each instance of a model, representing minimum, typical and maximum delays, different delays for rise and fall
transitions, or annotating delays using delay calculators or SDF files.

C model propagation delays can also be represented using the pin-to-pin path delays
in the Verilog shell module. This method provides the greatest amount of flexibility
and accuracy in modeling propagation delays. All path delays constructs can be used,
as well and Verilog timing constraints.

Example 7-8 shows adding pin-to-pin path delays to the scientific ALU shell module.

 Some Verilog simulators restrict the use of pin-to-pin path delays and SDF delay
back annotation to Verilog models which are represented with Verilog primitives and
net data types. To use path delays on a C model with these simulators, buffers must
be added to all input and output ports, with net data types connected to the inputs and
outputs of these buffers. Example 7-8 illustrates using buffers on all ports.

CD The source code for this example is on the CD accompanying this book.

- Application source file: Chapter.07/sci_alu_with_delays_vpi.c
- Verilog shell module: Chapter.07/sci_alu_with_delays_shell.v
- Verilog test bench: Chapter.07/sci_alu_with_delays_test.v
- Expected results log: Chapter.07/sci_alu_with_delays_test.log

---

Example 7-8: scientific ALU Verilog shell module with pin-to-pin path delays

```
`timescale 1ns / 100ps
module scientific_alu(result_out, exception, error,
                      a_in, b_in, opcode_in);
  output [63:0] result_out;
  output        exception, error;
  input  [63:0] a_in, b_in;
  input   [3:0] opcode_in;

  wire   [63:0] result_out, result_vector;
  wire   [63:0] a_in, a_vector;
  wire   [63:0] b_in, b_vector;
  wire    [3:0] opcode_in, opcode_vector;
  wire          exception, error;

  reg           exception_reg, error_reg;
  real          a, b, result; // real variables used in this module
```

```
// convert real numbers to/from 64-bit vector port connections
assign result_vector = $realtobits(result);
always @(a_vector)  a = $bitstoreal(a_vector);
always @(b_vector)  b = $bitstoreal(b_vector);

//call the PLI application which interfaces to the C model
initial
  $scientific_alu(a, b, opcode_vector,
                  result, exception_reg, error_reg);

specify
  (a_in, b_in *> result_out, exception, error) = (5.6, 4.7);
  (opcode_in  *> result_out, exception, error) = (3.4, 3.8);
endspecify

// add buffers to all ports, with nets connected to each buffer
// (this example uses the array of instance syntax in the
// from the IEEE 1364-1995 Verilog standard
buf result_buf[63:0] (result_out, result_vector);
buf excep_buf        (exception,  exception_reg);
buf error_buf        (error,      error_reg);
buf a_buf[63:0]      (a_vector, a_in);
buf b_buf[63:0]      (b_vector, b_in);
buf opcode_buf[3:0]  (opcode_vector, opcode_in);

endmodule
```

7.11 Summary

This chapter has presented several ways in which the VPI library can be used to interface a C language model with Verilog simulations. The VPI *simulation callback routine* provides a means to pass input changes to a C mode. The VPI routines to read and modify logic values allow information to be exchanged with the model. By creating a shell module which contains a system task that invokes the C model interface, the C model can be used in a Verilog design just as any other Verilog module.

This is the end of the discussion of the VPI portion of the Verilog PLI. The VPI library is the newest and most powerful portion of the PLI standard. Part Two of this book presents the TF and ACC portions of the PLI standard.

Part Two:

The TF/ACC Portion of the Verilog PLI Standard

CHAPTER 8   *Creating PLI Applications Using TF and ACC Routines*

*T*his chapter uses two short examples to introduce how the Verilog PLI works. The first example will allow a Verilog simulator to execute the ubiquitous *Hello World* C program. The second example uses the PLI to access specific activity within a Verilog simulation, by listing the name and current logic value of a Verilog net. The purpose of these examples is to introduce how to write a PLI application using the TF and ACC routine libraries in the Verilog PLI standard. Subsequent chapters in this part of the book build on the concepts presented in this chapter and show how to write more complex PLI applications.

8.1 The capabilities of the Verilog PLI

The Verilog **Programming Language Interface**, commonly referred to as the Verilog **PLI**, is a user-programmable procedural interface for Verilog digital logic simulators. The PLI gives Verilog users a way to extend the capabilities of a Verilog simulator by providing a means for the simulator to invoke other software programs, and then exchange information and synchronize activity with those programs.

The creators of the Verilog Hardware Description Language (HDL), which was first introduced in 1984, intended that the Verilog language should be focused on modeling digital logic. They felt that design verification tasks not directly related to modeling the design itself, such as reading and writing disk files, were already capably handled in other languages and should not be replicated in the Verilog HDL. Instead, the creators of Verilog added a procedural interface—the PLI—that provides a way for end-users of Verilog to write design verification tasks in the C programming language, and then have Verilog logic simulators execute those programs.

The capabilities of the PLI extend far beyond simple file I/O operations. The Verilog PLI allows Verilog designers to interface virtually any program to Verilog simulators. Some common engineering tasks that are required in the design of hardware, and for which the PLI is aptly fitted, include:

- C language models

    Abstract hardware models are sometimes represented in the C language instead of the Verilog HDL, perhaps to protect intellectual property or to optimize the performance of simulation. The PLI provides a number of ways to pass data from Verilog simulation to a C language model and from the C language model back into Verilog simulation.

- Access to programming language libraries

    The PLI allows Verilog models and test programs to access the libraries of the C programming language, such as the C math library. Through the PLI, Verilog can pass arguments to a C math function, and pass the return of the math function back to the Verilog simulator.

- Reading test vector files

    Test vectors are a common method of representing stimulus for simulation. A file reader program in C is easy to write, and, using the PLI, it is easy to pass values read by the C program to a Verilog simulation. The IEEE 1364-2001 Verilog standard adds significant file I/O enhancements to the Verilog HDL. Many file operations previously performed through the PLI can now be performed directly in the Verilog HDL. However, the PLI still provides greater and more versatile control of file formatting and file operations.

- Delay calculation

    Through the PLI, a Verilog simulator can call an ASIC or FPGA vendor's delay calculator program. The delay calculator can estimate the effects that fanout, voltage, temperature and other variances will have on the delays of ASIC cells or gates. Through the PLI, the delay calculator program can modify the Verilog simulation data structure so that simulation is using the more accurate delays. This usage of the PLI for dynamic delay calculation was common from the late 1980's to the mid 1990's, but has largely been replaced by static, pre-simulation delay estimation and SDF files.

- Custom output displays

    In order to verify design functionality, a logic simulator must generate outputs. The Verilog HDL provides formatted text output, and most simulators provide waveform displays and other graphical output displays. The PLI can be used to extend these output format capabilities. A video controller, for example, might generate video output for a CRT or other type of display panel. Using the PLI, an engineer can take the outputs from simulation of the video controller, and, while simulation is running, pass the data to a program which can display the data in the same form as the real display output. The PLI allows custom output programs to dynamically read simulation values while simulation is running.

- Co-simulation

    Complex designs often require several types of logic simulation, such as a mix of analog and digital simulations, or a mix of Verilog and VHDL simulations. Each simulator type could be run independently on its own regions of the design, but it is far more effective to simulate all regions of the design at the same time, using multiple simulators together. The PLI can be used as a communication channel for Verilog simulators to transfer data to and from other types of simulators.

- Design debug utilities

    The Verilog logic simulators on the market vary a great deal in what they provide for debugging design problems. There may be some type of information about a design that is needed, which cannot be accessed by the simulator's debugger. The Verilog PLI can access information deep inside a simulation data structure. It is often a very trivial task to write a short C program that uses the PLI to extract some specific information needed for debugging a design.

- Simulation analysis

    Most Verilog logic simulators are intended to simply apply input stimulus to a model and show the resulting model outputs. The PLI can be used to greatly enhance simulation by analyzing what had to happen inside a design to generate the output values. The PLI can be used to generate toggle check reports (how many times each node in a design changed value), code coverage reports (what Verilog HDL statements were exercised by the input tests), power usage (how much energy the chip consumed), etc.

The preceding examples are just a few of the tasks that can be performed using the Verilog PLI. The PLI provides full access to anything that is happening in a Verilog simulation, and allows external programs to modify the simulation. This open access enables truly unlimited possibilities. If an engineer can conceive of an application that requires interaction with Verilog simulations, and can write a program to perform the task, the PLI can be used to interface that application to a Verilog simulator.

8.2 General steps to create a PLI application

A PLI application is a user-defined C language application which can be executed by a Verilog simulator. The PLI application can interact with the simulation by both reading and modifying the simulation logic and delay values.

The general steps to create a PLI application are:

1. Define a *system task* or *system function* name for the application.

2. Write a C language *calltf routine* which will be executed by the simulator whenever simulation encounters the system task name or the system function name.

   Optionally, additional C language routines can be written which will be executed by the simulator for special conditions, such as when the simulator compiler or elaborator encounters the system task/function name.

3. *Register* the system task or system function name and the associated C language routines with the Verilog simulator. This registration tells the simulator about the new system task or system function name, and the name of the *calltf routine* associated with that task or system function (along with any other routines).

4. *Compile* the C source files which contain the PLI application routines, and link the object files into the Verilog simulator.

8.3 User-defined system tasks and system functions

In the Verilog language, a *system task* or a *system function* is a command which is executed by a Verilog simulator. The name of a system task or a system function begins with a *dollar sign* (*$*).

A *system task* is used like a programming statement that is executed from a Verilog *initial procedure* or *always procedure*. For example, to print a message at every positive edge of a clock, the *$display* system task can be used.

```
always @(posedge clock)
   $display("chip_out = %h", chip_out);
```

A *system function* is used like a programming function which returns a value. A function can be called anywhere that a logic value can be used. For example, to assign a random number to a vector at every positive edge of a clock, the *$random* system function can be used.

```
always @(posedge clock)
   vector <= $random();
```

The IEEE 1364 Verilog standard allows system tasks and system functions to be defined in three ways:

- A standard set of built-in system tasks and system functions.

  These are defined as part of the IEEE standard, such as *$display*, *$random*, and *$finish*. All IEEE compliant Verilog simulators will have these standard system tasks and system functions.

- Simulator specific system tasks and system functions.

  These are proprietary commands which are defined as part of a simulator, and may not exist in other simulators. Examples are *$save* to create a simulation check point file or *$db_settrace* to enable debug tracing.

- User-defined system tasks and system functions.

  These are created through the Programming Language Interface. A PLI application developer specifies the name and functionality of the system task/function.

User-defined system task or system function names begin with a dollar sign ($), just as the built-in system tasks and system functions. The user-defined system task/function name is then associated with a user-defined C application. When a Verilog simulator encounters the system task/function name, it will execute the C application associated with the name. This simple association allows users to extend the capability of a Verilog simulator in any manner desired. Chapter 9 discusses the concept of system tasks and system functions in more detail.

8.4 The $hello PLI application example

8.4.1 Step One: defining a $hello system task

The well known "Hello world" C program is a quick way to show how PLI applications are created. Though the C program itself is simple, it illustrates one of the most powerful features of the Verilog PLI—the ability for a Verilog designer to extend the Verilog language by having a Verilog simulation dynamically execute a user-defined program while simulation is running.

8.4.2 Step One: defining a $hello system task

The first step in creating a PLI application is create a new system task or system function name. A *$hello* user-defined system task will be created for this PLI application. An example of using *$hello* is:

```
module test;

  initial
    $hello();

endmodule
```

8.4.3 Step Two: writing a calltf routine for $hello

The second step in developing a PLI application is to write a C language routine which will be called when a Verilog simulator executes the *$hello* system task. In this example, the simulator will call a C language routine which prints the message:

```
Hello World!
```

The C language routine which will be called by the simulator is referred to as a *calltf routine*. This routine is a user-defined C function and can be any name. Example 8-1 lists the C source code for the *calltf routine* for *$hello*.

CD The source code for this example is on the CD accompanying this book.

- Application source file: Chapter.08/hello_tf.c
- Verilog test bench: Chapter.08/hello_test.v
- Expected results log: Chapter.08/hello_test.log

Example 8-1: *$hello* — a TF/ACC *calltf routine* for the PLI application

```
#include "veriuser.h"  /* IEEE 1364 PLI TF routine library */
#include "acc_user.h"  /* IEEE 1364 PLI ACC routine library */

int PLIbook_hello_calltf(int user_data, int reason)
{
  io_printf("\nHello World!\n\n");
  return(0);
}
```

In the above example:

- The header files *veriuser.h* and *acc_user.h* contain the TF and ACC libraries of functions provided by the PLI standard. These libraries are a collection of C functions which can be used by a PLI application to interact with a Verilog simulator. The *veriuser.h* and *acc_user.h* header files are part of the IEEE 1364 Verilog standard. This example does not use any routines from the ACC library, and so the acc_user.h file does not need to be included in this example.

- The **io_printf()** function used in the preceding example is very similar to the C `printf()` function. The difference is that `io_printf()` will write its message to the simulator's output channel and the simulator's output log file, whereas the C `printf()` writes its message to the operating system's standard out channel.

- The user_data and reason inputs are values which the simulator passes into all TF/ACC PLI applications. Most applications do not use these inputs, and simply ignore them. Chapter 9 explains the purpose of these inputs.

8.4.4 Step Three: Registering the $hello system task

The third step in creating a new PLI application is to tell the Verilog simulator about the new system task or system function name and the C routines which are associated with the application. This process is referred to a *registering* the PLI application.

The PLI IEEE 1364 standard provides an *interface mechanism* for this process. The interface mechanism defines:

- The type of application, which is a system task or system function.

- The system task or system function name.

- The name of the *calltf routine* and other C routines associated with the system task/function.

- Other information about the system task/function required by a simulator.

The TF and ACC interface mechanism for PLI applications is *not* part of the IEEE standard. The standard only defines what the interface mechanism needs to do. The TF and ACC interface mechanism is implemented in different ways in different Verilog simulators. Chapter 9 presents more details on how most Verilog simulators implement the interface mechanism.

8.4.5 Step Four: Compiling and linking the $hello system task

The final step in creating a new PLI application is to compile the C source code containing the application and linking the compiled files to the Verilog simulator. Once the application has been linked to the simulator, the simulator can invoke the *calltf routine* when simulation executes the *$hello* system task name. Compiling and linking is not part of the IEEE standard. This process is specific to both the simulator as well as the operating system on which simulations are run. Appendix A presents the steps for compiling and linking PLI applications for several major Verilog simulators.

8.4.6 Running simulations with the $hello system task

Once a PLI application has been registered with a Verilog simulator, the new system task or system function can be used in a Verilog model, just as with built-in system tasks and system functions. When the Verilog simulator encounters the user-defined system task, it will call the C routine which has been associated with that application.

The following Verilog model can be used to test the *$hello* PLI application:

```
module test;
   initial
     begin
       $hello();
       #10 $stop;
       $finish;
     end
endmodule
```

Figure 8-1 shows the results of simulating this Verilog model.

Figure 8-1: Simulation results using the *$hello* PLI application

```
Hello World!
```

8.5 The $show_value PLI application example

The *$hello* PLI application, though trivial from a C programming aspect, effectively illustrates the powerful capability provided by the Programming Language Interface—the ability to have a Verilog simulator call a user-defined C routine during simulation. The PLI provides far more capabilities, however. The C routine which is called by the simulator can use the TF and ACC libraries to read information from the simulation data structure, and to dynamically modify logic values and delay values in the data structure.

The system task *$show_value* illustrates using the PLI to allow a C routine to read current logic values within a Verilog simulation. The *$show_value* system task requires one argument, which is the name of a net or reg in the Verilog design. The *calltf routine* will then use the library of TF and ACC routines to read the current logic value of that net or reg, and print its name and logic value to the simulator's output screen. An example of using *$show_value* is:

```
module test;
  reg  a, b, ci;
  wire sum, co;
  ...
  initial
    begin
      ...
      $show_value(sum);
    end
endmodule
```

Two user-defined C routines will be associated with *$show_value*:

- A C routine to verify that *$show_value* has the correct type of argument.

- A C routine to print the name and logic value of the signal.

These programs are presented in the following sections.

8.5.1 Writing a checktf routine for $show_value

The PLI allows users to provide a C routine to verify that a system task or system function is being used correctly and has the correct types of arguments. This C routine is referred to as a ***checktf routine***.

The TF and ACC routines are two libraries of C functions which can access the arguments of a system task or system function and determine the Verilog data types of the objects listed in the arguments. This example uses some of these C functions. Only a brief explanation of each function is provided in this chapter. Subsequent chapters will present more details about each function.

The C source code for the *checktf routine* associated with *$show_value* is listed below. The C functions and constants from the PLI library are shown in bold text.

CD The source code for this example is on the CD accompanying this book.

- Application source file: `Chapter.08/show_value_acc.c`
- Verilog test bench: `Chapter.08/show_value_test.v`
- Expected results log: `Chapter.08/show_value_test.log`

Example 8-2: *$show_value — a TF/ACC checktf routine* for a PLI application

```
#include "veriuser.h"       /* IEEE 1364 PLI TF routine library  */
#include "acc_user.h"       /* IEEE 1364 PLI ACC routine library */

int PLIbook_ShowVal_checktf(int user_data, int reason)
{
  PLI_INT32 arg_type;
  handle    arg_handle;

  if (tf_nump() != 1)
    tf_error("$show_value must have 1 argument.");
  else if (tf_typep(1) == TF_NULLPARAM)
    tf_error("$show_value arg cannot be null.");
  else {
    arg_handle = acc_handle_tfarg(1);
    arg_type = acc_fetch_type(arg_handle);
    if (!(arg_type == accNet || arg_type == accReg))
      tf_error("$show_value arg must be a net or reg.");
  }
  return(0);
}
```

Observe the following key points in the above example:

- The function **tf_nump()** returns the number of arguments in the system task. This example uses tf_nump() to verify *$show_value* has one argument.

- The function **tf_typep()** returns a constant representing the general data type of a task/function argument. tf_typep() is used in this example to verify that the argument to *$show_value* is not null.

- The function **acc_handle_tfarg()** returns a pointer to a system task argument. The pointer is stored in a special C data type defined by the PLI, called a ***handle***.

- The function **acc_fetch_type()** returns a constant that more closely identifies the type of the argument. In this example, the test is checking that the argument is a Verilog net or reg data type.

- The function **tf_error()** prints an error message and causes simulation to abort.

- The data type **PLI_INT32** is a special data type defined in *veriuser.h* and *acc_user.h*, as a 32-bit C integer. The Verilog PLI uses several special data types instead of the C language **int, char** and other C data types to ensure portability across all operating systems and computer platforms (the ANSI C standard does not guarantee the bit width of int and char data types).

The fixed width PLI data types were added as part of the IEEE 1364-2001 standard. Prior to that, the PLI libraries used the regular C data types, such as byte, int and long, with the dangerous, non-portable assumption that an int would be 32 bits on any platform and operating system. Existing PLI applications that use the C data types instead of the new PLI data types will still compile and work correctly with the default definitions in the standard veriuser.h and acc_user.h header files. However, these older applications may need to be updated to the PLI data types in order to port correctly to new 64-bit operating systems. Section 9.4.4 on page 295 lists the fixed-width types defined in the IEEE 1364-2001 standard.

8.5.2 Writing the calltf routine for $show_value

The *calltf routine* is the C routine which will be executed when simulation encounters the *$show_value* system task during simulation.

The following example lists the C source code for the *calltf routine* associated with *$show_value*. Later chapters define the routines used in this example in full detail. The C functions, constants and data types from the PLI library are shown in bold text.

Example 8-3: *$show_value* — a TF/ACC *calltf routine* for a PLI application

```
#include "veriuser.h"
#include "acc_user.h"
int ShowValCall(int user_data, int reason)
{
  handle arg_handle;
  arg_handle = acc_handle_tfarg(1);
  io_printf("Signal %s has the value %s\n",
            acc_fetch_fullname(arg_handle),
            acc_fetch_value(arg_handle, "%b", null));
  return(0);
}
```

In the above example:

- The function **io_printf()** prints a formatted message. It is similar to the C printf() function, but prints the message to the simulation output channel and the simulation output log file.

- The function **acc_fetch_fullname()** returns the full hierarchical name of a Verilog object.

- The function **acc_fetch_value()** returns the logic value of a Verilog net, reg or variable. The value can be obtained in a variety of formats. In this example, the value is obtained as a string, with a binary representation of the value.

8.5.3 Registering the $show_value PLI application

The *$show_value* PLI application needs to be registered with the Verilog simulator, which informs the Verilog simulator about the PLI application and the names of the *calltf routine* and *checktf routine* C functions. The IEEE standard defines that all Verilog simulators should provide an interface mechanism to perform this registration, but the standard does not define how the interface mechanism should be implemented. Chapter 9 shows how most Verilog simulators implement this interface mechanism.

8.5.4 A test case for $show_value

The following Verilog HDL source code is a small test case for the *$show_value* application. The example lists the values of a simple 1-bit adder after a few input test values have been applied. Observe how *$show_value* is used as a Verilog programming statement within the test bench.

Example 8-4: *$show_value* — a Verilog HDL test case using a PLI application

```
`timescale 1ns / 1ns
module test;
   reg   a, b, ci, clk;
   wire sum, co;
   addbit i1 (a, b, ci, sum, co);
   initial
     begin
       clk = 0;
       a = 0;
       b = 0;
       ci = 0;
       #10 a = 1;
       #10 b = 1;

       $show_value(sum);
       $show_value(co);
       $show_value(i1.n3);

       #10 $stop;
       $finish;
     end
endmodule

/*** A gate level 1 bit adder model ***/
`timescale 1ns / 1ns
module addbit (a, b, ci, sum, co);
   input   a, b, ci;
   output sum, co;
   wire   a, b, ci, sum, co,
          n1, n2, n3;
```

```
xor     (n1, a, b);
xor #2 (sum, n1, ci);
and     (n2, a, b);
and     (n3, n1, ci);
or  #2 (co, n2, n3);
endmodule
```

8.5.5 Output from running the $show_value test case

Following is the output from simulating the test case for the *$show_value* application.

Figure 8-2: *$show_value* — sample simulation results

```
Signal test.sum has the value 1
Signal test.co has the value 0
Signal test.i1.n3 has the value 0
```

8.6 Summary

This chapter has presented two short examples which illustrate how PLI applications are created using the TF and ACC routine libraries in the PLI standard. The *$hello* PLI application showed how a Verilog simulation can execute another C program. The *$show_value* application illustrated how arguments can be passed from a Verilog simulation to a PLI application, and how the application can access information about the arguments.

The principles illustrated by the *$show_value* example can be readily expanded to provide much more powerful and useful capabilities. In Chapter 14, the functionality of *$show_value* will be extended to create an application called *$list_nets*, which automatically finds all nets in a module and prints the values of each net. Another example in the same chapter extends *$show_value* to create *$list_signals*, which automatically finds all nets, regs and variables in a module and prints the values of each one.

The next chapter presents much greater detail on how PLI application interact with Verilog simulations. Subsequent chapters in Part Two of this book will discuss the complete TF and ACC libraries, and include many examples of how the TF and ACC routines to create PLI applications which to interact with Verilog simulators.

CHAPTER 9 *Interfacing TF/ACC Applications to Verilog Simulators*

*D*on't skip this chapter! This chapter defines the terminology used by the Verilog PLI and how PLI applications which use the TF and ACC libraries are interfaced to Verilog simulators. All remaining chapters in Part Two of this book assume the principles covered in this chapter are understood. The general concepts presented are:

- PLI terms as used in this book
- Generations of the PLI standard
- System tasks and system functions
- How TF and ACC routines work
- A Complete PLI application example
- Interfacing PLI applications to Verilog simulators

9.1 General PLI terms as used in this book

The Verilog PLI has been in use since 1985, and, over the years, many reference manuals, technical papers and training courses have been written about the PLI. Unfortunately, these documents have not used consistent terms for describing the PLI. This inconsistency can make it difficult to cross reference information about the PLI in different documents.

This book strives to clearly define the PLI terms and consistently use this terminology. The general PLI terms used in this book are.

C program (often abbreviated to *program*)

A complete software program written in the C or C++ programming language. A C program must include a C *main* function.

C function

A function written in the C or C++ programming language. A C function does not include a C *main* function.

Verilog function

A Verilog HDL function, written in the Verilog Hardware Description Language. Verilog functions can only be called from Verilog source code.

User-defined system task or system function (abbreviated to *system task/function*)

A user-defined system task or system function is a construct that is used in Verilog HDL source code. The name of a user-defined system task or system function must begin with a dollar sign (*$*), and it is used in the Verilog language in the same way as a Verilog HDL standard system task or system function. When simulation encounters the user-defined system task or system function, the simulator will execute a *PLI routine* associated with the system task/function.

PLI routine

A C function which is part of a *PLI application*. The PLI routine is executed by the simulator when simulation encounters a user-defined system task or system function. The TF and ACC portion of the PLI standard define several PLI routine types: *calltf routines*, *checktf routines*, *sizetf routines*, *misctf routines* and *consumer routines*. These terms are defined in more detail later in this chapter.

PLI application (sometimes abbreviated to *application*)

A *user-defined system task or system function* with an associated set of one or more *PLI routines*.

PLI library

A library of C functions which are defined in the Verilog PLI standard. PLI library functions are called from *PLI routines*, and enable the *PLI routines* to interact with a Verilog simulation. The IEEE 1364 standard provides three PLI libraries, referred to as the *VPI library*, the *TF library* and the *ACC library*.

VPI routines, *TF routines* and *ACC routines*

C functions contained in the *PLI library*. The term *routine* is used for these functions, to avoid confusion with *Verilog functions* and *C functions*.

9.2 System tasks and system functions

The Verilog Hardware Description Language provides constructs called *"system tasks"* and *"system functions"*. A system task/function is not a hardware modeling construct. It is a command that is executed by a Verilog simulator.

The names of system tasks and system functions always begin with a dollar sign ($). A *system task* is analogous to a subroutine. When the task is called, the simulator branches to a program that executes the functionality of the task. When the task has completed, the simulation returns to the next statement following the task call. A *system function* returns a value. When the function is called, a simulator executes the program associated with the function, which returns a value into the simulation. The return value becomes part of the statement that called the function. The return value of a function may be defined to be an integer, a vector (with a specific bit size) or a floating point number.

9.2.1 Built-in system tasks and system functions

The IEEE 1364 Verilog standard defines a number of system tasks and system functions that are built into all Verilog simulators. Examples of the standard system tasks are *$display*, *$stop*, and *$finish*. Some of the standard system functions are *$time* and *$random*. These system tasks and system functions are part of the Verilog language, and the syntax and usage of these routines are covered in Verilog language books.

The IEEE 1364 standard also allows simulation vendors to add proprietary system tasks and system functions which are specific to a simulator product. For example, the Cadence Verilog-XL simulator adds the command *$db_settrace*, which is a system task that turns on Verilog statement execution tracing to aid in design debugging.

9.2.2 User-defined system tasks and system functions

The Verilog PLI provides a means for Verilog users to add additional system tasks and system functions. When simulation encounters a user-defined system task or system function, the simulator branches to a user-supplied C function, executes that function, and then returns back to executing the Verilog source at the point where the system task/function was called.

User-defined system task and system function names must begin with a dollar sign ($), and may only use the characters that are legal in Verilog names, which are:

a—z A—Z 1—9 _ $

Following are examples of legal user-defined system task and system function names:

```
$rand64        $cell_count        $GetVector
```

9.2.3 Overriding built-in system tasks and system functions

A user-defined system task or system function can be given the same name as a system task/function that is built into the simulator. This allows a PLI application developer to override the operation of a built-in system task/function with new functionality. For example, the built-in system function *$random* will return a 32-bit signed random number, with the random sequence defined by the simulator. If the verification of a design requires a special random number sequence, a random number generator could be written in the C language and then associated with a user-defined system function with the same *$random* name. The simulator would then call the user PLI application instead of the built-in random number generator.

9.2.4 System tasks are used as procedural programming statements

System tasks are procedural programming statements in the Verilog HDL. This means a system task may only be called from a Verilog **initial** procedure, an **always** procedure, a Verilog HDL **task** or a Verilog HDL **function**. A system task may not be called outside of a procedure, such as from a continuous assignment statement.

The following example calls a user-defined system task from an **always** procedure:

```
always @(posedge clock)
  $read_test_vector("vectors.pat", input_vector);
```

9.2.5 System functions are used as expressions

System functions are expressions in the Verilog HDL. This means a system function may be called anywhere a logic value may be used. The return value of the system function is considered the result of the expression. System functions may be called from a Verilog **initial** procedure, an **always** procedure, a Verilog HDL **task**, a Verilog HDL **function**, an **assign** continuous assignment statement, or as an operand in another expression.

The following examples call a user-defined system function several different ways:

```
always @(posedge clock)
  if (chip_out !== $pow(base,exponent))
    ...

initial
  $monitor("output = %f", $pow(base,exponent));

assign temp = i + $pow(base,exponent);
```

9.2.6 System task/function arguments

A system task or system function may have any number of arguments, including none. The user-supplied PLI application can read the values of the arguments of a system task/function. If the argument is a Verilog reg or variable (integer, real and time), then the PLI application can also write values into the task/function argument.

System task/function arguments are typically numbered from 1 to N, with the leftmost argument being argument number 1. For example:

```
module top (...);
  ...
  reg [15:0] in1;
  ...
  my_chip u1 (in1, out1);          argument 1 is a module
  ...                              instance name
  initial
    $cell_count(u1);
  ...                              argument 1 is a string
  always @(posedge clock)
    $read_vector_file("vectors.pat", in1);
  ...                                        argument 2 is a
endmodule                                    reg variable
```

9.3 Instantiated Verilog designs

In a Verilog HDL module, a reference to another module is referred to as a *module instance*. In the following Verilog source code, module **top** contains two *instances* of module **bottom.**

```
module top;
   reg   [7:0] in1;
   wire  [7:0] out1;
   bottom b1 (in1[7:4], out1[7:4]);
   bottom b2 (in2[3:0], out2[3:0]);
endmodule

module bottom (in, out);
   ...
endmodule
```

The Verilog PLI requires an instantiated Verilog data structure. This means that a Verilog simulator has created a data structure which contains the complete hierarchy tree that has been represented by one module containing instances of other modules. There is no limit to the number of levels of hierarchy in the Verilog language.

The PLI does not work directly with Verilog HDL source code—The PLI is simply a procedural interface provided by a simulator so that PLI applications can access the data structure of a simulation.

Multiple instances of a system task or system function

The system task or system function which invokes a PLI application can be instantiated any number of times in the Verilog source code. Consider the following Verilog HDL source code fragment:

```
module top;
   ...
   middle m1 (...);
   middle m2 (...);

   initial
      $my_app_1(in1, out1);

   always @(posedge clock)
      $my_app_1(in2, out2);

endmodule

module middle (...);
   ...
   bottom b1 (...);
   bottom b2 (...);

endmodule
```

```
module bottom (...);
  ...
  initial
    $my_app_2(in3, out3);

  always @(posedge clock)
    $my_app_2(in4, out4);

endmodule
```

In the preceding example, four different conditions are shown that can exist in a Verilog simulation.

- A single instance of a system task that is invoked one time

  In module **top**, the first *instance* of *$my_app_1* is in a Verilog initial procedure, which will be called once at the beginning of a simulation.

- A single instance of a system task that is invoked many times

  In module **top**, the second *instance* of *$my_app_1* is in a Verilog always procedure, which will be called every positive edge of the clock signal.

- Multiple instances of a system task that are each invoked one time

  In module **bottom**, the first *instance* of *$my_app_2* is in a Verilog initial procedure, which will be called once at the beginning of a simulation.

- Multiple instances of a system task that are each invoked many times

  In module **bottom**, the second *instance* of *$my_app_2* is in a Verilog always procedure, which will be called every positive edge of the clock signal.

Note that in the preceding Verilog source code example, module bottom is instantiated two times in module middle. The two instances of bottom, with two instances of *$my_app_2* inside bottom, creates four unique instances of *$my_app_2*. Module middle is instantiated twice in module top, resulting in eight unique instances of *$my_app_2*.

Thus, in the instantiated simulation data structure for the preceding Verilog source code, there are *two **instances*** of the *$my_app_1* system task, and ***eight instances*** of the *$my_app_2* system task. Each of these system task instances will have unique logic values for the inputs and outputs. One of the requirements of a PLI application is to allow for multiple unique instances of the application. This is an important consideration when an application allocates static variables or allocates system memory. Chapter 10 discusses these issues in more depth, in section 10.7 on page 325.

9.4 How PLI applications work

The PLI standard allows a Verilog user to create a user-defined system task or system function name and to associate one or more user-defined C routines with the system task/function name. When a Verilog simulator encounters the user-defined system task/function, it will execute the user-defined C routine associated with the name.

```
always @(posedge clock)
     result <= $pow(x,y);  ─────────> User-defined PLI routine
```

calculate x^y

9.4.1 The types of PLI routines

The TF and ACC portion of PLI standard defines several types of PLI routines which can be associated with a system task or system function. The type of the routine determines *when* the simulator will execute the routine. Some types of routines are runtime routines, which are invoked during simulation, and some types are elaboration or linking time routines, which are invoked prior to simulation. The types of PLI routines are:

- *calltf routines*
- *checktf routines*
- *sizetf routines*
- *misctf routines*

The purpose of these routines is defined in the following sections of this chapter.

9.4.2 Associating routine types with system task/functions

A system task/function can have multiple PLI routines associated with it, as long as each routine is a different type. A *$read_vector_file* application, for example, might have three routines associated with it—one which is called prior to simulation to perform syntax checking, one which is called when simulation first starts running to open the test vector file, and one which is called during simulation at every positive edge of clock to read vectors from the file.

Up to 4 different C routines may be associated with a system task/function in the TF and ACC portion of the PLI standard: the *calltf routine*, *checktf routine*, *sizetf routine* and *misctf routine*. The ACC library also provides for another type of PLI routine, called a *Value Change Link consumer routine*. This routine is not directly associated with a system task/function name. Instead, the *consumer routine* is called for logic value changes on specific objects with a simulation data structure.

The following sections of this chapter define in detail the purpose and usage of the *calltf routine*, *checktf routine*, *sizetf routine* and *misctf routine*. The *consumer routine* is discussed in Chapter 17.

 PLI routines are sometimes referred to as PLI programs or C programs. In proper C terminology, PLI routines are actually C *functions*. A *program* in the C language terminology requires a *main* function. PLI applications do not contain a main function. Instead, the routines which make up the PLI application are linked into a Verilog simulator, and the simulator contains the C main function. This allows the simulator to call the appropriate PLI routine whenever its associated system task/ function name is encountered.

9.4.3 C versus C++

Most Verilog simulators expect that PLI applications are written in ANSI C.

The PLI libraries are compliant with the ANSI C standard, but also include the appropriate prototypes and other information required for C++. This allows PLI applications to be developed in either C or C++. However, many Verilog simulators were written in the C language, and may or may not support linking C++ applications to the simulator. PLI application developers who wish to work with C++ should first check the limitations of the simulator product to which the applications will be linked. For maximum portability, PLI applications should be written using ANSI C.

9.4.4 Special PLI data types for portability

The TF and ACC libraries define several fixed-width data types, which are used by the routines in the libraries. This is because the ANSI C standard only guarantees the minimum width of C data types, which allows the width to be different on different computer architectures and operating systems. For example, a C **int** could be 16 bits wide, 32 bits wide or 64 bits wide, depending on what computer the application was compiled for. To provide portability across all operating systems, the TF and ACC libraries define special data types which are guaranteed to be a certain size. These data type definitions are made in the veriuser.h and acc_user.h header files, as follows:

Table 9-1: PLI fixed-width data types

PLI_INT32	32-bit signed integer
PLI_UINT32	32-bit unsigned integer
PLI_INT16	16-bit signed integer
PLI_UINT16	16-bit unsigned integer
PLI_BYTE8	8-bit signed integer
PLI_UBYTE8	8-bit unsigned integer

TIP

A PLI application should use the special PLI data types so that the application will work correctly on both 32-bit and 64-bit operating systems.

The veriuser.h and acc_user.h header file can be modified by simulator vendors to define the fixed width data types for a specific computer platform, operating system or C compiler. The default definitions, as defined in the IEEE 1364-2001 standard, are:

```
typedef int            PLI_INT32;
typedef unsigned int   PLI_UINT32;
typedef short          PLI_INT16;
typedef unsigned short PLI_UINT16;
typedef char           PLI_BYTE8;
typedef unsigned char  PLI_UBYTE8;
```

The fixed width PLI data types were added as part of the IEEE 1364-2001 standard. Prior to that, the PLI libraries used the regular C data types, such as byte, int and long, with the dangerous, non-portable assumption that an int would be 32 bits on any platform and operating system. Existing PLI applications that use the C data types instead of the new PLI data types will still compile and work correctly with the default definitions in the standard vpi_user.h header file. However, these older applications may need to be updated to the PLI data types in order to port correctly to new 64-bit operating systems.

9.5 calltf routines

The ***calltf routine*** is executed when simulation is running. For the *$pow* example that follows, at every positive edge of clock the *calltf routine* associated with *$pow* will be executed by the simulator. The routine is user-defined, and, for this example, will calculate a 32-bit number representing x to the power of y.

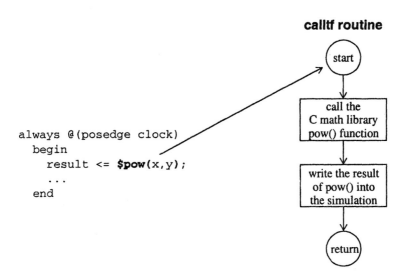

```
                                                    calltf routine

                                                        start

always @(posedge clock)                              call the
    begin                                         C math library
      result <= $pow(x,y);                         pow() function
      ...
    end                                           write the result
                                                    of pow() into
                                                   the simulation

                                                       return
```

9.6 checktf routines

The *checktf routine* is called by the simulator before simulation starts running—in other words, before simulation time 0. The routine is called at the earliest possible time after all simulation data structures required by the PLI are available. Generally this means after the design is fully instantiated, but no simulation events have occurred. The purpose of *checktf routine* is to verify that a system task/function is being used correctly (e.g.: to check that the call to the PLI application has the correct number of arguments, and that the arguments are the correct Verilog data types). For example, if the *$pow* system function were passed a module name as its second argument instead of a value, then the C math library would probably get an error when trying to use the name as an exponent value. Bad data can result in errors ranging from incorrect results to program crashes. Therefore, it is important to verify that each usage of a system task/function has valid arguments.

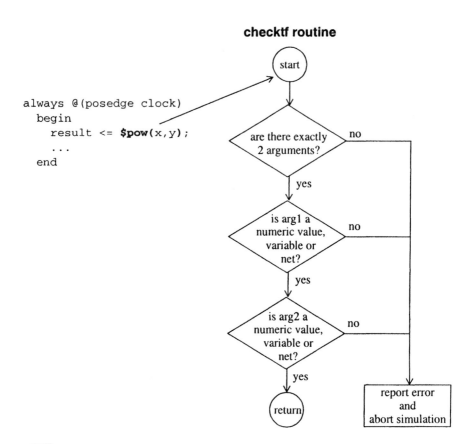

checktf routine

```
always @(posedge clock)
  begin
    result <= $pow(x,y);
    ...
  end
```

Use a *checktf routine* to improve the performance of PLI programs. Since this routine is only called one time prior to simulation time 0, the code to check the correctness of arguments is only executed once. The *calltf routine*, which may be invoked millions of times during a simulation, does not need to repeat the syntax checking, and therefore can execute more efficiently.

TIP

NOTE
The *checktf routine* will be called one time for each instance of a user-defined system task/function. If a design used the *$pow* user-defined system function in three different places, the *checktf routine* for *$pow* would be called three times. If a simulator allows system tasks and system functions to be invoked from the simulator's debug command line, then the *checktf routine* will be called prior to execution of the *calltf routine* for each interactive usage.

Limitations on checktf callbacks

The intent of the *checktf routine* is to verify the correctness of the arguments of an instance of a system task/function. The IEEE 1364 standard does not impose explicit restrictions on what TF and ACC routines will, or will not work in a *checktf routine*.

There are, however, considerations which should be observed. Since the *checktf routine* is called prior to simulation time 0, variables have not yet been assigned any values, and nets which are assigned values as soon as simulation starts may not yet reflect those values.

To ensure that a PLI application will be portable, only the following activities should be performed in a *checktf routine*:

- Accessing the arguments of an instance of a system task/function. Handles for the arguments can be obtained, and the properties of the arguments can be accessed to verify correctness. The values of literal numbers can be read, but variables and nets will not have been initialized. Note that Verilog `parameter` constants can be redefined for each instance of a Verilog module. The IEEE 1364 standard does not state whether the *checktf routine* will be called before or after parameter redefinitions have occurred. Reading parameter constant values from a *checktf routine* may not yield the same results on every simulator.

- Using TF and ACC routines such as `io_printf()`, which do not access objects in simulation. Attempting to traverse hierarchy to obtain handles beyond the system task arguments may not work correctly in all simulators. Attempting to write logic or delay values onto objects may result in PLI application errors.

 The checktf routine should only be used for syntax checking! Do not use this routine to perform run-time duties such as allocating memory or opening files. The PLI standard does ensure that activity performed by a checktf routine will remain in effect at simulation time. A *misctf routine* should be used instead of the *checktf routine* to perform run-time work at the very beginning of a simulation.

9.7 sizetf routines

The *sizetf routine* is only used with system functions which return scalar or vector values (system functions can also return real number values). A system function can return a user-specified number of bits, and the simulator compiler or elaborator may need to know how many bits to expect from the return, in order to correctly compile the statement from which the system function is called. A *sizetf routine* is called one time, before simulation time 0. The *sizetf routine* returns to the simulator how many bits wide the return value of system function will be.

 Even if a system function is used multiple times in the Verilog source code, the *sizetf routine* is only called once. The return value from the first call is used to determine the return size for all instances of the system function.

In the *$pow* example, the *calltf routine* will return a 32-bit value. Therefore, the *sizetf routine* associated with *$pow* needs to return a value of 32 to the simulator.

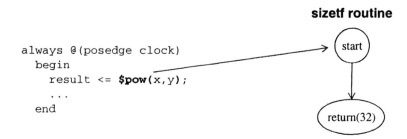

sizetf routine

Limitations on sizetf callbacks

The intent of the *sizetf routine* is to notify the simulator compiler of the return size for system functions. The simulator may invoke the *sizetf routine* very early in the elaboration/linking phase of a Verilog design, and, at this early stage, the Verilog hierarchy may not have been generated. In addition, the *sizetf routine* is only called one time for a system function name, and the return size applied to all instances of the system function. For these reasons, only standard C language statements and functions should be used in a *sizetf routine*. An error may result if any TF or ACC routines are called from a *sizetf routine*. Any memory or static variables allocated by a *sizetf routine* may not remain in effect for when simulation starts running.

9.8 misctf routines

The *misctf routine* is called for miscellaneous simulation events while the simulation is running. The IEEE 1364 standard defines the following miscellaneous events:

- End of Verilog source elaboration (just before the start of simulation time 0).

- Entering debug mode (such as when the *$stop* built-in system task is executed).

- End of simulation (such as when the *$finish* built-in system task is executed).

- Change of value on a user-defined system task argument.

- End of a simulation time step—the PLI *can* schedule additional simulation activity in the time step.

- End of a simulation time step—the PLI *cannot* schedule additional simulation activity in the time step.

- Simulation has reached a specified simulation time step.

In addition to the standard reasons for which the *misctf routine* will be called, the PLI standard allows simulation vendors to add other reasons that are proprietary to features of a simulator. For example, the Cadence Verilog-XL simulator has the ability to save the simulation state to a file and restart simulation from the save file. Since it might be important for a PLI application to know that a save or restart has occurred, Cadence adds save and restart reasons to the Verilog-XL implementation of the PLI.

Each time a *misctf routine* is called, the simulator passes to the routine a C constant which represents the reason for the call. The *misctf routine* can then check the value of the reason constant and act accordingly. The names of the reason constants are defined in the IEEE 1364 standard and are presented in Chapter 12.

A common usage of the *misctf routine* is to perform tasks at the very beginning and the very end of a simulation. A PLI application to read test vectors might need to open a test vector file at the start of simulation, and close the file at the end of simulation.

```
always @(posedge clock)
  $read_test_vector("vectors.pat", in1);
```

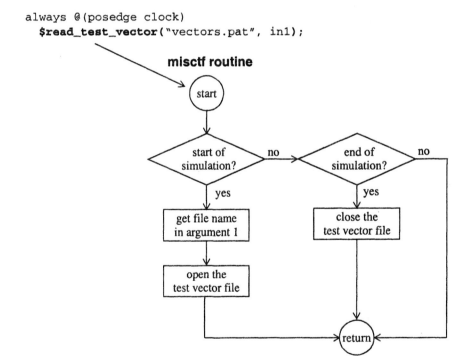

9.9 PLI routine inputs and outputs

PLI routines are C functions which are called by the Verilog simulator. When the function is called, the simulator will pass specific inputs to the function, and will receive a return value from the function.

The *sizetf routine*, *checktf routine* and *calltf routine* are defined in the PLI standard to have two C inputs and to return an integer value. The *misctf routine* is defined to have three C inputs and to return an integer value.

PLI routine inputs

The inputs that are passed to the PLI routines are:

- An integer *user_data* value

   The user_data value is defined along with the system task/function name (refer to section 9.12 on page 310 for more information on the user_data value).

- An integer *reason* value

   This value is generated by the PLI and is represented by constants which are defined in the *veriuser.h* PLI header file. For *misctf routines*, the reason input is used to specify why the routine was called.

 For the *sizetf routine*, *checktf routine* and *calltf routine*, the reason input serves no purpose, and can be ignored.

- An integer *paramvc* value (*misctf routines* only)

   This integer indicates which argument of a system task changed value. It is only used when a *misctf routine* is called with a reason of **REASON_PARAMVC**, which occurs when an argument of a system task changes value (see Chapter 12, section 12.5 on page 418 for more details on system task argument value change callbacks).

TF/ACC routine return values

Simulators prototype the *sizetf routine*, *checktf routine*, *calltf routine* and *misctf routine* as int functions. However, the only type of PLI routine that needs to return a value is the *sizetf routine*, which returns the bit-width of the system function return value. The return value from a *checktf routine*, *calltf routine* and *misctf routine* are ignored by the simulator.

It is a common practice to declare the *checktf routine*, *calltf routine* and *misctf routine* as void functions, since their return value is ignored, even though all of these routines are expected to be integer functions. However, declaring the functions as a different type from the PLI standard prototype may result in a C compiler warning message when the routine is compiled. To prevent this warning message, the pointer to the void function can be cast to a pointer to an int function. For example:

```
(int(*)())PLIbook_PowCalltf
```

Note, however, that some C compilers might not accept this cast.

9.10 A complete system function example — $pow

The following example illustrates all the parts of a complete system function, with the user-defined name of *$pow*.

- The system function will return a 32-bit value.

- The system function requires two arguments, a base value and an exponent value. Both arguments must be numeric integer values.

To implement the *$pow* functionality, four user-defined PLI routines are used:

- A *sizetf routine* to establish the return size of *$pow*.

- A *checktf routine* to verify that the *$pow* arguments are valid values.

- A *calltf routine* to calculate the base to the power of the exponent each time *$pow* is executed by the simulator.

- A *misctf routine* to print a message when simulation firsts starts running (immediately prior to simulation time 0).

The *$pow* system function does not really need the *misctf routine*, but to illustrate using this routine, this example uses a *misctf routine* to print a message at the start of simulation.

TIP

The *checktf routine*, *sizetf routine*, *calltf routine* and *misctf routine* are C functions. These functions may be located in separate files, or they can all be in the same file. Typically, smaller PLI applications place all the routines in a single file, while larger, more complex applications might break them into multiple files.

The *checktf routine*, *sizetf routine*, *calltf routine*, and *misctf routine* names can be any legal C function name. However, a typical Verilog HDL design may include several PLI applications. Using unique function name conventions can help avoid possible name conflicts when multiple PLI applications are linked together.

For this chapter, the *$pow* application has been simplified to only work with integer numbers. In practice, this application would probably accept both integer and floating point numbers for inputs, and return a floating point number for the result. Example 11-1 on page 348 of Chapter 11 presents another version of *$pow* that works with double-precision floating point numbers.

CD The source code for this example is on the CD accompanying this book.

- Application source file: `Chapter.09/pow_acc.c`
- Verilog test bench: `Chapter.09/pow_test.v`
- Expected results log: `Chapter.09/pow_test.log`

Example 9-1: *$pow* — a system function using TF/ACC routines

```
#include "veriuser.h"  /* IEEE 1364 PLI TF routine library  */
#include "acc_user.h"  /* IEEE 1364 PLI ACC routine library */

/**********************************************************************
 * Sizetf application
 **********************************************************************/
int PLIbook_pow_sizetf(int user_data, int reason)
{
  return(32);   /* $pow returns 32-bit values */
}

/**********************************************************************
 * checktf routine
 **********************************************************************/
int PLIbook_pow_checktf(int user_data, int reason)
{
  static PLI_INT32 valid_types[4] = {accReg, accIntegerVar,
                                     accConstant, 0};
  handle arg_handle;

  if (tf_nump() != 2)
    tf_error("$pow must have 2 arguments.\n");
  else if (tf_typep(1) == tf_nullparam)
    tf_error("$pow arg 1 cannot be null.\n");
  else if (tf_typep(2) == tf_nullparam)
    tf_error("$pow arg 2 cannot be null.\n");
  else {
    arg_handle = acc_handle_tfarg(1);
```

```
    if (!(acc_object_in_typelist(arg_handle, valid_types)) )
      tf_error("$pow arg1 must be number, variable or net.\n");
    arg_handle = acc_handle_tfarg(2);
    if (!(acc_object_in_typelist(arg_handle, valid_types)) )
      tf_error("$pow arg2 must be number, variable or net.\n");
  }
  return(0);
}

/**********************************************************************
 * calltf routine
 **********************************************************************/
#include <math.h>
int PLIbook_pow_calltf(int user_data, int reason)
{
  PLI_INT32 base, exp, result;

  base   = tf_getp(1);         /* read base value from tfarg 1 */
  exp    = tf_getp(2);         /* read exponent value from tfarg 2 */
  result = (PLI_INT32)pow( (double)base, (double)exp );
  tf_putp(0,result);           /* return result */
  return(0);
}

/**********************************************************************
 * misctf routine
 **********************************************************************/
int PLIbook_pow_misctf(int user_data, int reason, int paramvc)
{
  if (reason == reason_endofcompile)
    io_printf("\n$pow PLI application is being used.\n\n");
  return(0);
}
```

9.11 Interfacing PLI applications to Verilog simulators

As the previous sections in this chapter have shown, a PLI application may comprise:

- A system task or system function name
- A *calltf routine*
- A *checktf routine*
- A *sizetf routine*
- A *misctf routine*

After these items have been defined, the PLI application developer must carry out two basic steps:

1. Associate the system task/function name to the various application routines.

2. Link the applications into a Verilog simulator, so the simulator can call the appropriate routine when the system task/function name is encountered.

The PLI standard provides an **_interface mechanism_** to make the associations between the system task/function name and the application routines. There are two generations of the interface mechanism, one that was created for the older TF and ACC libraries, and a newer mechanism created for the VPI library. This chapter presents the TF and ACC interface mechanism. The newer VPI interface mechanism is presented in Part One of this book.

The TF and ACC interface mechanism is derived from the 1990 OVI PLI 1.0 standard. This older interface mechanism defines what all Verilog simulators should provide for interfacing PLI applications to a simulator, but the older interface does not define how the interface should be implemented. The IEEE 1364 standard includes this older interface mechanism to document the method for backward compatibility.

The TF/ACC interface mechanism is used to specify:

1. A system task/function **_name_**.

2. The application **_type_**, which is a _task_, _function_ or _real function_.

3. **_Pointers_** to the C functions for the _calltf routine, checktf routine, sizetf routine_ and _misctf routine_, if the routines exist (it is not required—and often not necessary—to provide each type of routine).

4. An integer **_user_data_** value, which the simulator will pass to the _calltf routine_, _checktf routine, sizetf routine_ and _misctf routine_ each time they are called.

9.11.1 The de facto standard veriusertfs array and s_tfcell structure

 The IEEE 1364 Verilog standard does not specify **_how_** a Verilog simulator should implement the TF and ACC interface mechanism—the standard only defines **_what_** the interface must specify. Every Verilog simulator provides a unique method for users to specify the interface items. However, most simulators use an array of s_tfcell structures, called a **_veriusertfs_** array, to specify the interface information. This section presents this de facto standard method.

The IEEE 1364 Verilog standard fully documents the VPI interface mechanism as the standard method for interfacing PLI applications. Prior to the IEEE standardization of Verilog, there was no standard method for how the information about a PLI application was specified. Different simulator companies found different ways to specify the system task/function name and the associated routines. In order to preserve backward compatibility with existing implementations, the IEEE 1364 standards committee chose to not standardize any of the existing methods of specifying the TF/ACC PLI application information. Instead, the standard only specifies what information needs to be provided to interface PLI applications written with the TF and ACC libraries.

Many Verilog simulators have adopted a similar method of specifying the information about a PLI application. This method uses a C array of **s_tfcell** structures. The array name is ***veriusertfs*** (which stands for *Verilog user tasks and functions*). The ver-iusertfs array is the original method implemented in the Cadence Verilog-XL product when the PLI was first introduced in the mid 1980's.

Each s_tfcell structure specifies the information about a single PLI application. There can be any number of structures in the *veriusertfs array*. A structure with the first field set to 0 is used to denote the end of the array. The s_tfcell structure is not defined in the IEEE standard. However, nearly all simulators use the same structure definition. The structure is shown below. The meaning of each field in the structure is explained in Table 9-2, on the following page.

```
typedef struct t_tfcell {
    short type;            /* one of the constants: usertask,
                              userfunction, userrealfunction */
    short data;            /* data passed to user routine */
    int (*checktf)();      /* pointer to the checktf routine */
    int (*sizetf)();       /* pointer to the sizetf routine */
    int (*calltf)();       /* pointer to the calltf routine */
    int (*misctf)();       /* pointer to the misctf routine */
    char *tfname;          /* name of the system task/function */
    int forwref;           /* usually set to 1 */
    char *tfveritool;      /* usually ignored */
    char *tferrmessage;    /* usually ignored */
} s_tfcell, *p_tfcell;
```

Table 9-2: Typical fields in the de facto standard veriusertfs array

veriusertfs Field	Definition
type	Defines the type of a PLI application as being either a system task or a system function. This field must be set to the C constant **usertask**, **userfunction** or **userrealfunction**.
user_data	Specifies an integer value—the value will be passed to the PLI application routines each time a routine is called.
checktf_app	Specifies a pointer to the C function that should be called by a simulator for the application's *checktf routine*.
sizetf_app	Specifies a pointer to the C function that should be called by a simulator for the application's *sizetf routine*.
calltf_app	Specifies a pointer to the C function that should be called by a simulator for the application's *calltf routine*.
misctf_app	Specifies a pointer to the C function that should be called by a simulator for the application's *misctf routine*.
"tf_name"	Specifies the name of the system task/function; must be a string beginning with a $.
forwref	Specifies instance name forward referencing as true (1) or false (0). This is not part of the IEEE standard and is ignored by most Verilog simulators. In the Cadence Verilog-XL simulator, setting to 1 makes Verilog-XL IEEE compliant by allowing module and primitive instance names to be used as system task/function arguments (a 0 makes instance names illegal in a system task/function argument).
tfveritool	This is not part of the IEEE standard and is ignored by most Verilog simulators.
tferrmessage	This is not part of the IEEE standard and is ignored by most Verilog simulators.

 NOTE The s_tfcell structure is not specified in the IEEE 1364 standard. Though many simulators use this structure, it is not required, and some simulators use different methods of specifying the PLI application information for the TF/ACC interface.

In most Verilog simulators, an array of s_tfcell structures, called a *veriusertfs array*, is used to define any number of PLI applications. A sample *veriusertfs array* is listed in example 9-2, with the entries for the *$show_value* application presented in Chapter 8 on page 280 and the *$pow* application presented in this chapter, in section 9.10 on page 303.

Example 9-2: example veriusertfs array for registering TF/ACC PLI applications

```
/* prototypes for the PLI application routines */
extern int PLIbook_ShowVal_checktf(), PLIbook_ShowVal_calltf();
extern int PLIbook_pow_checktf(), PLIbook_pow_sizetf(),
           PLIbook_pow_calltf(), PLIbook_pow_misctf();

/* the veriusertfs array */
s_tfcell veriusertfs[] =
{
    {usertask,                      /* type of PLI routine */
        0,                          /* user_data value */
        PLIbook_ShowVal_checktf,    /* checktf routine */
        0,                          /* sizetf routine */
        PLIbook_ShowVal_calltf,     /* calltf routine */
        0,                          /* misctf routine */
        "$show_value",              /* system task/function name */
        1                           /* forward reference = true */
    },


    {userfunction,                  /* type of PLI routine */
        0,                          /* user_data value */
        PLIbook_pow_checktf,        /* checktf routine */
        PLIbook_pow_sizetf,         /* sizetf routine */
        PLIbook_pow_calltf,         /* calltf routine */
        PLIbook_pow_misctf,         /* misctf routine */
        "$pow",                     /* system task/function name */
        1                           /* forward reference = true */
    },

    {0} /*** final entry must be 0 ***/
};
```

 The *veriusertfs array* is not specified in the IEEE 1364 standard. Though many simulators use this array, it is not required, and some simulators use different methods of specifying the PLI application information for the TF/ACC interface.

 Do not specify the veriusertfs array in the same file as the PLI application!

TIP
- Not all Verilog simulators use the *veriusertfs array* to specify PLI application information.
- The array is not standardized, and may be different in different simulators.
- The C language does not allow multiple global arrays with the same name. If two applications both contained a *veriusertfs array* definition, the applications could not be used together.

9.12 Using the user_data field

When a system task or system function is registered, a **user_data** value can be specified. The user_data is an integer value. This value is passed to the *calltf routine*, *checktf routine*, *sizetf routine* and *misctf routine* as a C function input each time the simulator calls one of these routines.

In the following example, two different system tasks, *$read_test_vector_bin* and *$read_test_vector_hex*, are associated with the same *calltf routine*. Assuming the de facto standard *veriusertfs* table is used, the registration for these applications is:

```
s_tfcell veriusertfs[] =
{
  {usertask,                      /* type of PLI routine */
     1,                           /* user_data value */
     0,                           /* checktf routine */
     0,                           /* sizetf routine */
     PLIbook_ReadVector_calltf,   /* calltf routine */
     0,                           /* misctf routine */
     "$read_test_vector_bin",     /* system task/function name */
     1                            /* forward reference = true */
  },

  {usertask,                      /* type of PLI routine */
     2,                           /* user_data value */
     0,                           /* checktf routine */
     0,                           /* sizetf routine */
     PLIbook_ReadVector_calltf,   /* calltf routine */
     0,                           /* misctf routine */
     "$read_test_vector_hex",     /* system task/function name */
     1                            /* forward reference = true */
  },
```

Both system tasks invoke the same *calltf routine*, but the user_data value for the two system task names is different. Therefore, the *calltf routine* can check the user_data value to determine which system task name was used to call the routine.

The user_data value is passed to the PLI routine as a C function input of type int. The following example illustrates reading the user_data value for the two system tasks registered in the previous example.

```
int PLIbook_ReadVector_calltf(int user_data)
{
  if (user_data == 1)
    /* read test vectors as binary values */
  else if (user_data == 2)
    /* read test vectors as hex values */
}
```

The user_data value listed in the register routine will be common to all instances of the registered system task. If, for example, the system task *$read_test_vector_hex* were used in two places in the Verilog design, then there would be two instances of the system task. Each task will be passed the same user_data value. Chapter 10, section 10.7 on page 325 discusses how to allocate data that is unique to each instance of a system task/function.

9.13 Compiling and linking PLI applications

After the C source files for the PLI applications have been defined and the interface information specified, the C source code must be compiled and linked into a Verilog simulator. This allows the simulator to call the appropriate PLI routine when the PLI application system task or system function is encountered by the simulator. The compiling and linking process is not part of the IEEE standard for the PLI. This process is defined by the simulator vendor, and is specific to both the simulator and the operating system on which the simulator is being run. Appendix A presents the instructions for compiling and linking PLI applications for several of the Verilog simulators which were available at the time this book was written.

9.14 Summary

This chapter has presented the major parts of PLI applications which are developed using the TF and ACC libraries. The main parts of an application are the name of a system task or system function, a ***checktf routine***, a ***calltf routine***, a ***sizetf routine***, and a ***misctf routine***. A system function called *$pow*, which calculates x^y and returns a 32-bit result, was used to illustrate the major parts of a PLI application. After defining the PLI application, the next step is to inform the simulator about the application. This is done through the ***PLI interface mechanism***. The IEEE 1364 standard defines what the interface mechanism must specify, but the TF/ACC libraries do not define how the interface should be implemented. Most, but not all, Verilog simulators use a veriusertfs array to specify the interface information.

The remaining chapters in this part of the book examine the TF and ACC libraries in detail, and present a number of examples of using the routines in these libraries. Appendix A presents how PLI applications written using the TF and ACC libraries are linked into different Verilog simulators.

CHAPTER 10 *How to Use the TF Routines*

T he TF routines are a library of 104 C functions that can interact with Verilog simulators. The discussion of TF routines is divided into four chapters: This chapter presents the TF routine library and how to use many of the TF routines. Chapter 11 presents how to read the values of the arguments of system tasks and system functions using TF routines. Chapter 12 presents how to use TF routines in conjunction with *misctf routines*, and Chapter 13 shows how to use the TF routines to interface C language models to Verilog simulators. Appendix B presents the complete syntax of the TF routine library.

The concepts presented in this chapter are:

- An overview of the library of TF routines

- Printing messages from PLI applications

- Obtaining information about system task/function arguments

- Using the TF work area

- Miscellaneous TF routines

10.1 The TF Library

The *TF* routines in the IEEE 1364 PLI standard are the oldest of three generations of the PLI standard (followed by the ACC routines and then the VPI routines). TF stands for *T*ask/*F*unction, because most of the TF routines deal with accessing information about system task/function arguments. There are also a number of utility routines in the TF library for printing messages, controlling simulation, etc.

The library of TF routines is defined in a C header file called *veriuser.h*, which is part of the IEEE 1364 standard. In addition to the TF function definitions, the header file also defines a large number of C constants and C structures used by the TF routines. A complete list of the routines in the TF library, along with any structure definitions and constant names used with each TF function, is provided in Appendix B.

All PLI applications that reference TF routines or constants must include the veriuser.h file. For example:

```
#include "veriuser.h"    /* IEEE 1364 PLI TF routine library */
```

10.1.1 Using TF routines in conjunction with the ACC and VPI routines

The TF library can be used entirely by itself to create complete and useful PLI applications. The TF library and the ACC library are also designed to be used together. The ACC routines complement and greatly extend the capabilities of the TF library.

The VPI library was designed to replace both the TF and ACC libraries with a more concise, more robust, and more versatile procedural interface. It is *not* intended to have TF routines and VPI routines used in the same application. However, there is no conflict between the two libraries, which means multiple PLI applications which use the different libraries can be linked into the same Verilog simulation.

The IEEE 1364 standard includes the TF and ACC libraries, in order to provide backward compatibility and portability of older PLI applications with modern Verilog simulators. The official policy of the IEEE 1364 standards committee is that as improvements and enhancements are added to the Verilog HDL, only the VPI library of the PLI will be expanded for those new features. There will be no new additions to the TF library.

10.1.2 Advantages of the TF library

The TF library has a major advantage over the ACC and VPI library, which offers a compelling reason to continue to develop and use PLI applications written solely with the TF library. The TF routines work directly with the arguments of system tasks and system functions. Unlike the VPI and ACC libraries, the TF routines cannot access the internal data structures of a simulation. This restricted access to the simulation data structure allows simulators to more efficiently optimize the simulation data structure, which can dramatically increase the run-time performance of a simulation.

When a PLI application includes routines from the VPI and ACC libraries, this level of optimization may not be possible.

 Not all simulators perform optimizations based on the type of PLI libraries which are used. Consult the documentation of specific simulators to determine if restricting PLI applications to only using the TF library can improve simulation performance.

10.2 System tasks and system functions

The Verilog PLI allows users to create user-defined system tasks and system functions which can be used in Verilog HDL source code. Refer back to Chapter 9 for more information creating user-defined system tasks and system functions.

10.2.1 System task/function arguments

A user-defined system task or system function can have any number of arguments, including none. With the TF library, the arguments are numbered from left to right, starting with argument number 1. In the following example:

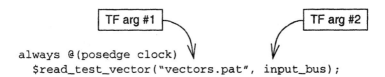

```
                  TF arg #1                          TF arg #2
always @(posedge clock)
  $read_test_vector("vectors.pat", input_bus);
```

- Task/Function argument number 1 is a string, with the value "vectors.pat".

- Task/Function argument number 2 is a signal, with the name input_bus.

Most of the TF routines in the PLI library are used for the purpose of accessing information about user-defined system task/function arguments.

The IEEE 1364-1995 standard for the TF routines, and the older OVI PLI 1.0 standard from which the IEEE standard was derived, refer to task/function *arguments* as *parameters*. While *parameter* is an alternate programming term for a function argument, it is too easily confused with the Verilog HDL keyword parameter. To avoid this confusion, the ACC and VPI generations of the PLI refer to task/function arguments as *arguments*. In the 1364-2001 standard, the word *parameter* has been changed to *argument* for TF routines, so that the standard uses consistent terminology for all routines. This book uses the term *argument* throughout.

10.2.2 Multiple instances of system tasks and system functions

The Verilog HDL source code can reference the same system task or system function any number of times. For example:

```
always @(posedge clock)
  $read_test_vector("A.dat", data_bus);

always @(negedge clock)
  $read_test_vector("B.dat", data_bus);
```

Just as a Verilog module can be used, or *"instantiated"*, many times in a design, every occurrence of a system task/function is a separate and unique *instance*. Each instance of *$read_test_vector* in the above example has different arguments. The PLI recognizes that each instance is unique, and keeps track of each instance. Therefore, at each positive edge of clock, the *calltf routine* associated with one instance of *$read_test_vector* will be invoked, and at the negative edge of clock the *calltf routine* associated with a different instance of *$read_test_vector* will be executed. It is important to understand that the Verilog simulator will call the same C functions for each instance of the system task/function, but the inputs and data associated with each call will be unique for each instance.

As an example, *$read_test_vector* might have a *calltf routine* named *gvCall()* associated with it. When the instance of *gvCall()* that is invoked at the positive edge of clock reads the value of task/function argument 1, it will see the string "A.dat". When the instance of *gvCall()* that is invoked at the negative edge of clock reads task/function argument number 1, it will see the string "B.dat".

10.2.3 Instance specific system task/function routines

A PLI *calltf routine*, *misctf routine*, *checktf routine* and *sizetf routine* are directly associated with the name of a system task or system function. (This association is part of the PLI interface mechanism, which was presented in Chapter 9.) Each of these routines can access the arguments of the system task/function which caused the routine to be called. If these routines call another C function, that function can also access the arguments of the system task/function.

The ACC library also provides for another type of PLI routine, called a *consumer routine*. Defining and using *consumer routines* is defined in Chapter 17. A *consumer routine* is *not* associated with the name of a system task or system function. Therefore, a *consumer routine* cannot directly access the arguments of system tasks/functions. The TF library also provides a means for PLI applications to indirectly access system task/function arguments. Indirect access is done by obtaining a pointer for an instance of a system task/function and then using an instance specific version of TF routines to access the arguments of that system task/function instance.

An instance pointer for a system task or system function is obtained using the **tf_getinstance()** routine. The syntax of this routine is:

PLI_BYTE8 ***tf_getinstance ( )**

The tf_getinstance() routine returns a pointer to the system task/function which called the PLI application.

There are instance-specific counterparts for many of the routines in the TF library. These instance-specific routines use the instance pointer to access the task/function arguments of a specific task/function instance. For example, the TF routine **tf_getp()** retrieves the value of an argument for the instance of the system task/function that called the PLI application. A counterpart to tf_getp() is **tf_igetp()**. This routine reads the value of an argument of a *different instance* of a system task/function, using the instance pointer of the other system task/function.

The typical usage of these routines is for a *calltf routine* or a *misctf routine* to obtain the pointer for the system task/function instance which called the routine. This handle is then saved in the user_data field of a *consumer routine*. This gives the consumer routine access to the instance handle, so that the consumer routine can access the arguments of the system task/function.

The instance specific versions of each TF routines are *not* presented in these chapters which discuss using the TF routines. The routines are listed in Appendix B, which presents the full syntax of the TF library. Functionally, the instance-specific versions of each routine work the same as the regular versions of the same routines. Example 13-7 on page 452 of Chapter 13 shows an example of one way in which the tf_getinstance() routine might be used. Chapter 18 also contains examples of using other instance specific routines.

The TF system task/function instance pointer is not the same as the ACC system task/function instance pointer. The ACC library also has instance specific versions of the routines which access the arguments of a system task or system function. However, the instance pointer returned from `tf_getinstance()` is different than the ACC handle returned from `acc_handle_tfinst()`. The instance pointer from a routine in one library should not be used with routines in the other library.

10.3 The PLI string buffer

A number of TF routines return pointers to C character strings, such as a file name or other name that is an argument to a system task/function. The string is stored in a temporary string buffer that is maintained by the PLI. This temporary buffer is limited in size, and will be overwritten by other calls to TF routines which return strings.

A PLI application should use the string pointer returned by a TF routine immediately. After another call is made to a TF routine which retrieves a string, there is no guarantee that the first string pointer will still be valid. If a string needs to be preserved, the PLI application should copy the string into application-allocated storage space. Following are two examples of using strings returned by TF routines. The TF routines used in these examples are described in more detail later in this chapter.

Read a string and use it immediately:

```
char *string_p;       /* string pointer only, no storage */
string_p = tf_strgetp(1, 'b');
io_printf("string_p points to %s\n", string_p);
```

Read a string and copy it to application-allocated storage for later use:

```
char *string_p;       /* string pointer only, no storage */
char *string_keep;    /* another string pointer only */
string_p = tfstrgetp(1, 'b'); /* get string from 1st arg */
string_keep = malloc(strlen(string_p)+1);
strcpy(string, string_p);  /* save string for later use */
```

10.4 Controlling simulation

There are two TF routines which can control the run-time execution of a simulation.

PLI_INT32 **tf_dostop ( )**

Executes the same functionality as the Verilog *$stop* built-in system task, which provides a means for a PLI application to send the simulation to the simulator's interactive debug environment. For example, a PLI application could be watching for a certain error condition to occur in a design (perhaps multiple drivers of a bus being enabled at the same time). When the error occurs, the PLI application could print a run-time warning message and then halt the simulation to allow an engineer to debug the problem using the simulator's debugger. When the user continues simulation from the interactive debug mode, the PLI application will resume execution at the statement which follows the call to `tf_dostop()`. The return value from `tf_dostop()` is not used, and should be ignored.

PLI_INT32 **tf_dofinish ( )**

Executes the same functionality as the Verilog *$finish* built-in system task, which provides a means for a PLI application to cause a simulation to exit, and in the process close any child processes or files that were opened by the simulator. An example of where `tf_dofinish()` can be used is with a PLI application to read test vectors from a file. When the file reader has reached the end of the test vector file, it can exit the simulation. The return value from `tf_dofinish()` is not used, and should be ignored.

10.5 Printing messages

The TF routines provide several ways to print messages.

void **io_printf (**
    PLI_BYTE8 *** format,**   /* character string with a formatted message */
              arg1...arg12) /* arguments to the formatted message string */

The `io_printf()` routine is used to print formatted text messages. The syntax for this routine is similar to the C `printf()` routine, but with a maximum of 12 arguments. The `io_printf()` routine prints the message to the simulation output channel, and to the simulation output log file. By using `io_printf()`, a PLI application can generate messages which are part of the simulation, without having to determine where the simulator directs its output.

The C printf() routine will print its message to the operating system's standard output channel. If the simulator uses some other output channel, such as a graphical window, then printing a message using the C printf() routine might not be seen by the simulator user. Nor will the message be recorded in any output files generated by the simulator.

Following is an example of using io_printf():

```
io_printf("Module %s has %d nets\n", module_name, num_nets);
```

void io_mcdprintf (

PLI_INT32 **mcd,**	/* multi-channel descriptor of open files */
PLI_BYTE8 * **format,**	/* character string with a formatted message */
arg1...arg12)	/* arguments to the formatted message string */

The io_mcdprintf() routine is used to write messages into files that were opened by the Verilog HDL *$fopen* system function. The *multi-channel descriptor* (mcd) returned by *$fopen* must be passed to the PLI application as an argument to a user-defined system task/function or its value obtained using an ACC routine. The syntax of tf_warning() is similar to the C printf() routine, but the maximum number of arguments that can be printed in a message is 12.

PLI_INT32 tf_warning (

PLI_BYTE8 * **format,**	/* character string with a formatted message */
arg1...arg5)	/* arguments to the formatted message string */

The tf_warning() routine prints a text message to the output channel used by the simulator that is running the PLI application. The formatting of the message is not defined by the IEEE 1364 standard, and different simulators will format the message in different ways. The syntax of tf_warning() is similar to the C printf() routine, but the maximum number of arguments that can be printed in a message is 5. The return value from tf_warning() is not used, and should be ignored. An example of using tf_warning() is:

```
tf_warning("Reached end-of-file in test vector file %s\n",
           file_name);
```

PLI_INT32 tf_error (

PLI_BYTE8 * **format,**	/* character string with a formatted message */
arg1...arg5)	/* arguments to the formatted message string */

The tf_error() routine prints a text message to the output channel of the simulator that is running the PLI application. The formatting of the message is not defined by the IEEE 1364 standard, and different simulators will format the message in different ways. The syntax of tf_error() is similar to the C printf() routine, but the maximum number of arguments that can be printed in a message is 5. The return value from tf_error() is not used, and should be ignored. An example of using tf_error() is:

```
tf_error("Could not open test vector file %s\n", file_name);
```

In addition to printing a message, tf_error() causes the simulation to abort if it is called from a *checktf routine*. This important feature allows PLI application developers to write syntax checking routines which abort a simulation before a *calltf routine* tries to use incorrect information (e.g.: if a task/function argument is expected to have a module instance name, but a user has specified a file name instead).

 Simulation will only abort if tf_error() is called from a *checktf routine*. tf_error() may also be used in a *calltf routine* or a *misctf routine* to print runtime error messages, but simulation will not be aborted.

 Some Verilog simulators do not adhere to the IEEE 1364 standard, in that they do not abort simulation when tf_error() is called from a *checktf routine*. For these simulators, the tf_dofinish() routine can be called after calling tf_error() to force an abort for a fatal error.

TIP

PLI_INT32 **tf_message (**

PLI_INT32 **level,**	/* a constant representing the error severity level; one of: **ERR_ERROR, ERR_SYSTEM, ERR_INTERNAL, ERR_WARNING, ERR_MESSAGE** */
PLI_BYTE8 *** facility,**	/* quoted string to be appended to the output message; must be 10 or less characters */
PLI_BYTE8 *** code,**	/* quoted string to be appended to the output message after facility; must be 10 or less characters */
PLI_BYTE8 *** format,**	/* quoted string of the formatted message */
arg1...arg5)	/* arguments to the formatted message string */

The tf_message() routine is essentially a combination of tf_warning() and tf_error(). The routine prints a text message with a maximum of 5 arguments. A severity level is specified that can be an error, warning or informational level. The severity level is specified as one of the constants: ERR_ERROR, ERR_SYSTEM, ERR_INTERNAL, ERR_WARNING, or ERR_MESSAGE. If tf_message() is called from a *checktf routine*, the first three severity levels will cause simulation to abort (like tf_error()). The tf_message() routine also allows multiple messages to be queued and then printed together (see the description of tf_text(), which follows). The tf_message() routine defines the error message format (whereas tf_warning() and tf_error() use the format defined by the simulator that is running the PLI application). The format is based on the Cadence Verilog-XL error/ warning message format, which consists of a text message and two short text codes. The codes are limited to a maximum of 9 characters each. The return value from tf_message() is not used, and should be ignored. An example of using tf_message() is:

```
tf_message(ERR_ERROR, "User", "TFARG",
           "Arg %d is illegal in $read_test_vector", argnum);
```

The above example will print a message similar to:

```
ERROR: Arg 2 is illegal in $read_test_vector [User-TFARG]
```

PLI_INT32 **tf_text (**
   PLI_BYTE8 *** format,**   /* character string with a formatted message */
       arg1...arg5) /* arguments to the formatted message string */

The tf_text() routine is used in conjunction with tf_message(). tf_text() allows multiple messages to be queued, which will all be printed when tf_message() is called. The return value from tf_text() is not used, and should be ignored. The following pseudo-code checks for three different types of errors, and uses tf_text() to queue any error messages. Then, if there were any errors, tf_message() is called to print the error messages and abort simulation.

```
bool err = FALSE;
if wrong number of arguments
  tf_text("$read_test_vector requires 2 args\n");
  err = TRUE;
if arg 1 is not a string
  tf_text("$read_test_vector arg 1 must be quoted file
name\n");
  err = TRUE;
if arg 2 is not a reg data type
  tf_text("$read_test_vector arg 2 must be a reg data
type\n");
  err = TRUE;
if (err)
  tf_message(ERR_ERROR, "User", "TFARG", "System task usage
error");
```

10.6 Checking system task/function arguments

Three TF routines provide information about system task/function arguments that can be useful for verifying the correctness of arguments or performing other duties that affect the system task/function arguments.

PLI_INT32 **tf_nump ( )**

The tf_nump() routine returns the number of arguments of the user-defined system task/function that called the PLI application. tf_nump() does not take any inputs. Two common ways to use tf_nump() are:

- In a *checktf routine*, to verify that a system task/function instance has the correct number of arguments. For example:

```
if (tf_nump() != 2 )
    tf_error("$read_test_vector has %d args, requires 2.\n",
            tf_nump());
```

- In any PLI application, to loop through the task/function arguments for processing. For example:

```
numargs = tf_nump();
for (i=1; i<= numargs; i++)
    . . .
```

 For faster program execution, do not use tf_nump() in a loop control test. In the
above example, tf_nump() is only called one time, and its return value is saved in
TIP a variable. If the loop had been coded as:
```
    for (i=1; i<=tf_nump(); i++)
```
then tf_nump() would be called every pass of the loop. The additional calls to the
TF routine are not necessary, yet require time to be executed.

In the Verilog HDL, it is legal for system tasks and system functions to have *null* arguments. The count returned by tf_nump() routine includes null arguments. The tf_typep() routine can be used to determine if an argument is null. The following examples will return the argument counts shown:

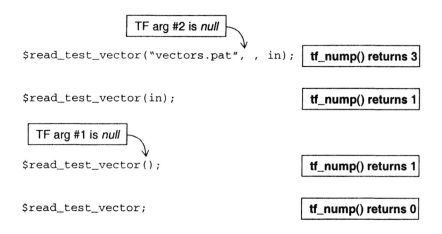

PLI_INT32 **tf_typep(n)**

    PLI_INT32   **n**)   /* index number of a system task/function argument */

tf_typep() returns a constant that represents the general data type of a specific system task/function argument. The input to tf_typep() is the index number of the system task/function argument of interest. The constants returned by tf_typep() are defined in the *veriuser.h* TF library. They are:

- **TF_NULLPARAM** — the argument is null.

- **TF_STRING** — the argument is a literal string.

- **TF_READONLY** — the argument is a literal integer number, an integer constant, or a Verilog net data type (wire, wand, wor, tri, triand, trior, trireg, supply0 or supply1).

- **TF_READONLYREAL** — the argument is a literal real number or a real constant.

- **TF_READWRITE** — the argument is a Verilog variable data type (reg, integer, time).

- **TF_READWRITEREAL** — the argument is a Verilog real variable data type (real, realtime).

Following are two examples of using tf_typep():

```
if (tf_typep(1) != TF_STRING)
   tf_error("$read_test_vector arg 1 must be a string\n");

if (tf_typep(2) != TF_READWRITE)
   tf_error("$read_test_vector arg 2 must be a reg type\n");
```

PLI_INT32 **tf_sizep (**
   PLI_INT32   **n**)   /* index number of a system task/function argument */

The `tf_sizep()` routine returns the number of bits in a system task/function argument. The input to `tf_sizep()` is the index number of the argument. There are many
ways the bit size of an argument can be useful. Following are two examples:

```
char *vector;
if (tf_sizep(1) != 1)
  tf_error("$my_app arg 1 must be scalar\n");
/* allocate memory for a string the size of TF arg 2 */
vector = malloc(tf_sizep(2) + 1);
```

10.7 The TF work area

The TF routines provide a *work area*, where a PLI application can store a pointer to
application data. Each instance of a system task/function is automatically allocated a
work area by the PLI, and this work area is shared by the *checktf routine*, *misctf routine* and *calltf routine* associated with that system task/function.

The work area is defined as a **PLI_BYTE** pointer. This pointer type is generally
equivalent to a `char*` pointer. The IEEE 1364-1995 and earlier generations of the TF
library defined the work area as a `char *` pointer. The IEEE 1364-2001 standard
changed to a `PLI_BYTE8 *` pointer in order to ensure portability across different operating systems.

PLI_INT32 **tf_setworkarea (**
   PLI_BYTE8 *** workarea**)   /* pointer to application-allocated storage */

PLI_BYTE8 ***tf_getworkarea ( )**

The `tf_setworkarea()` routine is used to write a pointer into the TF work area.
The return value from these routines is not used, and should be ignored. The
`tf_getworkarea()` routine retrieves the pointer from the work area.

There are two very powerful ways to use this work area, both of which greatly simplify writing PLI applications. The work area can be used to:

- Pass data from one PLI application routine to another application routine, such as
  from a *misctf routine* to a *calltf routine*.

- Store data that will be used in a future call to the same PLI application routine.

As an example of where both of the above capabilities might be needed, the *$read_test_vector* application might need to open a disk file from which to read vectors, and then read a line from the file each time the system task is called by simulation. The usage of the *$read_test_vector* example is:

```
always @(posedge clock)
  $read_test_vector("A.dat", data_bus);
```

Since the test vector file should only be opened once, rather than at every positive edge of clock, a convenient place to open the file is in the *misctf routine* for *$read_test_vector*. The *misctf routine* is called for miscellaneous reasons, such as **REASON_ENDOFCOMPILE**, which indicates the start of actual simulation. The test vector file can be opened at this time, and the file pointer saved in the TF work area for *$read_test_vector*. More information about *misctf routines* is presented in Chapter 12.

The *calltf routine* for *$read_test_vector* will perform the work of reading values from the file and applying them to simulation. The *calltf routine* will be invoked each time simulation encounters the system task, which, in the preceding example of *$read_test_vector*, is at each positive edge of clock. Since the *calltf routine* did not open the disk file, it does not have direct access to the file pointer. But, the *calltf routine* shares the work area with the *misctf routine*. Therefore, the *misctf routine* can store the file pointer at the beginning of simulation, and, each time the *calltf routine* is invoked, it can retrieve the file pointer from the work area.

Example 10-1, which follows, contains a partial *misctf routine* and *calltf routine* for the *$read_test_vector* example, which uses the TF work area to store a test vector file pointer. The complete code for *$read_test_vector* is listed at the end of this chapter, as example 10-3 on page 338.

Example 10-1: using the `tf_setworkarea()` and `tf_getworkarea()` routines

```
/*********************************************************************
 * misctf routine
 *********************************************************************/
int PLIbook_ReadVector_misctf(int user_data, int reason, int paramvc)
{
  FILE *in_file;

  ...

  if (reason == REASON_ENDOFCOMPILE) {   /* time to open vector file */
    in_file = fopen(tf_getcstringp(1), "r");

  tf_setworkarea((PLI_BYTE8 *)in_file);  /* save file ptr in workarea */
```

```
/*********************************************************************
 * calltf routine
 *********************************************************************/
int PLIbook_ReadVector_calltf(int user_data, int reason) {
  FILE *in_file;

  ...

  in_file = (FILE*)tf_getworkarea();          /* retrieve file pointer */

  /* read next test vector from file */
  ...
```

10.7.1 Using the TF work area instead of a static or global variable

A TF work area is automatically allocated and maintained for each instance of a user-defined system task/function. This means that if there are multiple instances of a system task/function, then each instance has its own unique work area.

For example, consider the Verilog HDL code:

```
always @(posedge clock)
  $read_test_vector("A.dat", data_bus);

always @(negedge clock)
  $read_test_vector("B.dat", data_bus);
```

In this example, the *$read_test_vector* instance that is called at the positive edge of clock needs to save a file pointer to *"A.dat"*, and the *$read_test_vector* that is called at the negative edge of clock needs to save a file pointer to *"B.dat"*.

A common solution in the C programming language to preserve data is to declare a static variable instead of an automatic variable. Sharing common data between C functions can be done using global variables.

These C programming techniques can cause serious problems in PLI applications!

The Verilog HDL allows a system task/function to be used multiple times in the Verilog source code, and each module which uses the task/function can be instantiated multiple times. Each occurrence in each instance of a module becomes a unique *instance* of the system task/function. However, the same C function will be called by the simulator for each instance.

Since both instances of *$read_test_vector* in the preceding example will invoke the same C function, a static or global variable cannot be used, because the two instances would share the same single copy of the variable. The variable can only hold a single file pointer, and there is no simple way of knowing which instance of *$read_test_vector* was the last one to store a value in the variable. Using static or global variables is a sure way to have problems when there are multiple instances of a system task or system function.

The TF work area avoids the problems of static and global variables. The PLI automatically allocates a unique TF work area for each instance of a system task or system function. If a Verilog design includes two instances of the *$read_test_vector* application, each instance will have a unique work area in which to save its file pointer.

10.7.2 Storing multiple values in the TF work area

The work area only stores a single pointer. This pointer can be used to store multiple values, by allocating a block of memory in which to store the values and saving the pointer to the memory block in the work area.

The following code fragment, example 10-2, stores a file pointer, a character pointer, and two integers in the TF work area. In this example, a structure called PLIbook_my_data is defined to contain the three values to be stored. At the start of simulation, the *misctf* routine performs three operations:

- Allocate memory for one PLIbook_my_data structure.

- Fill the fields in the structure (if needed at this time).

- Store a pointer to the allocated memory in the work area.

Example 10-2: storing multiple values in the TF work area

```
typedef struct PLIbook_my_data {
  FILE  *file_p;          /* pointer to a file */
  char  *tfinst_p;        /* a character pointer */
  int   temp1, temp2;     /* other integer data */
} PLIbook_my_data_s, *PLIbook_my_data_p;

void my_misc(int user_data, int reason, int paramvc)
{
  PLIbook_my_data_p data_p;
  switch(reason) {
    case REASON_ENDOFCOMPILE:  /* misctf called just before time 0 */
      data_p = (PLIbook_my_data_p)malloc(sizeof(PLIbook_my_data_s));
      data_p->file_p = fopen(tf_getcstringp(1), "r");
      data_p->tfinst_p = tf_getinstance();
      tf_setworkarea((PLI_BYTE8 *)data_p);  /* store ptr to data */
      break;
      . . .
}
```

10.8 Reading and using simulation times

The Verilog HDL represents simulation time as a 64 bit unsigned integer. Since the ANSI C standard does not provide an integer data type that is guaranteed to be 64 bits on all operating systems, the Verilog PLI uses a pair of C 32-bit integers to store the full 64 bits of simulation time. The lower 32 bits of Verilog time are placed in one integer, and the upper 32 bits of the Verilog time will be stored in a second integer, as shown in the following illustration:

```
PLI_INT32   high, low;
```

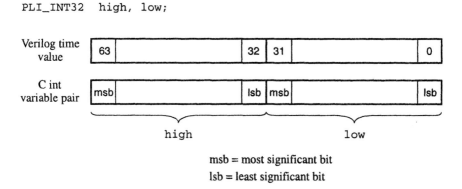

msb = most significant bit
lsb = least significant bit

Within each Verilog module, time is scaled to a simulation time unit and time precision, using the `timescale` compiler directive. The TF routines can retrieve the current scaled simulation time and information about a module's time scaling.

The `timescale` directive in Verilog indicates what time units are used in the modules which follow the directive. The time scale also indicates a time precision, which is how many decimal points of accuracy are permitted in the following modules. Delays with more resolution than the precision are rounded off. Each module in a design can have a different time scale, so one model can specify delays in nanoseconds, with two decimal points of precision, and another model can specify delays in microseconds, with three decimal points. A Verilog simulator will scale the times in each module to the simulator's internal time units.

The `timescale` directive is part of the Verilog language standard, and is not explained in this book. Refer to any good book on Verilog or the IEEE 1364 Verilog Language Reference Manual for more information on time scaling in Verilog. Page 7 lists a few sources for more information on the Verilog language.

10.8.1 Retrieving the current simulation time

PLI_INT32 **tf_gettime ( )**

The tf_gettime() routine retrieves the lower 32-bits of current simulation time as a 32-bit C integer.

PLI_INT32 **tf_getlongtime (**
  PLI_INT32 *** hightime)**   /* pointer to the upper 32-bits of the time */

This routine retrieves the current 64-bit simulation time. The lower 32-bits are returned, and the upper 32-bits are loaded into **hightime**.

double **tf_getrealtime ( )**

This routine retrieves the current simulation time as a real number.

PLI_BYTE8 ***tf_strgettime ( )**

This routine retrieves the current simulation time as a string.

The tf_gettime(), tf_getlongtime() and tf_getrealtime() routines return the current simulation time. For each of these routines, the time is automatically scaled from simulation time units to the time units and precision of the module which called the PLI application. Since tf_gettime() and tf_getlongtime() return the time as an integer, any decimal points of precision will be lost.

The tf_strgettime() routine returns the current simulation time as a pointer to a string, in the simulator's time units, which is usually the smallest time precision of all modules which make up the simulation.

10.8.2 Retrieving a future simulation time

PLI_INT32 **tf_getnextlongtime (**
  PLI_INT32 *** lowtime,**   /* pointer to the lower 32-bits of the time */
  PLI_INT32 *** hightime)**   /* pointer to the upper 32-bits of the time */

The tf_getnextlongtime() routine returns the next simulation time in which a simulation event is scheduled, and returns a flag indicating the significance of the time retrieved. The full 64-bits of simulation time is retrieved, with the lower 32-bits loaded into **lowtime**, and the upper 32-bits loaded into **hightime**. The simulation time is returned in simulation time units—it is not scaled to the time scale of the module which called the PLI application.

In order for tf_getnextlongtime() to determine when the next event is scheduled, it must be called at the very end of the current simulation time step, after all activity in the current time has been processed. The *misctf routine* can be scheduled to be called at the end of a time step, using the routine tf_rosynchronize() (refer to section 12.3 on page 408 for a description of using this routine). If tf_getnextlongtime() is not called at the very end of a time step, it retrieves the current simulation time, instead of the next simulation time. tf_getnextlongtime() returns a flag indicating the meaning of the simulation time retrieved, as follows:

0   tf_getnextlongtime() was called from a *misctf routine* at the end of a simulation time step (**REASON_ROSYNCH**). The time retrieved is the time the next simulation event is scheduled to occur.

1   tf_getnextlongtime() was called when there are no more simulation events scheduled (simulation is over). A 0 is returned for the simulation time.

2   tf_getnextlongtime() was **not** called from a *misctf routine* at the end of a simulation time step. The time retrieved is the current simulation time.

10.8.3 Retrieving time scale and precision values

PLI_INT32 **tf_gettimeunit ( )**

This routine returns an integer representing the time scale unit of the module containing the instance of the system task/function which called the PLI application.

PLI_INT32 **tf_gettimeprecision ( )**

This routine returns an integer representing the time scale precision of the module containing the instance of the system task/function which called the PLI application.

There is also an instance-specific version of these routines, namely **tf_igettimeunit()** and **tf_igettimeprecision()**. These instance-specific versions take as an input a pointer to an instance of a system task or function (obtained from tf_getinstance()), and return the time units and precision of the module containing that instance. If the instance pointer is null, then they return the simulator's internal time units. For most Verilog simulators, the internal time unit is the finest precision of all modules in the Verilog simulation.

All of these routines return an integer which represents a unit of time, as shown in Table 10-1, which follows. The time values are represented as the order of magnitude of 1 second, which is the exponent of 1 second times 10^n.

Table 10-1: Time unit values for tf_gettimeunit() and tf_gettimeprecision()

Exponent Value	Time Unit Represented
2	100 seconds (1×10^2)
1	10 seconds (1×10^1)
0	1 second (1×10^0)
-1	100 milliseconds (1×10^{-1})
-2	10 milliseconds (1×10^{-2})
-3	1 millisecond (1×10^{-3})
-4	100 microseconds (1×10^{-4})
-5	10 microseconds (1×10^{-5})
-6	1 microsecond (1×10^{-6})
-7	100 nanoseconds (1×10^{-7})
-8	10 nanoseconds (1×10^{-8})
-9	1 nanosecond (1×10^{-9})
-10	100 picoseconds (1×10^{-10})
-11	10 picoseconds (1×10^{-11})
-12	1 picosecond (1×10^{-12})
-13	100 femtoseconds (1×10^{-13})
-14	10 femtoseconds (1×10^{-14})
-15	1 femtosecond (1×10^{-15})

10.8.4 Utility routines to work with 64-bit time values

The TF library provides special routines to aid in manipulating 64-bit time values.

PLI_BYTE8 ***tf_longtime_tostr (**
    PLI_INT32   **lowtime,**   /* lower (right-most) 32-bits of a 64-bit integer */
    PLI_INT32   **hightime)**   /* upper (left-most) 32-bits of a 64-bit integer */

This routine converts a 64-bit simulation time that is represented as a pair of 32-bit C integers to a character string. The routine returns a pointer to the string, which is stored in temporary simulation-allocated storage.

void **tf_scale_longdelay (**

PLI_BYTE8	* **tfinst,**	/* pointer to an instance of a system task/function */
PLI_INT32	**lowdelay1,**	/* lower (right-most) 32-bits of first 64-bit time value */
PLI_INT32	**highdelay1,**	/* upper (left-most) 32-bits of first 64-bit time value */
PLI_INT32	**lowdelay2,**	/* lower (right-most) 32-bits of second 64-bit time value */
PLI_INT32	**highdelay2)**	/* upper (left-most) 32-bits of second 64-bit time value */

This routine scales a 64-bit time value that is stored as a pair of 32-bit C integers to the time scale of the module containing a specific instance of a system task/function. The value stored in arguments lowdelay1 and highdelay1 are scaled and deposited to lowdelay2 and highdelay2.

void **tf_scale_realdelay (**

PLI_BYTE8	* **tfinst,**	/* pointer to an instance of a system task/function */
double	**realdelay1,**	/* first time value */
double	* **realdelay2)**	/* pointer to second time value */

Scales a double precision real number time value to the time scale of the module containing a specific instance of a system task/function. The value in **realdelay1** is scaled and placed into **realdelay2**.

Scales a double precision real number time value to the time scale of the module containing a specific instance of a system task/function. The value stored in realdelay1 is scaled and deposited to realdelay2.

void **tf_unscale_longdelay (**

PLI_BYTE8	* **tfinst,**	/* pointer to an instance of a system task/function */
PLI_INT32	**lowdelay1,**	/* lower 32-bits of first 64-bit time value */
PLI_INT32	**highdelay1,**	/* upper 32-bits of first 64-bit time value */
PLI_INT32	* **lowdelay2,**	/* pointer to lower 32-bits of second 64-bit time value */
PLI_INT32	* **highdelay2)**	/* pointer to upper 32-bits of second 64-bit time value */

This routine converts a 64-bit time value expressed in simulation time units to the time scale of a specific instance of a system task/function. The time value stored in arguments lowdelay1 and highdelay1 are converted and deposited into lowdelay2 and highdelay2.

void **tf_unscale_realdelay (**

PLI_BYTE8	* **tfinst,**	/* pointer to an instance of a system task/function */
double	**realdelay1,**	/* real number value of first delay value */
double	* **realdelay2)**	/* pointer to real number value of second delay value */

Converts a real number value expressed in simulation time units to the time scale of a specific instance of a system task/function. The value stored in realdelay1 is converted and deposited to realdelay2.

10.9 Reading simulation invocation options

The PLI can access the invocation options used to invoke a Verilog simulator. Using the TF routines, a PLI application can access invocation options which begin with a plus sign (+). The ACC and VPI libraries can access any invocation option.

 The PLI can only access the invocation options used to invoke simulation. Some Verilog simulators use a separate command to compile Verilog HDL source code. When a simulator uses a separate compile command, the invocation options used to compile Verilog models cannot be accessed by the PLI.

PLI_BYTE8 ***mc_scan_plusargs (**
   PLI_BYTE8 *** plusarg)**   /* name of the invocation option */

The `mc_scan_plusargs()` routine scans the simulators invocations options for a specified user-defined invocation plus option. If the invocation option does not exist, the routine returns 0. If the option exists exactly as specified, the routine returns a null string. If the invocation option exits, and has additional characters beyond the specified option, the additional characters are returned in a string.

The following example performs a true/false test to see if simulation was invoked with a **+PLIbook_verbose** option:

```
if (mc_scan_plusargs("PLIbook_verbose"))
   io_printf("debug: value read from file is %s\n", vector);
```

Notice that the argument passed to mc_scan_plusargs() is a string containing the name of the invocation option without the plus sign.

The next example checks for the invocation option *+file+<file_name>* and prints the name of the file specified. If the invocation option is **+file+vector1.pat**, then the string *"vector1.pat"* will be printed.

```
if (file_name = mc_scan_plusargs("file+"))
   io_printf("Simulation invoked with +file+ name of %s\n",
             file_name);
else
   io_printf("Simulation was not invoked with +file+ \n");
```

Care must be taken when using mc_scan_plusargs() as a simple true/false test, without checking the return string. The following test,

```
if (mc_scan_plusargs("test1")
    /* do something */
```

will return a true (a non-null return) for both the invocation option of **+test1** and **+test10**. To determine which option was used, it is necessary to examine the return value from mc_scan_plusargs(). The former will return a pointer to a null string, and the latter will return a pointer to a string of "0".

10.10 Utility TF routines

There are several utility TF routines to aid in working with Verilog models and simulation.

Performing math operations on 64-bit values stored as a pair of integers

The routines tf_add_long(), tf_subtract_long(), tf_multiply_long(), tf_divide_long(), tf_compare_long() provide a means of performing math operations on 64-bit values, which are represented as a pair of 32-bit C integers.

PLI_INT32 **tf_add_long (**

PLI_INT32	*** low1,**	/* pointer to lower 32 bits of first operand */
PLI_INT32	*** high1,**	/* pointer to upper 32 bits of first operand */
PLI_INT32	**low2,**	/* lower 32 bits of second operand */
PLI_INT32	**high2)**	/* upper 32 bits of second operand */

The tf_add_long() routine adds two 64-bit values, and deposits the result back into the first operand.

PLI_INT32 **tf_subtract_long (**

PLI_INT32	*** low1,**	/* pointer to lower 32 bits of first operand */
PLI_INT32	*** high1,**	/* pointer to upper 32 bits of first operand */
PLI_INT32	**low2,**	/* lower 32 bits of second operand */
PLI_INT32	**high2)**	/* upper 32 bits of second operand */

The tf_subtract_long() routine subtracts two 64-bit values, and deposits the result back into the first operand.

void **tf_multiply_long (**

PLI_INT32	* **low1,**	/* pointer to lower 32 bits of first operand */
PLI_INT32	* **high1,**	/* pointer to upper 32 bits of first operand */
PLI_INT32	**low2,**	/* lower 32 bits of second operand */
PLI_INT32	**high2)**	/* upper 32 bits of second operand */

The `tf_multiply_long()` routine multiplies two 64-bit values, and deposits the result back into the first operand.

void **tf_divide_long (**

PLI_INT32	* **low1,**	/* pointer to lower 32 bits of first operand */
PLI_INT32	* **high1,**	/* pointer to upper 32 bits of first operand */
PLI_INT32	**low2,**	/* lower 32 bits of second operand */
PLI_INT32	**high2)**	/* upper 32 bits of second operand */

The `tf_divide_long()` routine divides two 64-bit values, and deposits the result into the first operand.

Converting 64-bit values stored as a pair of integers to/from doubles

Two TF routines are provided to aid in converting a 64-bit value that is stored in a pair of 32-bit C integers into a C double, and vice-versa.

void **tf_long_to_real (**

PLI_INT32	**low,**	/* lower (right-most) 32-bits of a 64-bit integer */
PLI_INT32	**high,**	/* upper (left-most) 32-bits of a 64-bit integer */
double	* **real)**	/* pointer to a double precision variable */

This routine converts a 64-bit value stored as a pair of 32-bit C integers to a C double precision value.

void **tf_real_to_long (**

double	**real,**	/* a double precision variable */
PLI_INT32	* **low,**	/* pointer to the lower 32-bits of a 64-bit integer */
PLI_INT32	* **high)**	/* pointer to upper 32-bits of a 64-bit integer */

The `tf_real_to_long()` routine converts a C double precision value to a 64-bit value stored in a pair of 32-bit C integers.

Getting the name of a module or scope containing a system task/function

PLI_BYTE8 *tf_mipname ( )

The tf_mipname() routine returns pointer to a string containing the full hierarchical path name of the *module instance* containing the system task/function which called the PLI application. If the system task/function call is located in a scope level within a module, such as a Verilog HDL function, the module instance containing the scope is still returned.

PLI_BYTE8 *tf_spname ( )

The tf_spname() routine returns a pointer to a string containing the full hierarchical path name of the Verilog *scope* containing the system task/function which called the PLI application. A scope may be a module instance, a named statement group, a Verilog task or a Verilog function.

10.11 A complete PLI application using TF Routines

This section contains the complete source code for the *$read_test_vector* PLI applications, which uses a *checktf routine*, a *calltf routine* and a *misctf routine*.

- The *checktf routine* verifies that the first system task/function argument is a string and that the second system task/function argument is writable.

- The *misctf routine* opens the test vector file specified in first system task/function argument.

- The *calltf routine* writes a 4-state value, represented as a string, onto the second system task/function argument.

The usage of *$read_test_vector* is as follows:

```
reg [19:0] input_vector;

always @(posedge clock)
  $read_test_vector("vector_file", input_vector);
```

 Most of the TF routines used in this example have been discussed in this chapter, but two routines are presented in the next chapter. The tf_getcstringp() routine is used to read the value of a string in Verilog and convert it to a C string, and the tf_strdelputp() routine is used to convert a C string into Verilog logic values.

> **CD** The source code for this example is on the CD accompanying this book.
>
> - Application source file: `Chapter.010/read_test_vector_tf.c`
> - Verilog test bench: `Chapter.010/read_test_vector_test.v`
> - Expected results log: `Chapter.010/read_test_vector_test.log`

Example 10-3: *$read_test_vector* — using the TF routines

```
#include <stdlib.h>    /* ANSI C standard library */
#include <stdio.h>     /* ANSI C standard input/output library */
#include "veriuser.h"  /* IEEE 1364 PLI TF routine library */
/*********************************************************************
 * checktf routine
 *********************************************************************/
int PLIbook_ReadVector_checktf(int user_data, int reason)
{
  if (tf_nump() != 2) /* check for two system task/function */
    tf_error("Usage: $read_test_vector(\"<file\",<reg_variable>);");
  if (tf_typep(1) != TF_STRING) /* check that first arg is a string */
    tf_error("$read_test_vector arg 1 must be a quoted file name");
  if (tf_typep(2) != TF_READWRITE) /* check that 2nd arg is reg type */
    tf_error("$read_test_vector arg 2 must be a register data type");
  return(0);
}


/*********************************************************************
 * misctf routine
 *    Use the misctf routine to open test vector file at the beginning
 *    of simulation, and save the file pointer in the work area for the
 *    instance of $read_test_vector that called the misctf routine.
 *********************************************************************/
int PLIbook_ReadVector_misctf(int user_data, int reason, int paramvc)
{
  FILE *in_file;
  char *file_name;

  if (reason == REASON_ENDOFCOMPILE) { /* time to open vector file */
    if ( (in_file = fopen((char *)tf_getcstringp(1),"r")) == NULL)
      tf_error("$read_test_vector cannot open file %s",
               tf_getcstringp(1));
    tf_setworkarea((PLI_BYTE8 *)in_file); /* save file pointer */
  }
  return(0);
}


/*********************************************************************
 * calltf routine
 *********************************************************************/
int PLIbook_ReadVector_calltf(int user_data, int reason)
{
```

```
FILE *in_file;
int   vec_size = (int)tf_sizep(2);      /* bit size of tfarg 2   */
char *vector = malloc(vec_size+1);      /* memory for vector string */
bool  VERBOSE = FALSE;                  /* flag for debug output */

if (mc_scan_plusargs("debug")) VERBOSE = TRUE; /* set verbose flag */

in_file = (FILE*)tf_getworkarea();      /* retrieve file pointer */

if ( (fscanf(in_file,"%s\n", vector)) == EOF) {    /* read a vector */
  tf_warning("$read_test_vector reached End-Of-File %s",
             tf_getcstringp(1));  /* get file name from task arg 1 */
  fclose(in_file);
  tf_dofinish();  /* exit simulation at end-of-file */
  return(0);
}
if (VERBOSE)
  io_printf("$read_test_vector: Value read from file=%s\n", vector);

/* write test vector value onto system task arg 2 */
if (!(tf_strdelputp(2,(PLI_INT32)vec_size,'b',vector,0,0)) )
  if (VERBOSE)
    tf_error("$read_test_vector could not write to arg 2 at time %s",
             tf_strgettime() );
  return(0);
}
```

10.12 Summary

This chapter has presented the routines contained in the PLI TF library. A complete PLI application, called *$read_test_vector*, was presented, which used many of these routines. This application reads a test vector file from a file, and passes the value to the Verilog simulator.

The next two chapters are a continuation of the discussion of the TF library. They present how to use the TF routines to read and modify the values of system task/function arguments and how to use the TF library with *misctf routines*. The *$read_stimulus* example presented in the Chapter 12 is a more elaborate test vector reader, which can read values in any radix, and can read both the vector value and the simulation time to apply the vector from the test vector file.

CHAPTER 11 *Reading and Writing Values Using TF Routines*

*T*his chapter presents the routines in the TF library that read and modify the values of system task and system function arguments. The TF library provides a number of powerful functions that automatically convert Verilog data types and logic values to C data types and back to Verilog data types and values.

The concepts presented in this chapter are:

- Working with 4-state logic values and 8-level signal strengths
- How the PLI writes values into Verilog simulations
- Returning values of system functions
- Reading and writing Verilog 2-state values
- Reading and writing Verilog 4-state values using C strings
- Reading and writing Verilog 2-state values encoded as 32-bit C integer values
- Reading Verilog logic strengths
- Reading and writing to Verilog memory arrays

11.1 Working with a 4-logic value, multiple strength level system

The Verilog HDL supports 4 logic values, **0**, **1**, **z** and **x** and multiple levels of signal strength. The C programming language cannot directly represent the same information. The TF library offers users several ways to handle the difference in data representation between Verilog and C:

- Convert Verilog logic values from 4-state to 2-state (0 and 1); z and x values are converted to 0, and strength levels are ignored.

- Convert Verilog logic values from 4-state values to C character strings ("0", "1", "z" and "x"), and ignore logic strength levels.

- Convert Verilog logic values from 4-state values to 2-bit encoded C values, referred to as an *aval/bval* pair, and ignore logic strengths.

- Convert Verilog logic values from 4-state values to 2-bit encoded C values, and convert logic strengths to two 32-bit C integer values, which represent the strength encoding defined in the Verilog HDL standard.

The TF routines which read and write Verilog logic values automatically convert Verilog logic values to and from the various C representations, making it easy to represent hardware logic in the C language.

11.2 How the PLI writes values into Verilog

The TF routines which write values into task/function arguments and return values of system functions are referred to as *"put"* routines.

11.2.1 Putting values into a system task/function argument

To "put" a value into a task/function argument requires passing the index number of the system task/function argument as one of the inputs to the TF routine.

System task/function arguments are numbered from 1 to N, with the left-most argument being argument being number 1. For example:

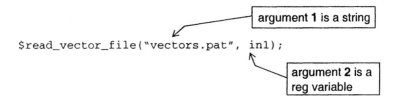

```
                                          ┌─ argument 1 is a string
$read_vector_file("vectors.pat", in1);
                                   └─ argument 2 is a
                                      reg variable
```

In the preceding example, the following TF routine will write c_value into the second task/function argument, which is input_vector:

```
                      ┌─ put 32-bit value on TF arg #2
tf_putp(2, c_value);
```

In order for the PLI to put a value into a system task/function argument, the argument must be *writable*. That is, the argument must be a Verilog HDL variable data type (reg, integer, time, real or realtime). The tf_typep() routine can be used to verify that a system task/function argument is writable. For example:

```
if (tf_typep(2) != TF_READWRITE)
    tf_error("$read_test_vector arg 2 must be a variable\n");
```

The tf_putp() and other TF put routines are discussed in more detail in subsequent sections of this chapter.

11.2.2 Returning system function values

To have the PLI return the value of a system function, the index number for the system task/function argument is set to **0**. The PLI application must have been declared as a type of *function* or *realfunction* in order to return a value using index number 0 (refer to Chapter 9, section 9.11 on page 305).

The TF library provides routines which return:

- A 2-state value of up to 32-bits wide, using the tf_putp() routine.

- A 2-state value of up to 64-bits wide, using the tf_putlongp() routine.

- A double precision floating point value, using the tf_putrealp() routine.

Using the ACC library or the VPI library, PLI applications can return system function values for vectors of any bit size, as well as signed integers.

As an example of returning a system function value, assume the Verilog HDL code contains the statement:

```
reg [31:0] result;

always @(posedge clock)
    result = $pow(a, b);
```

Then the following TF routine will return c_value as the system function return value:

```
                            put 32-bit value on system function return

tf_putp(0, c_value);
```

The return of a system function can only be written from a PLI *calltf routine*. The *calltf routine* is the C function that is called when the Verilog simulation executes the system function, and that is the only time the simulation expects a return value from the system function. It is an error to attempt to put a system function return value from a *checktf routine*, *misctf routine* or *sizetf routine*.

Example 9-1 on page 304 presented the *$pow* PLI application, which returns a 32-bit value. Example 11-1 on page 348 of this chapter illustrates returning a double-precision value for a *$realpow* system function.

11.3 Reading and writing 2-state values

Several of the TF routines allow PLI applications to read Verilog logic values from system task/function arguments and automatically convert the values into 2-state 32-bit C integers or C doubles for use in C programs. These routines automatically convert Verilog 4-state logic values into 2-state values in the C language. Logic X and Z Verilog values are converted to 0 in C, and Verilog strength levels are ignored. The routines require as an input the index number of the task/function argument to be read. A complementary set of routines write C values into a Verilog simulation, and automatically convert C data types into Verilog data types.

11.3.1 Reading 2-state values

PLI_INT32 **tf_getp (**
    PLI_INT32 **n)**   /* index number of a system task/function argument */

The `tf_getp()` routine returns the current value of argument **n** of the system task/function. This routine can access a Verilog vector of up to 32 bits, and returns the Verilog 2-state logic value as a single 32-bit integer. If the Verilog vector is wider than 32 bits, only the 32 least significant bits are read. An example of using `tf_getp()` is:

```
PLI_INT32 c_value;
c_value = tf_getp(1); /* read task/function argument 1 */
```

PLI_INT32 **tf_getlongp (**
    PLI_INT32 *** highvalue,**   /* pointer to the upper 32-bits of the value */
    PLI_INT32 **n)**   /* index number of a system task/function argument */

`tf_getlongp()` retrieves the value of argument **n** of the system task/function argument as a 64-bit integer. The routine accesses a Verilog vector of up to 64 bits, and returns Verilog 2-state logic value as a pair of 32-bit C integers, to form a 64-bit vector. The lower 32-bits of the 64-bit value are the return of tf_getlongp(). The upper 32-bits or the 64-bit value are placed into an integer variable pointed to by the first input to tf_getlongp(). If the Verilog vector is wider than 64 bits, only the 64 least significant bits are retrieved. The following example reads task/function argument number 4 as a 64-bit 2-state value:

```
PLI_INT32 high_order_bits, low_order_bits;
low_order_bits = tf_getlongp(&high_order_bits, 4);
```

double **tf_getrealp (**
  PLI_INT32   **n)**   /* index number of a system task/function argument */

The tf_getrealp() routine returns the current value of argument **n** of the system task/function as a C double (double-precision floating point number). The routine can be used for reading Verilog real values (literal values with a decimal point or Verilog real variables) or Verilog vector values. Real numbers in Verilog are stored using the ANSI definition of a C double precision number, so no conversion needs to be performed. Vectors are automatically converted into 2-state values and from integer to floating point. An example of using tf_getrealp() is:

```
double c_value;
c_value = tf_getrealp(3); /* read task/function arg #3 */
```

How Verilog vector widths are converted to C vector widths

The Verilog HDL can represent values in any vector width, from 1-bit wide (scalar) to 1 million bits wide. If a Verilog vector is less than the 32 or 64 bits that the tf_getp() and tf_getlongp() routines return, the upper bits of the C value are zero filled (not sign extended). If the Verilog vector size is greater than the 32 or 64 bit return value, then the upper bits (the most-significant bits) of the Verilog value are truncated, which means the value read into the PLI application is not accurate. The routine tf_sizep() can be used to determine the vector width of a system task/function argument and flag any potential problems. For example:

```
if (tf_sizep(1) > 64)
  tf_warning("$my_app does not support vectors greater than
    64 bits; some data may be lost\n");
```

11.3.2 Writing 2-state values

The TF library provides 3 routines which write 2-state values represented with C data types onto the arguments of a system task/function. These routines can also be used to write the return value of a system function into simulation.

PLI_INT32 **tf_putp (**
PLI_INT32	**n,**	/* index number of a system task/function arg., or 0 */
PLI_INT32	**value)**	/* a 32-bit integer value */

The `tf_putp()` routine deposits a 32-bit C integer value onto argument **n** of a system task/function. If **n** is 0, then the value is deposited as the return of a system function. An example of using `tf_putp()` to return a system function value is:

```
PLI_INT32 result
tf_putp(0, result);
```

PLI_INT32 **tf_putlongp (**
PLI_INT32	**n,**	/* index number of a system task/function arg., or 0 */
PLI_INT32	**lowvalue,**	/* lower (right-most) 32-bits of a 64-bit integer */
PLI_INT32	**highvalue)**	/* upper (left-most) 32-bits of a 64-bit integer */

`tf_putp()` deposits a 64-bit integer value, stored as a pair of 32-bit C integers, onto argument **n** of a system task/function. If **n** is 0, then the value is deposited as the return of a system function. An example of using `tf_putlongp()` to write a value into the second system task/function argument is:

```
PLI_INT32 high_bits, low_bits;
tf_putlongp(2, low_bits, high_bits);
```

PLI_INT32 **tf_putrealp (n, value)**
PLI_INT32	**n,**	/* index number of a system task/function arg., or 0 */
double	**value)**	/* a double precision real number */

The `tf_putp()` routine deposits a C double value onto argument **n** of a system task/function. If **n** is 0, then the value is deposited as the return of a system function. An example of using `tf_putlongp()` to write a value into the third system task/function argument is:

```
double c_value;
tf_putrealp(3, c_value);
```

How C vector widths are converted to Verilog vector widths

When using `tf_putp()` and `tf_putlongp()`, the bit width of a value in C might be different than the bit width of the task/function argument into which the value will be written. When a value is put into a task/function argument, the PLI follows the same rules as Verilog HDL unsigned assignment statements when there is a mismatch in the value widths, which are:

- If the C value is wider than the system task/function argument, the left-most bits of the C value are truncated.

- If the C value is narrower than the system task/function argument, the left-most bits of the task/function argument are zero filled.

11.3.3 Returning 2-state system function values

To have the PLI return the value of a system function, the index number for the system task/function argument is set to 0. In order to return a value using index number 0, the PLI application must have been declared as a type of *function* or *real-function* (refer back to Chapter 9, section 9.11 on page 305).

Using TF routines, the following types of C values can be put to a system function return value:

- A *32 bit C integer*, which is returned using `tf_putp()`. Within the Verilog simulation, the value is received as an unsigned vector of up to 32 bits.

- A pair of *32 bit C integers*, which is returned using `tf_putlongp()`. Within the Verilog simulation, the value is received as an unsigned vector of up to 64 bits.

- A *C double*, which is returned using `tf_putrealp()`. Within the Verilog simulation, the value is received as a double-precision real number.

The TF library cannot specify more that 64 bits as a system function return value. However, using the ACC library or the VPI library, PLI applications can return system function values for vectors of any vector size. The VPI routines can also return signed values to a system function.

 The return of a system function can only be written from a PLI *calltf routine*. The *calltf routine* is the C function that is called when the Verilog simulation executes the system function, and that is the only time the simulation expects a return value from the system function. It is an error to attempt to put a system function return value from a *checktf routine*, *misctf routine* or *sizetf routine*.

11.3.4 An example of reading and writing 2-state values

Example 9-1 on page 304 of Chapter 9 shows an example of using the tf_getp()
and tf_putp() routines to read the arguments of the *$pow* system function and
return a 32-bit value.

Using the TF library, the *$pow* function can easily be modified to read double-preci-
sion values as inputs, and return a double-precision value into simulation. Example
11-1 illustrates using tf_putp() or tf_getrealp() to read the arguments of a
$realpow system function. The tf_typep() routine is used to determine if the argu-
ments are integer or real number values. The tf_putrealp() routine is used to
write a real number as the *$realpow* system function return value.

> **CD** The source code for this example is on the CD accompanying this book.
>
> • Application source file: Chapter.011/realpow_tf.c
> • Verilog test bench: Chapter.011/realpow_test.v
> • Expected results log: Chapter.011/realpow_test.log

Example 11-1: *$realpow* — using TF routines to read/write 2-state values

```
/*************************************************************************
 * checktf routine
 *************************************************************************/
int PLIbook_realpow_checktf(int user_data, int reason)
{
  PLI_INT32 arg_type;

  if (tf_nump() != 2) {
    tf_error("$pow must have 2 arguments.\n");
    return(0);
  }
  arg_type = tf_typep(1);
  if (arg_type == TF_NULLPARAM)
    tf_error("$pow arg 1 cannot be null.\n");
  else
  if (    (arg_type != TF_READONLY)
       && (arg_type != TF_READONLYREAL)
       && (arg_type != TF_READWRITE)
       && (arg_type != TF_READWRITEREAL) ) {
    tf_error("$pow arg 1 must be number, variable or net.\n");
  }
  arg_type = tf_typep(2);
  if (arg_type == TF_NULLPARAM)
    tf_error("$pow arg 2 cannot be null.\n");
  else
  if (    (arg_type != TF_READONLY)
```

```
        && (arg_type != TF_READONLYREAL)
        && (arg_type != TF_READWRITE)
        && (arg_type != TF_READWRITEREAL) ) {
      tf_error("$pow arg 2 must be number, variable or net.\n");
  }
  return(0);
}

/************************************************************************
 * calltf routine
 ************************************************************************/
#include <math.h>
int PLIbook_realpow_calltf(int user_data, int reason)
{
  double    base, exp, result;
  PLI_INT32 arg_type;

  arg_type = tf_typep(1);
  if (    (arg_type == TF_READONLYREAL)
       || (arg_type == TF_READWRITEREAL) )
    base    = tf_getrealp(1);       /* read double value from tfarg 1 */
  else
    base    = (double)tf_getp(1);   /* read int value from tfarg 1 */

  arg_type = tf_typep(2);
  if (    (arg_type == TF_READONLYREAL)
       || (arg_type == TF_READWRITEREAL) )
    exp    = tf_getrealp(2);       /* read double value from tfarg 2 */
  else
    exp    = (double)tf_getp(2);   /* read int value from tfarg 2 */

  result = pow(base, exp);

  tf_putrealp(0,result);            /* return result */

  return(0);
}
```

11.4 Reading and writing 4-state logic values using C strings

Using strings in the C language is an easy way to represent the 4-state logic values of Verilog. There is a TF routine which allows PLI applications to read Verilog logic values from system task/function arguments and automatically convert the values into 4-state values represented as C strings. A set of TF routines will convert C strings back to Verilog 4-state values and write them onto system task function/arguments. These routines work with Verilog values 0, 1, z, x, but ignore the Verilog strength levels.

TIP The conversion from a Verilog logic value to a C string, and vice-versa, is expensive for simulation run-time performance. Using C strings to represent Verilog logic values is a simple way to represent 4-state logic in a printable form in the C language, and a simple way to work with Verilog vectors of any bit width. The best simulation performance will be achieved when Verilog values are represented in the C format that most closely resembles how the value is stored in a Verilog simulation.

11.4.1 Reading values as strings

PLI_BYTE8 ***tf_strgetp (**
   PLI_INT32   **n,**   /* index number of a system task/function argument */
   PLI_INT32   **format_char**)/* character in single quotes representing value format */

The tf_strgetp() routine returns the value of argument **n** of a system task/function. The routine automatically converts Verilog 4-state logic values into C character strings. The return from tf_strgetp() is a pointer to the C string. The string itself is stored in a temporary buffer within the PLI.

The string representation of the Verilog value is in an application-specified radix, which can be binary, octal, decimal or hexadecimal. The first input to tf_strgetp() is the index number of the system task/function argument to be read. The second input is a C character (in single quotes), called the *format* character, which indicates the radix in which to represent the value:

- `'b'` or `'B'` indicates the return string should be a binary representation.

- `'o'` or `'O'` indicates the return string should be an octal representation.

- `'d'` or `'D'` indicates the return string should be a decimal representation.

- `'h'` or `'H'` indicates the return string should be a hex representation.

The Verilog value which is read can be a literal number, a constant, a vector of any bit width, a real variable, or a string. The tf_strgetp() routine will convert the Verilog value into the C string using the following rules:

- A Verilog vector value will be represented in the C string exactly the same as the value in Verilog. The tf_strgetp() routine considers all vector values to be unsigned, so the value will always be a positive value (negative values are represented in their two's complement form).

- A Verilog literal integer or integer data type will be represented in the C string as a signed value if a decimal radix is used to read the value. If a binary, octal or hexadecimal radix is used to read the value, then the Verilog integer is represented as an unsigned vector, so the value will always be a positive value (negative values are represented in their two's complement form).

- A Verilog real value will be represented in the C string as a signed floating point value if a decimal radix is used to read the value. If a binary, octal or hexadecimal radix is used to read the value, then the real number will be rounded off and represented as an unsigned value. (Note: some simulators do not implement the conversion of real numbers to a decimal format correctly, and treat the decimal format the same as with binary, octal and hex conversions).

- A Verilog string value will be converted to a C string. The radix is ignored.

Following is an example of using `tf_strgetp()` to read the value of the second argument system task/function, and represent the value in hex.

```
char *i_str;
i_str = tf_strgetp(2,'h'); /* read arg 2 as 4-state string */
io_printf("Value of arg2 in PLI is %s\n", i_str);
```

 NOTE `tf_strgetp()` returns a pointer to a string, which is stored in a temporary string buffer. Because the buffer is temporary, the pointer will not remain valid. The PLI application should either use the string immediately, or copy the string into application-allocated storage.

11.4.2 Writing values represented with strings into Verilog

The TF library includes a set of routines which perform three key functions:

- Automatically convert logic values represented as C character strings into Verilog 4-state values.

- Write the value into a system task argument.

- Schedule a future time in simulation for the written value to take effect, similar to the propagation delay from an input change on a logic gate to the output change.

PLI_INT32 **tf_strdelputp (**

PLI_INT32	**n,**	/* index number of a system task/function argument */
PLI_INT32	**bit_length,**	/* number of bits to be written */
PLI_INT32	**format_char,**	/* character in single quotes indicating the value format, as **'b'**, **'B'**, **'o'**, **'O'**, **'d'**, **'D'**, **'h'** or **'H'** */
PLI_BYTE8	*** value,**	/* string representing the value to be written */
PLI_INT32	**delay,**	/* 32-bit integer delay before value is written */
PLI_INT32	**delay_type)**	/* code indicating delay method: **0** for inertial, **1** for modified transport, **2** for pure transport */

PLI_INT32 **tf_strlongdelputp (**

PLI_INT32	**n,**	/* index number of a system task/function argument */
PLI_INT32	**length,**	/* number of bits to be written */
PLI_INT32	**format_char,**	/* character in single quotes indicating the value format, as 'b', 'B', 'o', 'O', 'd', 'D', 'h' or 'H' */
PLI_BYTE8	*** value,**	/* string representing the value to be written */
PLI_INT32	**lowdelay,**	/* lower (right-most) 32-bits of a 64-bit time value */
PLI_INT32	**highdelay,**	/* upper (left-most) 32-bits of a 64-bit time value */
PLI_INT32	**mode)**	/* code indicating delay method: **0** for inertial, **1** for modified transport, **2** for pure transport */

PLI_INT32 **tf_strrealdelputp (**

PLI_INT32	**n,**	/* index number of a system task/function argument */
PLI_INT32	**length,**	/* number of bits to be written */
PLI_INT32	**format_char,**	/* character in single quotes indicating the value format, as 'b', 'B', 'o', 'O', 'd', 'D', 'h' or 'H' */
PLI_BYTE8	*** value,**	/* string representing the value to be written */
double	**delay,**	/* 32-bit integer delay before value is written */
PLI_INT32	**mode)**	/* code indicating delay method: **0** for inertial, **1** for modified transport, **2** for pure transport */

There are several inputs for `tf_strdelputp()`, `tf_strlongdelputp()` and `tf_strrealdelputp()`:

- **index** is the index number of the system task/function argument into which the value is to be written. It is illegal to specify an index of 0 to put the value as the return value of a system function using these routines.

- **length** is the bit-width of the Verilog vector value the string will be converted into, and should be set to the width of the task/function argument (which can be determined using `tf_sizep()`). The value represented as a C string will be written into the Verilog vector using the assignment rules of the Verilog language, which means that if the value is less than the maximum value which the vector stores, the upper bits of the vector are zero filled. If the value is greater than the maximum value which the vector stores, the most significant bits of the value are truncated.

- **format** is the radix format of the string representation of the value. The format is indicated as one of the following C characters:
  - `'b'` or `'B'` indicates the string is a binary representation.
  - `'o'` or `'O'` indicates the string is an octal representation.
  - `'d'` or `'D'` indicates the string is a decimal representation.
  - `'h'` or `'H'` indicates the string is a hexadecimal representation.

- **value** is a pointer to a string, or a literal string, with the logic value. The string may only contain characters which are legal for the format that is selected. An error status will be returned if the value is invalid, and the value may or may not be written into the simulation. The legal values are listed in table 11-1:

Table 11-1: Legal characters for the TF string put routines

Format	Legal Value Characters
binary	0, 1, x, X, z, Z
octal	0—7, x, X, z, Z
decimal	0—9
hexadecimal	0—9, a, A, b, B, c, C, d, D, e, E, f, F, x, X, z, Z

- **delay** is the amount of propagation delay to transpire between when the put routine is executed and when the logic change occurs in Verilog simulation.

- **mode** is the method of event scheduling that the simulation should use, represented by a literal integer of **0**, **1** or **2**, where **0** is *inertial delay*, **1** is *modified transport delay*, and **2** is *transport delay*. The delay modes define what the Verilog simulator should do if there is already a pending change scheduled for the same signal.

  - *Inertial delay*—all pending events on the signal are cancelled.

  - *Modified transport delay*—all pending events on the signal which are at a later time than this event are cancelled, which means the last event to be scheduled will always be the last event that occurs.

  - *Transport delay*—no pending events on the signal are cancelled, which means the last event to be scheduled may not be the last event which occurs.

The following examples illustrate several ways of writing values represented as strings into Verilog.

Write a literal C string with a hex value onto system task/function argument 2, as a 16-bit vector, and with no propagation delay:

```
tf_strdelputp(2, 16, 'H', "F5xZ", 0, 0);
```

Write a binary value stored in the string value_str onto system task/function argument 1, a 4-bit vector, and with a 25 time unit propagation delay, using inertial delay propagation:

```
char value_str[5];
strcpy(value_str, "1x0z");
tf_strdelputp(1, 4, 'b', value_str, 25, 0);
```

Write a hex value stored in the string `vector` onto task/function argument 2, using the bit-width of the argument, and with a 5.2 time unit propagation delay, using transport delay propagation:

```
double delay;
char *vector;
vector = malloc(tf_sizep(2) + 1);
strcpy(vector, "zzzzzzz");
delay = 5.2;
tf_strrealdelputp(2, tf_sizep(2), 'h', vector, delay, 2);
```

Specifying propagation delays when writing values:

`tf_strdelputp()` uses a 32-bit C integer value to specify the propagation delay time. `tf_strlongdelputp()` uses a 64-bit time value, specified using two 32-bit C integers for the high-order and low-order words of the time. `tf_strrealdelputp()` uses a double precision real number to specify the delay time.

To represent a 64 bit time value, the Verilog PLI uses a pair of 32-bit C integers to store the full 64 bits of simulation time. The lower 32 bits of Verilog time are placed in one integer, and the upper 32 bits of the Verilog time are stored in a second integer, as shown in the following illustration:

```
PLI_INT32   high, low;
```

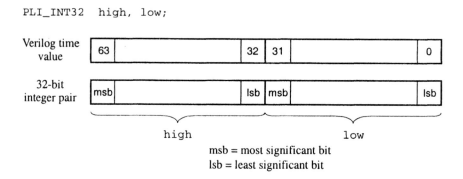

Within simulation, delays are scaled to a time unit specified in the Verilog models (using the `timescale` compiler directive). Time scaling permits each Verilog model to represent time in different units. For example, one model can represent time in nanoseconds with two decimal points of precision, and another model can represent delays in microseconds with no decimal points of precision.

The `tf_strdelputp()`, `tf_strlongdelputp()` and `tf_strrealdelputp()` routines schedule an event at a future simulation time, using the time scale of the Verilog module from which the PLI application was called. Therefore, a time value of 15 in one of these routines could represent 15 nanoseconds, 15 picoseconds, or some other time unit. The following example illustrates this automatic time scaling:

Assume a PLI application contained the following statement:

```
tf_strdelputp(1, 4, 'b', value_str, 15, 0);
```

In the following Verilog module, the delay of 15 in the PLI application would represent 15 nanoseconds.

```
'timescale 1ns/1ns
module test_chip1;
   ...
   always @(posedge clk)
     $read_test_vector("chip1.vectors", chip1_in);
   ...
```

In the next Verilog module, the same delay of 15 in the PLI application would represent 15 picoseconds.

```
'timescale 1ps/1fs
module test_chip2;
   ...
   always @(posedge clk)
     $read_test_vector(""chip2.vectors"", chip2_in);
   ...
```

Having the TF *put* routines use the calling module's time scale is generally desirable. It means the PLI application developer does not need to be concerned with time scales—the PLI automatically adjusts time to match the model in which the application is used. If needed, however, other TF routines allow the developer to read a module's time scale information, and to manually scale delays within a PLI application. Refer to Chapter 10, section 10.8 on page 329 for details on these routines.

Limitations on using TF string put routines

There are two important restrictions on using the tf_strdelputp(), tf_strlongdelputp() and tf_strrealdelputp() routines.

- These routines cannot be used to specify the return value of a system function. The Verilog standard requires that system function return values must be returned immediately, with no delay. Only tf_putp(), tf_putlongp() and tf_putrealp() can be used to return the value of a system function (the ACC and VPI libraries also have routines which can write the return value of system functions).

- These routines may not be used from a *misctf routine* that was called for REASON_ROSYNCH. At this read-only synchronization callback, the TF routines are prohibited from scheduling any new events in simulation.

No delay versus zero delay

The Verilog PLI makes a distinction between putting a value into simulation with no delay and putting a value into simulation with zero delay.

- The tf_putp(), tf_putlongp() and tf_putrealp() routines write a value into Verilog simulation instantly. When the PLI application returns back to simulation, any values written to a system task argument or system function return using these routines will already be in effect for Verilog HDL statements to use.

- The tf_strdelputp(), tf_strlongdelputp() and tf_strrealdelputp() routines schedule a value to be written into simulation. If a delay of zero is specified, the value is scheduled to be written into the task argument later in the current simulation time step. *When the system task returns back to simulation, the scheduled value will not yet have taken effect,* and other Verilog HDL statements scheduled to execute in the same simulation time step may or may not see the new value of the system task argument. Whether or not the new value is seen depends on where the value change which was scheduled by the PLI falls in the simulator's event queue in relation to other Verilog HDL events.

The following simple Verilog HDL source code illustrates the potential problem of putting a value into simulation using a delay of zero.

```
module test;
  reg [7:0] reg1, reg2;
  initial
    begin
      reg1 = 0; reg2 = 0;
      $put_value(reg1, reg2);
      $display("reg1=%d   reg2=%d", reg1, reg2);
      $strobe ("reg1=%d   reg2=%d", reg1, reg2);
      #1 $finish;
    end
endmodule
```

If the *calltf routine* for *$put_value* puts a value into reg1 using **tf_putp()**, then when *$put_value* returns to the simulation, and the *$display* statement prints the value of reg1, the <u>new</u> value will be printed.

If, however, the *calltf routine* for *$put_value* puts a value into reg2 using **tf_strdelputp()** with a delay of zero, then when *$put_value* returns to the simulation, and the *$display* statement prints the value of reg2, the <u>old</u> value might be printed. The old value is printed because the value written by the PLI has been scheduled to take place in the current time step, but it may not yet have taken effect. The *$strobe* statement which follows the *$display* will print the new value of both reg1 and reg2, because the definition of *$strobe* is to print its message at the end of the current simulation time step, after all value changes for that moment in time have taken effect.

11.5 Reading Verilog strings

In the Verilog HDL, a system task/function argument can be a string. For example:

```
$read_test_vector("vector_file.pat", input_vector);
```

A string in Verilog is represented as a vector containing the 8-bit ASCII code for each character, with no string termination character. The TF library provides a routine to read a Verilog string and convert it to a C string.

PLI_BYTE8 ***tf_getcstringp (**
  PLI_INT32   **n)**   /* index number of a system task/function argument */

tf_getcstringp() returns the value of argument **n** of a system task/function argument as a string, and returns *null* if an error occurs. The routine requires the index number of the task/function argument as an input, and expects that the value of the argument be a literal string or a vector with valid ASCII characters. The following example reads a file name stored in system task/function argument 1.

In a *checktf routine*:

```
if (tf_typep(1) != TF_STRING)
  tf_error("$read_vector arg 1 must be literal string\n");
```

In a *calltf routine*:

```
char *file_name;
file_name = tf_getcstringp(1);
```

 tf_getcstringp() returns a string pointer to a string, which is stored in a temporary string buffer. Because the buffer is temporary, the pointer will not remain valid. The PLI application should either use the string immediately, or copy the string into application-allocated storage. Refer to section 10.3 for details on PLI strings.

11.6 Reading and writing 4-state values using aval/bval encoding

There are three TF routines which read and write 4-state values of system task/function arguments, and represent the values as pairs of 32-bit C integers which encode the Verilog 4-state values. These routines also support Verilog real values and Verilog string values, and represent these values as a C double or C string, respectively (Verilog reals and strings use 2-state logic).

p_tfexprinfo **tf_exprinfo (**
    PLI_INT32   **n,**   /* index number of a system task/function argument */
    p_tfexprinfo   **info)**   /* pointer to application-allocated s_tfexprinfo structure */

The tf_exprinfo() routine reads detailed information about a system task/function argument (an *expression*). The information is retrieved into an **s_tfexprinfo** structure, which then contains the general data type of the argument, its current logic value and the number of bits in vectors. The routine returns the info pointer if successful and 0 if an error occurred.

PLI_INT32 **tf_evaluatep (**
    PLI_INT32   **n)**   /* index number of a system task/function argument */

tf_evaluatep() re-reads the logic value of a system task/function argument. The value of argument n must have been previously read using tf_exprinfo() before tf_evaluatep() may be used. The most current logic value is retrieved into the same C structure which was allocated for tf_exprinfo(). Therefore, the PLI application must maintain the s_tfexprinfo structure.

PLI_INT32 **tf_propagatep (**
    PLI_INT32   **n)**   /* index number of a system task/function argument */

tf_propagatep() writes a logic value to a system task/function argument. The value of argument n must have been previously read using tf_exprinfo() before tf_evaluatep() may be used. The new logic value to be written is stored in the same C structure which was allocated for tf_exprinfo(). Therefore, the PLI application must maintain the s_tfexprinfo structure. As soon as the value is written to the argument, the Verilog simulation will propagate the change to any Verilog expressions which read the value of that argument.

The **s_tfexprinfo** structure used by tf_exprinfo(), tf_evaluatep() and tf_propagatep() is defined in the veriuser.h file. The structure definition is listed below, followed by the explanation of the fields in structure.

```
typedef struct t_tfexprinfo {
  PLI_INT16  expr_type;    /* tf_nullparam,
                              tf_readonly,     tf_readonlyreal,
                              tf_readwrite,    tf_readwritereal,
                              tf_rwbitselect,  tf_rwpartselect,
                              tf_rwmemselect,  tf_string  */
  PLI_INT16  padding;
  struct t_vecval *expr_value_p;
  double     real_value;
  PLI_BYTE8 *expr_string;
  PLI_INT32  expr_ngroups;
  PLI_INT32  expr_vec_size;
  PLI_INT32  expr_sign;          /* may not be supported */
  PLI_INT32  expr_lhs_select;    /* may not be supported */
  PLI_INT32  expr_rhs_select;    /* may not be supported */
} s_tfexprinfo, *p_tfexprinfo;
```

 The sign, expr_lhs_select and expr_rhs_select fields are not supported by some Verilog simulators. For portability, an application should not rely on these fields.

The PLI application must first allocate memory for an s_tfexprinfo structure, and then pass a pointer to the memory to tf_exprinfo(). The routine will then fill in the fields of the structure. For example:

To allocate temporary storage that will be freed when the PLI application exits:

```
s_tfexprinfo  arg_info;
tf_exprinfo(n, &arg_info);
```

To allocate persistent storage that can be used each time the PLI application is called:

```
p_tfexprinfo  arg_info;
arg_info = (p_tfexprinfo)malloc(sizeof(s_tfexprinfo));
tf_exprinfo(n, arg_info);
```

 Using malloc(), rather than a local variable, allows the same s_tfexprinfo structure to be preserved from one call to the PLI application to another call. The PLI application must maintain a pointer to the structure so that it is available when needed. The TF work area is a good place to store the pointer, because the work area is both persistent and unique to each instance of a system task.

TIP

The fields in the s_tfexprinfo structure are explained in table 11-2, which follows.

Table 11-2: The `tf_exprinfo` structure

s_tfexprinfo field	Definition
expr_type	An integer constant which represents the general Verilog data type of the system task/function argument. The `expr_type` determines which of the remaining fields of the `s_tfexprinfo` structure will be used. The constants are: <table><tr><td>**TF_NULLPARAM**</td><td>the argument is null</td></tr><tr><td>**TF_STRING**</td><td>the argument is a string</td></tr><tr><td>**TF_READONLY**</td><td>the argument is a scalar net, vector net, net bit select, or net part select</td></tr><tr><td>**TF_READONLYREAL**</td><td>the arg is a constant real number</td></tr><tr><td>**TF_READWRITE**</td><td>the argument is a scalar reg, vector reg, integer, or time</td></tr><tr><td>**TF_READWRITEREAL**</td><td>the argument is a real variable</td></tr><tr><td>**TF_RWBITSELECT**</td><td>the argument is a bit select of a vector reg, integer, or time</td></tr><tr><td>**TF_RWPARTSELECT**</td><td>the argument is a part select of a vector reg, integer, or time</td></tr><tr><td>**TF_RWMEMSELECT**</td><td>the argument is a word select of a Verilog memory array</td></tr></table>
padding	
expr_value_p	if *expr_type* is TF_READONLY, TF_READWRITE, TF_RWBITSELECT, TF_RWPARTSELECT, or TF_RWMEMSELECT, this field contains a pointer to an array of `s_vecval` structures containing the 4-state logic value
real_value	if *expr_type* is TF_READONLYREAL, or TF_READWRITEREAL, then this field contains the real value
expr_string	if *expr_type* is TF_STRING, then this field contains a pointer to the string
expr_ngroups	if *expr_type* is TF_READONLY, TF_READWRITE, TF_RWBITSELECT, TF_RWPARTSELECT, or TF_RWMEMSELECT, then this field contains the number of elements in the array of `s_vecval` structures pointed to in the *expr_value_p* field

Table 11-2: The `tf_exprinfo` structure (continued)

s_tfexprinfo field	Definition
`expr_vec_size`	if *expr_type* is TF_READONLY, TF_READWRITE, TF_RWBITSELECT, TF_RWPARTSELECT, or TF_RWMEMSELECT, then this field contains the number of bits in task/function argument
`expr_sign`	*may* contains a flag indicating the sign type of the task/function argument, where 0 represents an unsigned value, and 1 represents a signed value. **Note:** some simulators do not use this field.
`expr_lhs_select` `expr_rhs_select`	if *expr_type* is TF_RWBITSELECT, TF_RWPARTSELECT, or TF_RWMEMSELECT, then these fields *may* contain the left-hand and right-hand indices of the bit or part select. **Note:** some simulators do not use these fields.

Representing 4-state logic values as encoded 32-bit C integers

The expression types **TF_READONLY**, **TF_READWRITE**, **TF_RWBITSELECT**, **TF_RWPARTSELECT** and **TF_RWMEMSELECT** use an *aval/bval* pair of 32-bit C integers to encode the 4 logic values of Verilog. The encoding uses 1 bit of each aval/bval pair to represent 1 bit of Verilog logic values. The encoding is:

Table 11-3: aval/bval logic value encoding

aval/bval pair	Verilog logic value represented
0/0	0
1/0	1
0/1	Z
1/1	X

The TF library defines C integers which are 32-bits wide, and therefore uses the aval/ bval pair to encode up to 32 bits of a Verilog vector. By using an array of aval/bval integer pairs, vector lengths of any size may be represented. The representation of a 40-bit vector in Verilog can be visualized as:

For the Verilog declaration: `reg [39:0] data;`

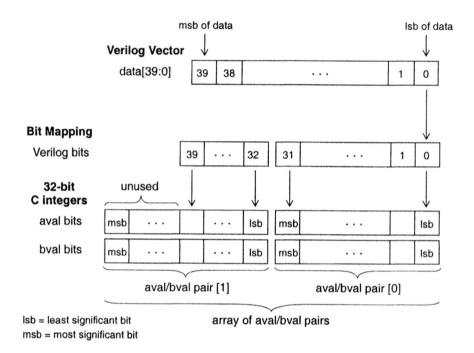

lsb = least significant bit
msb = most significant bit

The Verilog language supports any numbering convention for a vector's bit numbers. The least significant bit of the Verilog vector can be the smallest bit number, such as bit 0 (which is referred to as little endian convention). Or, the least significant bit of the Verilog vector can be the largest bit number, such as bit 39 (which is referred to as big endian convention). Verilog does not require there be a bit zero at all. Each of the following examples are valid 40-bit vector declarations in Verilog:

```
reg [39:0] data;    /* little endian -- LSB is bit 0  */
reg [0:39] data2;   /* big endian    -- LSB is bit 39 */
reg [40:1] data3;   /* little endian -- LSB is bit 1  */
```

The bit numbering used in Verilog does not affect the aval/bval representation of the Verilog vector. In the array of aval/bval pairs, the LSB of the Verilog vector will always be the LSB of the first 32-bit C integer in the array, and the MSB of the Verilog vector will always be the last bit in the array which is used. The following diagram illustrates the aval/bval array for a Verilog vector declared with a big endian convention.

For the Verilog declaration: **reg [1:40] data;**

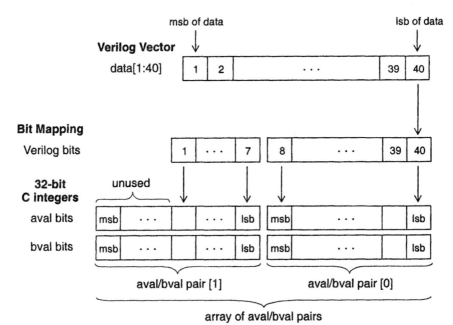

lsb = least significant bit
msb = most significant bit

The aval/bval pair is declared in an **s_vecval** structure, which is defined in the veriuser.h TF library. The structure definition is:

```
typedef struct t_vecval {
  PLI_INT32  avalbits;    /* aval/bval encoding: 0/0 == 0, */
  PLI_INT32  bvalbits;    /* 1/0 == 1, 0/1 == Z, 1/1 == X  */
} s_vecval, *p_vecval;
```

 NOTE The memory for the **s_vecval** structure is allocated and maintained by the simulator. The simulator will fill in the fields of the structure, and place a pointer to the s_vecval structure in the s_tfexprinfo structure that was allocated by the PLI application. The pointer to the s_tfexprinfo structure is passed to the simulator as an argument to tf_exprinfo().

To read a task/function argument's 4-state logic value using `tf_exprinfo()` involves three basic steps:

1. Allocate an `s_tfexprinfo` structure.

2. Call `tf_exprinfo()`, giving a pointer to the `s_tfexprinfo` structure as an argument to `tf_exprinfo()`.

3. Read the logic value of the task/function argument from the `s_tfexprinfo` structure. Use the `expr_type` field in the structure to determine what field in the structure contains the logic value. If the task/function argument is a scalar or vector signal, the logic value will be stored in an array of `s_vecval` structures, which is pointed to in the `expr_value_p` field of the `s_tfexprinfo` structure.

Example 11-2 lists a PLI application which uses `tf_exprinfo()` to read the 4-state value of any type of task/function argument, including any size of vector. Note that example 11-3 on page 366 lists a more efficient version of this example, called *$read_4state_value*.

CD The source code for this example is on the CD accompanying this book.

- Application source file: `Chapter.011/exprinfo_test_tf.c`
- Verilog test bench: `Chapter.011/exprinfo_test.v`
- Expected results log: `Chapter.011/exprinfo_test.log`

Example 11-2: *$exprinfo_test* — using `tf_exprinfo()` to read 4-state values

```
int PLIbook_exprinfoTest_calltf(int user_data, int reason)
{
  s_tfexprinfo info_s;     /* structure for tf_exprinfo() */
  p_vecval      val_array; /* pointer to value array in info struct */
  int i;

  tf_exprinfo(1, &info_s); /* read expression info for arg 1 */

  io_printf("Expression info:\n");
  switch (info_s.expr_type) {
    case TF_NULLPARAM:     io_printf(" type=TF_NULLPARAM\n"); break;
    case TF_STRING:        io_printf(" type=TF_STRING\n"); break;
    case TF_READONLY:      io_printf(" type=TF_READONLY\n"); break;
    case TF_READONLYREAL:  io_printf(" type=TF_READONLYREAL\n"); break;
    case TF_READWRITE:     io_printf(" type=TF_READWRITE\n"); break;
    case TF_READWRITEREAL: io_printf(" type=TF_READWRITEREAL\n");break;
    case TF_RWBITSELECT:   io_printf(" type=TF_RWBITSELECT\n"); break;
    case TF_RWPARTSELECT:  io_printf(" type=TF_RWPARTSELECT\n"); break;
    case TF_RWMEMSELECT:   io_printf(" type=TF_RWMEMSELECT\n"); break;
    default: io_printf(" type is unknown (%d)\n", info_s.expr_type);
  }
  io_printf(" ngroups = %d\n",      info_s.expr_ngroups);
```

```
io_printf(" vector size = %d\n", info_s.expr_vec_size);
io_printf(" sign = %d\n",        info_s.expr_sign);
io_printf(" LHS select = %d\n",  info_s.expr_lhs_select);
io_printf(" RHS select = %d\n",  info_s.expr_rhs_select);

switch (info_s.expr_type) {
  case TF_STRING:
    io_printf(" string value = %s\n", info_s.expr_string); break;
  case TF_READONLYREAL:
  case TF_READWRITEREAL:
    io_printf(" real value = %f\n", info_s.real_value); break;
  case TF_READONLY:
  case TF_READWRITE:
  case TF_RWBITSELECT:
  case TF_RWPARTSELECT:
  case TF_RWMEMSELECT:
    val_array = info_s.expr_value_p;
    io_printf(" vector value (in hex):\n");
    for (i=0; i<info_s.expr_ngroups; i++) {
      io_printf("  avalbits[%d] = %x\n", i, val_array[i].avalbits);
      io_printf("  bvalbits[%d] = %x\n", i, val_array[i].bvalbits);
    }
    break;
  }
  io_printf("\n\n");
  return(0);
}
```

11.6.1 Re-reading values previously read with tf_exprinfo()

PLI_INT32 **tf_evaluatep (**
  PLI_INT32   **n)**   /* index number of a system task/function argument */

The tf_evaluatep() routine can be used to re-read the logic value of a system task/ function argument that was previously read using tf_exprinfo(). The tf_evaluatep() routine retrieves the current value of the argument, and updates the value in the same structure that was used by tf_exprinfo(). Since only the value information in the structure is updated, tf_evaluatep() executes more efficiently than calling tf_exprinfo() a second time. *Note*: In order to use tf_evaluatep(), the PLI application must maintain the s_tfexprinfo structure.

The following PLI application example calls tf_exprinfo() from a *misctf routine* to read the value of the first task.function argument. The memory for the s_tfexprinfo is allocated using malloc so that the memory persists after the *misctf routine* has exited. The pointer to the structure is then stored in the PLI work area.

At some later simulation time, a *calltf routine* retrieves the structure pointer from the work area, and calls `tf_evaluatep()` to re-read the task/function argument's value. The PLI work area was presented in Chapter 10, section 10.7 on page 325, and the *misctf routine* is discussed in more detail in Chapter 12.

CD The source code for this example is on the CD accompanying this book.

- Application source file: `Chapter.011/evaluatep_test_tf.c`
- Verilog test bench: `Chapter.011/evaluatep_test.v`
- Expected results log: `Chapter.011/evaluatep_test.log`

Example 11-3: *$evaluatep_test* — using `tf_evaluatep()` to read 4-state values

```
/**********************************************************************
 * misctf routine
 *
 * The misctf routine is used to call tf_exprinfo() at the
 * beginning of simulation, so that the memory allocated by
 * tf_exprinfo() is only allocated one time for each instance of
 * $evaluatep_test.
 **********************************************************************/
int PLIbook_evaluatepTest_misctf(int user_data, int reason, int pvc)
{
  p_tfexprinfo info_p;      /* pointer to structure for tf_exprinfo() */
  p_vecval     val_array;  /* pointer to value array in info struct */
  int i;
  if (reason != REASON_ENDOFCOMPILE)
    return(0);  /* exit now if this is not the start of simulation */

  /* allocate memory for an s_tfexprinfo structure */
  info_p = (p_tfexprinfo)malloc(sizeof(s_tfexprinfo));
  tf_exprinfo(1, info_p);  /* read expression info for arg 1 */
  tf_setworkarea((PLI_BYTE8 *)info_p); /* save info pointer */

  io_printf("Expression info:\n");
  switch (info_p->expr_type) {
    case TF_NULLPARAM:    io_printf(" type=TF_NULLPARAM\n"); break;
    case TF_STRING:       io_printf(" type=TF_STRING\n"); break;
    case TF_READONLY:     io_printf(" type=TF_READONLY\n"); break;
    case TF_READONLYREAL: io_printf(" type=TF_READONLYREAL\n"); break;
    case TF_READWRITE:    io_printf(" type=TF_READWRITE\n"); break;
    case TF_READWRITEREAL:io_printf(" type=TF_READWRITEREAL\n"); break;
    case TF_RWBITSELECT:  io_printf(" type=TF_RWBITSELECT\n"); break;
    case TF_RWPARTSELECT: io_printf(" type=TF_RWPARTSELECT\n"); break;
    case TF_RWMEMSELECT:  io_printf(" type=TF_RWMEMSELECT\n"); break;
    default: io_printf(" type is unknown (%d)\n", info_p->expr_type);
  }
  io_printf(" ngroups = %d\n", info_p->expr_ngroups);
  io_printf(" vector size = %d\n", info_p->expr_vec_size);
  io_printf(" sign = %d\n", info_p->expr_sign);
```

```
   io_printf(" LHS select = %d\n", info_p->expr_lhs_select);
   io_printf(" RHS select = %d\n", info_p->expr_rhs_select);

 switch (info_p->expr_type) {
   case TF_STRING:
     io_printf(" string value = %s\n", info_p->expr_string); break;
   case TF_READONLYREAL:
   case TF_READWRITEREAL:
     io_printf(" real value = %f\n", info_p->real_value); break;
   case TF_READONLY:
   case TF_READWRITE:
   case TF_RWBITSELECT:
   case TF_RWPARTSELECT:
   case TF_RWMEMSELECT:
     val_array = info_p->expr_value_p;
     io_printf(" vector value (in hex):\n");
     for (i=0; i<info_p->expr_ngroups; i++) {
       io_printf("   avalbits[%d] = %x\n", i, val_array[i].avalbits);
       io_printf("   bvalbits[%d] = %x\n", i, val_array[i].bvalbits);
     }
     break;
 }
 io_printf("\n\n");
 return(0);
}
/***********************************************************************
 * calltf routine
 ***********************************************************************/
int PLIbook_evaluatepTest_calltf(int user_data, int reason)
{
 p_tfexprinfo info_p;        /* pointer to structure for tf_exprinfo() */
 p_vecval     val_array;     /* pointer to value array in info struct */
 int i;

 info_p = (p_tfexprinfo)tf_getworkarea(); /* retrieve info pointer */
 tf_evaluatep(1);                         /* re-read value of arg 1 */
 switch (info_p->expr_type) {
   case TF_STRING:
     io_printf(" string value = %s\n", info_p->expr_string); break;
   case TF_READONLYREAL:
   case TF_READWRITEREAL:
     io_printf(" real value = %f\n", info_p->real_value); break;
   case TF_READONLY:
   case TF_READWRITE:
   case TF_RWBITSELECT:
   case TF_RWPARTSELECT:
   case TF_RWMEMSELECT:
     val_array = info_p->expr_value_p;
     io_printf(" vector value (in hex):\n");
     for (i=0; i<info_p->expr_ngroups; i++) {
       io_printf("   avalbits[%d] = %x\n", i, val_array[i].avalbits);
       io_printf("   bvalbits[%d] = %x\n", i, val_array[i].bvalbits);
     }
```

```
    break;
  }
  io_printf("\n");
  return(0);
}
```

11.6.2 Writing to task/function arguments more than once

PLI_INT32 **tf_propagatep (**
  PLI_INT32   **n)**   /* index number of a system task/function argument */

The tf_propagatep() routine is used to write 4-state logic values to a system task/
function argument that was previously read using tf_exprinfo(). The value to be
written must be placed in the same s_tfexprinfo structure that was allocated for
tf_exprinfo(). Scalar and vector logic values are represented using aval/bval
pairs in an array of s_vecval structures, which were allocated by the simulator when
tf_exprinfo() was called.

tf_propagatep() can only modify the value of the *writable* Verilog data types supported by tf_exprinfo(). This means the *expr_type* must be TF_READWRITE,
TF_READWRITEREAL, TF_RWBITSELECT, TF_RWPARTSELECT or TF_RWMEMSELECT.

The tf_propagatep() routine can also be used to modify and propagate changes to
any Verilog memory word. This requires that the entire Verilog memory array be
accessed using tf_nodeinfo(), instead of tf_exprinfo(). The value to be modified is then contained in a character array pointed to in the s_tfnodeinfo structure
that is used by tf_nodeinfo(). Section 11.8 on page 378 discusses using
tf_nodeinfo() to access and modify the contents of Verilog memories.

 In order to use tf_propagatep() from one call of a PLI application to another,
the pointer to the s_tfexprinfo or s_tfnodeinfo structure must be
maintained by the application. Alternatively, the pointer to the field within the
structure which contains the value (or pointer to the value) to be modified can be
maintained. The PLI work area can be used to save the pointer.

Example 11-4 shows how the logic value of a Verilog signal can be read and then
modified using a combination of the routines tf_exprinfo(), tf_evaluatep(),
and tf_propagatep().

```
 ┌─────────────────────────────────────────────────────────────────────┐
 │  CD  The source code for this example is on the CD accompanying this book.  │
 │                                                                       │
 │  • Application source file:  Chapter.011/propagatep_test_tf.c         │
 │  • Verilog test bench:       Chapter.011/propagatep_test.v            │
 │  • Expected results log:     Chapter.011/propagatep_test.log          │
 └─────────────────────────────────────────────────────────────────────┘
```

Example 11-4: *$propagatep_test*—using `tf_propagatep()` to write 4-state values

```c
/*********************************************************************
 * misctf routine
 *
 * The misctf routine is used to call tf_exprinfo() at the
 * beginning of simulation, so that the memory allocated by
 * tf_exprinfo() is only allocated one time for each instance of
 * $read_4state_value.
 *********************************************************************/
int PLIbook_propagatepTest_misctf(int user_data, int reason, int pvc)
{
  p_tfexprinfo info_p;      /* pointer to structure for tf_exprinfo() */

  if (reason != REASON_ENDOFCOMPILE)
    return(0);  /* exit now if this is not the start of simulation */

  /* allocate memory for an s_tfexprinfo structure */
  info_p = (p_tfexprinfo)malloc(sizeof(s_tfexprinfo));

  tf_exprinfo(1, info_p);  /* read expression info for arg 1 */
  if (  (info_p->expr_type != TF_READWRITE)
     && (info_p->expr_type != TF_READWRITEREAL)
     && (info_p->expr_type != TF_RWBITSELECT)
     && (info_p->expr_type != TF_RWPARTSELECT)
     && (info_p->expr_type != TF_RWMEMSELECT) ) {
    io_printf("ERROR: Signal type not supported by $propagatep_test\n");
    tf_dofinish();
  }
  else
    tf_setworkarea((PLI_BYTE8 *)info_p); /* save info pointer */

  return(0);
}

/*********************************************************************
 * calltf routine
 *********************************************************************/
/* prototype for subroutine used by calltf routine */
void PLIbook_Print4stateValue();

int PLIbook_propagatepTest_calltf(int user_data, int reason)
{
  p_tfexprinfo info_p;      /* pointer to structure for tf_exprinfo() */

  info_p = (p_tfexprinfo)tf_getworkarea(); /* retrieve info pointer */
```

```
  io_printf("$propagatep_test called at time %d\n", tf_gettime());
  /* read current value of argument 1 */
  io_printf(" current value:\n");
  tf_evaluatep(1);
  PLIbook_Print4stateValue(info_p);

  /* modify value of argument 1 */
  switch (info_p->expr_type) {
    case TF_READWRITE:
    case TF_RWBITSELECT:
    case TF_RWPARTSELECT:
    case TF_RWMEMSELECT:
      info_p->expr_value_p[0].avalbits++;
      info_p->expr_value_p[0].bvalbits = 0;
      break;
    case TF_READWRITEREAL:
      info_p->real_value++;
      break;
  }
  tf_propagatep(1);

  /* read new value of argument 1 */
  io_printf(" new value:\n");
  tf_evaluatep(1);
  PLIbook_Print4stateValue(info_p);

  return(0);
}

void PLIbook_Print4stateValue(p_tfexprinfo info_p)
{
  int i;

  switch (info_p->expr_type) {
    case TF_READWRITEREAL:
      io_printf("  real value = %0.1f\n", info_p->real_value); break;
    case TF_READWRITE:
    case TF_RWBITSELECT:
    case TF_RWPARTSELECT:
    case TF_RWMEMSELECT:
      io_printf("  vector value (in hex):\n");
      for (i=0; i<info_p->expr_ngroups; i++) {
        io_printf("    avalbits[%d] = %x\n",
                  i, info_p->expr_value_p[i].avalbits);
        io_printf("    bvalbits[%d] = %x\n",
                  i, info_p->expr_value_p[i].bvalbits);
      }
      break;
  }
  return;
}
```

11.7 Reading 4-state logic values with strengths

The tf_nodeinfo() routine reads information about a system task/function argument (referred to as a *node*), including 4-state logic values and logic strengths.

p_tfnodeinfo **tf_nodeinfo (**
   PLI_INT32   **n,**   /* index number of a system task/function argument */
   p_tfnodeinfo **info)**   /* pointer to application-allocated s_tfnodeinfo structure */

The tf_nodeinfo() routine retrieves node information about a system task/function argument and places the information into an s_tfnodeinfo structure pointed to by info. The routine returns the value of info if successful, and 0 if an error occurred.

The tf_nodeinfo() routine is similar to tf_exprinfo(), but differs in these important ways:

- tf_nodeinfo() types more closely matches Verilog data types.

- tf_nodeinfo() can read the strength values of scalar nets.

- tf_nodeinfo() can read and modify the contents of entire Verilog memory arrays (tf_exprinfo() can only access word selects of a memory array).

- tf_nodeinfo() *cannot* read expressions in system task/function arguments (including literal values, bit and part selects of vectors, and strings).

- tf_nodeinfo() *cannot* be used with tf_evaluatep() to re-read an argument's value.

- tf_nodeinfo() *cannot* be used with tf_propagatep() to modify an argument's value. However, a specific word of a memory array that was accessed with tf_nodeinfo() can be passed to tf_propagatep().

 The Verilog language permits arbitrarily complex expressions to be used as system task/function arguments, but the IEEE 1364 Verilog standard does not clearly specify how tf_nodeinfo() should handle these expressions. To ensure portability to all simulators, a PLI application which uses tf_nodeinfo() should restrict the system task/function arguments that are accessed with the routine to the following: scalar or vector nets, scalar or vector regs, integer, time or real variables, and a word select of a one-dimensional array of reg, integer or time. *The* tf_nodeinfo() *routine may not return the same results on all simulators if a system task/function argument is a bit select or part select of a vector.*

tf_nodeinfo() uses an **s_tfnodeinfo** structure to receive the value and information of a task/function argument. This structure is defined in the veriuser.h file. The structure is listed below, and table 11-4, which follows, describes the fields of the structure.

```
typedef struct t_tfnodeinfo {
  PLI_INT16 node_type;  /* tf_null_node, tf_reg_node,
                           tf_integer_node, tf_real_node,
                           tf_time_node, tf_netvector_node,
                           tf_netscalar_node, tf_memory_node */
  PLI_INT16 padding;
  union {
    struct t_vecval     *vecval_p;
    struct t_strengthval *strengthval_p;
    PLI_BYTE8           *memoryval_p;
    double              *real_val_p;
  } node_value;
  PLI_BYTE8 *node_symbol;        /* signal name */
  PLI_INT32  node_ngroups;
  PLI_INT32  node_vec_size;
  PLI_INT32  node_sign;          /* may not be supported */
  PLI_INT32  node_ms_index;      /* may not be supported */
  PLI_INT32  node_ls_index;      /* may not be supported */
  PLI_INT32  node_mem_size;
  PLI_INT32  node_lhs_element;   /* may not be supported */
  PLI_INT32  node_rhs_element;   /* may not be supported */
  PLI_INT32 *node_handle;        /* not used */
} s_tfnodeinfo, *p_tfnodeinfo;
```

Table 11-4: The `tf_nodeinfo` structure

s_nodeinfo field	Definition	
node_type	A constant which represents the data type of the system task/function argument. The node_type determines which fields of the s_tfexprinfo structure will be used. The constants are:	
	TF_NULL_NODE	arg is null or is not a Verilog data type (e.g.: a number, expression or string)
	TF_REG_NODE	arg is a Verilog scalar or vector reg data type
	TF_INTEGER_NODE	arg is a Verilog integer data type
	TF_TIME_NODE	arg is a Verilog time data type
	TF_REAL_NODE	arg is a Verilog real data type
	TF_NETVECTOR_NODE	arg is a Verilog vector net data type
	TF_NETSCALAR_NODE	arg is a Verilog scalar net data type
	TF_MEMORY_NODE	arg is a word select of a one-dimensional array of a reg, integer or time types

Table 11-4: The `tf_nodeinfo` structure (continued)

s_nodeinfo field	Definition
`padding`	not used
`node_value`	A union of C data types that point to the logic value of the system task/function argument. The field within the union which points to the value is controlled by *node_type*:
	If *node_type* is TF_REG_NODE, TF_INTEGER_NODE, TF_TIME_NODE or TF_NETVECTOR_NODE, then `node_value.vecval_p` contains a pointer to an array of one or more `s_vecval` structures with the 4-state logic value
	If *node_type* is TF_NETSCALAR_NODE, then `node_value.strengthval_p` contains a pointer to one `s_strengthval` structure containing the logic strength
	If *node_type* is TF_MEMORY_NODE, then `node_value.memoryval_p` contains a pointer to a character array containing the 4-state logic value of the entire array
`node_symbol`	a pointer to a string containing the name of the signal in the system task/function argument
`node_ngroups`	If *node_type* is TF_REG_NODE, TF_INTEGER_NODE, TF_TIME_NODE or TF_NETVECTOR_NODE, this field contains the number of elements in the array of `s_vecval` structures pointed to by the *node_value.vecval_p* field
	If *node_type* is TF_MEMORY_NODE, then this field contains the number of characters in the character array pointed to by the *node_value.memoryval_p* field that represent one word in the Verilog memory array
`node_vec_size`	the number of bits in the system task/function argument
`node_sign`	a flag indicating the sign type of the task/function argument, 0 indicates an unsigned value, and 1 indicates a signed value. **Note:** some simulators do not use this field.
`node_ms_index` `node_ls_index`	if *expr_type* is **TF_REG_NODE** or **TF_NETVECTOR_NODE**, then these fields contain the bit number of the most significant bit and the least significant bit of the vector declaration. **Note:** some simulators do not use these fields.
`node_mem_size`	if *expr_type* is **TF_MEMORY_NODE**, then this field contains the number of addresses in the Verilog memory array.

Table 11-4: The `tf_nodeinfo` structure (continued)

s_nodeinfo field	Definition

`node_lhs_element`
 	
`node_rhs_element`
	if *expr_type* is **TF_REG_NODE** or **TF_NETVECTOR_NODE**, and the argument is a part select of a vector, then these fields contain the bit number of the most significant bit and the least significant bit of the part select. **Note:** some simulators do not use these fields.

`node_handle`	not used. **Note:** some simulators may store a pointer to the task/function argument signal in this field, but a PLI application should not rely on this data, as it may not be the same in all simulators.

 The `sign`, `node_ms_index`, `node_ls_index`, `node_lhs_element`, `node_rhs_element` and `node_handle` fields are not supported by some Verilog simulators. For portability, an application should not rely on these fields.

To read a task/function argument's 4-state logic value using `tf_nodeinfo()`, use the following basic steps:

1. Allocate an `s_tfnodeinfo` structure.

2. Call `tf_nodeinfo()` with a pointer to the `s_tfnodeinfo` structure.

3. Read the logic value from the `s_tfnodeinfo` structure. Use the `node_type` to determine what field of the `node_value` union contains the logic value, as described in table 11-4, above.

Reading vector and reg values with tf_nodeinfo()

Reading vector values and scalar `reg` values using `tf_nodeinfo()` is the same as with `tf_exprinfo()`. The logic value is retrieved into an array of `s_vecval` structures containing aval/bval pairs to encode the Verilog 4-state logic. The `node_ngroups` field indicates how many elements are in the array. The details on using `s_vecval` structures was presented earlier in this chapter, on page 361.

Reading strength values with tf_nodeinfo()

The logic value and strength value of scalar nets and bit selects of vector nets is represented as a single **s_strengthval** structure. The definition of this structure, as contained in the veriuser.h file, is:

```
typedef struct t_strengthval {
  PLI_INT32 strength0;
  PLI_INT32 strength1;
} s_strengthval, *p_strengthval;
```

The Verilog language has 8 strength levels for a logic zero, and 8 strength levels for a logic one. Each strength level is represented by a Verilog keyword, as follows:

Table 11-5: Verilog HDL strength levels and keywords

Strength Level	Strength Name	Specification Keyword	
7	Supply Drive	**supply0**	**supply1**
6	Strong Drive	**strong0**	**strong1**
5	Pull Drive	**pull0**	**pull1**
4	Large Capacitance	**large**	
3	Weak Drive	**weak0**	**weak1**
2	Medium Capacitance	**medium**	
1	Small Capacitance	**small**	
0	High Impedance	**highz0**	**highz1**

Within Verilog, the strength of a signal is stored as two 8-bit bytes, as shown in the diagram below:

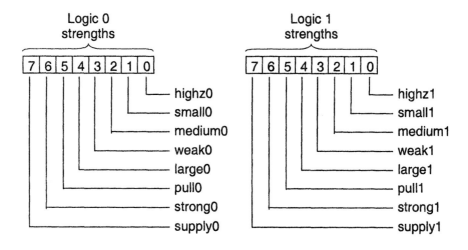

The full details on how the two strength bytes represent Verilog logic strength are defined in the Verilog language, and are outside the scope of this book. Refer to the list of Verilog HDL books on page 7 for suggestions on where to find more information on Verilog HDL strength levels.

In the PLI, the strength value returned by `tf_nodeinfo()` is represented by a pair of 32-bit C integers with a value from 00 (hex) to FF (hex). The value indicates which bits of the corresponding Verilog strength byte are set. Only 8 bits of each 32-bit C integer are used. Note that both integers represent bit 0, the highz bit, as the least significant bit of the 32-bit C integer.

Example 11-5 illustrates how to read the 4-state logic values and strength values of various Verilog HDL data types, using `tf_nodeinfo()`. This example does not read the values of a Verilog memory array. Reading memory values is presented next, in section 11.8 on page 378.

CD The source code for this example is on the CD accompanying this book.

- Application source file: `Chapter.011/nodeinfo_test_tf.c`
- Verilog test bench: `Chapter.011/nodeinfo_test.v`
- Expected results log: `Chapter.011/nodeinfo_test.log`

Example 11-5: *$nodeinfo_test* —using the `tf_nodeinfo()` routine

```
int PLIbook_nodeinfoTest_calltf(int user_data, int reason)
{
  s_tfnodeinfo    node_info;
  int             i;

  /* Get the nodeinfo structure for tfarg 1 */
  io_printf("Reading Node Information\n");
  if (!tf_nodeinfo(1, &node_info)) {
    tf_error("Error getting tf_nodeinfo for tfarg 1");
    return(0);
  }
  io_printf(" node_symbol is %s\n",
            node_info.node_symbol? node_info.node_symbol: "No Symbol");
  switch (node_info.node_type) {
    case TF_NULL_NODE:
      io_printf(" node_type = TF_NULL_NODE\n"); break;
    case TF_REG_NODE:
      io_printf(" node_type = TF_REG_NODE\n"); break;
    case TF_INTEGER_NODE:
      io_printf(" node_type = TF_INTEGER_NODE\n"); break;
    case TF_TIME_NODE:
      io_printf(" node_type = TF_TIME_NODE\n"); break;
```

```
    case TF_REAL_NODE:
      io_printf(" node_type = TF_REAL_NODE\n"); break;
    case TF_NETSCALAR_NODE:
      io_printf(" node_type = TF_NETSCALAR_NODE\n"); break;
    case TF_NETVECTOR_NODE:
      io_printf(" node_type = TF_NETVECTOR_NODE\n"); break;
    case TF_MEMORY_NODE:
      io_printf(" node_type = TF_MEMORY_NODE\n"); break;
    default:
      io_printf(" node_type = unknown (%d)\n\n", node_info.node_type);
  }
  io_printf(" node_ngroups = %d\n",      node_info.node_ngroups);
  io_printf(" node_vec_size = %d\n",     node_info.node_vec_size);
  io_printf(" node_sign = %d\n",         node_info.node_sign);
  io_printf(" node_ms_index = %d\n",     node_info.node_ms_index);
  io_printf(" node_ls_index = %d\n",     node_info.node_ls_index);
  io_printf(" node_mem_size = %d\n",     node_info.node_mem_size);
  io_printf(" node_lhs_element = %d\n", node_info.node_lhs_element);
  io_printf(" node_rhs_element = %d\n", node_info.node_rhs_element);

  switch (node_info.node_type) {
    case TF_REG_NODE:
      io_printf(" reg value (in hex):\n");
      for (i=0; i<node_info.node_ngroups; i++) {
        io_printf("    avalbits[%d] = %x\n",
                  i, node_info.node_value.vecval_p[i].avalbits);
        io_printf("    bvalbits[%d] = %x\n",
                  i, node_info.node_value.vecval_p[i].bvalbits);
      }
      break;
    case TF_INTEGER_NODE:
      io_printf(" integer value (in hex):\n");
      io_printf("    avalbits[0] = %x\n",
                node_info.node_value.vecval_p[0].avalbits);
      io_printf("    bvalbits[0] = %x\n",
                node_info.node_value.vecval_p[0].bvalbits);
      break;
    case TF_TIME_NODE:
      io_printf(" time value (in hex):\n");
      io_printf("    {avalbits[1],avalbits[0]} = %x%x\n",
                node_info.node_value.vecval_p[1].avalbits,
                node_info.node_value.vecval_p[0].avalbits);
      io_printf("    {bvalbits[1],bvalbits[0]} = %x%x\n",
                node_info.node_value.vecval_p[1].bvalbits,
                node_info.node_value.vecval_p[0].bvalbits);
      break;
    case TF_REAL_NODE:
      io_printf(" real value = %f\n",
                *node_info.node_value.real_val_p);
      break;
    case TF_NETSCALAR_NODE:
      io_printf(" scalar net value with strength (in hex):\n");
      io_printf("    strength0 = %x\n",
```

```
                    node_info.node_value.strengthval_p->strength0);
        io_printf("   strength1 = %x\n",
                    node_info.node_value.strengthval_p->strength1);
        break;
     case TF_NETVECTOR_NODE:
        for (i=0; i<node_info.node_ngroups; i++) {
          io_printf("   avalbits[%d] = %x\n",
                     i, node_info.node_value.vecval_p[i].avalbits);
          io_printf("   bvalbits[%d] = %x\n",
                     i, node_info.node_value.vecval_p[i].bvalbits);
        }
        break;
     case TF_MEMORY_NODE:
        io_printf(" memory arrays are not supported in this example\n");
        break;
   }
   io_printf("\n");
   return(0);
}
```

11.8 Reading from and writing into Verilog memory arrays

The **tf_nodeinfo()** routine can be used to both read and modify the contents of
Verilog memory arrays and variable arrays. In Verilog HDL terminology, a *memory
array* is a one dimensional array of the Verilog reg data type (regardless of the reg
vector width). A *variable array* is an array of Verilog integer or time data types.
In the IEEE 1364 Verilog standard, the description of the tf_nodeinfo() routine
does not distinguish between a memory array and a one-dimensional variable array.
The routine refers to either array type as a memory array.

TIP
Once a PLI application has obtained a pointer to a memory array using
tf_nodeinfo(), the values of the array can be both read and modified any
number of times during simulation. It is not necessary to call tf_nodeinfo()
each time access to the array is required. All that is necessary is to save the pointer to
the array which was returned from the first call to tf_nodeinfo().

NOTE
The IEEE 1364-2001 Verilog standard adds multi-dimensional arrays to the Verilog
language, as well as allowing arrays of any data type, including reals and wires. The
tf_nodeinfo() routine cannot access these new arrays. 1364-2001 also adds the
ability for the Verilog language to select bits or parts of a word within a array. Only
the VPI routines support multi-dimensional arrays and bit or part selects of array
words.

`tf_nodeinfo()` requires that a word-select of the array be specified in the system task/function argument. For example:

```
reg [23:0] RAM [0:3]; //array with 24-bit words, 4 words deep
initial
  $dump_mem_hex(RAM[2]);
```

Note that, although `tf_nodeinfo()` requires that a word select be specified, the routine ignores the value of the word select. The `tf_nodeinfo()` routine will always retrieve a pointer to the entire Verilog array, regardless of the word select which was specified (other routines, such as `tf_exprinfo()`, utilize the word select specified).

The logic values of the Verilog memory array are stored in an array of `bytes`, which is pointed to in the **node_value_p.memoryval_p** field. The `PLI_BYTE8` data type is used to ensure an 8-bit byte on all operating systems. On most operating systems the `PLI_BYTE8` data type is defined to a be a C `char` data type (Verilog standards prior to IEEE 1364-2001 used the `char` data type directly, but that type is not guaranteed to be 8 bits wide in the C standard).

This section shows how a Verilog memory array is mapped to a byte array in the C language.

Verilog array declaration syntax:

The syntax for declaring a one-dimensional array in Verilog is:

```
reg [<msb>:<lsb>] <memory_name> [<first_addr>:<last_addr>];

integer <array_name> [<first_address>:<last_address>];

time <array_name> [<first_address>:<last_address>];
```

The `integer` and `time` variables in Verilog have predefined word sizes of 32 bits and 64 bits, respectively. The word size of a `reg` data type can be any vector width, from 1-bit wide (scalar) to 1-million bits wide (the upper limit is actually defined by the simulator, but is 1 million bits in most simulators). The bit numbering of a word for a Verilog `reg` data type can use any numbering scheme. The following declarations all declare a 24-bit word size, and a 4-element array size:

```
reg [23:0] RAM1 [0:3];   //least-significant bit is bit 0
reg [0:23] RAM2 [0:3];   //least-significant bit is bit 23
reg [24:1] RAM3 [0:3];   //least-significant bit is bit 1
```

When an `integer` variable is used in an array, it is essentially the same as a `reg` declaration, with the msb as bit 31, and the lsb as bit 0. When a `time` variable is used in a memory array, it is essentially the same as a `reg` declaration, with the msb as bit 63, and the lsb as bit 0.

The array address numbering in Verilog can start and end with any address numbers, and do not need to have an address 0. The following examples all declare an array with 24-bit wide words, and 4 elements in the array:

```
reg [23:0] RAM1 [0:3];   //first word is address 0
reg [23:0] RAM2 [3:0];   //first word is address 3
reg [23:0] RAM3 [1:4];   //first word is address 1
```

11.8.1 Mapping Verilog Memory arrays into C arrays

To represent a Verilog memory array in storage represented in the C language involves three levels of mapping:

- Verilog array addresses need to be mapped to C array addresses.
- Verilog bit numbers within a word need to be mapped to a C representation.
- Verilog 4-state logic needs to be mapped to a C representation.

Mapping Verilog array numbering to C array numbering

The `tf_nodeinfo()` routine makes the bit numbering and address numbering used in Verilog transparent to the PLI application. This is done by mapping C array index numbers to the same position in a Verilog array, rather than mapping a C index number to a Verilog index number. There are two mappings which occur:

- The order of array addresses in the Verilog array are mapped into C array addresses.

    In the C language, arrays always begin with address 0. The `tf_nodeinfo()` routine maps a C array address 0 to the lowest word address in the Verilog array address. Note that, since Verilog can define array addressing in either ascending or descending order, the lowest word address number might be either the first address in the Verilog array, or it might be the last address in the Verilog array.

- The bytes within a Verilog array word are mapped to bytes in the C language.

    The `tf_nodeinfo()` routine divides a Verilog array word into 8-bit bytes, which are mapped to a byte array. The first byte in the C array is always the least-significant byte (the right-most byte) of the Verilog word, regardless of how the bits are numbered in Verilog.

The following diagrams illustrate how a Verilog array is mapped to a byte array.

A conceptual view of a Verilog array declaration

A conceptual view of a Verilog `reg` or variable array is:

`reg [msb:lsb] RAM [first_address:last_address];`

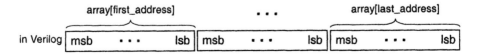

Two examples of Verilog memory declarations are:

`reg [23:0] RAM [0:3]; //lsb is lowest bit, ascending addresses`

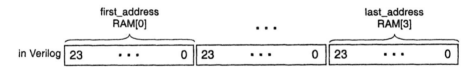

`reg [0:23] RAM [3:0]; //lsb is highest bit, descending addresses`

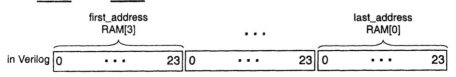

Mapping Verilog array addresses to C array addresses

The `tf_nodeinfo()` routine maps C array address 0 to the lowest number in the Verilog array address. Since Verilog can define array addressing in either ascending or descending order, the lowest address number might be either the first address of the Verilog array or it might be the last address in the Verilog array. The following diagrams show how two different Verilog memory declarations are mapped to a C array:

`reg [23:0] RAM [0:3];` //array with ascending address order

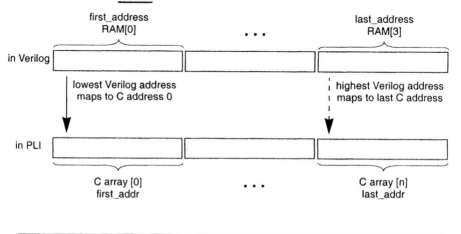

`reg [23:0] RAM [3:0];` //array with descending address order

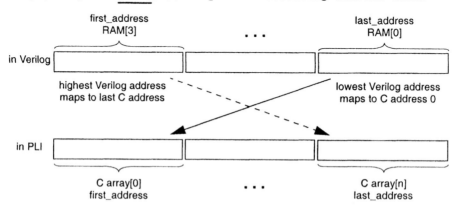

Mapping a Verilog array word into C bytes

The PLI represents each 8-bit byte of a Verilog array word as a **group**. Each group is stored as a PLI_BYTE8 data type, and the groups which make up a Verilog word are organized into a PLI_BYTE8 array. The least significant byte of a Verilog array word (which is the right-most byte) is stored in the first group in the C array.

The two diagrams which follow illustrates how two Verilog memory array declarations are represented as groups of 8-bits. *Note the reversed ordering that occurs in these diagrams:* In the Verilog HDL representation, the least-significant byte of a word is always the right most byte, but, in the PLI, the least-significant byte of the Verilog word becomes the first group in the C array. However, the bits within each byte remain in the same order as in the original Verilog word.

`reg [23:0] RAM [0:3]; //array with LSB the lowest bit number`

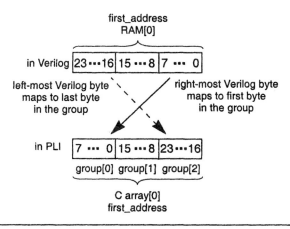

`reg [0:23] RAM [0:3]; //array with LSB the highest bit number`

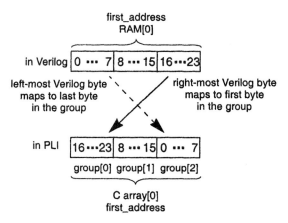

Mapping 4-state values into aval/bval pairs

Each bit in the Verilog array is represented by an aval/bval pair of bits, in order to encode the 4-state logic of Verilog. The encoding is listed in Table 11-6, and is the same encoding as used by `tf_exprinfo()` and other PLI routines which encode 4-state logic.

Table 11-6: aval/bval logic value encoding

aval/bval pair	Verilog logic value represented
0/0	0
1/0	1
0/1	Z
1/1	X

A Verilog array word is stored as a set of two 8-bit groups, with one set containing the aval values and the other set containing the bval values. Conceptually, the two sets of groups for a Verilog memory array can be thought of as follows:

```
reg [23:0] RAM [0:3]; //lsb is lowest bit, ascending addresses
```

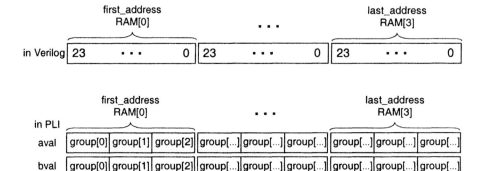

The conceptual view of a Verilog memory array shown above is represented as groups of bytes. The PLI stores the Verilog array using a `PLI_BYTE8` data type for each group, and uses an array of these bytes to store the entire Verilog memory. The aval and bval groups of each memory word are contained in the same array, as follows:

```
reg [23:0] RAM [0:3]; //lsb is lowest bit, ascending addresses
```

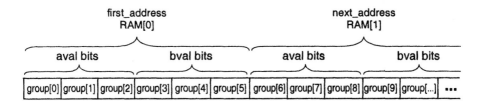

11.8.2 Verilog array information that is accessed by tf_nodeinfo()

After calling `tf_nodeinfo()` for a system task/function argument that contains a Verilog array word, the **s_tfnodeinfo** structure will contain the following information about the PLI_BYTE8 array which represents the Verilog array:

- **node_value.memoryval_p** contains a pointer to the start of the character array.

- **node_mem_size** contains the number of words represented by the Verilog array.

- **node_vec_size** contains the number of bits in each Verilog array word.

- **node_ngroups** contains the number of groups in each Verilog array word.

Using the pointer to the PLI_BYTE8 array, and knowing how many groups make up an array word, a PLI application can read the value of any word or discrete bit within the Verilog array.

Accessing one word of a Verilog array

A full word from a Verilog array can be accessed using the following formula:

Given the following call to `tf_nodeinfo()`:

```
s_tfnodeinfo node_info;
tf_nodeinfo(1, &node_info);  /* get info for tfarg 1 */
```

- The beginning of the character array and number of groups in a word are stored in the s_tfnodeinfo structure:

```
node_info.node_value.memoryval_p;

node_info.node_ngroups;
```

- Each Verilog array word is represented as an aval/bval pair of 8-bit groups, with two group sets required to represent one Verilog word. The number of groups in the character array which represent one Verilog array word can be calculated as:

```
int word_increment;
word_increment = node_info.node_ngroups * 2;
```

- To access a specific Verilog array word in the PLI_BYTE8 array, use the formula (assuming the desired word is stored in the variable memory_address):

```
PLI_BYTE8 *aval_ptr, *bval_ptr;
aval_ptr = node_info.node_value.memoryval_p
           + (word_increment * memory_address);
bval_ptr = aval_ptr + node_info.node_ngroups;
```

The following example will access the second word (address 1 in this example) of a RAM memory with 24 bit word widths.

reg [23:0] RAM [0:3]; //lsb is lowest bit, ascending addresses

s_tfnodeinfo node_info;

tf_node_info(1, &node_info);

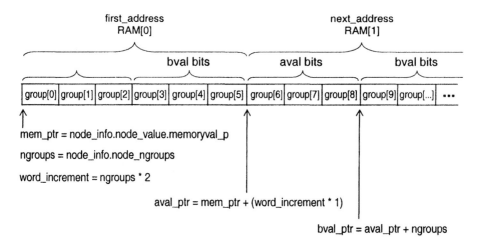

An example of reading values from a Verilog memory array

Example 11-6 illustrates two PLI applications, *$dump_mem_hex* and *$dump_mem_ascii*. The first application prints the aval/bval pair for each word in a memory array, in hexadecimal. The second application prints the value of just the aval bytes for each word in a memory array, in ASCII. Both applications call the same *checktf routine*, *misctf routine* and *calltf routine*. The user_data value associated with the PLI application is used to determine if the application routines were called by *$dump_mem_hex* (a user_data of 0) or *$dump_mem_ascii* (a user_data of 2).

CD The source code for this example is on the CD accompanying this book.

- Application source file: Chapter.011/dump_memory_tf.c
- Verilog test bench: Chapter.011/dump_and_fill_mem_test.v
- Expected results log: Chapter.011/dump_and_fill_mem_test.log

---

Example 11-6: *$dump_mem_??* — using tf_nodeinfo() with Verilog memories

```
#define HEX   0  /* values of user_data for system task names */
#define ASCII 2

/*********************************************************************
* checktf routine
*********************************************************************/
int PLIbook_DumpMem_checktf(int user_data, int reason)
{
  if (tf_nump() != 1)
    if (user_data == HEX)
      tf_error("Usage error: $dump_mem_hex(<memory_word_select>);");
    else
      tf_error("Usage error: $dump_mem_ascii(<memory_word_select>);");
  return(0);
}


/*********************************************************************
* misctf routine
*
* The misctf routine is used to call tf_nodeinfo() at the
* beginning of simulation, so that the memory allocated by
* tf_nodeinfo() is only allocated one time for each instance of
* $dump_mem_??.
*********************************************************************/
int PLIbook_DumpMem_misctf(int user_data, int reason, int paramvc)
{
  p_tfnodeinfo node_info;  /* pointer to structure for tf_nodeinfo() */

  if (reason != REASON_ENDOFCOMPILE)
    return(0);  /* exit now if this is not the start of simulation */
```

```
    /* allocate memory for an s_tfexprinfo structure */
    node_info = (p_tfnodeinfo)malloc(sizeof(s_tfnodeinfo));

    /* Get the nodeinfo structure for tfarg 1 */
    if (!tf_nodeinfo(1, node_info)) {
      tf_error("Err: $dump_mem_?? could not get tf_nodeinfo for tfarg 1");
      tf_dofinish(); /* about simulation */
      return(0);
    }
    else if (node_info->node_type != TF_MEMORY_NODE) {
      tf_error("Err: $dump_mem_?? arg is not a memory word -- aborting");
      tf_dofinish(); /* about simulation */
      return(0);
    }
    else
      tf_setworkarea((PLI_BYTE8 *)node_info); /* save info pointer */

    return(0);
}

/*************************************************************************
 * calltf routine
 *************************************************************************/
/* prototypes of functions invoked by the calltf routine */
void PLIbook_DumpMemHex();
void PLIbook_DumpMemAscii();

int PLIbook_DumpMem_calltf(int user_data, int reason)
{
    p_tfnodeinfo node_info;

    node_info = (p_tfnodeinfo)tf_getworkarea();

    io_printf("\nWithin PLI:\n");
    io_printf(" Memory array width=%d  depth=%d  ngroups=%d\n",
                node_info->node_vec_size,
                node_info->node_mem_size,
                node_info->node_ngroups);

    io_printf("\n\nnode_ms_index = %d    node_ls_index = %d\n\n",
                node_info->node_ms_index, node_info->node_ls_index);

    if (user_data == HEX)      /* application called by $dump_mem_hex */
      PLIbook_DumpMemHex(node_info);
    else                       /* application called by $dump_mem_ascii */
      PLIbook_DumpMemAscii(node_info);

    return(0);
}

/*************************************************************************
 * Function to dump each word of a Verilog array in hexadecimal
```

```
********************************************************************/
void PLIbook_DumpMemHex(p_tfnodeinfo node_info)
{
  PLI_BYTE8 *aval_ptr, *bval_ptr;
  int       word_increment, mem_address, group_num;

  io_printf(" Current memory contents of aval/bval groups in hex:\n");

  word_increment = node_info->node_ngroups * 2;
  for (mem_address = 0;
       mem_address < node_info->node_mem_size;
       mem_address++) {

    io_printf("   address %d:\t ", mem_address);

    /* set pointers to aval and bval words for the address */
    aval_ptr = node_info->node_value.memoryval_p
               + (mem_address * word_increment);
    bval_ptr = aval_ptr + node_info->node_ngroups;

    /* print groups in word in reverse order so will match Verilog:
       the highest group number represents the left-most byte of a
       Verilog word, the lowest group represents the right-most byte */
    for (group_num = node_info->node_ngroups - 1;
         group_num >= 0;
         group_num--) {
      io_printf("  group %d: %x/%x",
                group_num, aval_ptr[group_num], bval_ptr[group_num]);
    }
    io_printf("\n");
  }
  io_printf("\n\n");
  return;
}
/*********************************************************************
 * Function to dump each word of a Verilog array in ASCII
 *********************************************************************/
void PLIbook_DumpMemAscii(p_tfnodeinfo node_info)
{
  PLI_BYTE8 *aval_ptr;
  int       word_increment, mem_address, group_num;

  /* Read current memory values as a string using only aval bits */
  io_printf(" Current memory contents in ASCII are:\n");
  io_printf("  ");

  word_increment = node_info->node_ngroups * 2;
  for (mem_address = 0;
       mem_address < node_info->node_mem_size;
       mem_address++) {
    /* set pointer to aval word for the address */
    aval_ptr = node_info->node_value.memoryval_p
               + (mem_address * word_increment);

    /* print groups in word in reverse order so will match Verilog:
```

```
      the highest group number represents the left-most byte of a
      Verilog word, the lowest group represents the right-most byte */
   for (group_num = node_info->node_ngroups - 1;
        group_num >= 0;
        group_num--) {
      io_printf("%c", aval_ptr[group_num]);
   }
 }
 io_printf("\n\n");
 return;
}
```

11.8.3 Modifying values in Verilog memory and variable arrays

The `tf_nodeinfo()` routine retrieves a pointer to the actual storage within simulation of a one-dimensional Verilog `reg` or variable array. A PLI application can use this pointer to modify the values of the aval/bval pairs which represent the array.

 When a PLI application modifies the value of a Verilog array, the new value does not automatically propagate to Verilog statements which read the Verilog array.

Once a PLI application has modified an array value, the new value is available the next time a Verilog statement reads a word from the array. But, the changes in the Verilog array made by a PLI application do not automatically propagate to Verilog statements which are currently reading the Verilog array. The following two Verilog source code examples illustrate this difference:

Assuming an array declaration of:

```
reg [23:0] RAM [0:3]; //Verilog memory array
```

The following Verilog continuous assignment statement, which reads a word from the Verilog array, will not see a value change caused by a PLI application.

```
assign vector = RAM[1]; //continuously read address 1
```

The Verilog procedure listed next, which reads a word from the Verilog array at a positive edge of clock, will see the value change caused by a PLI application, the next time the statement is executed at a positive edge of clock.

```
always @(posedge clock)
   vector = RAM[1]; //read address 1 each posedge of clock
```

tf_propagatep() does not work with tf_nodeinfo()

 The description of `tf_propagatep()` in the IEEE 1364-1995 Verilog standard contains a serious errata. The IEEE 1364-2001 standard corrects this error.

The description in the IEEE 1364-1995 Verilog standard for the `tf_propagatep()` routine suggests that the routine can be used to propagate memory array contents to all Verilog HDL constructs which read values from the array. *This is an error.* The `tf_propagatep()` routine is intended to only propagate the value contained in a single word of an array (the word specified in the system task/function argument). The value to be propagated is the value pointed to in the `s_tfexprinfo` structure. The `tf_exprinfo()` routine must be used to setup the `s_tfexprinfo` structure. The value of the memory array word to be propagated is *not* the value pointed to in the `s_tfnodeinfo` structure.

 In some Verilog simulators, `tf_propagatep()` may cause some memory word values to propagate within simulation, other than the word contained in the `s_tfexprinfo` structure. This behavior is not part of the Verilog standard, and PLI applications should not depend on changes to memory array contents propagating within the Verilog simulation.

An example of writing values into a Verilog memory

Example 11-7 lists the C source code for a PLI application called *$fill_mem*, which fills every address of a Verilog array with the C language word index of each word in the array (that is, the first address of the Verilog array is loaded with the value of 0, the second address with the value of 1, etc.).

This example uses both `tf_exprinfo()` and `tf_nodeinfo()`, in order to illustrate the difference in using these routines.

- The `tf_exprinfo()` routine modifies the value of just the just memory word which is past in as a system task/function argument. The `tf_propagatep()` routine can be used in conjunction with `tf_exprinfo()` to cause the new value to propagate within simulation.

- The `tf_nodeinfo()` routine modifies the value of any memory location. The new values do not propagate within simulation. However, the new values are in the memory, and will be seen the next time a Verilog statement reads from the array.

> *CD* The source code for this example is on the CD accompanying this book.
>
> - Application source file: `Chapter.011/fill_memory_tf.c`
> - Verilog test bench: `Chapter.011/dump_and_fill_mem_test.v`
> - Expected results log: `Chapter.011/dump_and_fill_mem_test.log`

Example 11-7: *$fill_mem* — using `tf_nodeinfo()` to modify Verilog memories

```
/*****************************************************************
 * checktf routine
 *****************************************************************/
int PLIbook_FillMem_checktf(int user_data, int reason)
{
  if (tf_nump() != 2)
    tf_error("Usage: $fill_mem(mem_word_select,
word_select_address);");
  return(0);
}

/*****************************************************************
 * misctf routine
 *
 * The misctf routine is used to call tf_nodeinfo() at the
 * beginning of simulation, so that the memory allocated by
 * tf_nodeinfo() and tf_exprinfo() is only allocated one time for each
 * instance of $fill_mem.
 *****************************************************************/
typedef struct PLIbook_my_data {
  p_tfnodeinfo node_info;  /* pointer to structure for tf_nodeinfo() */
  p_tfexprinfo expr_info;  /* pointer to structure for tf_exprinfo() */
} PLIbook_my_data_s, *PLIbook_my_data_p;

int PLIbook_FillMem_misctf(int user_data, int reason, int paramvc)
{
  PLIbook_my_data_p info;  /* pointer to info structures */

  if (reason != REASON_ENDOFCOMPILE)
    return(0);  /* exit now if this is not the start of simulation */

  /* allocate memory for structure to store info structure */
  info = (PLIbook_my_data_p)malloc(sizeof(PLIbook_my_data_s));

  /* allocate memory for an s_nodeinfo and an s_tfexprinfo structure */
  info->node_info = (p_tfnodeinfo)malloc(sizeof(s_tfnodeinfo));
  info->expr_info = (p_tfexprinfo)malloc(sizeof(s_tfexprinfo));

  /* Get the nodeinfo structure for tfarg 1 */
  if (!tf_nodeinfo(1, info->node_info)) {
    tf_error("Error: $fill_mem could not get tf_nodeinfo for tfarg 1");
    tf_dofinish(); /* about simulation */
```

```
      return(0);
    }
    else if (info->node_info->node_type != TF_MEMORY_NODE) {
      tf_error("Error: $fill_mem arg is not a memory word -- aborting");
      tf_dofinish(); /* about simulation */
      return(0);
    }

    /* Get the exprinfo structure for tfarg 1 */
    if (!tf_exprinfo(1, info->expr_info)) {
      tf_error("Error: $fill_mem could not get tf_exprinfo for tfarg 1");
      tf_dofinish(); /* about simulation */
      return(0);
    }
    tf_setworkarea((PLI_BYTE8 *)info);   /* save info pointer */
    return(0);
}

/*********************************************************************
 * calltf routine
 *********************************************************************/
int PLIbook_FillMem_calltf(int user_data, int reason)
{
    int         depth, width, ngroups, word_increment, mem_address, i;
    PLI_BYTE8 *mem_ptr, *aval_ptr, *bval_ptr;
    PLIbook_my_data_p info;   /* pointer to info structures */

    info = (PLIbook_my_data_p)tf_getworkarea();

    mem_ptr     = info->node_info->node_value.memoryval_p;
    width       = (int)info->node_info->node_vec_size;
    depth       = (int)info->node_info->node_mem_size;
    ngroups     = (int)info->node_info->node_ngroups;

    /* Modify current memory values: set aval bits to memory address,
       set bval bits to 0 (2-state logic) */
    word_increment = ngroups * 2; /* 1 word = aval/bval group set */
    for (mem_address = 0;
         mem_address < 4; /* node_info->node_mem_size; */
         mem_address++) {
      aval_ptr = mem_ptr + (mem_address * word_increment);
      bval_ptr = aval_ptr + ngroups;
      aval_ptr[0] = mem_address;
      bval_ptr[0] = 0x0;
      for (i=1; i<ngroups; i++) {
        aval_ptr[i] = 0x0;
        bval_ptr[i] = 0x0;
      }
    }
    io_printf("  Memory contents have been modified by PLI\n");
    return(0);
}
```

11.8.4 Accessing one bit within a Verilog array word

A PLI application can access any discrete bit within a Verilog `reg` or variable array. This section shows two C coding methods which can be used to access the individual bits within a Verilog array.

- The first method uses a more obvious C coding style. This example is provided, in order to show how each bit is accessed.

- The second method uses a less intuitive, but more efficient C coding style.

A more obvious, but less efficient method

To access a discrete bit within a Verilog array word requires four basic steps:

1. Select a word from the `PLI_BYTE8` array which represents the Verilog array word to be modified, using the same method shown previously for reading and modifying words of Verilog reg and variable arrays (see page 385).

   ```
   PLI_BYTE8 *aval_ptr, *bval_ptr;
   aval_ptr = node_info.node_value.memoryval_p
               + (word_increment * memory_address);
   bval_ptr = aval_ptr + node_info.node_ngroups;
   ```

2. Determine which 8-bit group in the `PLI_BYTE8` array contains the desired bit in the word.

   ```
   int group_num;
   group_num = bit_num / 8;
   ```

   The desired bit number is divided by 8, because there are 8 bits in a group.

3. Determine which bit in the group contains the desired bit in the word.

   ```
   int group_bit;
   group_bit = bit_num % 8;
   ```

   The desired bit number is divided by 8 using the modulus operator, which returns the remainder of the division. The value of 8 is used because there are 8 bits in a group.

4. Set a mask to block out all unwanted bits in the group.

   ```
   PLI_BYTE8 mask;
   mask = 0x01;
   mask = mask << group_bit;
   ```

The mask is set to a hex value of 1, which sets the least-significant bit of the 8-bit byte. This bit is then shifted left to the desired bit within the group.

5. Read the value of the desired bit.

```
char aval_bit_value, bval_bit_value;
aval_bit_value = aval_ptr[group_num] & mask;
bval_bit_value = bval_ptr[group_num] & mask;
```

The value of the desired bit is accessed by logically *anding* the value of the group containing the bit with the mask value. Since only the desired bit in the mask is set, all other bits will be cleared. The resulting value of the aval/bval pair for the desired bit can then be mapped to the 4-state Verilog value.

The following C function shows how discrete bits of a memory word can be accessed using this more obvious, less efficient, C coding style. This C function is an excerpt from a complete example called *$dump_mem_bin*.

CD The source code for this example is on the CD accompanying this book.

- Application source file: `Chapter.011/dump_memory_tf.c`
- Verilog test bench: `Chapter.011/dump_and_fill_mem_test.v`
- Expected results log: `Chapter.011/dump_and_fill_mem_test.log`

Example 11-8: *$dump_mem_bin* — accessing a memory word bit; obvious method

```
void PLIbook_DumpMemBin(p_tfnodeinfo node_info)
{
  PLI_BYTE8 *aval_ptr, *bval_ptr;
  int       word_increment, mem_address, word_bit, group_num,
group_bit;
  PLI_BYTE8 aval_val, bval_val, bit_mask;

  io_printf(" Current memory contents in binary are:\n");

  word_increment = node_info->node_ngroups * 2;

  for (mem_address = 0;
       mem_address < node_info->node_mem_size;
       mem_address++) {

    io_printf("   address %d:\t ", mem_address);

    /* step 1: set pointers to aval and bval words for the address */
    aval_ptr = node_info->node_value.memoryval_p
               + (mem_address * word_increment);
```

```
  bval_ptr = aval_ptr + node_info->node_ngroups;
  for (word_bit = node_info->node_vec_size - 1;
       word_bit >= 0;
       word_bit--) {

    /* step 2: determine the group which contains the bit number */
    group_num = word_bit / 8;

    /* step 3: determine which bit in the group contains the bit */
    group_bit = word_bit % 8;

    /* step 4: set an 8-bit mask to block all unwanted bits in group */
    bit_mask = 0x01;  /* Set mask to most-signif. bit of 8-bit group */
    bit_mask = bit_mask << group_bit; /* Shift to bit to be modified */

    /* step 5: select desired aval and bval bits from the groups */
    aval_val = aval_ptr[group_num] & bit_mask;
    bval_val = bval_ptr[group_num] & bit_mask;

    /* translate aval/bval pair to 4-state logic value */
  if (!bval_val) {
      if (!aval_val)  io_printf("0");  /* aval/bval == 0/0 */
      else            io_printf("1");  /* aval/bval == 1/0 */
    }
    else {
      if (!aval_val)  io_printf("z");  /* aval/bval == 0/1 */
      else            io_printf("x");  /* aval/bval == 1/1 */
    }
  }
  io_printf("\n");
  }
  io_printf("\n");
  return;
}
```

A more efficient, but less obvious method

Drew Lynch, of Surefire Verification, who reviewed much of this book, suggests the following, more efficient C coding method to access the bits of a Verilog array word. This method uses a combination of the C shift and logical-and operators to select a single bit from a memory word, instead of the mathematical operators and intermediate variables used in the previous example.

```
aval_ptr = node_info->node_value.memoryval_p
           + (mem_address * word_increment);
bval_ptr = aval_ptr + node_info->node_value.ngroups;

aval_bit_value =
        (aval_ptr[word_bit >> 3]) & (1 << (word_bit & 0x7));
bval_bit_value =
        (bval_ptr[word_bit >> 3]) & (1 << (word_bit & 0x7));
```

      select group within word   select bit within group

The first operation in this expression, (bval_ptr[word_bit >> 3]), selects a specific 8-bit group from the aval or bval bits of a Verilog array word. Shifting the bit-select value right three times is equivalent to dividing the bit-select value by 8, using integer division. The word is divided by 8, because there are 8-bits in a group.

The second operation, (1 << (word_bit & 0x7)), selects a specific bit from an 8-bit group. By logically-anding the desired bit-select with a value of hex 7, a value of 0 through 7 is derived. This is equivalent to the modulus operation of (word_bit % 8). A literal value of 1 is then shifted left by the result of this modulus operation. This shift will select a specific bit within an 8-bit group.

The final operation in this expression is to logically-and the selected bit within the selected group. This operation will return 1 if that bit is set, and 0 if that bit is not set.

The following C function shows how a discrete bit of a memory word can be accessed using this more efficient C coding style. This C function can replace the less efficient example listed in Example 11-8 on page 395.

CD The source code for this example is on the CD accompanying this book.

- Application source file: Chapter.011/dump_memory_tf.c
- Verilog test bench: Chapter.011/dump_and_fill_mem_test.v
- Expected results log: Chapter.011/dump_and_fill_mem_test.log

---

Example 11-9: *$dump_mem_bin* — accessing a memory word bit; efficient method

```
void PLIbook_DumpMemBin(p_tfnodeinfo node_info)
{
  PLI_BYTE8 *aval_ptr, *bval_ptr;
  int       word_increment, mem_address, word_bit;
  PLI_BYTE8 aval_val, bval_val;
```

```
io_printf(" Current memory contents in binary are:\n");

word_increment = node_info->node_ngroups * 2; /* 1 word = aval/bval
pair */
for (mem_address = 0;
     mem_address < node_info->node_mem_size;
     mem_address++) {
  /* set pointers to aval and bval words for the address */
  aval_ptr = node_info->node_value.memoryval_p
             + (mem_address * word_increment);
  bval_ptr = aval_ptr + node_info->node_ngroups;
  io_printf("   address %d:\t ", mem_address);

  /* print groups in word in reverse order so will match Verilog:
     the highest group number represents the left-most byte of a
     Verilog word, the lowest group represents the right-most byte */
  for (word_bit = node_info->node_vec_size - 1;
       word_bit >= 0;
       word_bit--) {
  aval_val = (aval_ptr[word_bit >> 3]) & (1 << (word_bit & 0x7));
  bval_val = (bval_ptr[word_bit >> 3]) & (1 << (word_bit & 0x7));

    /* translate aval/bval pair to 4-state logic value */
  if (!bval_val) {
      if (!aval_val)  io_printf("0");  /* aval/bval == 0/0 */
      else            io_printf("1");  /* aval/bval == 1/0 */
    }
    else {
      if (!aval_val)  io_printf("z");  /* aval/bval == 0/1 */
      else            io_printf("x");  /* aval/bval == 1/1 */
    }
  }
  io_printf("\n");
}
return;
}
```

Modifying the bits of a Verilog array

A bit of a Verilog reg or variable array can be modified by setting the aval and bval values for the desired bit. The pointer to the desired bit is obtained in the same manner as when reading a bit value. The PLI stores the value of a Verilog array word as 8-bit groups. Therefore, an 8-bit mask value can be logically and'ed or or'ed with the group, to set of clear a specific bit.

The following steps can be used to set a 4-state logic value in the aval/bval group pair:

1. Select a word from the PLI_BYTE8 array which represents the Verilog array word to be modified, using the same method shown previously for reading and modifying words of Verilog reg and variable arrays.

```
PLI_BYTE8 *aval_ptr, *bval_ptr;
aval_ptr = node_info.node_value.memoryval_p
           + (word_increment * memory_address);
bval_ptr = aval_ptr + node_info.node_ngroups;
```

2. Determine which 8-bit group in the PLI_BYTE8 array contains the desired bit in the Verilog word.

```
int group_num;
group_num = bit_num / 8;
```

3. Determine which bit in the 8-bit group contains the desired bit in the word.

```
int group_bit;
group_bit = bit_num % 8;
```

4. Set a mask to block out all unwanted bits in the group.

```
PLI_BYTE8 mask;
mask = 0x01;
mask = mask << group_bit;
```

The mask is set to a hex value of 1, which sets the least-significant bit of the 8-bit character. The least-significant bit is then shifted left to the desired bit within the group.

5. Modify desired aval and bval bits from the groups. To clear a bit: logically AND group with the inverse of the mask . To set a bit, logically OR group with the mask. For example:

```
switch (bit_val) {
   case '0': aval_ptr[group_num] = aval_ptr[group_num] & ~bit_mask;
             bval_ptr[group_num] = bval_ptr[group_num] & ~bit_mask;
             break;
   case '1': aval_ptr[group_num] = aval_ptr[group_num] |  bit_mask;
             bval_ptr[group_num] = bval_ptr[group_num] & ~bit_mask;
             break;
   case 'z': aval_ptr[group_num] = aval_ptr[group_num] & ~bit_mask;
             bval_ptr[group_num] = bval_ptr[group_num] |  bit_mask;
             break;
   case 'x': aval_ptr[group_num] = aval_ptr[group_num] |  bit_mask;
             bval_ptr[group_num] = bval_ptr[group_num] |  bit_mask;
             break;
}
```

TIP The preceding steps use a less efficient coding style, in order to explicitly show the steps required to modify a specific bit of a Verilog array. A more efficient method of modifying a specific bit would be to use logical operators in a compound expression, instead of using mathematical operators and intermediate variables. The concepts shown in example 11-9 on page 397 can be applied to this example, in order to make the code more efficient.

11.9 Summary

This chapter has presented the routines contained in the PLI TF library, which are used to read and write the logic values of system task and system function arguments, and to specify the return value of a system function. The Verilog HDL provides a number of logic value constructs which do not have a direct counterpart in the C language, such as 4-state logic and vectors with arbitrary bit widths The TF routines provide several options for converting values from Verilog to C and from C to Verilog.

The next chapter is a continuation of the discussion of the TF library, and presents how to use the TF library with *misctf routines*.

CHAPTER 12 *Synchronizing to Simulations Using Misctf Routines*

*T*he *misctf routine* offers several very useful features that can enhance the functionality of PLI applications. There are a number of TF routines which work specifically with the *misctf routine*. The concepts presented in this chapter are:

- The purpose of the *misctf routine*
- Automatic callbacks for simulation events
- Application scheduled callbacks at the end of a simulation time step
- Application scheduled callbacks at future simulation times
- Task/function argument value change callbacks
- Simulation save and restart callbacks

12.1 The purpose of the misctf routine

The PLI *misctf routine* is unique from the *calltf routine*, *checktf routine* and *sizetf routine*. The *misctf routine* is called for a number of miscellaneous reasons during simulation, which are not directly related to the simulation having encountered or executed the user-defined system task/function name. As an example, if the Verilog HDL source code contained the user-defined system task:

```
always @(posedge clock)
  $read_test_vector("vectors.pat", input_bus);
```

The **checktf routine** will be called when the simulation compiler or elaborator encounters the *$read_test_vector* task, and the **calltf routine** will be called at every positive edge of clock when the simulator executes the *$read_test_vector* task. The **misctf routine** will be called for miscellaneous *simulation events* that might occur during simulation, such as:

- Elaboration or loading of the simulation is complete, and simulation is about to start running.
- The simulation executed a break point and entered interactive debug mode.
- The simulation finished and is exiting.
- The state of the simulation is being saved.
- Simulation is being restarted from a saved state.

In addition to the simulation events that will call a *misctf routine*, PLI applications can schedule callbacks to the *misctf routine* for other specific reasons, such as:

- Simulation is finished processing the current simulation time step.
- Simulation has reached a specific future simulation time step.
- An argument of a user-defined system task has changed value.

12.1.1 The misctf reason input

Whenever a *misctf routine* is called, it is passed three C function inputs:

1. **user_data**: The integer value that was specified when the user-defined task/function name was registered using the PLI interface mechanism. Refer to chapter 9, sections 9.9 and 9.11 for more details on how to use the user_data value.

2. **reason**: An integer constant that is generated by the Verilog simulator to indicate why the simulator called the *misctf routine*. The reason constants are defined in the veriuser.h file.

3. **paramvc**: An integer value which indicates the index number of an argument to a system task that changed value. The *paramvc* input is covered in section 12.5.

Table 12-1 lists the reasons for which a *misctf routine* can be called, and which the IEEE 1364 standard says every Verilog simulator should support.

Table 12-2 lists additional reasons the *misctf routine* may be called, which the IEEE 1364 standard indicates are optional events that may or may not be occur in a specific simulator. The constants for these reasons are included in the standard IEEE veriuser.h header file. The IEEE standard also allows simulators to add additional reason constants unique to that product, which are not in the IEEE veriuser.h file.

All reason constants defined in `veriuser.h` are defined in both all lower case letters and in all upper case letters. There is no difference between the two.

Table 12-1: IEEE 1364 standard *misctf* callback reasons

Reason Constant	Description
`reason_endofcompile` `REASON_ENDOFCOMPILE`	end of Verilog source elaboration/start of execution
`reason_finish` `REASON_FINISH`	simulation is finished and is preparing to exit
`reason_paramvc` `REASON_PARAMVC`[†]	change of value on a user-defined system task/function argument
`reason_synch` `REASON_SYNCH`[†]	end of a time step flagged by `tf_synchronize()`
`reason_rosynch` `REASON_ROSYNCH`[†]	end of a time step flagged by `tf_rosynchronize()`
`reason_reactivate` `REASON_REACTIVATE`[†]	a simulation time step has been reached which was flagged using `tf_setdelay()`

Table 12-2: IEEE 1364 optional *misctf* callback reasons

Reason Constant	Description
`reason_paramdrc` `REASON_PARAMDRC`[†]	a strength change occurred on the driver of a user-defined system task/function argument
`reason_force` `REASON_FORCE`	a procedural `force` or procedural continuous `assign` statement was executed
`reason_release` `REASON_RELEASE`	a procedural `release` or procedural `deassign` was executed
`reason_disable` `REASON_DISABLE`	a procedural `disable` statement was executed
`reason_interactive` `REASON_INTERACTIVE`	a simulation breakpoint was executed, such as the *$stop* built-in system task
`reason_scope` `REASON_SCOPE`	simulation changed interactive debug scope, such as by executing the *$scope* built-in system task
`reason_startofsave` `REASON_STARTOFSAVE`	simulation has started executing a checkpoint save, such as the *$save* built-in system task

Table 12-2: IEEE 1364 optional *misctf* callback reasons (continued)

Reason Constant	Description
reason_save **REASON_SAVE**	simulation has finished executing a checkpoint save
reason_restart **REASON_RESTART**	simulation has executed a checkpoint restart, such as the *$restart* built-in system task
reason_reset **REASON_RESET**	simulation has started executing a reset back to time 0, such as executing the *$reset* built-in system task
reason_endofreset **REASON_ENDOFRESET**	simulation has finished executing a reset to time 0

[†] indicates that the *misctf* routine will only be called for this reason if requested by the PLI application. The *misctf* routine will be called automatically for all other reasons, if that event occurs in simulation.

Example 12-1 illustrates using the reason input. *Notice that the reason the misctf routine is called is passed into the routine as the second C function input.*

Example 12-1: using the *misctf* reason input

```
int PLIbook_ReadVector_misctf(int data, int reason, int paramvc)
{
  FILE *in_file;
  char *file_name = (char *)tf_getcstringp(1);
  if (reason == REASON_ENDOFCOMPILE) { /* time to open vector file */
    if ( (in_file = fopen(file_name,"r")) == NULL)
      tf_error("$read_test_vector cannot open file %s", file_name);
    tf_setworkarea((PLI_BYTE8 *)in_file); /* save file pointer */
  }
  if (reason == REASON_FINISH) {  /* time to close vector file */
    in_file = (FILE *)tf_getworkarea();  /* retrieve file pointer */
    fclose(in_file);
  }
  return(0);
}
```

12.2 Automatic callbacks for simulation events

The *misctf routine* will be called automatically for a number of miscellaneous simulation events. These callbacks to the *misctf routine* allow a PLI application to perform specific operations, or to process data at those times.

For most automatic callback reasons, the *misctf routine* will not be called until the *calltf routine* for that instance of the system task has been executed at least once. For certain automatic callback reasons, however, the *misctf routine* will be called for each instance of the system task in the design, even if the system task is not executed during simulation. These are: REASON_ENDOFCOMPILE, REASON_STARTOFSAVE, REASON_SAVE, REASON_RESTART, REASON_RESET and REASON_ENDOFRESET.

 The *misctf routine* is called **automatically** for most reasons, many of which may not be needed by a particular PLI application. It is imperative to check the reason input, and only perform operations for the desired reasons.

TIP
The *misctf routine* has an inherent performance inefficiency, because it is called automatically for reasons that a PLI application may not need. It is not required for a PLI application to specify a *misctf routine*, and the run-time execution of a simulation can be improved if no *misctf routine* is specified. However, there are powerful features in the TF part of the PLI which require using the *misctf routine*.

The **REASON_ENDOFCOMPILE** *misctf* callback reason is poorly named. The callback does not occur at the end of compilation, but, instead, the callback occurs just before simulation time zero, when simulation is invoked and the simulation data files are loaded. A better term than end-of-compile would be start-of-simulation, indicating that the Verilog source code has been loaded, and simulation is about to start running.

There are a number of ways the REASON_ENDOFCOMPILE can be used. PLI applications can take advantage of the fact that start-of-simulation only occurs once in a simulation. This is a good time to perform any operations that need to be executed just once during the simulation, such as allocating memory for the PLI application, opening disk files, and opening graphics windows.

TIP
Use the TF work area to save memory pointers, file pointers and other information that needs to be preserved during simulation. See section 10.7 on page 325 of Chapter 10 for more information on the TF work area.

The **REASON_FINISH** callback indicates that the simulator is about to exit, which is generally caused by a *$finish* built-in system task having been executed in the Verilog source code, or a tf_dofinish() routine having been executed in a PLI application.

Many Verilog simulators also provide an exit command from the simulator's interactive debug environment. The REASON_FINISH callback allows a PLI application to clean up before simulation exits. For example, if the PLI application had allocated memory, opened disk files, or opened graphics windows, then the application has the opportunity to free the memory and close the files and windows at the time the simulator exits.

The **REASON_INTERACTIVE** callback occurs whenever the simulator encounters a breakpoint and enters the simulator's interactive debug environment. The *$stop* built-in system task and tf_dostop() routine are two methods of specifying breakpoints, and most simulators have additional ways to specify breakpoints.

Example 12-2, which is listed below, is an excerpt from a PLI application called *$read_stimulus*, which reads stimulus from a disk file. The stimulus includes both a test vector value and a simulation time to apply the vector. The complete *$read_stimulus* application is listed in example 12-7 on page 425. This example uses the *misctf routine* automatic callbacks to:

- Allocate a block or memory for data storage and open a file at the start of simulation (REASON_ENDOFCOMPILE).

- Free the data structure memory and close the file at the end of simulation (REASON_FINISH).

- Print a status message each time simulation halts at a breakpoint (REASON_INTERACTIVE).

- Save the data structure memory if the simulation state is saved (REASON_SAVE).

- Restore the data structure memory and file read position if the simulation is restarted from a saved state (REASON_RESTART).

Note that some of the automatic callback reasons used in this example are not part of the required callbacks in the IEEE standard. They are, however, part of the recommended optional callbacks specified in the standard. It is not a portability problem to use these optional callback reasons in a PLI application, since the constants which represent the optional reasons are defined in the IEEE standard. If a simulator has not implemented the optional callback features, the callback will simply never occur with that simulator.

Example 12-2: partial-code for using automatic *misctf routine* callbacks

```
int PLIbook_ReadVector_misctf(int user_data, int reason, int paramvc)
{
  switch(reason) {
    case REASON_ENDOFCOMPILE: /* misctf called at start of simulation */
      /* allocate a memory block for data storage */
      data_p = (p_stim_data)malloc(sizeof(s_stim_data));
```

```
        /* store a pointer to the application data in the work area */
        tf_setworkarea((PLI_BYTE8 *)data_p);
        /* add code to open stimulus file and save pointer to file */
        break;

    case REASON_REACTIVATE:  /* misctf called by tf_setdelay */
        /* get the pointer to the data structure from the work area */
        data_p = (p_stim_data)tf_getworkarea();
        /* add code to read next line from file and apply to simulation */
        break;

    case REASON_FINISH:    /* misctf called at end of simulation */
        /* add code to close files and fee any allocated memory */
        break;

    case REASON_SAVE:  /* misctf called end of simulation */
      /* add code to save PLI application data */
      break;

    case REASON_RESTART:  /* misctf called end of simulation */
      /* add code to retrieve saved PLI application data */
      break;
  }
  return(0);
}
```

Checktf routines versus reason_endofcompile callbacks

A *checktf routine* and a *misctf routine* at a REASON_ENDOFCOMPILE callback are both called prior to simulation time zero, and both callbacks occur for each instance of a system task/function. The IEEE standard does not state when in the elaboration/linking phase the *checktf routine* will be called. Potentially, a *checktf routine* could be called before the simulator has finished building its simulation data structure, which means some elaboration-time activity—such as parameter redefinitions—may not be completed before the *checktf routine* is called. On the other hand, a *misctf routine* callback for REASON_ENDOFCOMPILE will always occur immediately before simulation starts running, after the simulator has completed building its internal data structures.

To be portable to all Verilog simulators, PLI applications should limit activity in a *checktf routine* to verifying the correctness of system task functions only. Activity such as allocating memory, opening files, and other start-of-simulation actions should not be performed during compile/load time, but instead should be postponed until the *misctf routine* is called at the end of elaboration, which is the start of simulation.

12.3 Application scheduled callbacks at the end of a time step

A PLI application can schedule callbacks to the *misctf routine* at the end of the current simulation time step. This capability allows PLI applications to synchronize with the activity in simulation. The callback to the *misctf routine* can be scheduled from either a *calltf routine* or a *misctf routine* (but not a *sizetf routine* or *checktf routine*). The TF library allows two types of synchronization callbacks within the current simulation time step:

- *Read-write synchronization* will call the *misctf routine* at the end of all known events in the current simulation time. A PLI application is allowed to write logic values into the simulation in the current time (e.g.: a PLI application can add more events in the current simulation time after all known events have been processed).

- *Read-only synchronization* will call the *misctf routine* at the end of all events in the current simulation time. A PLI application is *not* allowed to write logic values into the simulation in the current or a future simulation time.

There are a number of reasons a PLI application might need to synchronize activity with the end of a simulation time step. One example would be to communicate value changes to a C language model, and have the C model pass value changes for that time step back to simulation. This type of activity would utilize a read-write synchronization.

Another example would be to wait until all activity in a simulation time step is stable before reading logic values from the simulation. This type of activity would utilize a read-write synchronization. In the following Verilog HDL source code:

```
always @(a or b)              always @(a or b)
    $my_strobe(sum);              sum = a + b;
```

parallel (concurrent) activity

The *calltf routine* for *$my_strobe* will be called every time a or b changes value, at which time the PLI application might need to read the value of argument 1 (the sum signal). At the same moment in simulation time, however, the sum variable is also scheduled to change value in the simulation. This is a classic race condition in Verilog simulation that is caused by the concurrent activity of a value being read and written in the same simulation time step. The Verilog standard states that most concurrent activity can be processed in any order by the simulation, which means that the *calltf routine* can be executed either before or after the sum signal changes at the positive edge of the clock. The outcome of this race condition is unpredictable.

The race condition in the above example can be resolved by synchronizing the *$my_strobe* PLI application to the very end of the simulation time step in which the change on a or b occurs. When a read-only synchronization callback to the *misctf routine* occurs, any and all statements for that time step will have been executed, and the values of the arguments to the system task/function will be at their most current value. For the *$my_strobe* example, when the *calltf routine* is invoked when a or b changes, it can schedule a read-write synchronize callback to the *misctf routine* at the end of that time step. The *misctf routine* can then read the value of sum, and be assured that it has the most current value for that moment in time.

The TF library provides two routines to synchronize to the end of a simulation time step, one for read-write synchronization, and one for read-only synchronization.

PLI_INT32 **tf_synchronize ( )**

The `tf_synchronize()` routine schedules a callback to the *misctf routine* at the end of the current simulation time step in a ***read-write*** mode. The reason constant passed to the *misctf routine* is **REASON_SYNCH**. In this mode, the *misctf routine* can read the values of objects, and can also modify the values of objects. The routine returns 0 if successful, and 1 if an error occurred.

PLI_INT32 **tf_rosynchronize ( )**

`tf_rosynchronize()` schedules a callback to the *misctf routine* at the end of the current simulation time step in a ***read-only*** mode. The reason constant passed to the *misctf routine* is **REASON_ROSYNCH**. In this mode, the *misctf routine* is only permitted to read values. It is not allowed to modify values in either the current or a future simulation time. The routine returns 0 if successful, and 1 if an error occurred.

Verilog event scheduling

The IEEE 1364 Verilog standard contains a generalized description of an event scheduling algorithm for Verilog simulators. This algorithm is complex, and beyond the scope of this book. The following paragraphs present a simplified version of the IEEE algorithm, in order to show the differences between **tf_synchronize()** and **tf_rosynchronize()** simulation callbacks. In essence, there are four distinct regions of events within a simulation time step in Verilog, which are referred to a *slots* in the following figure. Within each slot, certain types of simulation events are scheduled to be executed. ***The Verilog standard allows simulators to optimize many of the events within a slot in any order, and to interleave the types of events within each slot in any order.***

Figure 12-1: Organization of events in a Verilog simulation time step

Current Simulation Time:
  Slot 1: Active Events

events within
a slot may be
interleaved
in any order!

  Evaluate right-hand side of nonblocking assignments
  Evaluate right-hand-side & change left-hand-side of blocking assignments
  Evaluate right-hand-side & change left-hand-side of continuous assignments
  Evaluate changes on primitive inputs, & schedule changes to outputs
  Print outputs from scheduled $display and $write tasks
  Call PLI calltf routines for scheduled system tasks and system functions

  Slot 2: Nonblocking Assignment Update Events
  Change left-hand side of nonblocking assignments

  Slot 3: Read-write Synchronization
  Call misctf routines which were scheduled using tf_synchronize()

  Slot 4: Read-only Synchronization
  Print outputs from scheduled $strobe and $monitor tasks
  Call misctf routines which were scheduled using tf_rosynchronize()

Next Simulation Time:
  . . .

> NOTE ► *Some simulators reverse the order of nonblocking assignment update events and read-write synchronization!* The IEEE 1364 description of event scheduling is much more complex than illustrated above, and allows simulators latitude on the ordering of some types of events.

The IEEE 1364 Verilog standard is purposely written in such a way as to allow simulators a certain degree of latitude on how to implement their internal event scheduling algorithms. This allows simulator vendors the flexibility needed to create competitive products. The latitude is also necessary to allow simulators the freedom to order events differently as required for the behavior of different types of hardware. CMOS hardware does not behave exactly the same as ECL hardware, for example. The flexibility provided in the Verilog standard is one of the strengths of Verilog. But, for PLI application developers, the latitude in event scheduling permitted by the Verilog standard can be a source of frustration. It means that when synchronizing PLI applications within a simulation time step, *misctf routine* callbacks scheduled by tf_synchronize() may occur in a different order in different simulators. There is no simple fix for this; it is just the nature of the Verilog language. Note, however, that the tf_rosynchronize() does not have this ambiguity. The read-only synchronization events will occur after all other types of simulation events within a time step, in all Verilog simulators.

In the preceding illustration, a simulation will first execute all scheduled events in slot 1, the active events. Processing an active event may cause another active event to be scheduled, or it may cause a nonblocking assignment update event to be scheduled. Consider the following Verilog HDL code:

```
always @(a or b)
  begin
    sum = a + b;
    $my_strobe(sum, zero_flag);
  end

assign zero_flag = ~|sum;
```

In this example, when either a or b changes, the statements sum = a + b; and $my_strobe(sum, zero_flag); are scheduled as active events. Because these events are within a sequential begin—end statement group, the Verilog standard guarantees that these active events will be executed in the order they are listed. However, when sum changes, the statement assign zero_flag = ~|sum; is also scheduled as an active event. The assign statement is a concurrent (parallel) process, and simulators are allowed to interleave these concurrent events between sequential events. That is, the new active event to assign zero_flag can be scheduled to occur either before or after the active event to execute the *calltf* routine for *$my_strobe*. For simulation results, the affects of the event ordering should not matter—everything appears to have happened at the same time in simulation time.

This latitude to interleave active events provided by the IEEE standard may affect the results of PLI *calltf routines*. In one simulator, the *calltf routine* might be executed before the concurrent continuous assignment statement has been executed, and therefore will see the value of zero_flag before it changes. In another simulator, however, the assignment to zero_flag might be interleaved before the *calltf routine*, and therefore the *calltf routine* will see value of zero_flag after the assignment is executed. This situation is very easy to rectify. Instead of immediately reading the value of zero_flag, the *calltf routine* can request that the simulation call the *Misctf routine* using the reason **tf_rosynchronize()**. This callback will be scheduled in the read-write synchronization event list, after all active events have been processed.

Simulation will not proceed to slot 2 until it has executed all scheduled active events within slot 1, which includes executing the *calltf* routine for *$my_strobe* in the preceding example. Simulators then move the events in slot 2 to the active event list, and begin processing those events. As these new active events are processed, they may cause additional active events, nonblocking assignment update events or read-write synchronization events to be scheduled.

Once all the events that were moved from slot 2 to the active event list have been processed, simulators will move the events in slot 3 to the active event list, and begin processing those. Once again, these new active events may cause new active events, nonblocking assignment update events or read-write synchronization events to be scheduled.

Most Verilog simulators will execute all the originally scheduled events in slots 2 and 3, by moving those events to the active event slot. The simulator will then return to slot 2, and move any new events that have been scheduled to the active event list and process those new events. Then the simulator will move any new slot 3 events to the active list and process those. This loop through slots 1, 2 and 3 will continue until all three slots are completely empty.

It is critical to note that when simulation returns to a slot that was previously processed, only *new* events will be processed—events which have already been executed have been removed from that slot's event list, and will not be executed a second time. Assuming there are no zero-delay infinite loops in the Verilog code (which is possible), simulation will eventually complete all events in slots 1, 2 and 3, and then proceed to slot 4.

In the read-write synchronization slot, which may be either slot 2 or 3, any registered *misctf routine* callbacks which were requested using **tf_synchronize()** will be invoked. In this mode, the PLI is allowed to schedule new events in the current simulation time step. These new events will be scheduled as active events in slot 1. The simulation callback routine can also register another *misctf routine* callback using **tf_synchronize()**. The new callback will be scheduled as a new read-write synchronization event, which will not be executed until after the simulator revisits the read-write synchronization slot. Any scheduled active events and nonblocking assignment update events will be executed before the read-write synchronization slot is revisited.

Once all events in slots 1, 2 and 3 have been processed, simulation will proceed to slot 4, where any registered misctf routine callbacks which were requested using **tf_rosynchronize()** will be invoked. In the read-only synchronize mode, the PLI is allowed to read information from the simulation, but the PLI may *not* schedule new events in the current simulation time step. Slot 4 represents the true end of the current simulation time step. When the *misctf routine* is invoked with REASON_ROSYNCH, the routine may not modify or schedule the modification of any values in simulation. However, the *misctf routine* is allowed to schedule another read-only synchronize callback to itself, using tf_rosynchronize(). The VPI library, however, is only prohibited from scheduling new events at the current time. VPI routines can still schedule events at future simulation times.

Read-write synchronization ambiguity

 A critical latitude permitted in the IEEE 1364 Verilog standard is that slot 2 and slot 3 in the illustration show in Figure 12-1 can be reversed. This ambiguity means that in some simulators, a read-write synchronization will occur after nonblocking assignments have been updated. In other Verilog simulators, a read-write synchronization will occur before nonblocking assignments are updated.

Slot 2 in the illustration shown in Figure 12-1 on page 410 may be either the non-blocking assignment update events or the read-write synchronization events. If it is the latter, then the PLI application will be able to see the results of all active events, but the left-hand side of the nonblocking assignment update events will not yet have been processed. Therefore the PLI application will not see the results of those events. The following Verilog code illustrates this potential problem:

```
always @(posedge clock)
  begin
    sum <= a + b;       //non-blocking assignment to sum
    $my_strobe(sum);
  end
```

In this example, using a read-write synchronization will not resolve the ambiguity possible between different simulators. A simulator might process the read-write syn-chronize events before processing the nonblocking assignment update events, in which case a PLI application called as a read-write synchronize event will see the value of sum before it changes. A coding style that can be used to reduce this ambiguity is for the first call to a *misctf routine* for **REASON_SYNCH** can request a second call-back for **REASON_SYNCH**. This second callback will occur after all scheduled nonblocking assignment update events have been processed. However, there is no way to ensure that new nonblocking assignment update events have not been sched-uled and not yet processed. The only sure way to ensure that a PLI application has been called after all nonblocking assignment update events have been processed is to request that the misctf routine be called in read-only mode, using `tf_rosynchronize()`.

 The VPI library may provide more control over the order of read-write syn-chronization callbacks. Refer to Section 6.7.1 on page 218 for a discussion of synchronizing to simulation time using the VPI library.

Example 12-3 illustrates using `tf_rosynchronize()` to implement *$my_strobe*. In this example, the *calltf routine* for *$my_strobe* will be executed as an active event. However, at the same moment in simulation time the signal sum may also be scheduled to change, possibly as an active event (by a blocking assignment) or a nonblocking assignment update event (by a nonblocking assignment). Therefore, the *calltf routine* does not read the value of sum. Instead, the *calltf routine* schedules a *read-only synchronize* callback to a *misctf routine* at the end of the current simulation time. The value of sum is then read from the *misctf routine*.

CD The source code for this example is on the CD accompanying this book.

- Application source file: `Chapter.012/my_strobe_tf.c`
- Verilog test bench: `Chapter.012/my_strobe_test.v`
- Expected results log: `Chapter.012/my_strobe_test.log`

Example 12-3: *$my_strobe* — using the `tf_rosynchronize()` routine

```
/*************************************************************************
 * calltf routine
 *************************************************************************/
int PLIbook_MyStrobe_calltf(int user_data, int reason) {
  tf_rosynchronize();
  return(0);
}

/*************************************************************************
 * misctf routine
 *************************************************************************/
int PLIbook_MyStrobe_misctf(int user_data, int reason, int paramvc)
{
  if (reason == REASON_ROSYNCH) {
    io_printf("Status of tfarg 1 is %s:\n", tf_strgetp(1,'b'));
  }
  return(0);
}
```

12.4 Application-scheduled callbacks at a future simulation time

The TF library provides three routines that allow a PLI application to schedule its *misctf routine* to be called at a future simulation time step.

PLI_INT32 **tf_setdelay (**
    PLI_INT32   **delay**   /* a 32-bit time value greater than or equal to 0 */

PLI_INT32 **tf_setlongdelay (**
   PLI_INT32   **lowdelay,**   /* lower (right-most) 32-bits of a 64-bit time value */
   PLI_INT32   **highdelay)**   /* upper (left-most) 32-bits of a 64-bit time value */

PLI_INT32 **tf_setrealdelay (**
   double   **realdelay)**   /* a real time value greater than or equal to 0.0 */

The `tf_setdelay()`, `tf_setlongdelay()`, and `tf_setrealdelay()` routines allow a PLI application to schedule its *misctf routine* to be called at a future simulation time step. The simulation time is specified as a 32-bit C integer, a pair of 32-bit C integers (for a 64-bit time value) or as a C double. The callback is scheduled as an active event, which will occur in the first slot of simulation activity at the future time step. The time value is a relative delay from the current simulation time. The delay is automatically scaled to the time scale of the Verilog module which contains the system task/function which called the PLI application. The routines will return a 1 (true) if successful, and a 0 (false) if an error occurred.

Any number of callbacks can be scheduled. If multiple callbacks are scheduled in the same time step, the *misctf routine* will be called once for each scheduled callback. Callbacks can also be scheduled in the current time step (using a delay of zero). When the *misctf routine* is invoked by a callback scheduled using any of these routines, the simulator will pass a reason value of **REASON_REACTIVATE**.

The TF library can also remove any callbacks scheduled with `tf_setdelay()`, `tf_setlongdelay()`, and `tf_setrealdelay()` which have not yet transpired.

PLI_INT32 **tf_clearalldelays ( )**

`tf_clearalldelays()` will remove all pending (not yet executed) callbacks for a specific instance of a *misctf routine*. Suppose a Verilog model contained two instances of a system task named *$read_stimulus*, and each instance had scheduled future callbacks. If a PLI application for one of the *$read_stimulus* instances called `tf_clearalldelays()`, only the pending callbacks for that instance would be removed. The pending callbacks for the other instance would not be affected. The return value from `tf_clearalldelays()` is not used.

An example of scheduling callbacks at a future time

Example 12-4, which follows, illustrates one usage of `tf_setdelay()`. The example implements a system task called `$read_stimulus`, which uses the *misctf routine* to read one line at a time from a stimulus vector file. Each line contains a simulation time and a test vector, as follows:

```
time        vector
----    ----------------
 10     11111111xxxxxxxx
 17     00000000zzzzzzzz
 30     1000000011011101
```

The *misctf routine* for *$read_stimulus* will need to process the following loop as long as there are more test vectors in the file:

1. Read a time and the next test vector from the file.

2. Schedule test vector to be applied to simulation at the desired time.

3. Schedule the *misctf routine* to called again in the same future time step in which the test vector is applied.

The usage model for *$read_stimulus* is:

```
reg   [15:0] input_vector;

initial
    $read_stimulus("read_stimulus.pat", input_vector);
```

Notice that *$read_stimulus* is called from a Verilog `initial` procedure, which means the *calltf routine* for *$read_stimulus* will only be invoked one time throughout simulation. The *calltf routine* calls `tf_setdelay(0)`, which schedules a callback to the *misctf routine* with `REASON_REACTIVATE`. The *misctf routine* will then read the first line from the file, and schedule the test vector to be applied at the simulation time specified by the delay (using `tf_strdelputp()`). After scheduling the vector to be applied, the *misctf routine* calls `tf_setdelay()` with a delay time in order to schedule a future callback to itself, again with `REASON_REACTIVATE`. When the last line of the file is read, the *misctf routine* will cause the simulation to exit. The *misctf routine* is also used to open the test vector file at the beginning of simulation (for `REASON_ENDOFCOMPILE`). Example 12-7 on page 425, listed at the end of this chapter, is a more robust version of the *$read_stimulus* example, with added capabilities for supporting different vector formats and delay specifications.

CD The source code for this example is on the CD accompanying this book.

-   Application source file: `Chapter.012/read_stimulus_short_tf.c`
-   Verilog test bench: `Chapter.012/read_stimulus_short_test.v`
-   Expected results log: `Chapter.012/read_stimulus_short_test.log`

Example 12-4: *$read_stimulus* — using the `tf_setdelay()` routine

```
/****************************************************************/
/* calltf routine                                               */
/****************************************************************/
int PLIbook_ReadStimulusShort_calltf(int user_data, int reason)
{                   /* call the misctf routine at the end of current  */
  tf_setdelay(0); /* time step, with REASON_REACTIVATE; the misctf */
                  /* routine reads the stimulus file.               */
  return(0);
}


/****************************************************************/
/* misctf routine                                               */
/****************************************************************/
int PLIbook_ReadStimulusShort_misctf(int user_data, int reason
                                     int paramvc)
{
  PLI_INT32  delay;
  FILE       *file_p;       /* pointer to the test vector file      */
  PLI_BYTE8  vector[1024];  /* stimulus vector -- hard coded limit  */

  switch(reason) {
    case REASON_ENDOFCOMPILE: /* misctf called at start of simulation*/
      file_p = fopen(tf_getcstringp(1), "r");
      /* store a pointer to the file in the work area */
      tf_setworkarea((char *)file_p);
      break;

    case REASON_REACTIVATE:  /* misctf called by tf_setdelay */
      /* get the file pointer from the work area */
      file_p = (FILE *)tf_getworkarea();
      /* read next line from the file */
      if ( (fscanf(file_p,"%d %s\n", &delay, (char *)vector)) == EOF) {
        io_printf("$read_stimuulus reached end-of-file\n");
        return(0);  /* exit without scheduling another callback */
        break;
      }
      /* schedule the vector to be applied after the delay period */
      tf_strdelputp(2, tf_sizep(2), 'b', vector, delay, 2);
      /* call this routine back after delay time */
      tf_setdelay(delay);
      break;
  }
  return(0);
}
```

12.5 System task/function argument value change callbacks

A PLI application can schedule with the simulator to call the *misctf routine* whenever an argument to a system task changes value.

PLI_INT32 **tf_asynchon ( )**

The tf_asynchon() routine enables asynchronous callbacks to the *misctf routine* whenever an argument of a system task changes value or strength. The reason constant which the simulator will pass to the *misctf routine* is **REASON_PARAMVC**. The simulator also passes a third C function input to the *misctf routine*, called the paramvc input, which contains the index of the system task argument which changed value. Using this index number, a PLI application can then read the new value of the task argument, or do any other processing desired at the time of the value change. The routine returns 1 if successful, and 0 if an error occurred.

 Asynchronous callbacks may only be used with user-defined system tasks. The routine tf_asynchon() is ignored if it is called by a PLI application that is defined as a system function.

PLI_INT32 **tf_asynchoff ( )**

tf_asynchoff() disables the argument change callbacks for the current instance of the system task. The return value is not used, and should be ignored.

The following example illustrates the basic usage of tf_asynchon(). In this example, the asynchronous callbacks are enabled by the *calltf routine*. The PLI application is invoked by *$my_monitor1*, as follows:

```
initial
  $my_monitor1(clock, d, q);
```

CD The source code for this example is on the CD accompanying this book.

- Application source file: Chapter.012/my_monitor1_tf.c
- Verilog test bench: Chapter.012/my_monitor1_test.v
- Expected results log: Chapter.012/my_monitor1_test.log

Example 12-5: *$my_monitor1* — asynchronous argument value change callbacks

```
/******************************************************************
 * calltf routine
 ******************************************************************/
int PLIbook_MyMonitor1_calltf(int user_data, int reason)
{
  tf_asynchon();  /* enable asynchronous misctf callbacks */
  return(0);
}
/******************************************************************
 * misctf routine
 ******************************************************************/
int PLIbook_MyMonitor1_misctf(int user_data, int reason, int paramvc)
{
  if (reason == REASON_PARAMVC) {
    io_printf("At %s: tfarg %d changed, new value is %s:\n",
              tf_strgettime(), paramvc, tf_strgetp(paramvc,'b'));
  }
  return(0);
}
```

In the preceding example, if two task arguments change at the same moment in simulation time, the *misctf routine* could be called twice, asynchronously. This may be desirable in some applications, but in other applications it may be desirable to only process data at the end of the simulation time step, when all activity for that moment of time is complete. The synchronous callback to the *misctf routine* can be combined with asynchronous callbacks to accomplish this. Synchronous callbacks at the end of the time step are scheduled using tf_synchronize() and tf_rosynchronize(), which were discussed in section 12.3 on page 408. In order to pass information about which system task arguments changed value (which result in asynchronous callbacks of the *misctf routine*) to the synchronous callback of the *misctf routine*, it is necessary to temporarily save those task arguments which changed value. The PLI provides a mechanism to automatically save this information.

Simulation maintains two internal flags for each system task argument. The flags are referred to as the **PVC** flags (for **Parameter Value Change**). One of the flags is the *current PVC flag*, and the other is the *saved PVC flag*. Each time an asynchronous callback occurs, the simulation sets the current PVC flag for the task argument that changed value. By itself, the current PVC flag has little value. However, a *misctf routine* can move or copy the current PVC flag to the saved PVC flag. Then, at the end of the simulation time step, the *misctf routine* can examine all saved PVC flags to determine which task arguments changed value sometime during the current time step.

 PVC flags are not enabled by a simulator until tf_asynchon() has been called to enable asynchronous callbacks to the *misctf routine*.

PLI_INT32 **tf_copypvc_flag (**
    PLI_INT32   **n**)   /* index number of a system task/function arg, or **−1** */

tf_copypvc_flag() is used to copy the *current* PVC flag of a specific system task argument to the *saved* PVC flag for that argument. The current PVC flag is not cleared by the copy. tf_copypvc_flag() must be called from a *misctf routine* that was called with REASON_PARAMVC, which is when the current PVC flag is active. The input provided to tf_copypvc_flag() is the index number of a task argument, typically the argument that just changed value. The return value of tf_copypvc_flag() is the value of the PVC flag which was just copied. If the input value to tf_copypvc_flag() is **−1**, then the routine returns the result of a logical OR of all *saved* PVC flags.

PLI_INT32 **tf_movepvc_flag (**
    PLI_INT32   **n**)   /* index number of a system task/function arg, or **−1** */

The tf_movepvc_flag() routine serves a dual role:

- If tf_movepvc_flag() is called from a *misctf routine* that was called with REASON_PARAMVC, the routine will move the *current* PVC flag of a specific system task argument to the *saved* PVC flag for that argument. The current PVC flag is cleared by the move. The input provided to tf_movepvc_flag() is the index number of a task argument, typically the argument that just changed value. The return value of tf_movepvc_flag() is the value of the PVC flag which was just moved. If the input value to tf_movepvc_flag() is **−1**, then the routine returns the result of a logical OR of all *saved* PVC flags.

- If tf_movepvc_flag(-1) is called from a *misctf routine* that was called with REASON_SYNCH or REASON_ROSYNCH, the routine will setup for all *saved* PVC flags to be examined by tf_getpchange().

PLI_INT32 **tf_testpvc_flag (**
    PLI_INT32   **n**)   /* index number of a system task/function arg, or **−1** */

The tf_testpvc_flag() routine can be used to see if the *saved* PVC flag of a specific system task argument is set. The input provided to tf_testpvc_flag() is the index number of a task argument. The return is the value of the PVC flag. The tf_testpvc_flag() can also be used to test if any saved PVC flag is set, by specifying a **−1** for **n**. When the input value is **−1**, then tf_testpvc_flag() will return the result of a logical OR of all saved PVC flags.

PLI_INT32 **tf_getpchange (**

    PLI_INT32   **n)**   /* index number of a system task/function argument */

tf_getpchange() is used to scan all saved PVC flags and return the index number of each system task argument with PVC flag that is set. tf_getpchange() should be used in a *misctf routine* that was called with REASON_SYNCH or REASON_ROSYNCH. The input to tf_getpchange() is the index number of a system task argument. The routine returns the index number of the *next* task argument with a saved PVC flag that is set. If there are no task arguments with a greater index number than the input that has a saved PVC flag set, then tf_getpchange() will return 0. There are two mandatory rules for using tf_getpchange():

1. tf_movepvc_flag(-1) must be called prior to calling tf_getpchange().

2. The first call to tf_getpchange() must have **0** as the input value.

Example 12-6 implements *$my_monitor2*. This example enables asynchronous argument value change callbacks, and then, in any time step that one or more task arguments change value, prints a summary of all changes. The printing of the summary is synchronized to the end of the simulation time step in which the changes occurred.

```
initial
  $my_monitor2(clock, d, q);
```

CD The source code for this example is on the CD accompanying this book.

- Application source file: Chapter.012/my_monitor2_tf.c
- Verilog test bench: Chapter.012/my_monitor2_test.v
- Expected results log: Chapter.012/my_monitor2_test.log

Example 12-6: *$my_monitor2* — synchronized analysis of argument value changes

```
/*********************************************************************
 * calltf routine
 ********************************************************************/
int PLIbook_MyMonitor2_calltf(int user_data, int reason)
{
  tf_asynchon();  /* enable asynchronous misctf callbacks */
  return(0);
}
/*********************************************************************
 * misctf routine
 ********************************************************************/
int PLIbook_MyMonitor2_misctf(int user_data, int reason, int paramvc)
{
  PLI_INT32 arg_num;
```

```
  if (reason == REASON_PARAMVC) {
    /* io_printf("At %s: change detected on tfarg %d\n",
             tf_strgettime(), paramvc); */
    tf_copypvc_flag(paramvc);
    tf_rosynchronize(); /* schedule a callback at end of time step */
  }

  if (reason == REASON_ROSYNCH) {
    io_printf("Reached end of time step %s:\n", tf_strgettime());
    if (tf_movepvc_flag(-1)) { /* only print if something changed */
      arg_num = 0;
      while (arg_num = tf_getpchange(arg_num) ) {
        io_printf("  tfarg %d changed to %s\n",
                  arg_num, tf_strgetp(arg_num,'b'));
      }
    io_printf("\n");
    }
  }
  return(0);
}
```

12.6 Simulation save and restart checkpoint files

Some Verilog simulators provide the means to save the simulation state, and restart simulation from the saved state. Save and restart are sometimes referred to as checkpointing. A *$save* built-in system task typically creates the checkpoint file, and a *$restart* built-in system task resumes simulation from the saved state in the checkpoint file.

If a simulator supports save and restart capabilities, then the *misctf routine* will be called automatically with **REASON_SAVE** at the start of a save state, and **REASON_RESTART** at the end of a simulator restart from a saved file.

PLI_INT32 **tf_write_save (**
    PLI_BYTE8 *** blockptr,**   /* pointer to a block of application-allocated memory */
    PLI_INT32   **blocklength)** /* length of the block of memory in bytes */

The tf_write_save() routine may only be called from the *misctf routine* of a PLI application. The routine instructs the simulator to add an application-specified block of memory to the save file being created by the simulator. The block of memory can contain any information important to the PLI application. The data stored in the PLI work area is not automatically saved. The application should copy that information into the block of memory to be saved before calling tf_write_save(). The inputs to `tf_write_save()` are a pointer to the application-specified block of memory, and the length of the memory block. The routine will return a non-zero value if successful, and 0 if an error occurred.

PLI_INT32 **tf_read_restart (**
    PLI_BYTE8 ***blockptr,**   /* pointer to a block of memory */
    PLI_INT32   **blocklength**) /* length of the block of memory in bytes */

The `tf_read_restart()` routine may only be called from the *misctf routine* of a PLI application. The routine retrieves from the simulator the data that was saved by `tf_write_save()`. Before calling `tf_read_restart()`, the PLI application must first allocate memory to receive the restored data. A pointer to the allocated memory, and the length of the saved data to be restored is passed to `tf_read_restart()`. The routine will return 1 if successful, and 0 if an error occurred.

`tf_read_restart()` retrieves the saved data as an input stream, and automatically maintains a pointer to how much of the saved block of data has been retrieved. This means the block of saved data can be restored using one call to `tf_read_restart()`, or using multiple calls. Often, the length of the block of data that was saved is not known when the restart occurs. An easy way to pass the length of the saved data from the save operation to the restart operation, is to make the first piece of data saved an integer which contains the save block length. Upon restart, `tf_read_restart()` can be called twice: First, with a block length equal to one integer, in order to retrieve the total saved block length, and then a second time, to retrieve the remainder of the saved block.

 It is mandatory that all data be retrieved when ever a restart occurs. Because `tf_read_restart()` retrieves the data as a stream, if one PLI application does not retrieve its data, another PLI application, which might have saved data afterwards, will not be able to retrieve the correct data. The PLI standard guarantees that all calls to the *misctf routine* on a restart will occur in the same order that calls occurred on a save. A PLI application should never assume it will be the only PLI application which saved data, and can therefore ignore retrieving all saved data.

 The VPI library provides a more flexible method of saving and restoring PLI application data when simulations execute a save or restart.
TIP

When simulation is restarted, any pointers which were saved may no longer be valid. A file pointer, for example, can no longer be used, because the file may have been closed between the save and restart operations. Any application-allocated memory may have been freed or relocated, making pointers to the memory invalid. The PLI application must define a scheme to re-open files if needed, and to re-allocate memory if needed.

Save and restart capabilities are not implemented in some simulators. To allow PLI applications to be portable, however, the IEEE 1364 standard includes the `tf_write_save()` and `tf_read_restart()` routines as part of the PLI standard. Simulators which do not have save and restart capability will return an error status when the routines are called. The application should check the error status to determine if the application data was saved or restored by the simulator.

Example 12-7 on page 425 illustrates how to use save and restart with the *$read_stimulus* stimulus pattern file reader. In this example, the *misctf routine* will append the position indicator of the next line in the stimulus file to be read to the simulation save file. When simulation is restarted, the stimulus file is re-opened, and the C file pointer is reset to the saved position indicator. The example also adds a character string to the simulation save file, in order to illustrate how the data can be retrieved when simulation is restarted. Note that the example also saves the file pointer and stimulus pattern string pointers, but these pointers may no longer be valid when simulation is restarted. Therefore these pointers are recreated on a restart.

12.7 A complete example of using misctf routine callbacks

Example 12-7, which follows, is a more extensive stimulus file reader than was presented in example 12-4 earlier in this chapter. The *$read_stimulus_ba* PLI application listed below utilizes several types of *misctf routine* callbacks:

- `REASON_ENDOFCOMPILE` is used to both allocate memory that the *$read_stimulus_ba* application requires, and to open the file containing the stimulus patterns.

- `REASON_REACTIVATE` is used to retrieve a stimulus pattern and a simulation time from the file, and then schedule the simulator to apply the test pattern and the designated simulation time.

- `REASON_ROSYNCH` is used to terminate simulation when *$read_stimulus_ba* has reached the end of the stimulus pattern file. Instead of terminating simulation immediately on end of file, a read-only synchronous callback is scheduled, to allow any other activity in the current simulation time step to complete first.

- `REASON_FINISH` is used to close the stimulus file and free the memory that was allocated at the start of simulation.

- `REASON_SAVE` is used to save the current position from which the *$read_stimulus_ba* will retrieve the next stimulus pattern from the file.

- `REASON_RESTART` is used to retrieve the file position indicator, so that simulation can resume at the next test stimulus pattern in the file.

This example also extends the file handling capability by permitting stimulus values to be specified as either hex patterns or binary patterns, and by allowing the simulation times to be specified as relative delays or as absolute time from when the PLI application was invoked. This is accomplished by having several versions of the system task name, all of which call the same PLI application, but with different user-data values:

- *$read_stimulus_ba* has a user-data of *0*

- *$read_stimulus_br* has a user-data of *1*

- *$read_stimulus_ha* has a user-data of *2*

- *$read_stimulus_hr* has a user-data of *3*

The usage for this example is:

```
initial
  $read_stimulus_<base><delay_type>("file_name",
verilog_reg);
```

where:

```
<base>       is b or h (for binary or hex vectors).
<delay_type> is a or r for absolute or relative times.
"file_name"  is the name of the file to be read, in quotes.
verilog_reg  is a verilog variable data type of the same bit
             width as the patterns to be read.
```

For example:

```
initial
  $read_stimulus_ba("read_stimulus.pat", input_vector);
```

CD   The source code for this example is on the CD accompanying this book.

- Application source file: `Chapter.012/read_stimulus_file_tf.c`
- Verilog test bench: `Chapter.012/read_stimulus_file_test.v`
- Expected results log: `Chapter.012/read_stimulus_file_test.log`

Example 12-7: *$read_stimulus_ba* — using several *misctf routine* callbacks

```
#include <stdio.h>      /* ANSI C standard input/output library */
#include "veriuser.h"   /* IEEE 1364 PLI TF routine library   */

/* prototypes of sub-functions */
char *PLIbook_reason_name(PLI_INT32 reason);
```

```
/******************************************************************/
/* structure definition for data used by the misctf routine      */
/******************************************************************/
typedef struct PLIbook_stim_data {
  FILE *file_ptr;       /* pointer to the test vector file  */
  long  file_position;  /* position within file of next byte to read */
  char *vector;         /* pointer to stimulus vector */
  char  dummy_msg[20];  /* dummy message field to show save/restart */
} PLIbook_stim_data_s, *PLIbook_stim_data_p;

/******************************************************************/
/* checktf routine                                               */
/******************************************************************/
int PLIbook_ReadStimulus_checktf(int user_data, int reason)
{
  bool  err = FALSE;

  if (tf_nump() != 2) {
    tf_text("$read_stimulus_?? requires 2 arguments\n");
    err = TRUE;
  }
  if (tf_typep(1) != tf_string) {
    tf_text("$read_stimulus_?? arg 1 must be a string\n");
    err = TRUE;
  }
  if (tf_typep(2) != tf_readwrite) {
    tf_text("$read_stimulus_?? arg 2 must be a register data type");
    err = TRUE;
  }
  if (err)
    tf_message(ERR_ERROR, "", "", ""); /* print stored messages */
  return(0);
}

/******************************************************************/
/* calltf routine                                                */
/******************************************************************/
int PLIbook_ReadStimulus_calltf(int user_data, int reason)
{                  /* call the misctf routine at the end of current  */
  tf_setdelay(0);  /* time step, with REASON_REACTIVATE; the misctf */
                   /* routine reads the stimulus file.              */
  return(0);
}

/******************************************************************/
/* misctf routine                                                */
/******************************************************************/
int PLIbook_ReadStimulus_misctf(int user_data, int reason, int paramvc)
{
  PLI_INT32 delay;
  PLI_BYTE8 base;
  PLIbook_stim_data_p data_p;

  bool  debug = FALSE;
```

```
if(mc_scan_plusargs("read_stimulus_debug")) debug = TRUE;

if (debug)
    io_printf("** read_stimulus misctf called for %s at time %s **\n",
              PLIbook_reason_name(reason), tf_strgettime());

switch(reason) {
  case REASON_ENDOFCOMPILE: /* misctf called at start of simulation */

      /* Check work area to see if application data is already
      /* allocated and the vector file opened. This check is necessary
      /* because REASON_RESTART can occur before REASON_ENDOFCOMPILE */
      data_p = (PLIbook_stim_data_p)tf_getworkarea();
      if (data_p)
        return(0); /* abort if work area already contains data */

      /* allocate a memory block for data storage */
      data_p=(PLIbook_stim_data_p)malloc(sizeof(PLIbook_stim_data_s));

      /* store a pointer to the application data in the work area */
      tf_setworkarea((PLI_BYTE8 *)data_p);

      /* fill in the application data fields */
      data_p->file_ptr = fopen((char *)tf_getcstringp(1), "r");
      if ( !data_p->file_ptr ) {
        tf_error("$read_stimulus_?? could not open file %s",
                 (char *)tf_getcstringp(1));
        tf_dofinish(); /* exit simulation if cannot open file */
        break;
      }
      data_p->file_position = ftell(data_p->file_ptr);
      data_p->vector = (char *)malloc(((int)tf_sizep(2) * 8) + 1);
      strcpy(data_p->dummy_msg, "Hello world");
      break;

  case REASON_REACTIVATE:  /* misctf called by tf_setdelay */

      /* get the pointer to the data structure from the work area */
      data_p = (PLIbook_stim_data_p)tf_getworkarea();

      /* read next line from the file */
      if ( (fscanf(data_p->file_ptr,"%d %s\n",
            &delay, data_p->vector)) == EOF) {
        tf_rosynchronize();  /* if end of file, schedule a callback */
        break;               /* at the end of the current time step */
      }
      if (debug)
        io_printf("** values read from file: delay = %d  vector = %s\n",
                  delay, data_p->vector);

      /* set flag for test vector radix */
      switch(user_data) {
        case 0:
        case 1: base = 'b'; break; /* vectors are in binary format */
        case 2:
        case 3: base = 'h'; break; /* vectors are in hex format */
      }
```

```
      /* convert absolute delays to relative to current time */
      switch(user_data) {
        case 0:
        case 2: /* using absolute delays; convert to relative */
                delay = delay - tf_gettime();
      }

      /* schedule the vector to be applied after the delay period */
      tf_strdelputp(2, tf_sizep(2), base, (PLI_BYTE8 *)data_p->vector,
                    delay, 0);

      /* schedule reactive callback to this routine after delay time */
      tf_setdelay(delay);
      break;
    case REASON_ROSYNCH:  /* misctf called at end of time step */
      io_printf("\n$read_stimulus_?? has encountered end-of-file.\n");
      tf_dofinish;
      break;

    case REASON_FINISH:    /* misctf called at end of simulation */
      io_printf("\nPLI is processing finish at simulation time %s\n\n",
                tf_strgettime());
      /* get the pointer to the application data from the work area */
      data_p = (PLIbook_stim_data_p)tf_getworkarea();
      /* close file */
      if (data_p->file_ptr) fclose(data_p->file_ptr);
      /* de-allocate storage */
      free(data_p);
      break;

    case REASON_SAVE:  /* misctf called for $save */

      /* get the pointer to the application data from the work area */
      data_p = (PLIbook_stim_data_p)tf_getworkarea();

      /* save current file position in the application data */
      data_p->file_position = ftell(data_p->file_ptr);

      /* add application data to simulation save file */
      tf_write_save((PLI_BYTE8 *)data_p, sizeof(PLIbook_stim_data_s));
      if (debug)
        io_printf("\nPLI data saved (last file position was %ld)\n\n",
                  data_p->file_position);
      break;

    case REASON_RESTART:  /* misctf called end of simulation */

      /* re-allocate memory for PLI application data */
      data_p=(PLIbook_stim_data_p)malloc(sizeof(PLIbook_stim_data_s));

      /* save new application data pointer in work area */
      tf_setworkarea((PLI_BYTE8 *)data_p);

      /* retrieve old application data from save file */
      if (!tf_read_restart((PLI_BYTE8 *)data_p,
          sizeof(PLIbook_stim_data_s)) ) {
```

```
            tf_error("\nError retrieving PLI data from save file!\n");
            return(0);
        }
        if (debug) {
            io_printf("\nPLI data retrieved from save file.\n");
            /* test to see if old application data was restored */
            io_printf("  dummy message = %s, file position = %ld\n\n",
                      data_p->dummy_msg, data_p->file_position);
        }
        /* re-open test vector file */
        data_p->file_ptr = fopen((char *)tf_getcstringp(1), "r");
        if ( !data_p->file_ptr ) {
            tf_error("$read_stimulus_?? could not re-open file %s",
                     (char *)tf_getcstringp(1));
            tf_dofinish(); /* exit simulation if cannot open file */
        }

        /* re-position file to next test vector to be read */
        if ( fseek(data_p->file_ptr, data_p->file_position, SEEK_SET)) {
            tf_error("$read_stimulus_?? could not reposition file");
            tf_dofinish(); /* exit simulation if reposition open file */
        }
        break;
    }
    return(0);
}
/***********************************************************************/
/* Function to convert reason integer to reason name                   */
/***********************************************************************/
char *PLIbook_reason_name(PLI_INT32 reason)
{
    char str[25];
    switch (reason) {
        case REASON_ENDOFCOMPILE : return("REASON_ENDOFCOMPILE"); break;
        case REASON_FINISH       : return("REASON_FINISH"); break;
        case REASON_INTERACTIVE  : return("REASON_INTERACTIVE"); break;
        case REASON_SYNCH        : return("REASON_SYNCH"); break;
        case REASON_ROSYNCH      : return("REASON_ROSYNCH"); break;
        case REASON_REACTIVATE   : return("REASON_REACTIVATE"); break;
        case REASON_PARAMVC      : return("REASON_PARAMVC"); break;
        case REASON_PARAMDRC     : return("REASON_PARAMDRC"); break;
        case REASON_SAVE         : return("REASON_SAVE"); break;
        case REASON_RESTART      : return("REASON_RESTART"); break;
        case REASON_RESET        : return("REASON_RESET"); break;
        case REASON_ENDOFRESET   : return("REASON_ENDOFRESET"); break;
        case REASON_FORCE        : return("REASON_FORCE"); break;
        case REASON_RELEASE      : return("REASON_RELEASE"); break;
    }
    return("Non-standard or Unknown Reason");
}
```

12.8 Summary

This chapter has presented the TF library routines which allow PLI applications to synchronize with other types of activity that can occur during a simulation. The TF routines use the *misctf routine* to do this synchronization. Some types of callback to the *misctf routine* are automatic. The PLI application does not request the callback. Other types of callbacks are application defined, and only occur if the PLI application requests the callback.

The next chapter applies the concepts presented in this chapter and the previous chapter, to show how the TF routines can be used to interface a hardware model written in the C programming language into a Verilog simulation.

CHAPTER 13 *Interfacing to C Models Using TF Routines*

*O*ne of the ways the TF library can be used is to create an interface to hardware models written in the C programming language. The TF library provides routines to read and modify logic values within a simulation, and to synchronize activity with logic value changes and with simulation time. This chapter shows several ways in which a C model can be interfaced to a Verilog simulation using the TF library (Chapter 7 presents using the VPI library for interfacing to C models, and Chapter 18 shows how to use the ACC library to accomplish this same task).

The concepts presented in this chapter are:

- Representing hardware models in C
- Verilog HDL shell modules
- Combinational logic interfaces to C models
- Sequential logic interfaces to C models
- Synchronizing with the end of a simulation time step
- Synchronizing with a future simulation time step
- Multiple instances of a C model
- Creating instance specific storage within C models
- Representing propagation delays in C models

TIP

One reason for representing hardware models in the C language is to achieve faster simulation performance. The C programming language allows a very abstract, algorithmic representation of hardware functionality, without representing detailed timing, multi-state logic, hardware concurrency, and other hardware specific details offered by the Verilog language.

The PLI can be a means to access the efficiency of a highly abstract C model. However, a poorly written PLI application can become a bottleneck that offsets much of the efficiency gains. Care must be taken to write PLI applications that execute as efficiently as possible. Some guidelines that can help maximize the efficiency and run-time performance of PLI applications are:

- Good C programming practices are essential. General C programming style and techniques are not discussed within the scope of this book.
- Consider every call to a PLI routine as expensive, and try to minimize the number of calls.
- Routines which convert logic values from a simulator's internal representation to C strings, and vice-versa, are very expensive in terms of performance. Best efficiency is attained when the value representation in C is as similar as possible to the value representation in Verilog.
- Use the Verilog language to model the things hardware description languages do well, such as representing hardware parallelism and hardware propagation times. Simulator vendors have invested a great deal in optimizing a simulator's algorithms, and that optimization should be utilized.

NOTE

The objective of this book is to show several ways in which the TF library can be used to interface to C models. Short examples are presented that are written in a relatively easy to follow C coding style. In order to meet the book's objectives, the examples do not always follow the guidelines of efficient C coding and prudent usage of the PLI routines. It is expected that, if these examples are adapted for other applications, the coding style will also be modified to be more efficient and robust.

13.1 How to interface C models with Verilog simulations

The power and flexibility of the C programming language and the Verilog PLI provide a wide variety of methods that can be used to interface a Verilog simulation with a C language model. All methods have three essential concepts in common:

- Value changes which occur in the Verilog simulator must be passed to the C model.
- Value changes within the C model must be passed to the Verilog simulation.
- Simulated time in both the Verilog simulation and the C model must remain synchronized.

This chapter will present some of the more common methods of interfacing a Verilog simulation with a C model. The methods presented are by no means the only ways this interface can be accomplished, and may not always be the most efficient methods. However, the methods presented have many advantages, including simplicity to implement, portability to many types of Verilog simulators, and the ability to use the C model any number of times and anywhere in the hierarchy of a Verilog design.

The fundamental steps that are presented in this chapter are:

1. Create the C language model as an independent block of pure C code that does not use the PLI routines in any way. The C model will have inputs and outputs, but it will not know the source of the inputs or the destination of the outputs. The C code to implement the model might be in the form of a C function with no main function, or it might be a complete C program with its own main function.

2. Create a Verilog HDL **shell module**, also called a *wrapper module* or *bus functional module* (BFM), to represent the inputs and outputs of the C language model. This module will be written completely in the Verilog language, but will not contain any functionality. To represent the functionality of the model, the shell module will call a PLI application.

3. Create a PLI application to serve as an interface between the C model and the Verilog shell module. The PLI application is a communication channel, which:

    - Uses PLI routines to retrieve data from the Verilog HDL shell module, and pass the data to the C model via standard C programming.

    - Uses standard C programming to receive data from the C model, and pass the data to the Verilog shell module via PLI routines.

The following diagram shows how the blocks which are created in these three steps interact with each other.

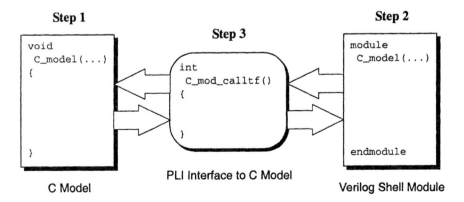

This chapter presents steps 2 and 3 of this interface method in detail. Step 1 is to model some desired functionality or algorithm in the C language. This step is pure C programming, which does not directly involve the Verilog language or the Verilog PLI. This chapter does not cover how to implement ideas in the C language, The focus is on how to interface that implementation with a Verilog simulation. To maintain this focus, the C model example presented in this chapter will be a practical example, but relatively simple to model in C. The C model example used will illustrates all of the important concepts of integrating C models into a Verilog simulation.

13.2 Creating the C language model

A hardware model can be represented in the C programming language in two basic forms, either as a C function or as an independent C program.

13.2.1 Using functions to represent the C model

When the C model is represented as a C function, that function can be linked into the Verilog simulator, together with the PLI application that serves as the interface to the model. The PLI application can then call the function when needed, passing inputs to the function, and receiving outputs from the function. One advantage of representing a C model as a function is the simplicity of passing values to and from the model. Another advantage is ease of porting to different operating systems, since the C model is called directly from the PLI application as a C function. A disadvantage of using a function to represent the C model is that the C model must contain additional code to allow a Verilog design to instantiate the C model multiple times. The model needs to specifically create unique storage for each instance.

13.2.2 Using independent programs to represent the C model

When the C model is represented as an independent program, which means it has its own C *main* function, then the Verilog simulation and the C model can be run as parallel processes on the same or on different computers. The PLI application which serves as an interface between the simulation and the model will need to create and maintain some type of communication channel between the two programs. This communication can be accomplished several ways, such as using the exec command in the C standard library. On Unix operating systems, the fork or vfork commands with either Unix pipes or Unix sockets is an efficient method to communicate with the C model program. On PC systems running a DOS or windows operating system, the spawn command can be used to invoke the C model program and establish two-way communications between the PLI application and the C model process.

One of the advantages of representing the C model as an independent model is the ability to have parallel processes running on the same computer or separate computers. Another advantage is that when a Verilog design instantiates multiple instances of the C model, each instance will be a separate process with its own memory storage. The major disadvantage of independent programs when compared to using a C function to represent the C model is that the PLI interface to invoke and communicate with the separate process is more complex, and might be operating system dependent.

David Roberts, of Cadence Design Systems, who reviewed many of the chapters of this book, has provided a full example of representing a C model as a separate C program. This example is included with the CD that accompanies this book.

13.3 A C model example

The C model used for the different PLI interfaces shown in this chapter is a scientific Arithmetic Logic Unit, which utilizes the C math library. The C model is represented as a C function which will be called from the PLI interface mechanism. This model is written entirely with standard C library routines and C data types, without reference to any PLI routines or PLI data types. This same example is also used in other chapters, to show how a PLI interface to C models can be created, using the VPI and ACC libraries of the PLI.

The inputs and outputs of the scientific ALU C model are shown below, and Table 13-1 shows the operations which the ALU performs.

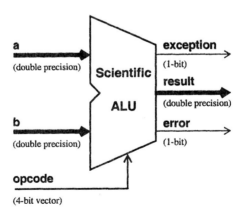

exception is set to 1 whenever an operation results in a value which is out of range of the double-precision result.

error is set to 1 whenever an input to an operation is out of range for the operation.

Table 13-1: Scientific ALU C model operations

Opcode	C Math Library Operation
0	pow(a,b) — returns a to the power of b
1	sqrt(a) — returns the square root of a
2	exp(a) — returns the natural exponent of a
3	ldexp(a,b) — returns a * (2 to the power of b)
4	fabs(a) — returns the absolute of a
5	fmod(a,b) — returns the floating remainder of a / b
6	ceil(a) — returns smallest whole number not less than a
7	floor(a) — returns largest whole number not more than a
8	log(a) — returns the natural log of a
9	log10(a) — returns the base 10 log of a
A	sin(a) — returns the sine of a
B	cos(a) — returns the cosine of a
C	tan(a) — returns the tangent of a
D	asin(a) — returns the arcsine of a
E	acos(a) — returns the arccosine of a
F	atan(a) — returns the arctangent of a

The source code for the scientific ALU is listed in Example 13-1. This version of the ALU generates a result, but does not store the result. A latched version of the ALU which stores the operation result is listed later in this chapter.

CD The source code for this example is on the CD accompanying this book.

- Application source file: `Chapter.13/sci_alu_comb_calltf_tf.c`
- Verilog shell module: `Chapter.13/sci_alu_comb_calltf_shell.v`
- Verilog test bench: `Chapter.13/sci_alu_comb_calltf_test.v`
- Expected results log: `Chapter.13/sci_alu_comb_calltf_test.log`

Example 13-1: scientific ALU C model — combinational logic version

```
#include <stdio.h>
#include <math.h>
#include <errno.h>
void PLIbook_ScientificALU_C_model(
        double  a,       /* input */
        double  b,       /* input */
        int     opcode,  /* input */
        double *result,  /* output from ALU */
        int    *excep,   /* output; set if result is out of range */
        int    *err)     /* output; set if input is out of range */
{
  switch (opcode) {
    case 0x0: *result = pow    (a, b);      break;
    case 0x1: *result = sqrt   (a);         break;
    case 0x2: *result = exp    (a);         break;
    case 0x3: *result = ldexp  (a, (int)b); break;
    case 0x4: *result = fabs   (a);         break;
    case 0x5: *result = fmod   (a, b);      break;
    case 0x6: *result = ceil   (a);         break;
    case 0x7: *result = floor  (a);         break;
    case 0x8: *result = log    (a);         break;
    case 0x9: *result = log10  (a);         break;
    case 0xA: *result = sin    (a);         break;
    case 0xB: *result = cos    (a);         break;
    case 0xC: *result = tan    (a);         break;
    case 0xD: *result = asin   (a);         break;
    case 0xE: *result = acos   (a);         break;
    case 0xF: *result = atan   (a);         break;
  }
  *excep = (errno == ERANGE);  /* result of math func. out-of-range */
  #ifdef WIN32  /* for Microsoft Windows compatibility */
  *err   = (_isnan(*result) ||  /* result is not-a-number, or */
            errno == EDOM);     /* arg to math func. is out-of-range */
  #else
  *err   = (isnan(*result) ||   /* result is not-a-number, or */
            errno == EDOM);     /* arg to math func. is out-of-range */
  #endif
  if (*err) *result = 0.0;      /* set result to 0 if error occurred */
  errno = 0;                    /* clear the error flag */
  return;
}
```

13.4 Creating a Verilog shell module

A *shell module* allows a Verilog design to reference a C model using standard Verilog HDL syntax. The shell module is a Verilog module which has the same input and output ports as the C model, but the module has no functionality modeled within. To represent the module's functionality, the shell module invokes a PLI application, which in turn invokes the C model. A shell module is sometimes referred to as a *wrapper module*, because the module is wrapped around the call to a PLI application. The shell module for a combinational logic version of the scientific ALU is listed below.

CD The source code for this example is on the CD accompanying this book.

- Application source file: `Chapter.13/sci_alu_comb_calltf_tf.c`
- Verilog shell module: `Chapter.13/sci_alu_comb_calltf_shell.v`
- Verilog test bench: `Chapter.13/sci_alu_comb_calltf_test.v`
- Expected results log: `Chapter.13/sci_alu_comb_calltf_test.log`

Example 13-2: Verilog shell module for the scientific ALU C model

```
`timescale 1ns / 1ns
module scientific_alu(a_in, b_in, opcode,
                      result_out, exception, error);
  input   [63:0] a_in, b_in;
  input   [3:0] opcode;
  output  [63:0] result_out;
  output          exception, error;

  real            a, b, result; // real variables used in this module
  reg             exception, error;

  // convert real numbers to/from 64-bit vector port connections
  assign result_out = $realtobits(result);
  always @(a_in) a  = $bitstoreal(a_in);
  always @(b_in) b  = $bitstoreal(b_in);

  //call the PLI application which interfaces to the C model
  //using combinational logic input sensitivity
  always @(a or b or opcode)
    $scientific_alu(a, b, opcode, result, exception, error);

endmodule
```

 NOTE In this example, the primary inputs and outputs of the model are double-precision floating point values, represented as Verilog `real` data types. The Verilog language does not permit real numbers to be connected to module ports. However, the language provides built-in system functions which convert real numbers to 64-bit vectors, and vice-versa, so the real values can be passed through a module port connection. These built-in system functions are *$realtobits()* and *$bitstoreal()*.

The Verilog shell module that represents the C model can be instantiated in a design in the same way as any other Verilog module. For example:

```
module chip (...)
  ...
  scientific_alu u1 (a, b, opcode, result, excep, err);
  ...
endmodule
```

Creating a shell module to represent the C model is not mandatory. The PLI application could be called directly from any place in a Verilog design. However, there are important advantages to using a shell module to represent the C model:

- The shell module provides a simple method to encapsulate the C model.

- The shell module can be instantiated anywhere in a Verilog design hierarchy.

- The shell module can be instantiated any number of times in a Verilog design.

- The shell module can add Verilog HDL delays to the C model, which can accurately represent rise and fall delay delays, state-dependent delays, and timing constraints such as setup times.

- Delays within a shell module can be annotated using delay calculators or SDF files for additional delay accuracy for each instance of the shell module. Section 13.10 later in this chapter discusses how delays can be represented in the Verilog shell module.

13.5 Creating a combinational logic interface to a C model

In a combinational logic model, the outputs of the model continuously reflect the input values of the model. The inputs are asynchronous—when any input changes value, the model outputs are re-evaluated to reflect the input change.

The TF library works with the arguments of a system task. In the discussion of the Verilog shell module, in the previous section of this chapter, it was recommended that a system task be created to represent the C model interface, and that this task list all of the C model inputs and outputs as arguments. The reason for listing all inputs and outputs as task arguments is because this gives a PLI application easy access to these signals using the TF library.

For a combinational logic model, the PLI application which represents the interface to the C model must be called whenever one of the inputs to the C model changes value. There are two ways this can be accomplished:

- The simplest method is to use the Verilog HDL to invoke the *calltf routine* of the PLI application each time an input to the C model changes. This requires using a Verilog `always` procedure with a combinational logic sensitivity list. For the scientific ALU model example, the Verilog shell module would contain:

```
always @(a or b or opcode)
    $scientific_alu(a, b, opcode, result, exception, error);
```

- An alternative method is to use the TF routines to invoke the *misctf routine* of the PLI application each time an argument to a system task changes. This requires using a Verilog `initial` procedure to instantiate the system task which represents the interface to the C model. For the scientific ALU model example, the Verilog shell module would contain:

```
initial
    $scientific_alu(a, b, opcode, result, exception, error);
```

Examples of using each of these methods to represent a combinational logic interface to a C model are shown in sections 13.5.1 and 13.5.2.

13.5.1 Using the Verilog HDL to represent a combinational logic interface

When the Verilog HDL is used to invoke a system task whenever an input changes, the *calltf routine* of the PLI application will be invoked for each input change. In the example:

```
always @(a or b or opcode)
    $scientific_alu(a, b, opcode, result, exception, error);
```

The *calltf routine* for *$scientific_alu* will need to:

- Read the values of the C model inputs from the system task arguments

- Call the C model and pass the input values

- Write the outputs of the C model onto the system task arguments

Example 13-3 illustrates a complete PLI application for *$scientific_alu*, using the *calltf routine* to read and modify the values of the system task.

CD The source code for this example is on the CD accompanying this book.

- Application source file: `Chapter.13/sci_alu_comb_calltf_tf.c`
- Verilog shell module: `Chapter.13/sci_alu_comb_calltf_shell.v`
- Verilog test bench: `Chapter.13/sci_alu_comb_calltf_test.v`
- Expected results log: `Chapter.13/sci_alu_comb_calltf_test.log`

Example 13-3: combinational logic C model interface using a *calltf routine*

```
/***********************************************************************
 * calltf routine: Serves as an interface between Verilog simulation
 * and the C model.  Called whenever the C model inputs change value,
 * reads the input values, and passes the values to the C model, and
 * writes the C model outputs into simulation.
 ***********************************************************************/
int PLIbook_ScientificALU_calltf(int user_data, int reason)
{
  #define ALU_A        1   /* system task arg 1 is ALU A input       */
  #define ALU_B        2   /* system task arg 2 is ALU B input       */
  #define ALU_OP       3   /* system task arg 3 is ALU opcode input  */
  #define ALU_RESULT   4   /* system task arg 4 is ALU result output */
  #define ALU_EXCEPT   5   /* system task arg 5 is ALU exception output */
  #define ALU_ERROR    6   /* system task arg 6 is ALU error output  */

  double  a, b, result;
  int     opcode, excep, err;

  /* Read current values of C model inputs from Verilog simulation */
  a       = tf_getrealp(ALU_A);
  b       = tf_getrealp(ALU_B);
  opcode = (int)tf_getp(ALU_OP);

  /****** Call C model ******/
  PLIbook_ScientificALU_C_model(a, b, opcode, &result, &excep, &err);

  /* Write the C model outputs onto the Verilog signals */
  tf_putrealp(ALU_RESULT, result);
  tf_putp    (ALU_EXCEPT, (PLI_INT32)excep);
  tf_putp    (ALU_ERROR,  (PLI_INT32)err);

  return(0);
}

/***********************************************************************
 * checktf routine: Verifies that $scientific_alu() is used correctly.
 *   Note: For simplicity, only limited data types are allowed for
 *   task arguments.  Could add checks to allow other data types.
 ***********************************************************************/
int PLIbook_ScientificALU_checktf(int user_data, int reason)
{
  if (tf_nump() != 6)
    tf_error("$scientific_alu requires 6 arguments");
  else {
    if (tf_typep(ALU_A) != TF_READWRITEREAL)
      tf_error("$scientific_alu arg 1 must be a real variable\n");
    if (tf_typep(ALU_B) != TF_READWRITEREAL)
      tf_error("$scientific_alu arg 2 must be a real variable\n");
    if (tf_typep(ALU_OP) != TF_READONLY)
      tf_error("$scientific_alu arg 3 must be a net\n");
    else if (tf_sizep(ALU_OP) != 4)
```

```
            tf_error("$scientific_alu arg 3 must be a 4-bit vector\n");
      if (tf_typep(ALU_RESULT) != TF_READWRITEREAL)
         tf_error("$scientific_alu arg 4 must be a real variable\n");
      if (tf_typep(ALU_EXCEPT) != TF_READWRITE)
         tf_error("$scientific_alu arg 5 must be a reg\n");
      else if (tf_sizep(ALU_EXCEPT) != 1)
         tf_error("$scientific_alu arg 5 must be scalar\n");
      if (tf_typep(ALU_ERROR) != TF_READWRITE)
         tf_error("$scientific_alu arg 6 must be a reg\n");
      else if (tf_sizep(ALU_ERROR) != 1)
         tf_error("$scientific_alu arg 6 must be scalar\n");
   }
   return(0);
}
```

13.5.2 Using a misctf routine to represent a combinational logic interface

The tf_asynchon() routine provides a simple method of creating a combinational logic interface to a C model. This routine schedules asynchronous callbacks to the *misctf routine* whenever an argument of a system task changes value. The *misctf routine* can read the input values from the system task arguments, and pass the input values to the C model. The outputs of the C model are then passed back to the Verilog simulation by writing the results onto the system task arguments in the Verilog shell module. Chapter 12 presented how to use the tf_asynchon() routine.

By using *misctf routine* callbacks when a system task argument changes value, the *calltf routine* does not need to be invoked for every value change. In the Verilog shell module, the system task which represents the C model interface can be called from a Verilog initial procedure, so that the *calltf routine* is only invoked one time, instead of repeatedly. For example:

```
initial
   $scientific_alu(a, b, opcode, result, exception, error);
```

Using tf_asynchon() is an easy way to accurately represent a combinational logic interface, but there are two cautions which must be observed when using this routine: First, the *misctf routine* is called for any task argument change, including changes to the C model outputs, and, second, the *misctf routine* is called automatically for many miscellaneous reasons which are not required by the C model interface. The *misctf routine* must filter out these extra calls.

TIP   The VPI and ACC libraries offer a more efficient method of representing a combinational logic interface to a C model. These libraries provide a means to have a PLI application called only when specific signals change value. By placing value change flags on just the inputs to a C model, the PLI application will only be called when an input changes value. However, the t f_asynchon() routine allows an interface to be created very quickly and simply, which can be advantageous. Also, some simulators execute the TF routines more efficiently than the ACC and VPI routines, which, for these simulators, may make using t f_asynchon() a more efficient method, despite the extra calls to the *misctf routine*.

The basic steps involved with using t f_asynchon() and the *misctf routine* to implement a combinational logic interface are:

1. Create a PLI application system task to represent the interface between the Verilog shell module and the C model. The system task is invoked from the shell module, and all of the C model inputs and outputs are listed as arguments to the system task

2. In the *calltf routine* associated with the system task, invoke the t f_asynchon() routine, to enable asynchronous callbacks to the *misctf routine*.

3. In the *misctf routine*, which is called whenever a system task argument changes value, read the values of all inputs and pass the values to the C model. The output values of the C model are returned to the same *misctf routine*, which then writes the values to the system task arguments that represent the outputs of the C model.

The following example implements a combinational logic interface for the scientific ALU C model. The *misctf routine* has checks to avoid calling the C model when it was called for reasons are not due to value changes on the inputs of the C model.

CD   The source code for this example is on the CD accompanying this book.

- Application source file: `Chapter.13/sci_alu_comb_misctf_tf.c`
- Verilog shell module: `Chapter.13/sci_alu_comb_misctf_shell.v`
- Verilog test bench: `Chapter.13/sci_alu_comb_misctf_test.v`
- Expected results log: `Chapter.13/sci_alu_comb_misctf_test.log`

Example 13-4: combinational logic C model interface using a *misctf routine*

```
/**********************************************************************
 * calltf routine: turns on asynchronous callbacks to the misctf
 * routine whenever an argument to the system task changes value.
 **********************************************************************/
int PLIbook_ScientificALU_calltf(int user_data, int reason)
{
  tf_asynchon();
  return(0);
}

/**********************************************************************
 * misctf routine: Serves as an interface between Verilog simulation
 * and the C model.  Called whenever the C model inputs change value,
 * reads the input values, and passes the values to the C model, and
 * writes the C model outputs into simulation.
 **********************************************************************/
int PLIbook_ScientificALU_misctf(int user_data, int reason, int
paramvc)
{
  #define ALU_A      1  /* system task arg 1 is ALU A input         */
  #define ALU_B      2  /* system task arg 2 is ALU B input         */
  #define ALU_OP     3  /* system task arg 3 is ALU opcode input    */
  #define ALU_RESULT 4  /* system task arg 4 is ALU result output   */
  #define ALU_EXCEPT 5  /* system task arg 5 is ALU exception output */
  #define ALU_ERROR  6  /* system task arg 6 is ALU error output    */

  double  a, b, result;
  int     opcode, excep, err;

  /* abort if misctf was not called for a task argument value change */
  if (reason != REASON_PARAMVC)
    return(0);

  /* abort if task argument that changed was a model output */
  if (paramvc > ALU_OP) /* model outputs are after model inputs */
    return(0);

  /* Read current values of C model inputs from Verilog simulation */
  a      = tf_getrealp(ALU_A);
  b      = tf_getrealp(ALU_B);
  opcode = (int)tf_getp(ALU_OP);

  /****** Call C model ******/
  PLIbook_ScientificALU_C_model(a, b, opcode, &result, &excep, &err);

  /* Write the C model outputs onto the Verilog signals */
  tf_putrealp(ALU_RESULT, result);
  tf_putp     (ALU_EXCEPT, (PLI_INT32)excep);
  tf_putp     (ALU_ERROR,  (PLI_INT32)err);

  return(0);
}
```

13.6 Creating a sequential logic interface to a C model

In a sequential logic model, the outputs of the model change synchronously with an input strobe, such as a positive edge of a clock. There may also be one or more asynchronous inputs, such as a reset signal. As with representing a combinational logic interface, a sequential logic interface can be represented using either the Verilog HDL or within the PLI application.

13.6.1 Using the Verilog HDL to represent a sequential logic interface

When the Verilog HDL is used to represent the sequential logic of the C model interface, the interface will be represented using a *calltf routine*. The Verilog HDL will invoke the system task at each clock cycle, which will result in the *calltf routine* being called synchronously at that time. The Verilog shell module will contain an always procedure with a sequential logic sensitivity list. For example:

```
always @(posedge clock)
    $scientific_alu(clock, a, b, op, result, excep, err);
```

An asynchronous reset can be represented in the Verilog procedure as follows (this example models an active low reset):

```
always @(posedge clock or negedge rst)
    $scientific_alu(clock, rst, a, b, op, result, excep, err);
```

When the Verilog HDL is used to synchronize the call to the PLI application to a clock, the *calltf routine* does not need to be any different than with combinational logic. When a clock change occurs, the *calltf routine* will be invoked, and all inputs to the C model will be read at that time. An example of this method is not shown here, since it is virtually the same as the example of using a *calltf routine* with the combinational logic that was listed previously in example 13-3.

13.6.2 Using the misctf routine to represent a sequential logic interface

The TF routines do not provide a convenient method of synchronizing activity to only one input change, such as the clock line. However, the asynchronous callbacks for any input change using `tf_asynchon()` can be used to create a synchronous, sequential logic interface. The process simply involves adding additional filters in the *misctf routine*, so that value changes on input signals other than the clock line are ignored. The basic steps involved with using `tf_asynchon()` to implement a synchronous sequential logic interface are very similar to implementing a combinational logic interface. The one difference is that C model input values are only read and passed to the C model when the clock input changes.

Example 13-5 implements a sequential logic interface for the scientific ALU C model using the *misctf routine* and asynchronous callbacks for all task argument changes. This example ignores all task argument changes, except for change which represents the clock input.

Note that Verilog uses 4-state logic, so there are 12 possible transitions on the clock signal. Any transition will cause the *misctf routine* to be called. The following example checks that the value of the clock is a logic 1, with the assumption if the clock just changed value, and it is now a logic 1, it must have been a positive edge of the clock. By checking for a logic one, all negative going transitions on the clock line are ignored, as well as transitions from 0 to Z and 0 to X. The positive going transitions that this example will interpret as a positive edge of clock are transitions from 0 to 1, Z to 1, and X to 1. This filtering of transitions is not quite the same as the posedge keyword in the Verilog language, where 0 to Z and 0 to X are also treated as positive transitions.

CD The source code for this example is on the CD accompanying this book.

- Application source file: `Chapter.13/sci_alu_sequential_tf.c`
- Verilog shell module: `Chapter.13/sci_alu_sequential_shell.v`
- Verilog test bench: `Chapter.13/sci_alu_sequential_test.v`
- Expected results log: `Chapter.13/sci_alu_sequential_test.log`

Example 13-5: sequential logic C model interface using TF routines

```
/******************************************************************
 * calltf routine: turns on asynchronous callbacks to the misctf
 * routine whenever an argument to the system task changes value
 ******************************************************************/
int PLIbook_ScientificALU_calltf(int user_data, int reason)
{
  tf_asynchon();
  return(0);
}


/******************************************************************
 * misctf routine: Serves as an interface between Verilog simulation
 * and the C model.  Called whenever the C model inputs change value,
 * ignores all changes except a positive edge of clock, on the positive
 * edge of clock, reads the input values, and passes the values to the
 * C model, and writes the C model outputs into simulation.
 ******************************************************************/
```

```
int PLIbook_ScientificALU_misctf(int user_data, int reason, int
paramvc)
{
  #define ALU_CLOCK  1  /* system task arg 1 is ALU clock input     */
  #define ALU_A      2  /* system task arg 2 is ALU A input         */
  #define ALU_B      3  /* system task arg 3 is ALU B input         */
  #define ALU_OP     4  /* system task arg 4 is ALU opcode input    */
  #define ALU_RESULT 5  /* system task arg 5 is ALU result output   */
  #define ALU_EXCEPT 6  /* system task arg 6 is ALU exception output */
  #define ALU_ERROR  7  /* system task arg 7 is ALU error output    */

  double  a, b, result;
  int     opcode, excep, err, clock;

  /* abort if misctf was not called for a task argument value change */
  if (reason != REASON_PARAMVC)
    return(0);

  /* abort if task argument that changed was not the clock input */
  if (paramvc != ALU_CLOCK)
    return(0);

  /* Read current values of C model inputs from Verilog simulation */
  clock  = (int)tf_getp(ALU_CLOCK);
  if (clock != 1) /* abort if not a positive edge of the clock input */
    return(0);
  a      = tf_getrealp(ALU_A);
  b      = tf_getrealp(ALU_B);
  opcode = (int)tf_getp(ALU_OP);

  /****** Call C model ******/
  PLIbook_ScientificALU_C_model(clock, a, b, opcode,
                                &result, &excep, &err);

  /* Write the C model outputs onto the Verilog signals */
  tf_putrealp(ALU_RESULT, result);
  tf_putp    (ALU_EXCEPT, (PLI_INT32)excep);
  tf_putp    (ALU_ERROR,  (PLI_INT32)err);

  return(0);
}
```

13.7 Synchronizing with the end of a simulation time step

Within a simulation, several inputs to the C model might change at the same moment in simulation time. The *calltf routine* or the *misctf routine* will be called for each input change, which means these routines may be called before all input value changes have occurred for that time step.

With a combinational logic interface, the *calltf routine* or *misctf routine* will be called for every input change. This is the correct functionality for combinational logic. At the completion of a simulation time step, the outputs from the C model represent the most current input values. However, by synchronizing the call to the C model with the end of the simulation time step in which changes occur, the multiple calls to the C model within a time step could be optimized to a single call. With a sequential logic interface synchronized to a clock, when the *calltf routine* or *misctf routine* is called at a clock change, other input changes at that moment in simulation time may or may not have occurred. It may be desirable to ensure that the C model is not called until all inputs have their most current value for the time step in which the clock changes.

By using the `tf_synchronize()` routine, both combinational logic and sequential logic C model interfaces can be synchronized to the end of a current simulation time step. This is done by using the call to the *calltf routine* or *misctf routine* when a value change occurred, to schedule a synchronous callback to the *misctf routine* for the end of the current time step. Note that `tf_synchronize()` does not guarantee that all changes in the current simulation time step have occurred. The `tf_synchronize()` routine schedules a read/write synchronization callback to the *misctf routine* after all known events in a time step have been processed by the simulator. However, other PLI applications or simulation activity may schedule new events in the same simulation time step. The events caused by other PLI applications could affect the inputs of the C model after the `tf_synchronize()` callback has occurred.

 The IEEE 1364 Verilog standard allows PLI read-write synchronization events to be processed either before or after nonblocking assignments. This ambiguity can lead to different results from different simulators. The reason the Verilog standard allows this latitude in event scheduling is discussed in Chapter 12, on page 221.

If the C model interface requires absolute assurance that no other inputs will change in the current simulation time step, the PLI application must use the `tf_rosynchronize()` read-only synchronization callback, instead of a read/write synchronization. In a read-only synchronization, the *misctf routine* is not allowed to write return values into the simulation using any of the TF put routines or the `acc_set_value()` routine. The VPI routines, however, can return values into simulation at a future time from a read-only simulation callback.

Example 13-6 modifies the combinational logic interface, which was presented previously in Example 13-4. This modified version schedules a *misctf routine* callback for two different reasons:

- At a value change of a task argument, a synchronize callback is scheduled at the end of the current time step.

- At a synchronize callback at the end of a simulation time step, the values of all C model inputs are read and passed to the C model.

This example uses the TF work area to store a flag to indicate when a synchronous callback to the *misctf routine* has already been scheduled for the current simulation time step. This flag prevents more than one synchronous callback to the *misctf routine* being requested in the same time step. Since the TF work area is unique for each instance of a system task, each instance of the C model will have a unique flag.

CD The source code for this example is on the CD accompanying this book.

- Application source file: `Chapter.13/sci_alu_synchronized_tf.c`
- Verilog shell module: `Chapter.13/sci_alu_synchronized_shell.v`
- Verilog test bench: `Chapter.13/sci_alu_synchronized_test.v`
- Expected results log: `Chapter.13/sci_alu_synchronized_test.log`

Example 13-6: C model interface synchronized to the end of a time step

```
/*********************************************************************
 * calltf routine: turns on asynchronous callbacks to the misctf
 * routine whenever an argument to the system task changes value
 *********************************************************************/
int PLIbook_ScientificALU_calltf(int user_data, int reason)
{
  tf_asynchon();
  tf_setworkarea(NULL); /* set work area to null */
  return(0);
}

/*********************************************************************
 * misctf routine: Serves as an interface between Verilog simulation
 * and the C model.  The misctf routine performs different operations
 * depending on the reason it is called:
 * - For a value change callback: schedules a callback to the misctf
 *   application synchronized to the end of a time step. Only schedules
 *   one callback for a time step.
 * - For a synchronize callback: reads the input values, and passes
 *   the values to the C model, and writes the C model outputs into
 *   simulation.
 *********************************************************************/
```

```
int PLIbook_ScientificALU_misctf(int user_data, int reason, int
paramvc)
{
  #define ALU_A       1  /* system task arg 1 is ALU A input      */
  #define ALU_B       2  /* system task arg 2 is ALU B input      */
  #define ALU_OP      3  /* system task arg 3 is ALU opcode input */
  #define ALU_RESULT  4  /* system task arg 4 is ALU result output */
  #define ALU_EXCEPT  5  /* system task arg 5 is ALU exception output */
  #define ALU_ERROR   6  /* system task arg 6 is ALU error output */

  double  a, b, result;
  int     opcode, excep, err;

  /* check if misctf was called for a task argument value change */
  if (reason == REASON_PARAMVC) {
    /* abort if task argument that changed was a model output */
    if (paramvc > ALU_OP) /* model outputs are after model inputs */
      return(0);

    /* If the TF work area is null, then no misctf synchronize    */
    /* callback has been scheduled for this time step (the work area */
    /* is set to non-null by this routine, and is set to null by the */
    /* misctf after a synchronize callback is processed.          */
    if (tf_getworkarea() == NULL) {
    /* Schedule a synchronize callback to misctf for this instance */
      tf_synchronize();
      tf_setworkarea("1"); /* set work area to non-null */
    }
    return(0);
  }

  /* check if misctf was called for end-of-time step synchronize */
  if (reason == REASON_SYNCH) {
    /* Read current values of C model inputs from Verilog simulation */
    a      = tf_getrealp(ALU_A);
    b      = tf_getrealp(ALU_B);
    opcode = (int)tf_getp(ALU_OP);

    /****** Call C model ******/
    PLIbook_ScientificALU_C_model(a, b, opcode, &result, &excep, &err);

    /* Write the C model outputs onto the Verilog signals */
    tf_putrealp(ALU_RESULT, result);
    tf_putp     (ALU_EXCEPT, (PLI_INT32)excep);
    tf_putp     (ALU_ERROR,  (PLI_INT32)err);
    tf_setworkarea(NULL); /* set work area to null */
  }
  return(0);
}
```

13.8 Synchronizing with a future simulation time step

In certain C model applications, it may be necessary to synchronize C model activity with future simulation activity. The `tf_setdelay()` routine and variations of this routine can be used to schedule a call to the *misctf routine* for a specific amount of time in the future, relative to the current simulation time.

The `tf_getnextlongtime()` routine returns the future simulation time in which the next simulation event is scheduled to occur. This provides a way for a PLI application to synchronize activity for when the Verilog simulator is processing simulation events.

These TF routines for synchronizing with future simulation times were presented in more detail in Chapter 12.

13.9 Allocating storage within a C model

Special attention and care must be taken when a C model uses static variables or allocates memory.

The Verilog language can instantiate a model any number of times. Each instance of the Verilog shell module creates a unique instance of the system task which invokes the PLI interface to the C model. Therefore, the *calltf routine* and *misctf routine* which are invoked by a system task instance will both be unique to a task instance, and any memory which is allocated by the *calltf routine* and *misctf routine* will also be unique for each instance of the system task.

When a C model is represented as an independent program, multiple instances of the model are not a problem, as each instance will invoke a new process with unique storage for each process.

When the C model is represented as a C function, however, multiple instances of the model will share the same function. The C function must allow for the possibility of multiple instances, and provide unique storage for each instance.

Example 13-7 presents a latched version of the scientific ALU, which can store the result of a previous operation indefinitely. Example 13-8 presents a combinational logic interface to this latched model. This example interface is based on the interface method shown previously in example 13-4, which used the *misctf routine*. The same principles apply to the interface method which uses the *calltf routine* as the interface to the C model. This example allocates unique storage within the C model for each instance of the C model. The instance pointer of the system task which represents the C model interface is used to identify which storage area belongs to which instance of the model. This instance pointer is obtained using tf_getinstance() and is passed to the C model as an input to the model function.

CD The source code for this example is on the CD accompanying this book.

- Application source file: Chapter.13/sci_alu_latched_tf.c
- Verilog shell module: Chapter.13/sci_alu_latched_shell.v
- Verilog test bench: Chapter.13/sci_alu_latched_test.v
- Expected results log: Chapter.13/sci_alu_latched_test.log

Example 13-7: scientific ALU C model with latched outputs

```
/*************************************************************************
 * Definition for a structure to store output values when the ALU is
 * latched.  When enable is 1, the ALU returns the currently calculated
 * outputs, and when 0, the ALU returns the latched previous results.
 *************************************************************************/
#include <stdlib.h>
#include <stdio.h>
 typedef struct PLIbook_SciALUoutputs   *PLIbook_SciALUoutputs_p;
 typedef struct PLIbook_SciALUoutputs {
   char *instance_p; /* shows which task instance owns this space */
   double result;    /* stored result of previous operation */
   int    excep;
   int    err;
   PLIbook_SciALUoutputs_p next_ALU_outputs; /* next stack location */
 } PLIbook_SciALUoutputs_s;

  /* declare global stack pointer */
  static PLIbook_SciALUoutputs_p ALU_outputs_stack = NULL;


/*************************************************************************
 * C model of a Scientific Arithmetic Logic Unit.
 *     Latched outputs version.
 *************************************************************************/
#include <math.h>
#include <errno.h>
void PLIbook_ScientificALU_C_model(
        int     enable,     /* input; 0 = latched */
        double  a,          /* input */
        double  b,          /* input */
        int     opcode,     /* input */
        double *result,     /* output from ALU */
        int    *excep,      /* output; set if result is out of range */
        int    *err,        /* output; set if input is out of range */
        char   *instance_p) /* input; pointer to system task instance */
{
```

```
PLIbook_SciALUoutputs_p ALU_outputs;
/* Locate the output storage in the stack for this model instance  */
/* If no storage is found, then allocate a storage block and add   */
/* the storage to the stack.                                       */
ALU_outputs = ALU_outputs_stack; /* top-of-stack is in global var. */
while (ALU_outputs && (ALU_outputs->instance_p != instance_p))
  ALU_outputs = ALU_outputs->next_ALU_outputs;
/* If no storage area found for this model instance, create one */
if (ALU_outputs == NULL) {
  ALU_outputs =
    (PLIbook_SciALUoutputs_p)malloc(sizeof(PLIbook_SciALUoutputs_s));
  ALU_outputs->instance_p = instance_p; /* set owner of this space */
  ALU_outputs->next_ALU_outputs = ALU_outputs_stack;
  ALU_outputs_stack = ALU_outputs; /* save new top-of-stack */
}
if (enable) { /* ALU is not latched, calculate outputs and store */
  switch (opcode) {
    case 0x0: ALU_outputs->result = pow   (a, b);       break;
    case 0x1: ALU_outputs->result = sqrt  (a);          break;
    case 0x2: ALU_outputs->result = exp   (a);          break;
    case 0x3: ALU_outputs->result = ldexp (a, (int)b);  break;
    case 0x4: ALU_outputs->result = fabs  (a);          break;
    case 0x5: ALU_outputs->result = fmod  (a, b);       break;
    case 0x6: ALU_outputs->result = ceil  (a);          break;
    case 0x7: ALU_outputs->result = floor (a);          break;
    case 0x8: ALU_outputs->result = log   (a);          break;
    case 0x9: ALU_outputs->result = log10 (a);          break;
    case 0xA: ALU_outputs->result = sin   (a);          break;
    case 0xB: ALU_outputs->result = cos   (a);          break;
    case 0xC: ALU_outputs->result = tan   (a);          break;
    case 0xD: ALU_outputs->result = asin  (a);          break;
    case 0xE: ALU_outputs->result = acos  (a);          break;
    case 0xF: ALU_outputs->result = atan  (a);          break;
  }
  ALU_outputs->excep = (errno == ERANGE);/* result out-of-range */
  #ifdef WIN32  /* for Microsoft Windows compatibility */
   ALU_outputs->err  = (_isnan(*result) || /* not-a-number, or */
                        errno == EDOM);   /* arg out-of-range */
  #else
   ALU_outputs->err  = (isnan(*result) ||  /* not-a-number, or */
                        errno == EDOM);   /* arg out-of-range */
  #endif
  if (ALU_outputs->err) ALU_outputs->result = 0.0;
  errno = 0;                               /* clear error flag */
}
/* return the values stored in the C model */
*result = ALU_outputs->result;
*err    = ALU_outputs->err;
*excep  = ALU_outputs->excep;
return;
}
```

Example 13-8: combinational logic interface to latched scientific ALU C model

```
/*********************************************************************
 * calltf routine: turns on asynchronous callbacks to the misctf
 * routine whenever an argument to the system task changes value
 *********************************************************************/
int PLIbook_ScientificALU_calltf(int user_data, int reason)
{
  tf_asynchon();
  return(0);
}


/*********************************************************************
 * misctf routine: Serves as an interface between Verilog simulation
 * and the C model.  Called whenever the C model inputs change value,
 * reads the input values, and passes the values to the C model, and
 * puts the C model outputs into simulation.  Passes the instance
 * pointer of the Verilog system task which represents the C model
 * to serve as a unique flag within the C model.
 *********************************************************************/
int PLIbook_ScientificALU_misctf(int user_data, int reason, int
paramvc)
{
  #define ALU_ENABLE 1   /* system task arg 1 is ALU enable input    */
  #define ALU_A      2   /* system task arg 2 is ALU A input         */
  #define ALU_B      3   /* system task arg 3 is ALU B input         */
  #define ALU_OP     4   /* system task arg 4 is ALU opcode input    */
  #define ALU_RESULT 5   /* system task arg 5 is ALU result output   */
  #define ALU_EXCEPT 6   /* system task arg 6 is ALU exception output */
  #define ALU_ERROR  7   /* system task arg 7 is ALU error output    */

  double  a, b, result;
  int     opcode, excep, err, enable;
  char    *instance_p;

  /* abort if misctf was not called for a task argument value change */
  if (reason != REASON_PARAMVC)
    return(0);

  /* abort if task argument that changed was a model output */
  if (paramvc > ALU_OP) /* model outputs are after model inputs */
    return(0);

  enable = (int)tf_getp(ALU_ENABLE);
  a      = tf_getrealp(ALU_A);
  b      = tf_getrealp(ALU_B);
  opcode = (int)tf_getp(ALU_OP);

  /* Obtain the instance pointer for this system task instance */
  instance_p = (char *)tf_getinstance();

  /****** Call C model ******/
  PLIbook_ScientificALU_C_model(enable, a, b, opcode,
                                &result, &excep, &err, instance_p);

  /* Write the C model outputs onto the Verilog signals */
```

```
tf_putrealp(ALU_RESULT, result);
tf_putp     (ALU_EXCEPT, (PLI_INT32)excep);
tf_putp     (ALU_ERROR,  (PLI_INT32)err);

return(0);
}
```

13.10 Representing propagation delays in a C model

Propagation delays from an input change to an output change in a C model can be represented in two ways:

- Using delays in the PLI interface.
- Using delays in the Verilog shell module.

Delays in the PLI interface are represented by specifying a delay value with the `tf_strdelputp()`, `tf_strlongdelputp()` and `tf_strrealdelputp()` routines, which write values onto the system task arguments at a future simulation time step. Either inertial or transport event propagation can be represented, depending on the requirements of the C model. However, using the `tf_strdelputp()` and related routines has disadvantages. The logic values in the PLI application must be converted to strings, in order to put the value onto an argument. This conversion to a string, which the simulator must then convert into a Verilog logic value, is not efficient for simulation run-time performance. Also, the TF routines do not offer a great deal of flexibility on creating delays which are different for each instance of a model, representing minimum, typical and maximum delays, different delays for rise and fall transitions, or annotating delays using delay calculators or SDF files.

C model propagation delays can also be represented using the pin-to-pin path delays in the Verilog shell module. This method provides the greatest amount of flexibility and accuracy in modeling propagation delays. All path delay constructs can be used, as well and Verilog timing constraints.

Example 13-9 illustrates adding pin-to-pin path delays to the scientific ALU shell module.

 Some Verilog simulators restrict the use of pin-to-pin path delays and SDF delay back annotation to models which are represented with Verilog primitives and net data types. To use path delays on a C model with these simulators, buffers must be added to all input and output ports, with net data types connected to the inputs and outputs of these buffers. Example 13-9 illustrates using buffers in this way.

CD The source code for this example is on the CD accompanying this book.

- Application source file: Chapter.13/sci_alu_with_delays_tf.c
- Verilog shell module: Chapter.13/sci_alu_with_delays_shell.v
- Verilog test bench: Chapter.13/sci_alu_with_delays_test.v
- Expected results log: Chapter.13/sci_alu_with_delays_test.log

Example 13-9: scientific ALU Verilog shell module with pin-to-pin path delays

```
`timescale 1ns / 100ps
module scientific_alu(a_in, b_in, opcode_in,
                      result_out, exception, error);
  output [63:0] result_out;
  output        exception, error;
  input  [63:0] a_in, b_in;
  input   [3:0] opcode_in;

  wire   [63:0] result_out, result_vector;
  wire   [63:0] a_in, a_vector;
  wire   [63:0] b_in, b_vector;
  wire    [3:0] opcode_in, opcode_vector;
  wire          exception, error;
  reg           exception_reg, error_reg;
  real          a, b, result; // real variables used in this module

  // convert real numbers to/from 64-bit vector port connections
  assign result_vector = $realtobits(result);
  always @(a_vector)  a = $bitstoreal(a_vector);
  always @(b_vector)  b = $bitstoreal(b_vector);

  //call the PLI application which interfaces to the C model
  initial
    $scientific_alu(a, b, opcode_vector,
                    result, exception_reg, error_reg);

  specify
    (a_in, b_in *> result_out, exception, error) = (5.6, 4.7);
    (opcode_in  *> result_out, exception, error) = (3.4, 3.8);
  endspecify

  // add buffers to all ports, with nets connected to each buffer
  // (this example uses the array of instance syntax in the
  // from the IEEE 1364-1995 Verilog standard
  buf result_buf[63:0] (result_out, result_vector);
  buf excep_buf        (exception,  exception_reg);
  buf error_buf        (error,      error_reg);
  buf a_buf[63:0]      (a_vector, a_in);
  buf b_buf[63:0]      (b_vector, b_in);
  buf opcode_buf[3:0]  (opcode_vector, opcode_in);
endmodule
```

13.11 Summary

This chapter has presented a few ways in which the TF library can be used to interface a C language model with Verilog simulations. By creating a shell module which contains the system task that invokes the C model interface, the C model can be used in a Verilog design, just as any other Verilog module. The interface between the shell module and the C model can done through a *calltf routine*, using the Verilog HDL to control when the *calltf routine* is invoked. The interface can also be done through the *misctf routine*, using the TF library to control when the *misctf routine* is invoked. The tf_asynchon() routine provides a simple means of using the *misctf routine* to pass input changes to a C model between the C model and the Verilog shell module. The TF routines to read and modify logic values allow information to be exchanged in a variety of formats.

CHAPTER 14 *How to Use the ACC Routines*

T his chapter introduces the ACC portion of the PLI standard, and shows how to use the ACC routines to access information within a simulation data structure. Two complete PLI applications, *$show_all_nets* and *$show_all_signals*, will be created in this chapter, to illustrate how the ACC routines work. The remaining chapters in this part of the book then build on the principles presented in this chapter by explaining the routines within the ACC library in much more detail.

The concepts presented in this chapter are:

- An overview of how ACC routines work

- Advantages of the ACC library

- Creating a complete PLI application using the ACC library

- Obtaining handles to Verilog HDL objects

- Accessing properties of Verilog HDL objects

- Reading values of Verilog HDL objects

14.1 Specification of $show_all_nets and $show_all_signals

To show how the ACC routines are used, two PLI applications will be created. These examples will be built up, one step at a time, as this chapter progresses.

The first example presented is an application called *$show_all_nets*. The usage of this application is:

```
$show_all_nets(<module_instance_name>);
```

This PLI application will:

1. Access the first argument of the system task, which is the name of a module instance.

2. Print the hierarchical path and name of that module, along with the current simulation time.

3. Search for all net signals in the module, and print the data type and current logic value of each net.

This chapter will first illustrate a *checktf routine* for *$show_all_nets*, which verifies that the argument provided as an input is a valid module instance name. Then a *calltf routine* will be created to perform the functionality of the system task.

The second example is a PLI application called ***$show_all_signals***. This application prints the current value of all net, reg and variable data types in a module. The usage of this application is:

```
$show_all_signals(<module_instance_name>);
```

To illustrate some additional ways to use the ACC routines, two enhancements to the *$show_all_signals* example will be presented. These are:

-   Use no argument or a null argument to *$show_all_signals*, to represent the module instance containing the *$show_all_signals* system task.

-   Allow multiple arguments to *$show_all_signals*, so the values of signals in several modules can be printed with one call to *$show_all_signals*.

14.2 The ACC routine library

"ACC" stands for "access", and the library of ACC routines are often referred to as *access routines*. The ACC routines are the second of three primary generations of the PLI functions (the TF routines were the first generation, and the VPI routines are the newest). The primary purpose of the ACC routines is to provide a PLI application access to the internal data structures of a simulation. The ACC routines provide a consistent layer between a user's PLI application and the underlying data structures of a simulation. The PLI application does not need to know the specifics about how the simulator stores its data, and the same PLI application will work with many different simulators.

The ACC routines treat Verilog HDL constructs as *objects*, and many of the ACC routines provide ways to locate any specific object or type of objects within a simulation data structure. Other ACC routines can then read and modify information about each object.

The ACC library can be divided into five basic groups of routines:

- *handle* routines obtain a handle for one specific Verilog HDL object.

- *next* routines locate and return handles for a specific type of Verilog object.

- *fetch* routines access information about an object.

- *set* routines modify information about an object.

- *miscellaneous* routines perform a variety of operations.

The library of ACC routines is defined in a C header file called **acc_user.h**, which is part of the IEEE 1364 standard. This header file also defines a number of C constants and C structures used by the ACC routines. All PLI applications that use ACC routines must include the acc_user.h file.

The ACC library is designed to work with the TF library. An example of including the header files for these two libraries is:

```
#include "veriuser.h"    /* IEEE 1364 TF PLI library */
#include "acc_user.h"    /* IEEE 1364 ACC PLI library */
```

The ACC routines are intended to complement and expand the capabilities of the TF routines, rather than replace the TF routines. While there is some overlap in the functionality provided by the two libraries, there are a number of capabilities which are unique to the TF library—for example: `tf_nump()`, `io_printf()`, `tf_setworkarea()` and `tf_getworkarea()`.

 The VPI library was designed to replace both the TF and ACC libraries with a more concise, more robust, and more versatile procedural interface. The IEEE 1364 standard includes the TF and ACC libraries, in order to provide backward compatibility and portability of older PLI applications with modern Verilog simulators. The official policy of the IEEE 1364 standards committee is that, as improvements and enhancements are added to the Verilog language, only the VPI library of the PLI will be expanded to support those new features. The TF and ACC libraries will be maintained, but not enhanced, in future versions of the IEEE 1364 standard.

14.3 Advantages of the ACC library

The ACC routines provide direct access to much of a Verilog simulation data structure. This is in contrast to the TF library, which only provides access to the arguments of system tasks and system functions. This direct access allows a PLI application to more fully analyze and interact with a Verilog simulation.

Advantages of ACC routines, compared to TF routines

An important advantage the direct access which ACC routines provide can be seen in the example PLI applications presented in this chapter. These applications print the current logic values of all signals in a module. To implement this functionality with TF routines, every signal name would need to be listed as an argument to the system task. If a module had dozens of signals, listing each one as an argument would be very awkward. Using ACC routines, only one argument needs to be passed to the system task—the name of a module instance. From this starting point, the ACC routines can find all the signals in the module, and directly access the information of those signals.

TIP
Using ACC routines can have a negative impact on simulation run-time performance, because ACC routines have arbitrary access to the simulation data structure. Using only TF routines can improve the performance of some simulators.

The ACC routines provide arbitrary, run-time access to information within the simulation data structure. The information that ACC routines will access during simulation cannot be predicted at elaboration/linking time. This can prevent a compiler or elaborator from effectively optimizing the simulation data structure for maximum simulation performance. The TF routines restrict access to the arguments of a system task or system function. This restricted access can be determined at elaboration/linking time, and can therefore be optimized by the compiler or elaborator.

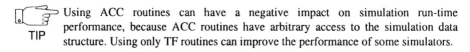
NOTE Some simulators, such as Cadence NC-Verilog™ and Synopsys VCS™, require special configurations or invocation options to enable some or all of the ACC routines. Refer to Appendix A for more information about specific simulators.

Advantages of ACC routines, compared to VPI routines

The VPI library is designed to replace the ACC library, and also extend the capabilities of the PLI. In the author's opinion, the VPI library is a better choice for creating new PLI applications. However, there are some advantages which the ACC library offers, compared to the VPI library.

First and foremost, the ACC library is supported by virtually every major Verilog simulator, whereas the VPI standard is only supported by a few leading simulators. The more widespread support of ACC routines makes a PLI application portable to a variety of simulators and engineering environments.

Second, it is often much easier and faster to develop PLI applications using the ACC library, compared to the VPI library. The ACC library is a much larger library, and often has predefined routines which take care of much of the work that a PLI application needs to accomplish. In the smaller VPI library, the PLI application developer must code much of the corresponding functionality by hand. However, the larger ACC library also tends be less efficient for simulation run-time performance. When using ACC routines, PLI application developers may tend to implement poorly structured C code, which also impacts simulation performance and becomes difficult to maintain.

14.4 Verilog HDL objects

The ACC routines treat Verilog HDL constructs as **objects**, and many of the ACC routines provide ways to locate any specific object or type of object within a simulation data structure. Other ACC routines can then read and modify information about each object. The simple Verilog HDL example which follows has several objects which can be accessed by the library of ACC functions.

```
module test;
  reg  [1:0] test_in;
  wire [1:0] test_out;
  buf2 u1 (test_in, test_out);
  initial
    begin
      test_in = 3;
      #50 $display("in=%d, out=5d", test_in, test_out);
    end
endmodule

module buf2 (in, out);
  input  [1:0] in;
  output [1:0] out;
  wire   [1:0] in, out;
  buf #5 n0 (out[0], in[0]);
  buf #7 n1 (out[1], in[1]);
endmodule
```

In this Verilog HDL example, the objects that a PLI application can access include:

- A top-level module, with the definition name *"test"*. Within this module are:
  - A `reg` signal, with a vector size of 2 and the name *"test_in"*. The signal will have a logic value which can be read and modified by the PLI application.
  - A `wire` net, with a vector size of 2 and the name *"test_out"*. The net reflects a resolved logic value which can be read by the PLI application.
  - A module instance, with the definition name *"buf2"* and the instance name *"u1"*. Within this module are:
  - Two ports, with the names *"in"* and *"out"*. Each port has a vector size and direction.
  - Two `wire` nets, with vector sizes and names. The nets reflect a resolved logic value which can be read by the PLI application.
  - Two primitive instances, with the definition name *"buf"* and the instance names *"n0"* and *"n1"*. Each primitive has a delay value which can be read and modified by the PLI.
  - Terminals on each primitive instance, with bit-selects of specific nets connected to the terminals, such as *"out[0]"* and *"in[0]"*.

14.4.1 The ACC handle data type

The ACC routines use a special data type, called a *handle*, to access Verilog HDL objects. The handle data type is defined in the ACC library (in the *acc_user.h* header file). The declaration type for variables to store a handle is **handle** (spelled with all lower case letters). An example declaration for two handle variables is:

```
handle  module_handle, net_handle;
```

There are more than 45 ACC routines that locate objects within a simulation data structure and return handles for the objects. Other ACC routines are used to access information about an object, using the object's handle as a reference point. The information that can be accessed depends on the type of the object, but might include the object's name and current logic value.

The object oriented method of accessing information used by the ACC routines is very similar to the object oriented method used by the VPI portion of the PLI standard, which was presented in Part One of this book. The only real difference is that the ACC routines have a more limited list of what Verilog HDL constructs are considered objects. For example, Verilog procedures and procedural statements are not objects in the ACC library, and therefore the ACC routines cannot access the procedural portions of a Verilog design. Verilog memory arrays are also not objects, and cannot be accessed. In the VPI library, virtually everything that exists in the Verilog language is considered an object.

 Do not share handles between VPI routines and ACC routines! The VPI routines in the PLI standard also use the concept of a handle for referencing Verilog objects. The IEEE 1364 standard does not guarantee that a handle which is obtained with the VPI library will be the same as a handle which is obtained with the ACC library.

14.5 ACC handle routines

The *$show_all_nets* application shown in this chapter will need to obtain a handle to the first system task/function argument, in order to access all signals within a module.

ACC *handle* routines return a handle for a single object. *These ACC routines are object-specific.* There is a different routine for each type of object that can be accessed using handle routines. Some examples of ACC handle routines are:

- **acc_handle_tfarg()** obtains a handle for an object named in a system task/function argument.
- **acc_handle_terminal()** obtains a handle for an object connected to the terminal of a primitive.
- **acc_handle_port()** obtains a handle for a module port.

There are 25 ACC handle routines. This chapter introduces these routines. More details on these routines are presented in the next chapter.

Using ACC handle routines

A system task or system function can have any number of arguments, including none. The arguments are numbered from left to right, starting with argument number 1. In the following example:

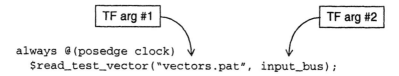

```
always @(posedge clock)
    $read_test_vector("vectors.pat", input_bus);
```

- Task/Function argument number 1 is a string, with the value "vectors.pat".
- Task/Function argument number 2 is a signal, with the name input_bus.

For the *$show_all_nets* application, the first system task argument will be a module instance name. For example:

```
always @(posedge clock)
      $show_all_nets(top.i1);
```

The ACC handle routine that is used to obtain a handle for a system task/function argument is `acc_handle_tfarg()`. The syntax of this routine is:

handle **acc_handle_tfarg (**
    PLI_INT32   **n)**   /* position number of a system task/function argument */

This routine returns a handle for the object named as an argument in the system task/ function which called the PLI application. Arguments are numbered from left to right, beginning with 1.

A handle for the module instance that is named in the first system task argument can be obtained, using the following C code:

```
handle tfarg_handle;
tfarg_handle = acc_handle_tfarg(1);
```

14.6 ACC next routines

The ACC library provides a set of routines to make it easy to access all occurrences of a specific type of object. These routines are referred to as *ACC next routines*. As with the ACC handle routines, *the ACC next routines are object-type specific*, so there is a different ACC next routine for most types of Verilog objects which the ACC routines can access. A few examples are:

- **acc_next_net()** obtains handles for all nets within a module.
- **acc_next_port()** obtains handles for all ports within a module.
- **acc_next_primitive()** obtains handles for all primitive instances within a module.

There are 22 ACC next routines. How these next routines are used is presented in this chapter, and the full list and syntax of the routines is presented in the next chapter.

There are two important terms used with ACC next routines:

- *target objects* are the type of objects for which the ACC next routine will obtain handles. For acc_next_net(), the target objects will be Verilog nets.

- *reference objects* are where the ACC next routine will search for the target objects. For example, to find all nets within a module, the reference object is the module.

Most ACC next routines require two inputs:

1. A handle for a reference object.

2. A handle to the previous target object found.

Using ACC next routines

The *$show_all_nets* application will need to access all nets within a module. The acc_next_net() routine will be used to access these nets. The syntax for this routine is:

handle **acc_next_net (**
    handle   **module,**   /* handle for a module */
    handle   **prev_net)**   /* handle for the previous net found; initially **null** */

The reference object for this routine is a handle for a module instance, and the target handle will be for a net. The basic usage of the routine is:

```
next_net_handle = acc_next_net(module_handle, previous_net_handle)
```

The ACC next routines return the target object handles one handle at a time. In order to locate all of a specific object, the next routine must be placed in a loop. All ACC next routines follow the same rules for how the target object handles are retrieved:

- To locate the first of the target objects within a reference object, the handle for the previous target found must be set to **null**.

- To locate the next target object within a reference object, the previous target handle found must be set to the previous target found.

- When the ACC next routine cannot find any more of target objects, the routine returns a **null**. This return value can be used to terminate the loop.

The *calltf routine* for the *$show_all_nets* application will need to find all net signals in a module instance. This can be done using acc_next_net(), as follows:

```
handle mod_h, net_h;
```

```
mod_h = acc_handle_tfarg(1);

net_h = null; /* initialize the target handle to null */
while ( (net_h = acc_next_net(mod_h, net_h)) != null) {
   /* perform desired operations on the net handle */
}
```

The **null** used in the above example is defined in the IEEE 1364 standard *acc_user.h* file as a long 0. Another common C coding style is to terminate the loop when the value assigned to the target handle is 0, instead of explicitly comparing the assigned value to null. For example:

```
net_h = null; /* initialize the target handle to null */
while ( net_h = acc_next_port(mod_h, net_h) ) {
   /* perform desired operations on the port handle */
}
```

 The **null** (all lower case letters) that is used by ACC routines is defined in the *acc_user.h* ACC library file. This is not the same null as the standard C language **NULL** (all capital letters) defined in *stdlib.h* library.

14.7 Accessing object types and fulltypes

Every Verilog object which can be accessed by ACC routines has a *type* property and a *fulltype* property. These properties identify what Verilog object is referenced by a Verilog handle.

acc_fetch_type() retrieves the *type* property of an object, which identifies the general type of an object. The syntax of this routine is:

PLI_INT32 **acc_fetch_type (**
   handle   **object)**   /* handle for an object */

The type property is an integer constant, such as accModule, accPort, accNet, accPrimitive, etc. This property can be used many different ways—one common usage is to verify that a handle which was obtained references the type of object expected. For example, the *$show_all_nets* application requires that the first task/ function argument be a module instance. The PLI application could verify that the argument is correct, using the following code fragment:

```
handle tfarg_handle;
tfarg_handle = acc_handle_tfarg(1);
if (acc_fetch_type(tfarg_handle) != accModule)
  /* report error that argument is not correct */
```

The acc_fetch_fulltype() routine retrieves the *fulltype* property of an object.

PLI_INT32 **acc_fetch_fulltype (**
   handle   **object**)   /* handle for an object */

The fulltype property provides more detailed information about an object. For example, if a PLI application has obtained a handle for a Verilog net, the constant returned for acc_fetch_type(net_handle) is **accNet**, while the constants returned for acc_fetch_fulltype(net_handle) include **accWire, accWor, accWand**, etc. For a Verilog module, acc_fetch_type(module_handle) returns the constant **accModule**, and acc_fetch_fulltype(module_handle) returns one of the constants: **accTopModule, accModuleInstance, accCellInstance**.

The type and fulltype properties are represented by constants with integer values. Printing the type of an object directly would print the integer value of the constant, not the name of the constant. A useful routine for debugging problems in a PLI application is **acc_fetch_type_str()**. This routine takes a type or fulltype constant value as its input, and returns a pointer to a string which contains the actual name of the constant. The syntax of this routine is:

PLI_BYTE8 ***acc_fetch_type_str (**
   PLI_INT32   **type**)   /* any type or fulltype constant */

An example of using the acc_fetch_type_str() routine to print the name of a type constant is:

```
PLI_INT32 object_type;
object_type = acc_fetch_type(tfarg_handle);
 if (object_type != accModule) {
   tf_error("Tfarg type of %s is illegal.\n",
            acc_fetch_type_str(object_type));
 }
```

14.8 Accessing the names of objects

Many Verilog objects have one or more **name** properties, which can be accessed using ACC routines.

The name properties which an object can have are:

- A *local name*. For objects such as nets, the local name is the *declaration name* of the object within a module. Note that in Verilog, a net can be implicitly declared by simply referencing the name. For a module or primitive, the local name is the *instance name* within the module that the module or primitive is instantiated.

- A *hierarchical path name*, which is the Verilog HDL design hierarchy path to an object, starting with the top of the design hierarchy.

- A *definition name*, which is the *definition name* of a Verilog module or primitive.

The following Verilog HDL source code fragment illustrates the difference between *name*, *full name* and *definition name*.

```
module test;
   wire  a, b, ci, sum, co;

   addbit u1 (a, b, ci, sum, co);
endmodule
```

local name: "u1"
full name: "test.u1"
definition name: "addbit"

```
module addbit (a, b, ci, sum, co);
   input   a, b, ci;
   output  sum, co;

   wire  a, b, ci, sum, co;

   xor      g1 (n1, a, b);
   xor #2 g2 (sum, n1, ci);
   and      g3 (n2, a, b);
   and      g4 (n3, n1, ci);
   or  #2 g5 (co, n2, n3);
endmodule
```

local name: "sum"
full name: "test.u1.sum"

local name: "g1"
full name: "test.u1.g1"
definition name: "xor"

The routines which retrieve an object's name are **acc_fetch_name()**, **acc_fetch_fullname()** and **acc_fetch_defname()**. The syntax of these three routines is:

PLI_BYTE8 ***acc_fetch_name (**
   handle   **object)**   /* handle for an object */

PLI_BYTE8 ***acc_fetch_fullname (**
   handle   **object)**   /* handle for an object */

PLI_BYTE8 ***acc_fetch_defname (**
   handle   **object)**   /* handle for a module or primitive */

14.9 The ACC string buffer

ACC routines that retrieve strings, such as `acc_fetch_name()` and `acc_fetch_value()`, will retrieve the string into a temporary string buffer, and returns a pointer to the string. Other ACC routines which return pointers to strings also share this string buffer. This temporary buffer is limited in size, and when the buffer is full, it wraps around to the beginning of the buffer. The buffer can hold multiple strings, but, when it is full, previous strings will be overwritten. A PLI application should use the string pointer returned by an ACC routine immediately. After another call is made to an ACC routine which retrieves a string, there is no guarantee that the first string pointer will still be valid. If a string needs to be preserved, the PLI application should copy the string into application-allocated storage space. Following are two examples of using strings in the PLI.

Read a string and use it immediately:

```
PLI_BYTE8 *string_p;        /* pointer only, no storage */
string_p = acc_fetch_name(net_handle);
io_printf("string_p points to %s\n", string_p);
```

Read a string and copy it to application-allocated storage for later use:

```
char *string_p;        /* string pointer only, no storage */
char *string_keep;     /* another string pointer */
string_p = (char *)acc_fetch_name(net_handle);
string_keep = malloc(strlen(string_p)+1);
strcpy(string, string_p);   /* save string for later use */
```

A PLI application can reset the ACC string buffer pointer to the beginning of the buffer using the routine **acc_reset_buffer()**. Since the buffer automatically wraps around to the beginning, there is little, if any, need to use the `acc_reset_buffer()` routine.

14.10 Reading the logic values of Verilog objects

The ACC routine **acc_fetch_value()** retrieves the value of Verilog objects which contain a logic value. The Verilog language uses 4-state logic values, comprising logic 0, 1, Z and X. The acc_fetch_value() routine automatically converts Verilog 4-state logic into various C data types for representation in PLI applications. The simplest way to represent 4-state logic in C is to use character strings, and this is the method that is used in the *$show_all_nets* application. Chapter 16 presents reading and writing Verilog logic values in more detail.

PLI_BYTE8 ***acc_fetch_value (**

handle	**object,**	/* handle for a net, reg or variable */
PLI_BYTE8	*** format_str,**	/* character string controlling the radix of the retrieved value: **"%b", "%o", "%d", "%h", "%v"** or **"%%"** */
p_acc_value	**value)**	/* pointer to an application-allocated s_acc_value structure to receive the value; only used if format_str is "%%" */

This routine has three inputs, but only the first two are used when retrieving values as a string. The third input can be set to **null** if it is not used.

The format string controls how the Verilog logic value should be represented in the C string. A format of **"%b"** indicates the logic value should be represented using binary numbers using the characters ('0', '1', 'z', and 'x'). A **"%h"** format indicates the value should be represented using hexadecimal numbers, using the characters ('0' through 'F', 'z', and 'x'). A **"%o"** format indicates an octal representation, and a **"%d"** format represents a decimal representation. Other formats are available, which are discussed in Chapter 16.

Once a handle for a net has been obtained, the value for the net can be retrieved and printed as a string, using acc_fetch_value(), as follows:

```
io_printf("  net %s value is %s\n",
          acc_fetch_name(net_h),
          acc_fetch_value(net_h, "%b", null));
```

TIP

Using C strings to represent 4-state logic is a simple method for reading and printing a Verilog logic value. However, the automatic conversion from Verilog values to C strings can be expensive for the run-time performance of a PLI application. If a PLI application will access a large number of values, or if the application will be called many times during a simulation, it is better to use a more efficient format for reading values. Chapter 16 presents all the formats for reading logic values that are available using ACC routines, and discusses performance considerations.

14.11 A complete PLI application using ACC routines

Example 14-1 lists a complete *checktf routine* and *calltf routine* for the *$show_all_nets* PLI application. Note the mixture of ACC routines and TF routines. These two libraries are designed to complement each other. Retrieving the current simulation time, for example, is done using tf_strgettime(). There is no routine in the ACC library to retrieve the current simulation time.

CD The source code for this example is on the CD accompanying this book.

- Application source file: Chapter.014/show_all_nets_acc.c
- Verilog test bench: Chapter.014/show_all_nets_test.v
- Expected results log: Chapter.014/show_all_nets_test.log

Example 14-1: *$show_all_nets* — using ACC routines in a PLI application

```
#include "veriuser.h"   /* IEEE 1364 PLI TF routine library  */
#include "acc_user.h"   /* IEEE 1364 PLI ACC routine library */
/************************************************************************
 * checktf routine
 ************************************************************************/
int PLIbook_ShowNets_checktf(int user_data, int reason)
{
  acc_initialize();
  if (tf_nump() != 1)
    tf_error("$show_all_nets must have 1 argument.");
  else if (tf_typep(1) == TF_NULLPARAM)
    tf_error("$show_all_nets arg cannot be null.");
  else if (acc_fetch_type(acc_handle_tfarg(1)) != accModule)
    tf_error("$show_all_nets arg must be a module instance.");
  acc_close();
  return(0);
}

/************************************************************************
 * calltf routine
 ************************************************************************/
int PLIbook_ShowNets_calltf(int user_data, int reason)
{
  handle module_handle, net_handle;
  acc_initialize();
  module_handle = acc_handle_tfarg(1);
  io_printf("\nAt time %s, nets in module %s (%s):\n",
            tf_strgettime(),
            acc_fetch_fullname(module_handle),
            acc_fetch_defname(module_handle));
  net_handle = null;      /* start with known value for target handle */
```

```
  while (net_handle=acc_next_net(module_handle,net_handle)) {
    io_printf("  %-13s %-13s  value is  %s (hex)\n",
              acc_fetch_type_str(acc_fetch_fulltype(net_handle)),
              acc_fetch_name(net_handle),
              acc_fetch_value(net_handle, "%h", null));
  }
  acc_close();
  return(0);
}
```

Example 14-2, which follows, lists a simple Verilog HDL design to test *$show_all_nets*. Example 14-3 which follows shows the output of running simulation with this test design.

Example 14-2: *$show_all_nets* — Verilog HDL test case for the PLI application

```
`timescale 1ns / 1ns
module top;
  reg  [2:0] test;
  tri  [1:0] results;

  addbit i1 (test[0], test[1], test[2], results[0], results[1]);

  initial
    begin
      test = 3'b000;
      #10 test = 3'b001;

      #10 $show_all_nets(top);
      #10 $show_all_nets(i1);

      #10 $stop;
      #10 $finish;
    end
endmodule

/*** A gate level 1 bit adder model ***/
`timescale 1ns / 1ns
module addbit (a, b, ci, sum, co);
  input  a, b, ci;
  output sum, co;

  wire  a, b, ci, sum, co, n1, n2, n3;

  xor      (n1, a, b);
  xor #2 (sum, n1, ci);
  and      (n2, a, b);
  and      (n3, n1, ci);
  or  #2 (co, n2, n3);
endmodule
```

Example 14-3: *$show_all_nets* — simulation results

```
At time 20, nets in module top (top):
  accTri          results           value is  1 (hex)

At time 30, nets in module top.i1 (addbit):
  accWire         a                 value is  1 (hex)
  accWire         b                 value is  0 (hex)
  accWire         ci                value is  0 (hex)
  accWire         sum               value is  1 (hex)
  accWire         co                value is  0 (hex)
  accWire         n1                value is  1 (hex)
  accWire         n2                value is  0 (hex)
  accWire         n3                value is  0 (hex)
```

14.12 Accessing handles for reg and variable data types

The Verilog HDL defines two general data type groups, **nets** and **variables**. The variable data type group includes the Verilog keywords **reg**, **integer**, **time** and **real**. The PLI treats the reg data type as a unique object, and groups the integer, time and real data types into an object class called **variables**.

Most Verilog HDL objects which can be accessed by the ACC routines have either an object-specific *handle* routine, or an object-specific *next* routine to access a specific type of object. However, there are no object-specific routines for accessing the reg and variable data types. Instead, there is a generic ACC *next* routine which can be used to obtain handles for a number of different types of objects.

acc_next() is used to obtain handles for multiple types of objects. The syntax of this routine is:

handle **acc_next (**
PLI_INT32	*** type_list,**	/* pointer to a static array with a list of type or fulltype constants */
handle	**scope,**	/* handle for the scope in which to scan for objects */
handle	**prev_object)**	/* handle for the previous object found; initially **null** */

This routine requires three inputs:

1. A pointer to a list of *type* or *fulltype* constants. This list must be a static PLI_INT32 array, with **0** in the last element of the array.

2. A handle for the reference object, which determines where the routine will search for the destination objects.

3. A handle to the last target object found.

By providing a list of constants, acc_next() can be used in three general contexts:

- acc_next() can obtain handles for a specific type of object that does not have an object-specific ACC next routine by providing the *type* constant of that object.

- acc_next() can obtain handles for several different types of objects at the same time by providing a list of multiple object *type* constants.

- acc_next() can obtain handles for only certain objects within a larger class of objects by providing a list of one or more *fulltype* constants. For example, acc_next_net() will access all nets of any net type within a reference module. If only wired-logic net types were desired, acc_next() can be used to access only those specific net fulltypes.

The following example illustrates using acc_next() to obtain handles for only wired logic nets, and exclude other types of nets:

```
handle mod_h, net_h;

static PLI_INT32 wired_nets[5] = {accWand, accWor,
                                  accTriand, accTrior, 0};

/* add code to get a module handle */

net_h = null; /* initialize the target handle to null */
while (net_h = acc_next(wired_nets, mod_h, port_h) ) {
  /* perform desired operations on the net handle */
}
```

The acc_next() routine only supports a subset of the object type and fulltype constants. Object types which are not supported by acc_next() can only be accessed using an object-specific next routine. The supported constants are:

- Verilog data type constants: accNet, accReg, accIntegerVar, accTimeVar, accRealVar, and accNamedEvent.

- Verilog net fulltypes: accWire, accTri, accWand, accTriand, accWor, accTrior, accTri0, accTri1, accTrireg, accSupply0, and accSupply1.

- Verilog module type and fulltype constants: accModule, accTopModule, accModuleInstance, and accCellInstance.

- Verilog primitive fulltypes: accCombPrim, accSeqPrim, accAndGate, accNandGate, accNorGate, accOrGate, accXorGate, accXnorGate, accBufGate, accNotGate, accBufif0Gate, accBufif1Gate, accNotif0Gate, accNotif1Gate, accNmosGate, accPmosGate, accCmos-Gate, accRnmosGate, accRpmosGate, accRcmosGate, accRtranGate, accRtranif0Gate, accRtranif1Gate, accTranGate, accTranif0Gate, accTranif1Gate, accPullupGate, accPulldownGate.

 NOTE The reg, integer, time, real and event data types do not have a corresponding object-specific ACC next routine. The generic acc_next() routine must be used to obtain handles for these objects.

14.12.1 A complete PLI application for $show_all_signals

Example 14-4 lists the complete C code for the *$show_all_signals* PLI application. This application uses the generic acc_next() routine to obtain handles for all signals in a module, including the data types of net, reg, integer, time, and real.

CD The source code for this example is on the CD accompanying this book.

- Application source file: Chapter.014/show_all_signals1_acc.c
- Verilog test bench: Chapter.014/show_all_signals1_test.v
- Expected results log: Chapter.014/show_all_signals1_test.log

---

Example 14-4: *$show_all_signals*, version 1 — using the acc_next() routine

```
#include "veriuser.h"  /* IEEE 1364 PLI TF routine library */
#include "acc_user.h"  /* IEEE 1364 PLI ACC routine library */
/******************************************************************
 * checktf routine
 ******************************************************************/
int PLIbook_ShowSignals1_checktf(int user_data, int reason)
{
  acc_initialize();
  if (tf_nump() != 1)
    tf_error("$show_all_signals must have 1 argument.");
  else if (tf_typep(1) == TF_NULLPARAM)
    tf_error("$show_all_signals arg cannot be null.");
  else if (acc_fetch_type(acc_handle_tfarg(1)) != accModule)
    tf_error("$show_all_signals arg must be a module instance.");
  acc_close();
  return(0);
}
```

```
/*****************************************************************************
 * calltf routine
 *****************************************************************************/
int PLIbook_ShowSignals1_calltf(int user_data, int reason)
{
  handle module_h, signal_h;
  static PLI_INT32 signal_types[6] = {accNet, accReg, accIntegerVar,
                                      accTimeVar, accRealVar, 0 };
  acc_initialize();
  module_h = acc_handle_tfarg(1);
  io_printf("\nAt time %s, signals in module %s (%s):\n",
            tf_strgettime(),
            acc_fetch_fullname(module_h),
            acc_fetch_defname(module_h));
  signal_h = null;    /* start with known value for target handle */
  while (signal_h = acc_next(signal_types, module_h, signal_h)) {
    io_printf("  %-13s %-13s  value is  %s (hex)\n",
              acc_fetch_type_str(acc_fetch_fulltype(signal_h)),
              acc_fetch_name(signal_h),
              acc_fetch_value(signal_h, "%h", null));
  }
  acc_close();
  return(0);
}
```

Example 14-5 lists Verilog source code for testing *$show_all_signals*. This test is similar to the test for *$show_all_nets*, but the lower level adder model has been changed from a gate level model to an RTL model in order to use more data types in the Verilog source code. Example 14-6, which follows, shows the simulation results from running a simulation with *$show_all_signals*.

Example 14-5: *$show_all_signals1* — Verilog HDL test case for the PLI application

```
`timescale 1ns / 1ns
module top;
  integer    test;
  tri   [1:0] results;

  addbit i1 (test[0], test[1], test[2], results[0], results[1]);

  initial
    begin
      test = 3'b000;
      #10 test = 3'b001;

      #10 $show_all_signals1(top);
      #10 $show_all_signals1(i1);
```

```
      #10 $stop;
      #10 $finish;
    end
endmodule

/*** An RTL level 1 bit adder model ***/
'timescale 1ns / 1ns
module addbit (a, b, ci, sum, co);
  input   a, b, ci;
  output sum, co;

  wire   a, b, ci;
  reg    sum, co;

  always @(a or b or ci)
    {co, sum} = a + b + ci;

endmodule
```

Example 14-6: *$show_all_signals1* — simulation results

```
At time 20, signals in module top (top):
  accTri         results          value is  1 (hex)
  accIntegerVar test             value is  00000001 (hex)

At time 30, signals in module top.i1 (addbit):
  accWire        a                value is  1 (hex)
  accWire        b                value is  0 (hex)
  accWire        ci               value is  0 (hex)
  accRegister    sum              value is  1 (hex)
  accRegister    co               value is  0 (hex)
```

14.13 Obtaining handles to the current hierarchy scope

The Verilog language allows a system task or system function to be invoked from any hierarchy scope. A *scope* in the Verilog HDL is a level of design hierarchy, and can be represented by several constructs:

- Module instances
- Named statement groups
- Verilog HDL tasks
- Verilog HDL function

The following example calls the *$show_all_signals* from a named statement group:

```
module top;
  ...
  always @(posedge clock)
    begin: local
      integer i;
      reg      local_bus;
      ...
      $show_all_signals;
    end
endmodule
```

A useful enhancement to the *$show_all_signals* example is to allow either no system task argument or a null system task argument to represent the module instance which called the *$show_all_signals* system task. The difference between no argument and a null argument is shown in the following two examples.

No system task/function arguments:

```
$show_all_signals;
```

A null system task/function argument:

```
$show_all_signals();
```

The following Verilog source code shows the enhanced usage possibilities for the *$show_all_signals* example:

```
module top;
  ...
  addbit i1 (a, b, ci, sum, co);    // instance of an adder
  ...
  always @(posedge clock)
    $show_all_signals;        // list signals in this module
    $show_all_signals(i1);    // list signals in instance i1
endmodule

module addbit (a, b, ci, sum, co);
  ...
  always @(sum or co)
    $show_all_signals();      // list signals in this module
endmodule
```

In order to access all signals in a module, the *$show_all_signals* application will need a handle for a module instance, but now the name of the module instance is not passed to the PLI application as an argument to *$show_all_signals*. The ACC library provides two ways to obtain the handle for the module which called a PLI application.

acc_handle_calling_mod_m returns the handle for the module from which a PLI application was called. The syntax for this routine is:

handle **acc_handle_calling_mod_m**

Note that in the ACC library, this routine is defined as a macro, not a function. Therefore, it should not be called with parentheses at the end of the name.

The *$show_all_signals* PLI application can be easily enhanced to work with no argument or a null argument using the code:

```
if (tf_nump() == 0)   /* no task/function arguments */
  mod_h = acc_handle_calling_mod_m;
else if (tf_typep(1) == tf_nullparam) /* null argument */
  mod_h = acc_handle_calling_mod_m;
else     /* a task/function argument exists */
  mod_h = acc_handle_tfarg(1);
```

In the Verilog HDL, a level of design hierarchy can be an object other than a module. The following example calls the *$show_all_signals* from a named statement group:

```
module top;
   . . .
   always @(posedge clock)
     begin: local
       integer i;
       reg      local_bus;
       . . .
       $show_all_signals;      // list signals in this scope
     end
endmodule
```

In the above example, the *$show_all_signals* applications should search for signal names in the local hierarchy scope, which is a named statement group, instead of a module instance. The name of any type of hierarchy scope can be passed to the system task as a task/function argument, but, in this example, *$show_all_signals* is being called with no arguments. To obtain the local hierarchy scope without being passed, the scope name requires two steps:

1. Call the **acc_handle_tfinst()** routine to obtain a handle for the system task/function which called the PLI application. This routine does not require any inputs.

2. Call the **acc_handle_scope()** routine to obtain a handle for the scope containing a reference handle. For this example, the reference handle will be the system task handle.

Example 14-7 contains the complete listing of the enhanced *$show_all_signals*, with the ability to use either no arguments or null arguments to represent the local design hierarchy scope.

CD The source code for this example is on the CD accompanying this book.

- Application source file: Chapter.014/show_all_signals2_acc.c
- Verilog test bench: Chapter.014/show_all_signals2_test.v
- Expected results log: Chapter.014/show_all_signals2_test.log

---

Example 14-7: *$show_all_signals*, version 2—obtaining a handle for the local scope

```
/*********************************************************************
 * checktf routine
 *********************************************************************/
int PLIbook_ShowSignals2_checktf(int user_data, int reason)
{
  acc_initialize();
  if (tf_nump() == 0)
    return(0); /* no arguments is OK, skip remaining checks */
  if (tf_nump() > 1)
    tf_error("$show_all_signals must have 0 or 1 argument.");
  else if (tf_typep(1) == TF_NULLPARAM)
    return(0); /* null argument is OK, skip remaining checks */
  else if (acc_fetch_type(acc_handle_tfarg(1)) != accModule)
    tf_error("$show_all_signals arg must be a module instance.");
  acc_close();
  return(0);
}

/*********************************************************************
 * calltf routine
 *********************************************************************/
int PLIbook_ShowSignals2_calltf(int user_data, int reason)
{
  handle module_h, signal_h;
  static PLI_INT32 signal_types[6] = {accNet, accReg, accIntegerVar,
                                      accTimeVar, accRealVar, 0 };
  acc_initialize();
  if (tf_nump() == 0)
    module_h = acc_handle_scope(acc_handle_tfinst());
  else if (tf_typep(1) == tf_nullparam)
    module_h = acc_handle_scope(acc_handle_tfinst());
  else
    module_h = acc_handle_tfarg(1);
```

```
io_printf("\nAt time %s, signals in module %s (%s):\n",
          tf_strgettime(),
          acc_fetch_fullname(module_h),
          acc_fetch_defname(module_h));
signal_h = null;    /* start with known value for target handle */
while (signal_h = acc_next(signal_types, module_h, signal_h)) {
  io_printf("  %-13s %-13s  value is  %s (hex)\n",
            acc_fetch_type_str(acc_fetch_fulltype(signal_h)),
            acc_fetch_name(signal_h),
            acc_fetch_value(signal_h, "%h", null));
}
acc_close();
return(0);
}
```

14.14 Obtaining handles to multiple task/function arguments

Another useful modification to the *$show_all_signals* application is to allow multiple
hierarchy scopes to be specified at the same time. For example:

```
$show_all_signals(i1, ,top.local);
```

In the above example, there are three system task/function arguments, the second
argument being null, to indicate the local hierarchy scope.

Example 14-8 illustrates using a C `for` loop to access each system task/function argument.

TIP

This example uses a more structured programming style by creating a separate
function called *get_all_signals()* to do the work of searching for all signals and
printing the current logic value. By moving this application logic into a separate C
function, the *calltf routine* is kept shorter and easier to read, plus there is less
duplication of code. Using structured programming techniques makes it easier to
maintain or enhance the functionality of the PLI application.

CD The source code for this example is on the CD accompanying this book.

- Application source file: `Chapter.014/show_all_signals3_acc.c`
- Verilog test bench: `Chapter.014/show_all_signals3_test.v`
- Expected results log: `Chapter.014/show_all_signals3_test.log`

Example 14-8: *$show_all_signals*, version 3—obtaining handles for multiple tfargs

```
#include "veriuser.h"   /* IEEE 1364 PLI TF routine library  */
#include "acc_user.h"   /* IEEE 1364 PLI ACC routine library */
/********************************************************************
 * checktf routine
 ********************************************************************/
int PLIbook_ShowSignals3_checktf(int user_data, int reason)
{
  int i, numargs;

  acc_initialize();
  numargs = (int)tf_nump();
  if (numargs == 0)
    return(0); /* no arguments is OK, skip remaining checks */
  for (i = 1; i <= numargs; i++) {
    if (tf_typep(i) == TF_NULLPARAM)
      break;  /* null argument is OK, skip other checks for this arg */
    else if (acc_fetch_type(acc_handle_tfarg(i)) != accModule)
      tf_error("$show_all_signals arg must be a module instance.");
  }
  acc_close();
  return(0);
}


/********************************************************************
 * calltf routine
 ********************************************************************/
void PLIbook_GetAllSignals();   /* prototype function used by calltf */

int PLIbook_ShowSignals3_calltf(int user_data, int reason)
{
  handle module_h;
  int i, numargs;
  acc_initialize();
  numargs = (int)tf_nump();
  if (numargs == 0) {
    module_h = acc_handle_scope(acc_handle_tfinst());
    PLIbook_GetAllSignals(module_h);
  }
  else
    for (i = 1; i <= numargs; i++) {
      if (tf_typep(i) == tf_nullparam)
        module_h = acc_handle_scope(acc_handle_tfinst());
      else
        module_h = acc_handle_tfarg(i);
      PLIbook_GetAllSignals(module_h);
    }
  acc_close();
  return(0);
}
```

```
void PLIbook_GetAllSignals(handle module_h)
{
  handle signal_h;
  static PLI_INT32 signal_types[6] = {accNet, accReg, accIntegerVar,
                                      accTimeVar, accRealVar, 0 };
  io_printf("\nAt time %s, signals in module %s (%s):\n",
            tf_strgettime(),
            acc_fetch_fullname(module_h),
            acc_fetch_defname(module_h));
  signal_h = null;     /* start with known value for target handle */
  while (signal_h = acc_next(signal_types, module_h, signal_h)) {
    io_printf("  %-13s %-13s  value is  %s (hex)\n",
              acc_fetch_type_str(acc_fetch_fulltype(signal_h)),
              acc_fetch_name(signal_h),
              acc_fetch_value(signal_h, "%h", null));
  }
  return;
}
```

14.15 Summary

The ACC routines in the PLI standard provide direct access to what is happening within a Verilog simulation. This access is done using an object oriented method, where most Verilog HDL constructs that can exist in a simulation data structure are treated as objects. The ACC routines use *handles* to reference these objects. A large number of ACC routines are provided to obtain handles for the various types of Verilog HDL objects, and other ACC routines retrieve data about the objects, such as the name of an object or the vector size of a net.

This chapter has focused on how to create PLI applications using the ACC library. The following three chapters will present more detail on the syntax and usage of the 103 ACC routines in the PLI standard. These chapters include several additional examples of PLI applications which use the ACC routines.

CHAPTER 15 *Details on the ACC Routine Library*

*T*he ACC library provides 103 C functions that can interact with Verilog simulators. The previous chapter provided an overview of the ACC library, and how ACC routines are used in PLI applications. This chapter presents a more detailed description of the ACC library, and how ACC routines access information from Verilog simulations.

The concepts presented in this chapter are:

- Initializing and configuring ACC routines
- ACC error handling
- ACC object diagrams and object relationships
- Using ACC *handle* routines
- Using ACC *next* routines
- Miscellaneous ACC routines

15.1 PLI application performance considerations

The run-time performance of a simulator can be impacted in either a positive way or a negative way by PLI applications. Often, a complex algorithm can be represented in the C language, using C language data types, much more efficiently than in the hardware-centric Verilog HDL language. The C language can be used for an abstract representation of a design, when 4-state logic, logic transitions, simulation time, and other details are not required, but which a hardware description language must be able to represent. The abstraction that C offers often makes it possible to greatly increase the run-time performance of simulation.

However, a poorly thought out PLI application can actually decrease the run-time performance of a simulation. Each call to a routine in the PLI library will take time to be executed. It is important to architect a PLI application to minimize the number of times ACC routines are used.

The following guidelines can help in planning an efficient PLI application:

- Good C programming practices are essential. C programming style and techniques are not discussed within the scope of this book.

- Consider every call to an ACC routine as expensive, and try to minimize the number of calls.

- ACC routines which obtain object handles using an object's name are less efficient than routines which obtain object handles based on a relationship to another object.

- Routines which convert logic values from a simulator's internal representation to C strings, and vice-versa, are less efficient that using other C data types. Strings are a convenient means of representing 4-state values for printing, but strings should be used prudently.

- When the same object must be accessed many times during a simulation, the object handle can be obtained once, and saved in application-allocated storage. Using a pointer to the storage, a PLI application has immediate access to the object handle, without having to call an ACC routine to obtain the handle each time it is needed.

- Use the ACC library to access the unique abilities of hardware description languages, such as representing hardware parallelism and hardware propagation times. Simulator vendors have invested a great deal in optimizing a simulator's algorithms, and that optimization should be utilized in a PLI application.

When developing a PLI application, one primary consideration should be how often a PLI application will be called during a simulation. It is well worth the effort to optimize the performance of an application that is invoked every clock cycle, but may not be as important for an application that is only invoked once during a simulation.

 The objective of this book is to show how the routines in the ACC library are used. Short examples of using many of these routines are shown in the context of complete PLI applications. In order to meet the book's objectives, the examples presented in this book do not always follow the guidelines of efficient C coding and prudent usage of the ACC routines. It is expected that when parts of these example PLI applications are adapted for other applications, the coding style will also be modified to be more efficient and robust.

15.2 Initializing and configuring ACC routines

Several ACC routines can be configured for how the routines access information in the Verilog simulation data structure. In addition, some ACC routines need to allocate and initialize memory for their operation. Three ACC routines are used to initialize and configure ACC routines.

15.2.1 Initializing and closing the ACC environment

Two special ACC routines, `acc_initialize()` and `acc_close()`, are used to initialize and maintain the environment used by the ACC library.

PLI_INT32 **acc_initialize ( )**

The `acc_initialize()` routine performs two primary operations: allocate and initialize any memory that is needed by the ACC routines, and set all ACC configurations to their default values. The routine returns true if it was successful, and false if an error occurred.

void **acc_close ( )**

`acc_close()` frees any memory that was allocated by `acc_initialize()`, and resets all ACC configurations back to their default values.

An example of using `acc_initialize()` and `acc_close()` is:

```
my_calltf_app()
{
  /* declarations */
  acc_initialize();
  /* use routines from ACC library */
  acc_close();
}
```

When to use acc_initialize() and acc_close()

TIP Simulation run-time performance might be improved through judicious usage of `acc_initialize()` and `acc_close()`. PLI applications which do not use any ACC routines do not need to call `acc_initialize()` and `acc_close()`.

Some—but not all—Verilog simulators do not need `acc_initialize()` and `acc_close()`. In many simulators, the configuration of ACC routines is automatically set to default values when the PLI application is entered. Any memory required by ACC routines is automatically allocated when needed, and automatically freed when not needed. Refer to the documentation of a specific simulator to see if `acc_initialize()` and `acc_close()` can be omitted for that simulator.

Calling `acc_initialize()` and `acc_close()` when the routines are not required may slightly slow down the run-time performance of a simulator. If a PLI application is invoked many times during a simulation, such as at every clock cycle, this performance cost can become expensive. For those simulators where the routines are not needed, initializing and closing the ACC routines is an unnecessary cost to simulation performance.

 The IEEE standard recommends that `acc_initialize()` be called at the beginning of every PLI application which uses the ACC library, and that `acc_close()` be called at the end of the application. Following the IEEE recommendation will ensure that a PLI application is well behaved and portable to all Verilog simulators.

Three possible ways to utilize `acc_initialize()` and `acc_close()` are:

- Follow the IEEE recommendation, and use `acc_initialize()` and `acc_close()` in all ACC based PLI applications. This provides maximum portability of the PLI application, with minimum optimization.

- Use conditional compilation to include or exclude `acc_initialize()` and `acc_close()`, depending on the simulator for which the PLI application is being compiled. This provides better optimization, but complicates developing and compiling PLI applications.

- Only use `acc_initialize()` and `acc_close()` in PLI applications which use ACC routines that are affected by `acc_configure()` (excluding warning and error configuration, which affect all ACC routines). The ACC routines which can be configured are the routines which are most likely to need initialization. This approach provides a reasonable compromise between portability and performance.

15.2.2 Configuring the ACC environment

Several ACC routines can be configured for how they operate. A special ACC routine is used to configure all of these routines.

PLI_INT32 **acc_configure (**
    PLI_INT32   **config,**   /* one of the constants listed in following table */
    PLI_BYTE8 ***value)**   /* configuration value as a character string */

The `acc_configure()` routine configures specific ACC routines. The routine returns 1 if successful, and 0 if an error occurred. The name of each configuration is represented by a constant, and the setting of the configuration is represented as a string. Table 15-1 lists the configuration constants and the ACC routines affected. The specific configurations are presented in the sections of this book that discuss the affected routines.

Table 15-1: ACC configuration constants

Configuration Constant	Description and Routines Affected
accDevelopmentVersion	documents the version of the PLI standard for which the PLI application was developed no ACC routines affected
accDisplayErrors	enables or disables printing error messages for run-time errors caused by ACC routines all ACC routines
accDisplayWarnings	enables or disables printing warning messages for run-time warnings caused by ACC routines all ACC routines
accDefaultAttr0	controls the default return value if an attribute is not found `acc_fetch_attribute()`

accEnableArgs	controls which input arguments must be specified
	
	`acc_handle_modpath()`
	`acc_handle_tchck()`
	`acc_handle_scope()`

accPathDelimStr	controls the delimiter used in path names
	
	`acc_fetch_name()`
	`acc_fetch_fullname()`
	`acc_fetch_attribute()`
	`acc_fetch_attribute_int()`
	`acc_fetch_attribute_str()`

Table 15-1: ACC configuration constants (continued)

Configuration Constant	Description and Routines Affected

accMinTypMaxDelays	controls whether min and max delays are used
	
	`acc_fetch_delays()`
	`acc_append_delays()`
	`acc_replace_delays()`

accPathDelayCount	controls the number of path delays read or modified
	
	`acc_fetch_delays()`
	`acc_append_delays()`
	`acc_replace_delays()`
	`acc_fetch_pulsere()`
	`acc_append_pulsere()`
	`acc_replace_pulsere()`

accToHiZDelay	controls how turn-off delays are specified
	
	`acc_append_delays()`
	`acc_replace_delays()`

accMapToMipd	controls how interconnect delays are mapped to input ports `acc_replace_delays()`

15.3 ACC routine error handling

The ACC routines have built-in error handling for when a call to a routine cannot perform its operation. For example, an ACC routine to obtain a handle for an object
using the object's name will fail if an object of that name does not exist.

The ACC library provides three actions which should occur when an ACC routine
cannot perform its operation:

- Set a global error flag, called **acc_error_flag**.
- Display an error message to the simulator's output channel.
- Return an exception value, if the ACC routine returns values.

15.3.1 The global ACC error flag

The global **acc_error_flag** is set to 0 if a call to an ACC routine is successful, and
is set to a non-zero value if the call was unsuccessful. The status of the error flag is

updated each time an ACC routine is called. Every ACC routine shares the global `acc_error_flag`. The following C code fragment illustrates using the error flag.

```
handle net_handle, module_handle;
char *net_name;

/* add code to read the name of a net from a file */

net_handle = acc_handle_by_name(net_name, module_handle);
if (acc_error_flag) {
   io_printf("Net %s was could not be found\n", net_name);
}
else {
   ... /* use the net handle obtained */
```

15.3.2 Exception return values

ACC routines which return values have a defined exception value. Most ACC routines follow the same convention for the exception value. However, there are a few routines which do not adhere to the convention. The exception values used by most ACC routines are listed in Table 15-2. If an ACC routine follows a different convention, its exception value is noted in the description of the routine.

Table 15-2: Exception return values for most ACC routines

Type of ACC Routine	Exception Value
ACC routines which return integer values	0
ACC routines which return double-precision values	0.0
ACC routines which return boolean values	false
ACC routines which return **handle** values	null
ACC routines which return pointers to character strings	null

15.3.3 Enabling and disabling ACC error and warning messages

By default, whenever an ACC routine is not successful, an ACC error message is printed in the output channel of the simulator. This is the same output channel used by routines such as `io_printf()` — it is not necessarily the operating system's *stdout* or *stderr* message channels. The ACC library provides a means to disable this automatic error message generation, using `acc_configure()`, as follows:

```
acc_configure(accDisplayErrors, "false");
```

Setting the configuration of `accDisplayErrors` to "true" will re-enable automatic error message generation, as will re-initializing the ACC environment.

Some simulators may also generate ACC warning messages, such as when an invalid input is provided to the ACC routine. Typically, a warning message indicates that a less severe error occurred with the ACC routine. The `acc_error_flag` is not set when a warning message occurs.

The end-user of a PLI application may not be interested in these non-fatal warnings. But, these warnings can be of interest to a PLI application developer. By default, the generation of ACC warning messages is *disabled*. The `acc_configure()` routine is used to enable warning messages, as follows:

```
acc_configure(accDisplayWarnings, "true");
```

TIP
Warning messages can provide valuable information when developing or debugging a PLI application. Often a warning message can indicate a potential problem with an application that might not be obvious as an application is being tested. It is a good technique to enable warning messages until it is certain that a PLI application is working correctly, and then disable warning messages for better run-time performance of the PLI application. The TF and ACC routines allow user-defined invocation options to be created, and many PLI applications use invocation options as a way to turn displaying ACC warning messages on or off.

15.4 Using ACC object diagrams

The IEEE 1364 Verilog standard defines what Verilog HDL constructs are considered objects in the ACC library. However, in the ACC portion of the IEEE 1364 standard, the relationships between objects are not defined. Object relationships are only defined in the VPI portion of the standard.

This book adds *object diagrams* for each object which the ACC routines can access. These ACC diagrams are not part of the IEEE 1364 standard. They are the invention of the book's author. The ACC object diagrams document:

- The *properties* of an object. For example, a net object has *name*, *vector size*, and *logic value* properties (as well as several other properties).

- The *relationships* of the object. Relationships indicate how an object is connected to, or contained within, other objects within a Verilog data structure. For example, a net is contained within a module, and may also be connected to other objects, such as a module port or primitive terminal.

The object diagrams for an object are based on enclosures with arrows. The type of object is listed within each enclosure, and the relationships to other objects are shown as arrows between the enclosures. The properties of the object are listed in a table below the diagrams. The specific ACC routines required to access the properties of an object or to traverse to a related object are shown in the diagrams.

The complete set of ACC object diagrams are in Appendix C. This section shows how to read the ACC object diagrams, Section 15.7, later in this chapter, shows how to utilize the diagrams to traverse from any point in a Verilog design hierarchy to any other point.

Following is a partial object diagram for a Verilog module object:

Figure 15-1: ACC object diagram for Verilog modules (partial)

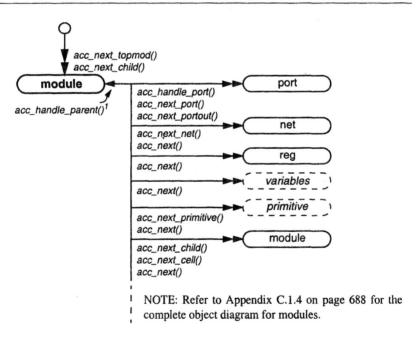

NOTE: Refer to Appendix C.1.4 on page 688 for the complete object diagram for modules.

acc_fetch_type()	returns **accModule**
acc_fetch_fulltype()	returns **accTopModule, accModuleInstance, accCellInstance**
acc_fetch_name()	returns the instance name of a module
acc_fetch_fullname()	returns the hierarchical path name of a module
acc_fetch_defname()	returns the definition name of a module
acc_fetch_timescale_info()	returns the timescale of a module
acc_fetch_location()	returns source file name & line no. of module instance

15.4.1 Object diagram symbols

A object diagram contains four primary symbols and four font type faces:

- A *solid circle*, such as ○, designates either the top level of a Verilog hierarchy tree, or the hierarchy scope from which the PLI application was called.

- A *solid enclosure*, such as ⬭**module**⬭ or ⬭ port ⬭, designates a Verilog object. The name of the object is shown within the enclosure. The font used for the name has significance:

  - A **non-italicized, bold font** designates that this object is being defined in this diagram. In the diagram for module objects, the name, **module**, is in bold.

  - A non-italicized, non-bold font designates that this object is being referenced in this diagram, but is not being defined. The definition will appear in a different diagram. For example, in the diagram for module objects, the name port, is not bolded.

- A *small dotted enclosure*, such as ⌐ *variables* ⌐, designates a reference to a named class of Verilog objects. A class of objects is a group of several objects which have something in common. The name of the object class is shown within the enclosure, using an *italicized, non-bold font*. The specific objects within a class are listed in the diagram for the class definition.

- A *large dotted enclosure*, which has smaller enclosures within it, designates the definition of a class of Verilog objects. The large enclosure contains all of the objects which make up the class. A class of objects may or may not have a name, but only a named class can be referenced in another diagram. The name of the class is shown at the top of the large enclosure, using an *italicized, bold font*. Two examples of object class definitions are:

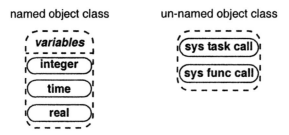

15.4.2 Traversing object relationships

The relationship of one object to another object is shown as arrows in the object diagrams. Each arrow indicates the relationship from a reference object (the originating end of the line) to a target object (the terminating end of the line). The type of arrow at the target object indicates the type of relationship. Most objects can be both a reference object and a target object, depending on which direction the Verilog hierarchy is being traversed.

There are two types of object relationships possible in a Verilog design for which ACC routines can obtain handles:

- *One-to-one* relationships are represented by a line which terminates with a single arrow in a object diagram. A one-to-one relationship indicates that a given object is related to, at most, one of another type of object. In the module object diagram shown in Figure 15-1 on page 495, there is a single arrow going from ⊂ port ⊃ back to ⊂ **module** ⊃ , which indicates a given port is only contained within one module.

- *One-to-many* relationships are represented by a line which terminates with a double arrow in a object diagram. A one-to-many relationship indicates that a given object is related to any number of another type of object. In the module object diagram shown in Figure 15-1 on page 495, there is a double arrow going from ⊂ **module** ⊃ to ⊂ port ⊃ , which indicates there may be any number of ports within a module.

In most object relationships, the connecting line both originates and terminates with a single or double arrow. In certain relationships, however, the connecting line has no arrow on the originating end, at the reference object. This indicates that the relationship is one-way. That is, the target object can be accessed from the reference object, but it is not possible to get back to the reference object from the target object.

Several ACC routines provide a means for a PLI application to locate and retrieve handles for Verilog objects. This chapter will introduce these routines, and subsequent chapters in this part of the book will present several examples of using these routines in practical PLI applications.

There are two general groups of ACC routines used to obtain handles for objects, which are referred to as *ACC handle routines* and *ACC next routines*.

- The *ACC handle routines* obtain handles for one object of a specific object type. For example, the routine acc_handle_port() will obtain a handle for one specific port within a module.

- The *ACC next routines* obtain handles for all objects of a specific object type. For example, the routine acc_next_port() will obtain handles for all of the ports within a module.

The description and usage of these routines are presented in sections 15.5 and 15.6, which follow.

15.5 Using ACC handle routines

An ACC **handle** is an abstraction used to reference an object within the simulation data structure. ACC routines use this abstraction to access information about the object. This layer of abstraction allows PLI applications to be portable to any Verilog simulator, because the abstract handle is a layer between the PLI application and the internal data structures of the simulator.

ACC **handle routines** return a handle for a single object. Generally, these routines are used when there is a one-to-one relationship shown in the object diagram, which is represented by a single arrow at the terminating end of a relationship line. *ACC handle routines are object-specific*, so there is a different routine for each type of object that can be accessed using handle routines. There are 25 handle routines in the ACC library:

```
acc_handle_by_name()              acc_handle_scope()

acc_handle_calling_mod_m          acc_handle_path()

acc_handle_condition()            acc_handle_pathin()

acc_handle_conn()                 acc_handle_pathout()

acc_handle_datapath()             acc_handle_port()

acc_handle_hiconn()               acc_handle_simulated_net()

acc_handle_interactive_scope()    acc_handle_tchk()

acc_handle_itfarg()               acc_handle_tchkarg1()

acc_handle_loconn()               acc_handle_tchkarg2()

acc_handle_modpath()              acc_handle_terminal()

acc_handle_notifier()             acc_handle_tfarg()

acc_handle_object()               acc_handle_tfinst()

acc_handle_parent()
```

The ACC handle routines are straightforward and simple to use. The complete syntax for each ACC handle routine is shown in Appendix C, and is not duplicated in this section. This chapter includes examples of using many of these ACC handle routines, and provides additional description on a few of the routines.

Most ACC handle routines require as an input a handle for a reference object. The routines return a handle for a specific target object. The ACC object diagrams show which ACC handle routine is used to obtain a handle for a specific object. As an example, assume that a PLI application had already obtained a handle for a port, and needs to locate the module which contains that port. The *reference* point will be the port, and the *target* will be a Verilog module. In the object diagram for modules, which was shown in Figure 15-1 on page 495, there is a single arrow from ⬭ port ⬭ to ⬭ **module** ⬭ , indicating a one-to-one relationship. Next to the target object (the module), the ACC routine to obtain a handle for the module is shown as being **acc_handle_parent()**. Using this information, the following C code fragment can be used to obtain the module handle from the port handle:

```
handle   port_handle, module_handle;

/* add code to obtain handle for a port */

module_handle = acc_handle_parent(port_handle);
```

15.6 Using ACC next routines

Many objects in Verilog have a one-to-many relationship with other objects. These relationships are represented by a double arrow terminating at the target object in the object diagrams. The *ACC next routines* are used to obtain handles for all of objects in this type of relationship. As with the ACC handle routines, the ACC next routines are object-type specific, so there is a different ACC routines for most types of Verilog objects for which next routines can obtain handles. There are 22 ACC next routines:

```
acc_next()                        acc_next_net()

acc_next_bit()                    acc_next_output()

acc_next_cell()                   acc_next_parameter()

acc_next_cell_load()              acc_next_port()

acc_next_child()                  acc_next_portout()

acc_next_driver()                 acc_next_primitive()

acc_next_hiconn()                 acc_next_scope()

acc_next_input()                  acc_next_specparam()

acc_next_load()                   acc_next_tchk()

acc_next_loconn()                 acc_next_terminal()

acc_next_modpath()                acc_next_topmod()
```

The complete syntax for each ACC next routine is shown in Appendix C, and is not duplicated in this section. Chapter 14 discussed how ACC next routines are used (refer back to section 14.6 on page 466). This chapter shows how the ACC object diagrams relate to the ACC next routines.

Most of the ACC next routines require two inputs:

1. A handle for the *reference object*, which determines where the routine will search for the destination objects.

2. A handle for the *previous target object* found.

The ACC object diagrams show which ACC next routine should be used to traverse from one object to another. As an example, suppose a PLI application needed to obtain handles for all ports in a module. In the object diagram for modules shown in Figure 15-1 on page 495, there is a double arrow from ⬭ **module** to ⬭ port , indicating a one-to-many relationship. The ACC routines listed next to the target objects (the ports) show that **acc_next_port()** can be used to access all ports within that module. For example:

```
port_h = null; /* initialize the target handle for null */
while ( port_h = acc_next_port(mod_h, port_h) ) {
   /* perform desired operations on the port handle */
}
```

Obtaining handles for just one object, using ACC next routines

On occasion, a PLI application might need to obtain a handle for just the first target object in a one-to-many relationship. Or, an application may only need to determine if any target objects exist, without obtaining any target object handles.

For example, an application might need to determine if a Verilog module contains instances of other modules, but the handles for all of the module instances are not needed. The object diagram for modules (refer back to Figure 15-1 on page 495) shows a double arrow from ⬭ **module** to ⬭ module , indicating a one-to-many relationship from a module to all module instances within that module. The diagram shows that **acc_next_child()** can be used to access all module instances within a module. Since this example PLI application does not need the child module handles which would be returned from acc_next_child(), the application can simplify the way the next routine is used in two ways:

• The ACC next routine does not need to be called in a loop to access all handles.

• The previous target handle can be replaced with **null**, since the routine will not be called to find a second target object.

The following code fragment illustrates using acc_next_child() to perform a true/false test, where a true is returned if a module represents the bottom of the Verilog hierarchy (the module does not contains any module instances).

```
int is_bottom_module(handle this_module_h)
{
  if (acc_next_child(this_module_h, null))
    return(0); /* a module instance was found */
  else
    return(1); /* no child module instances */
}
```

A similar situation that occasionally arises in PLI application is when only the first target object of a one-to-many relationship is desired. Once again, since the second or subsequent target objects will not be accessed, the ACC next routine does not need to be called in a loop, and the previous target handle does not need to be provided. The following code fragment illustrates using an ACC next routine to obtain only the first port of a module.

```
handle first_port_handle;

first_port_handle = acc_next_port(module_handle, null);
```

15.7 Traversing Verilog hierarchy using object relationships

By following the object's object diagrams and using the appropriate ACC handle and ACC next routines, the Verilog design hierarchy can be traced from one object to any other object anywhere in the Verilog design. Traversing the design hierarchy often requires traversing from one object, to another object, to another object, until the desired destination is attained. For example, suppose a PLI application had obtained a handle for a module, and the application needs to locate every module output, where the output is also connected to a module path delay. The following Verilog source code shows the starting and ending objects for which handles are desired:

```
                              ┌────────────────┐
                              │ start with a handle │
                              │ to a module     │
                              └────────────────┘
module dff (clk, d, q, qb);
    input  clk, d;                 ┌──────────────────────────┐
    output q, qb;                  │ obtain handles for all output │
                                   │ ports which are also     │
    ff_prim g1 (q, d, clk,);       │ connected to a module path │
    not g2 (qb, q);                │ delay (port q in this example) │
                                   └──────────────────────────┘
    specify
       (d *> q) = 2.5;
    endspecify
endmodule
```

By following the connections in the object diagrams, the following object connections can be traversed:

1. The partial *module* diagram shown in Figure 15-2, which follows, shows a one-to-many connection from ⬭ **module** ⬭ to ⬭ mod path ⬭. The diagram also shows that the ACC next routine `acc_next_modpath()` is used to traverse from the module to the module path target. Since this is a next routine, a handle for each module path in the module can be obtained.

Figure 15-2: ACC object diagram for Verilog modules (partial)

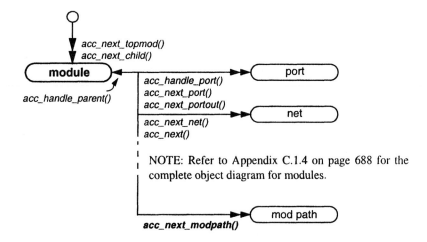

2. The *mod path* object diagram in Figure 15-3 shows a one-to-many connection from ⬭ **mod path** ⬭ to ⬭ **path term** ⬭. This connection can be traversed with specific ACC routines to obtain handles for the path input terminals or output terminals. Since this example application is looking for the output ports connected to module paths, the output terminal will be the connection which needs to be followed. In addition, the diagram shows two ACC routines to obtain handles for the path output terminal. The ACC handle routine, `acc_handle_pathout()`, will return only a single handle, which is the first output of the path. The ACC next routine, `acc_next_output()`, will return handles for all outputs in the path (if the routine is called in a loop).

Figure 15-3: ACC object diagram for module paths (partial)

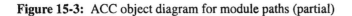
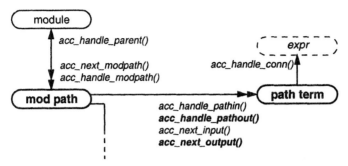

NOTE: Refer to Appendix C.1.12 on page 697 for the complete object diagram for module paths.

Since the objective for this example PLI application is to find all ports connected to path outputs, the acc_next_output() routine will be the right choice for this example. Using this routine, a handle for each path output in the module path can be obtained. Note that this call to acc_next_output() must be done for each module path in the module, because the acc_next_modpath() routine will return the module path handles one at a time. Nested loops will be used, in order to obtain handles for each path output for each module path. If C while loops are used, the nested loops can be coded as:

```
modpath_h = pathterm_h = null;
while (modpath_h=acc_next_modpath(module_h, modpath_h)) {
  while (pathterm_h=acc_next_output(modpath_h,pathterm_h)) {
    /* continue traversing to port driven by this output */
  }
}
```

3. The ⬭ **path term** object is part of the ***mod path*** object diagram, shown in Figure 15-3 on the previous page. The ***path term*** object shows a one-to-one connection to an object group called ⟨___ *expr* ___⟩ (expr stands for expression). The diagram indicates that this object relationship is traversed using acc_handle_conn().

The object diagram for the expression group, shown in Figure 15-4, below, lists many types of Verilog objects, including nets, variables, and several other objects. These are the types of objects which can be used in connections and statements in many places in the Verilog language. For example, the connection to a module port could be any of the objects listed in the expression class. The objects referenced within the expression group are defined in other diagrams.

Figure 15-4: ACC object diagram of expressions (partial)

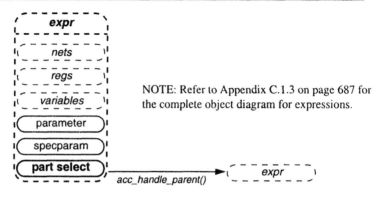

NOTE: Refer to Appendix C.1.3 on page 687 for the complete object diagram for expressions.

The type of object within the expression class can be determined using `acc_fetch_type`(*expression_handle*). For this illustration, it will be assumed that the type returned is **accNet**. Therefore, the object diagram for nets will show the next relationship in the connection from a module path to a port.

4. The *nets* object diagram, shown in Figure 15-5 on the following page, shows two types of connections from `nets` to `ports`.

Figure 15-5: ACC object diagram for nets (partial)

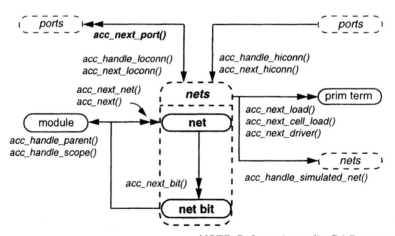

NOTE: Refer to Appendix C.1.7 on page 692 for the complete object diagram for nets.

One connection to ports is accessed using `acc_next_port()`. The other connection has no terminating arrow, which indicates that the ACC routines cannot traverse hierarchy in that direction. The connection accessed by `acc_next_port()` is the connection from a net within a module to a port of that module. The other connection is from a port of a *module instance* within the current module to a net. The following Verilog model illustrates the difference between the two types of port objects:

```
module chip (clock, in1, out1, out2);

    wire out1, out2;

    dff u1 (clock, in1, out1, out2);

endmodule


module dff (clk, d, q, qb);
    ...
endmodule
```

> module ports are accessed from an internal net using `acc_next_port()`

> *module instance* ports cannot be accessed from an internal net using ACC routines

Example summary

In the example described in this section, four object connections were followed to traverse the Verilog hierarchy from a module object to the output ports connected to a path delay. The connection path is from ⟨ **module** ⟩ to ⟨ **mod path** ⟩ to ⟨ **path term** ⟩ to ⟨ *nets* ⟩ (in the expression group) to ⟨ *ports* ⟩.

The C source code listed in Example 15-1 illustrates how these multiple connections through the Verilog hierarchy are traversed. The PLI application is *$list_pathout_ports*. The usage is:

```
$list_pathout_ports(<module_instance_name>);
```

CD The source code for this example is on the CD accompanying this book.

- Application source file: `Chapter.015/list_pathout_ports_acc.c`
- Verilog test bench: `Chapter.015/list_pathout_ports_test.v`
- Expected results log: `Chapter.015/list_pathout_ports_test.log`

Example 15-1: *$list_pathout_ports* — traversing Verilog hierarchy

```
int PLIbook_ListPorts_calltf(int user_data, int reason)
{
  handle module_h, modpath_h, pathterm_h, net_h, port_h;

  acc_initialize();

  module_h = acc_handle_tfarg(1);
  io_printf("\nOutput ports with path delays in module %s:\n",
            acc_fetch_defname(module_h));

  modpath_h = pathterm_h = net_h = port_h = null;
  while (modpath_h = acc_next_modpath(module_h, modpath_h)) {
    while (pathterm_h = acc_next_output(modpath_h, pathterm_h)) {
      net_h = acc_handle_conn(pathterm_h);
      while (port_h = acc_next_port(net_h, port_h)) {
        io_printf("  Port %s\n", acc_fetch_name(port_h));
      }
    }
  }
  acc_close();
  return(0);
}
```

15.8 Traversing hierarchy across module ports

Verilog hierarchy connections can be traversed across module boundaries by following the connections within and without module ports. A module port has two connections, as shown in the following diagram.

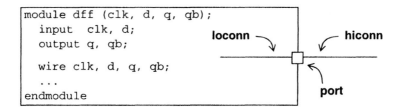

The *loconn* (hierarchically lower connection) is the signal *inside* the module that is connected to a port. In the object diagrams, the internal connection is shown as an *expression*, because a number of different types of objects can be connected to the port. Within a module, the Verilog HDL syntax restricts the connection to a port to the Verilog data types of *nets*, *regs* and *variables* (except real variables). The connection can also be a scalar signal (1-bit wide), a vector, a bit or part select of a vector, or a concatenation of any these signals. For the loconn connection, any of these expression types can be connected to an output port, but only expressions using the *net* class of data types can be connected to an input or inout port.

The ***hiconn*** (hierarchically higher connection) is the signal *outside* the module that is connected to a port. External to a module, the expression connected to the port can be a scalar signal, a vector, a bit or part select of a vector, or a concatenation of these signals. Any of these expression types can be connected to an input port, but only expressions using the *net* class of data types can be connected to an output or inout port. (The Verilog language also allows the *hiconn* connection of an input port to be a constant, a literal value, an operation, or the return of a function call, but the ACC routines cannot access these types of objects).

The Verilog HDL syntax allows concatenation multiple internal signals to be connected to *loconn* of a module port. Each signal can have a different data type, and the bits of the *loconn* connections can have different port directions. The ACC routines refer to a port with a mix of signal directions connected to it as a *mixed I/O* port.

The object diagram for *ports*, shown in Figure 15-6, shows a one-to-many connection to either the *hiconn* or *loconn* connection to the port. The connecting object is an *expr* object group, to allow for the different types of objects which may be connected to the port. Using the handle for the *expr* object, the type property of the object can be accessed using either `acc_fetch_type()` or `acc_fetch_fulltype()` to determine what is connected to the port. The object diagram for that type of object can then be used to continue traversing the Verilog design hierarchy. Accessing the object type property is presented previously, in Chapter 14, section 14.7 on page 468.

Figure 15-6: ACC object diagram for ports (partial)

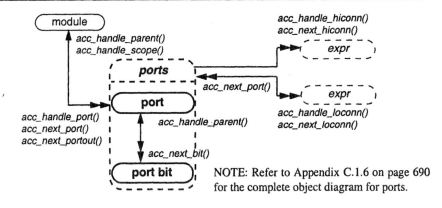

NOTE: Refer to Appendix C.1.6 on page 690 for the complete object diagram for ports.

Both ACC handle routines and ACC next routines can be used to obtain the handles for the hiconn and loconn connections of a port. This provides flexibility for different PLI applications. Table 15-3, on the following page, shows what type of object will be returned, based on different types of port connections and the ACC routine used. The terms used in this table are defined after the table.

Table 15-3: Port connection handles

Routine	Port Property	Connection Handle
acc_handle_loconn()	scalar port	scalar connection
acc_handle_hiconn()	expanded vector port	vector connection
	unexpanded vector port	vector connection
	bit select of a port	bit select of connection
acc_next_loconn()	scalar port	scalar connection
	expanded vector port	bit select of connection
	unexpanded vector port	vector connection
	bit select of a port	illegal
acc_next_hiconn()	scalar port	scalar connection
	expanded vector port	vector connection
	unexpanded vector port	vector connection
	bit select of a port	illegal

Bit-select, part-select and concatenation ports

When a bit-select or part-select of a vector signal is connected to a module port, the port type inherits this property. If a concatenation of several signals is connected to a port, this property is also reflected in the port type. The type property of the port is accessed, using acc_fetch_fulltype(), which was presented previously, in Chapter 14, section 14.7 on page 468.

The ACC constants which represent ports are:

• **accScalarPort** — The port has a scalar signal attached to it.

```
module m1 (a);
   input a;
   wire  a;
```

- **accVectorPort** — The port has a vector signal attached to it.

```
module m1 (a);
  input [7:0] a;
  wire  [7:0] a;
  . . .
```

- **accBitSelectPort** — The port has a bit-select of a vector signal attached to it.

```
module m1 (a[0]);
  input [7:0] a;
  wire  [7:0] a;
  . . .
```

- **accPartSelectPort** — The port has a part-select of a vector signal attached to it.

```
module m1 (a[7:4], a[3:0]);
  input [7:0] a;
  wire  [7:0] a;
  . . .
```

- **accConcatPort** — The port has a concatenation of signals attached to it.

```
module m1 (.a{n,m});
  input  [3:0] n;
  output [3:0] m;
  wire   [3:0] n, m;
  . . .
```

Expanded and unexpanded vectors

A module port may be any vector size. If a vector port has a special *expanded* property, then the ACC routines are allowed to access discrete bits of the vector. If the port is *unexpanded*, then certain ACC routines, such as acc_next_loconn(), may be restricted to only accessing the complete vector. The expanded or unexpanded property is inherited from the type of loconn signal connected to the port. The routine acc_object_of_type() is used to determine if a port is expanded or unexpanded. This routine is presented in Chapter 16, in section 16.2 on page 538.

Expanded vectors are vectors for which the simulator allows access to individual bits within the vector. Access to the bits of a vector is the default in the Verilog language, and can also be explicitly declared in the Verilog source code, using the scalared keyword. For example:

```
wire scalared [63:0] data_bus;
```

Unexpanded vectors are vectors for which the simulator may prohibit access to individual bits within the vector. This is a feature which some simulators use to optimize run-time performance. Vector only access must be explicitly declared in the Verilog source code, using the `vectored` keyword. For example:

```
wire vectored [31:0] address_bus;
```

An example of traversing hierarchy across module ports

The following example accesses all ports of a module and lists the port name, port size, port direction, and type of object connected as the *loconn* and *hiconn*. The name, size and direction are all properties of the module port.

> **CD** The source code for this example is on the CD accompanying this book.
>
> • Application source file: `Chapter.015/port_info_acc.c`
> • Verilog test bench: `Chapter.015/port_info_test.v`
> • Expected results log: `Chapter.015/port_info_test.log`

Example 15-2: *$port_info* — traversing Verilog hierarchy across module ports

```
int PLIbook_PortInfo_calltf(int user_data, int reason)
{
  handle mod_h, port_h, loconn_h, hiconn_h;

  acc_initialize();
  acc_configure(accDisplayWarnings, "true");

  mod_h = acc_handle_tfarg(1);

  io_printf("\nModule %s:\n", acc_fetch_defname(mod_h));
  io_printf("  Instance name: %s\n", acc_fetch_fullname(mod_h));
  switch ( acc_fetch_fulltype(mod_h) ) {
    case accTopModule:
      io_printf("  Module type: top-level\n");
      break;
    case accModuleInstance:
      io_printf("  Module type: module instance\n");
      break;
    case accCellInstance:
      io_printf("  Module type: cell module\n");
      break;
    default:
      io_printf("  Module type: unknown\n");
  }

  io_printf("  Ports:\n");
```

```
port_h = null;
while (port_h = acc_next_port(mod_h, port_h)) {
   io_printf("   %-8s", acc_fetch_name(port_h) );
   io_printf("%2d-bit ", acc_fetch_size(port_h) );
   switch ( acc_fetch_direction(port_h) ) {
     case accInput:
       io_printf("input   ");
       break;
     case accOutput:
       io_printf("output  ");
       break;
     case accInout:
       io_printf("inout   ");
       break;
     case accMixedIo:
       io_printf("mixed input/output  ");
       break;
     default:
       io_printf("unknown direction  ");
   }

   if (acc_object_of_type(port_h, accExpandedVector))
     io_printf("  Expanded=true   ");
   else
     io_printf("  Expanded=false  ");
   if (acc_object_of_type(port_h, accUnExpandedVector))
     io_printf("  Unexpanded=true\n");
   else
     io_printf("  Unexpanded=false\n");

   loconn_h = acc_handle_loconn(port_h);
   io_printf("      Loconn type = %s\n",
             acc_fetch_type_str(acc_fetch_fulltype(loconn_h)));

   hiconn_h = acc_handle_hiconn(port_h);
   if (hiconn_h)
     io_printf("      Hiconn type = %s\n\n",
               acc_fetch_type_str(acc_fetch_fulltype(hiconn_h)));
   else
     io_printf("      Hiconn type = none\n\n");
} /* end of next port_h loop */
acc_close();
return(0);
}
```

15.9 Identifying modules and library cells

In the Verilog language, modules are building blocks which make up a larger design. A design hierarchy tree is formed when one module instantiates another module.

A PLI application often needs to distinguish one level of a design hierarchy from others. An ASIC library vendor, for example, may need to locate the cells from the ASIC library within the rest of the Verilog hierarchy, in order to calculate cell delays or power usage. Figure 15-7 shows a simple hierarchy tree. A PLI application might wish to traverse from the top of the design, and locate the module instances of *cellA*, *cellB*, and *cellC*. To access *cellA* and *cellB*, the PLI application will need to traverse through *modA*. The same PLI application might wish to consider *cellC* a leaf in the hierarchy tree, and ignore the hierarchy below that leaf (the instance of *modB* in this example).

Figure 15-7: Example Verilog design hierarchy tree

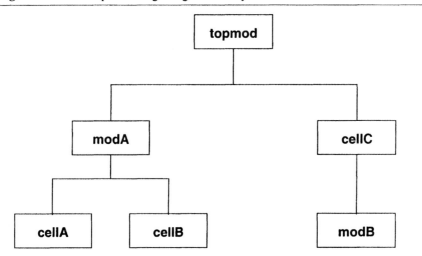

For convenience in developing PLI applications, the Verilog language provides a special flag which can be placed in the Verilog source code. This flag designates that the modules which follow the flag are to be treated as cell modules. The flag has no effect on simulation. It is strictly a flag which is used by certain ACC routines to distinguish one type of module from another. The `` `celldefine`` compiler directive is used to set the cell module flag, and the `` `endcelldefiine`` directive is used to terminate the flag. All modules which are compiled between the two directives will be flagged as cell modules.

The IEEE 1364 standard states that any Verilog module which is loaded into simulation using a library scan option should automatically be flagged as a cell module. In most Verilog simulators, the library scan options are -v, -y, and the 'uselib compiler directive. Some simulators provide special invocation options to enable or disable this automatic flagging of modules loaded from a library as being a cell modules.

The ACC library provides two different ACC next routines to find module instances within a module:

handle **acc_next_child (**
    handle   **module,**   /* handle for a module, or **null** */
    handle   **prev_child)** /* handle for the previous child found; initially **null** */

The acc_next_child() routine returns handles for all instances of any module type within a reference module. If the module handle is null, then the next top-level module is returned. In the diagram shown above, if acc_next_child() were passed a reference handle for **topmod**, it would retrieve handles for **modA** and **cellC**. Those are the modules which are instantiated within topmod.

handle **acc_next_cell (**
    handle   **module,**   /* handle for a module */
    handle   **prev_cell)** /* handle for the previous cell found; initially **null** */

The acc_next_cell() routine returns handles for all instances of modules which are flagged as cells. This routine automatically traverses all levels of hierarchy, starting with the reference module. In the hierarchy tree shown in Figure 15-7 on page 512, if acc_next_cell() were given a reference handle for **topmod**, it would retrieve handles for **cellA, cellB** and **cellC**. Those are the cell modules which are instantiated within *and below* topmod.

Example 15-3 lists a short PLI application which looks for all top-level modules in a design, and then searches all levels of hierarchy below that top level and prints the name of every cell module used in the design.

CD The source code for this example is on the CD accompanying this book.

- Application source file: Chapter.015/list_cells_acc.c
- Verilog test bench: Chapter.015/list_cells_test.v
- Expected results log: Chapter.015/list_cells_test.log

Example 15-3: *$list_cells* — working with module cells

```
·int PLIbook_ListCells_calltf()
{
  handle mod_h, cell_h;
  int    cell_cnt = 0;

  acc_initialize();
  acc_configure(accDisplayWarnings, "true");

  mod_h = acc_handle_tfarg(1);
  io_printf("\nCells in Module %s, instance %s:\n",
            acc_fetch_defname(mod_h), acc_fetch_fullname(mod_h));
  cell_h = null; /* start with null (no cells found yet) */
  while (cell_h = acc_next_cell(mod_h,cell_h)) {
    io_printf("  %s (%s)\n",
              acc_fetch_fullname(cell_h), acc_fetch_defname(cell_h));
    cell_cnt++;
  }
  io_printf("Total cells in this hierarchy tree = %d\n\n", cell_cnt);
  acc_close();
  return(0);
}
```

15.10 Accessing loads and drivers

The ACC library provides three routines to find the drivers and loads of a Verilog net. These routines also distinguish between standard modules and cell modules.

handle **acc_next_driver (**
handle	**net,**	/* handle for a scalar net or bit-select of a vector net */
handle	**prev_driver)**	/* handle for the previous driver found; initially **null** */

This routine returns the handle for the next *primitive terminal driver* on a net.

handle **acc_next_load (**
handle	**net,**	/* handle for a scalar net or bit-select of a vector net */
handle	**prev_load)**	/* handle for the previous load found; initially **null** */

This routine returns the handle for the next *primitive terminal load* on a net. If there are multiple loads within a module, all loads will be returned .

handle **acc_next_cell_load (**
    handle   **net,**   /* handle for a scalar net or bit-select of a vector net */
    handle   **prev_load)**   /* handle for the previous load found; initially **null** */

This routine returns the handle for the next *primitive terminal of a cell load* on a net. Loads that are not in cell modules are not returned. Only the first load within a cell will be returned if there are multiple loads within the cell.

Both acc_next_load() and acc_next_cell_load() return primitive terminals as the loads of a net. The difference in the routines is which primitive terminals are considered to be loads. If a net fans out to multiple loads within a module, acc_next_load() returns a handle for each load. acc_next_cell_load() is more discriminatory. This routine only returns the primitive terminals which load a net, if the primitive is within a module that was flagged as a cell, and only one load per cell module will be returned. The following diagram illustrates the difference between the loads returned from each of these routines.

 acc_next_driver(), acc_next_load() and acc_next_cell_load() only locate primitive input terminals as the drivers or loads of a net. The VPI routines will locate other types of drivers and loads, such as procedural assignments and continuous assignment statements.

Figure 15-8: Difference of acc_next_load() and acc_next_cell_load()

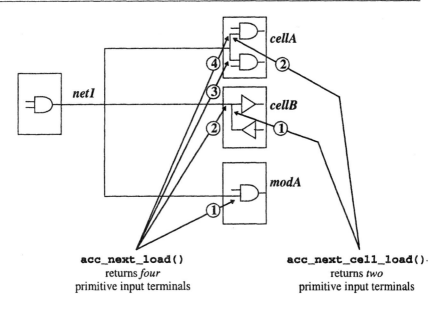

15.11 Accessing model timing

The ACC library provides several routines designed specifically to provide access to the timing information within Verilog simulation. This timing information is specific to each *instance* of a module, rather than to each definition of a module. By reading and/or modifying the timing of each instance, PLI applications can greatly increase the timing accuracy of a simulation.

The PLI can access the following timing constructs within a Verilog simulation:

- Primitive delays

- Module path delays

- Timing constraint checks

- Module inter-connect delays

The ACC library provides several routines to obtain handles for these timing objects, which are presented on the following pages. There are also special ACC routines for reading and modifying the values of the delays on each object. Reading and modifying delays is presented in the next chapter.

15.11.1 Obtaining handles for primitives

The Verilog language allows a unique set of delays to be specified on each instance of a primitive. The syntax for specifying primitive delays is part of the Verilog language, and is not presented in this book. From the perspectives of the PLI, there are only two major considerations for accessing primitive delays:

- Verilog primitives can be two-state or three-state devices. A two-state primitive has timing for output transitions from 0 to 1 and from 1 to 0. A three-state primitive has timing for transitions from 0 to 1, 1 to 0, and from any value to hi-impedance.

- Primitive delays can be represented as a single value for each possible transition, or as a minimum:typical:maximum set of delays for each transition.

To access the delays of a primitive, a handle for the primitive must be obtained. The routines which can obtain handles for primitives are:

- **`acc_next_primitive()`** returns handles for all primitive instances within a module instance.

- **`acc_handle_object()`** returns a handle for a primitive instance, using the instance name of the primitive, searching in the design hierarchy scope in which the PLI is currently operating.

- **acc_handle_by_name()** returns a handle for a primitive instance, using the instance name of the primitive, searching in the design hierarchy scope specified as an input to the routine.

The syntax and usage of the ACC next routines was presented in Chapter 14, in section 14.6 on page 466. The syntax and usage of acc_handle_object() and acc_handle_by_name() is discussed in more detail later in this chapter, in section 15.14 on page 528.

15.11.2 Obtaining handles for module paths

The primary Verilog construct for representing the timing of a module is a *module path delay*. Module path delays are also referred to as *pin-to-pin delays*, or simply as *path delays*. These delays are specified within the specify block of a module. The full syntax for specifying module path delays is part of the Verilog language, and is outside the scope of this book. The PLI standard has special terms for the various components of a module path delay, which are presented in the following paragraphs. The components of a module path delay which are identified by the PLI are:

- The *module path*
- The path *input terminals*
- The path *output terminals*
- The *data path*
- The data path *source terminals*
- The data path *destination terminals*
- The path *conditional expression*

The following diagrams illustrate these components of a module path delay.

Module path delay with single inputs and outputs:

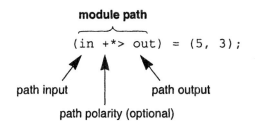

Module path delay with multiple inputs and outputs:

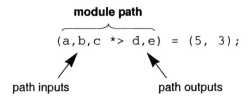

Edge sensitive module path delay:

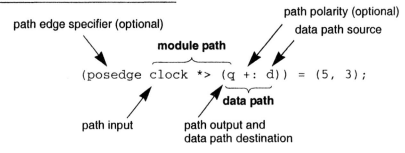

Conditional module path delay:

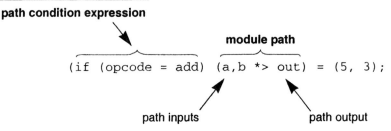

There are several ACC handle and ACC next routines used to obtain handles for the different components of a module path. The ACC handle and ACC next routines which obtain handles for module paths and the components of a module path are:

- **acc_next_modpath()** returns handles for all module paths within a module.

- **acc_handle_modpath()** returns a handle for a specific module path within a module instance, using a description of the input and output of the path.

- **acc_handle_pathin()** returns a handle for the first input of a module path.

- **acc_handle_pathout()** returns a handle for the first output of a module path.

- **acc_handle_condition()** returns a handle for the condition expression of a module path or data path.

- **acc_next_input()** returns handles for all inputs of a module path, or all source terminals of a data path.

- **acc_next_output()** returns handles for all outputs of a module path, or all destination terminals of a data path.

- **acc_handle_conn()** returns a handle for the net connected to a module path terminal or data path terminal.

The general syntax and usage of the ACC handle routines and ACC next routines was presented in Chapter 14, in sections 14.5 and 14.6.

The ACC routines listed above allow a PLI application to obtain the module path timing information for any module instance in a design. Example 15-1 on page 506, presented earlier in this chapter, illustrates accessing all module paths in a module, and all output terminals of each module path. After a handle is obtained, other ACC routines can then fetch information about the path, such as the polarity and condition edges. Additional ACC routines can read and modify the delay values of each module path, for each instance of the module containing the path. The routines to read and modify module path delays are presented in the next chapter.

Configuring acc_handle_modpath()

The routine acc_handle_modpath() uses handles for the path input and output terminals to obtain a handle for the module. This routine can be configured to describe the terminals in different ways. The syntax of this routine is:

handle **acc_handle_modpath (**
handle	**object,**	/* handle for a module */
PLI_BYTE8	*** src_name,**	/* name of net connected to path source, or **null** */
PLI_BYTE8	*** dest_name,**	/* name of net connected to path destination, or **null** */
handle	**src_handle,**	/* handle for net connected to path source, or **null** */
handle	**dest_handle)**	/* handle for net connected to path destination, or **null** */

The acc_handle_modpath() routine uses the acc_configure() routine to control which arguments are used to describe the path terminals. The syntax for acc_configure() was presented earlier in this chapter, in section 15.2 on page 489.

- acc_configure(accEnableArgs, "no_acc_handle_modpath") configures acc_handle_modpath() to use only the src_name and dest_name arguments. A literal string or pointer to a string is provided for the arguments. The src_handle and dest_handle arguments are ignored, and can be set to null or dropped from the argument list. This is the default configuration.

- acc_configure(accEnableArgs, "acc_handle_modpath") configures acc_handle_modpath() to use the src_name and dest_name arguments, if a literal string or pointer to a string is provided for the arguments. If the name arguments are set to null, then the src_handle and dest_handle arguments are used.

Memory allocation for acc_next_input() and acc_next_output()

The routines acc_next_input() and acc_next_output() require special attention in a PLI application. Some simulators may need to allocate memory in order to process these routines. This memory will automatically be freed once the PLI application has retrieved the last terminal, which is indicated when the ACC next routine returns a null. However, if a PLI application does not call these routines in a loop until the routine returns a null, then the memory is not released. The PLI application must then manually free the memory by calling **acc_release_object()**.

PLI_INT32 **acc_release_object (**
    handle   **object)**   /* handle for a module path or data path terminal */

This routine returns 1 if successful and 0 if an error occurred.

15.11.3 Obtaining handles for module inter-connect paths

A module inter-connect path is a connection from the output port of one module to the input port of another module.

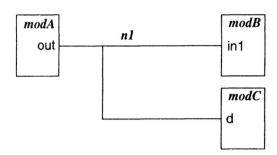

In this illustration, there is a single net **n1**, but two inter-connect paths, **out** to **in1**, and **out** to **d**.

The acc_handle_path() routine obtains a handle for an inter-connect path, using handles for an output port and the input port. The syntax for this routine is:

handle **acc_handle_path (**
handle	**source,**	/* handle for a scalar output or inout port, or bit-select of a vector output or inout port */
handle	**destination**)	/* handle for a scalar input or inout port, or bit-select of a vector input or inout port */

There is no construct in the Verilog language to represent inter-connect delays. However, using an inter-connect path handle, a PLI application can place delays on the inter-connection and read the value of those delays. Reading and modifying delays is presented in Chapter 14.

15.11.4 Obtaining handles for timing constraint checks

Verilog modules can specify timing constraints between one or more input ports of the module, such as setup and hold times. These constraints are specified in a Verilog specify block, and use a special form of built-in system tasks. The names of these tasks are: *$setup*, *$hold*, *$setuphold*, *$skew*, *$recovery*, *$period*, and *$width*. The PLI uses specific terminology to identify the arguments of a Verilog timing constraint.

Example timing check:

```
$setup(data, posedge clock && (reset), 10, err_flag);
```

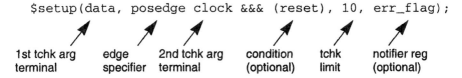

| 1st tchk arg
terminal | edge
specifier | 2nd tchk arg
terminal | condition
(optional) | tchk
limit | notifier reg
(optional) |

The PLI can access the timing constraint checks for each instance of a module, and can modify the constraint values of each instance. Handles for the optional condition and notifier fields of a timing check can also be accessed, as well as the type of edge specifier.

The Verilog HDL syntax of timing constraint checks is not consistent. Some checks have two terminals, like the *$setup* timing check shown above; and some checks have just one terminal. Most timing checks have a single tchk limit, as shown above, but some timing checks specify two limits. To allow for these variations of syntax in the Verilog language, the ACC library provide specific routines to access the individual fields of each timing check:

- **acc_next_tchk()** returns handles for all timing checks within a module instance.

- **acc_handle_tchk()** returns a handle for a specific timing check, using a description of the timing check arguments.

- **acc_handle_tchkarg1()** returns a handle for the first terminal of a timing check.

- **`acc_handle_tchkarg2()`** returns a handle for the second terminal of a timing check.

- **`acc_handle_conn()`** returns a handle for the net connected to a timing check argument.

The general syntax and usage of the ACC handle routines and ACC next routines was presented in Chapter 14, in sections 14.5 and 14.6. After a handle is obtained, other ACC routines can then fetch information about the constraint, such as the polarity and condition edges. Additional ACC routines can read and modify the values of each timing constraint, for each instance of the module containing the constraint. The routines to read and modify module constraint values are presented in the next chapter.

 The IEEE 1364-2001 standard adds additional timing constraint checks to the Verilog language, and adds additional arguments to some existing timing constraint checks. These enhancements increase the accuracy for modeling deep-submicron technologies. The ACC library cannot access these new timing constraint checks and arguments. Only the VPI library supports these enhancements.

Configuring acc_handle_tchk()

The routine `acc_handle_tchk()` uses a description of the timing check arguments to obtain a handle for the timing check. This routine can be configured to describe the timing check arguments in different ways. The syntax for `acc_handle_tchk()` is:

handle **acc_handle_tchk (**

handle	**object,**	/* handle for a module */
PLI_INT32	**type,**	/* constant representing timing check type. One of: **accHold, accNochange, accPeriod, accRecovery, accSetup, accSkew** or **accWidth** */
PLI_BYTE8	*** name1,**	/* name of net connected to the 1st tchk arg, or **null** */
PLI_INT32	**edge1,**	/* constant representing the edge of the 1st timing check argument (constant names are listed below) */
PLI_BYTE8	*** name2,**	/* name of net connected to the 2nd tchk arg, or **null** */
PLI_INT32	**edge2,**	/* constant representing the edge of the 2nd tchk arg (constant names are listed below) */
handle	**conn1,**	/* handle for net connected to 1st tchk arg, or **null** */
handle	**conn2)**	/* handle for net connected to 2nd tchk arg, or **null** */

edge1 and **edge2** identifiers are one of the following:

- One of the constants: `accNoedge, accPosedge, accNegedge`

- List of constants separated by +: `accEdge01, accEdge0x, accEdgex1`

- List of constants separated by +: `accEdge10, accEdge1x, accEdgex0`.

`acc_handle_tchk()` uses the `acc_configure()` routine to control which arguments are used to describe the path terminals. The syntax for `acc_configure()` was presented earlier in this chapter, in section 15.2 on page 489.

- `acc_configure(accEnableArgs, "no_acc_handle_tchk")` configures the `acc_handle_tchk()` routine to use only the name1 and name2 arguments. A literal string or pointer to a string is provided for the arguments. The conn1 and conn2 arguments are ignored, and can be set to `null` or dropped from the argument list. This is the default configuration.

- `acc_configure(accEnableArgs, "acc_handle_tchk")` configures the `acc_handle_tchk()` routine to use the name1 and name2 arguments, if a literal string or pointer to a string is provided for the arguments. If the name arguments are set to `null`, then the conn1 and conn2 arguments are used.

15.12 Counting the number of objects

The ACC library provides a convenient routine for counting how many of a specific type of object exist within a reference object.

PLI_INT32 **acc_count (**

handle	*** next_routine,** /* name, without parenthesis, of any ACC *next* routine,
	except acc_next_topmod */
handle	**object)** /* reference object for the ACC *next* routine */

The `acc_count()` routine returns the number of objects which were located by an ACC next routine. The inputs to `acc_count()` are the name of a next routine, and a reference point in which that routine should search for its target objects. The handles for the objects are not returned.

An example of using `acc_count()` is shown in example 15-4. This example implements a *$count_loads* system function, which counts how many cell loads are driven by an output port of a module, and then returns that count back to the simulation as a system function return. *$count_loads* is passed a module port name as its input.

An example of using *$count_loads* might be:

```
if ($count_loads(chip_out) > 5)
    $display("Warning: output port has too many loads!");
```

CD The source code for this example is on the CD accompanying this book.

- Application source file: `Chapter.015/count_loads_acc.c`
- Verilog test bench: `Chapter.015/count_loads_test.v`
- Expected results log: `Chapter.015/count_loads_test.log`

Example 15-4: *$count_loads* — using the `acc_count()` routine

```
/**********************************************************************
 * Sizetf application
 **********************************************************************/
int PLIbook_CountLoads_sizetf(int user_data, int reason)
{
  return(32);  /* $count_loads returns a 32-bit integer value */
}


/**********************************************************************
 * checktf routine
 **********************************************************************/
int PLIbook_CountLoads_checktf(int user_data, int reason)
{
  static PLI_INT32 valid_args[4] = {accNet, accReg, accRegBit, 0};
  PLI_INT32 direction;
  handle tfarg_h, port_h;

  acc_initialize();
  if (tf_nump() != 1)
    tf_error("$count_loads must have 1 argument.");
  else if (tf_typep(1) == TF_NULLPARAM)
    tf_error("$count_loads arg cannot be null.");
  /* acc_handle_tfarg() returns a loconn handle, not a port handle */
  else if (tf_sizep(1) != 1)
    tf_error("$count_loads arg must be scalar or a bit-select.");
  else {
    tfarg_h = acc_handle_tfarg(1);
    if (!acc_object_in_typelist(tfarg_h, valid_args)) {
      tf_error("$count_loads arg must be a net or reg signal");
      return(0);
    }
    port_h = acc_next_port(tfarg_h, null);
    if (port_h == null) {
      tf_error("$count_loads arg is not connected to a module port.");
      return(0);
    }
    direction = acc_fetch_direction(port_h);
    if (   direction != accOutput
        && direction != accInout)
      tf_error("$count_loads arg must be an output or inout port.");
  }
}
```

```
  acc_close();
  return(0);
}

/***********************************************************************
 * calltf routine
 ***********************************************************************/
int PLIbook_CountLoads_calltf(int user_data, int reason)
{
  handle loconn_h, port_h, hiconn_h;
  int    load_count;

  acc_initialize();
  acc_configure(accDisplayWarnings, "true");

  /* acc_handle_tfarg() may return loconn handle, not port handle */
  loconn_h = acc_handle_tfarg(1);
  if (acc_fetch_type(loconn_h) == accPort)
    port_h = loconn_h;
  else
    port_h = acc_next_port(loconn_h, null);
  hiconn_h = acc_handle_hiconn(port_h);
  load_count = acc_count(acc_next_cell_load, hiconn_h);
  tf_putp(0, load_count);
  acc_close();
  return(0);
}
```

15.13 Collecting and maintaining lists of object handles

Sometimes a PLI application may need to use the same ACC next routine to find the same target objects multiple times in the same call to a PLI application. For example, an application might need to find all of some type of object in a module (such as all the primitives) in order perform some type of operation. Then later, in the same call to the PLI application, it might be necessary to again find all of the same objects in the same module, in order to perform some other type of operation.

Sometimes an instance of a PLI application might be called many times during a simulation. For example, an application that prints the logic value of all drivers of a net might be called every clock cycle. In this case, the same ACC next routine might be called millions of times during a simulation. Each call will need to locate and return the same set of objects.

Repeatedly calling the same ACC next routine to locate the same objects can have a negative impact on simulation run-time performance. The ACC library provides a pair of routines to help make a PLI application more efficient when the same ACC next routine needs to be called multiple times.

handle ***acc_collect (**

handle	*** next_routine,**	/* name, without parenthesis, of any ACC *next* routine,
		except acc_next_topmod */
handle	**object,**	/* reference object for the ACC *next* routine */
PLI_INT32	*** count)**	/* the number of objects collected */

The **acc_collect()** routine is provided the name of an ACC next routine, a handle for a reference object, and a pointer to an integer variable. The routine will execute the specified ACC next routine and collect a list of all of the target handles. The list of handles is stored in an array of handles, and a pointer to the array is returned. The total number of objects found is placed the integer variable pointed to as the third argument to acc_collect().

Once the array of object handles has been collected, a PLI application can reference the array as often as needed, without having to call the ACC next routine again and again. The array is stored in persistent memory, which is allocated by acc_collect(). Therefore, as long as the pointer for the array is saved by the PLI application, the array can be used as often as needed. If the PLI application saves the pointer to the array, then the array can be used in future calls to the PLI application as well. The TF work area is the proper place to preserve the pointer to the array of object handles. The work area is specific to each instance of a system task/function, and is shared by the *checktf routine, calltf routine* and *misctf routine*. Refer to section 10.7 on page 325 of Chapter 10, for details on using the TF work area.

Because the memory for the array of object handles allocated by acc_collect() is persistent, the PLI application must explicitly release the memory when the array is no longer required. The acc_free() routine is used to release the memory which was allocated by acc_collect(). If acc_free() is not called, the memory allocated for the array of handles will not be released until the simulation exits. The syntax for acc_free() is:

void **acc_free (**

| *handle* | *** array_ptr)** | /* pointer to an array of handles */ |

The following C code shows how a list of handles for all signals in a module can be created, using acc_collect(). The array of signal handles is generated at the start of simulation, when the *misctf routine* is called for REASON_ENDOFCOMPILE, and the array memory is freed at the end of simulation, when the *misctf routine* is called for REASON_FINISH. The pointer to the array is stored in the TF work area, so that the *calltf routine* can access the array whenever needed.

```
┌─────────────────────────────────────────────────────────────────────┐
│  CD  The source code for this example is on the CD accompanying this book. │
│                                                                         │
│  • Application source file:  Chapter.015/display_all_acc.c             │
│  • Verilog test bench:       Chapter.015/display_all_test.v            │
│  • Expected results log:     Chapter.015/display_all_test.log          │
└─────────────────────────────────────────────────────────────────────┘
```

Example 15-5: *$display_all_nets* — using the `acc_collect()` routine

```c
/**********************************************************************
 * structure definition for data to be passed from misctf to calltf
 **********************************************************************/
typedef struct PLIbook_NetData {
  handle  module_h;
  handle *net_array;
  int     net_count;
} PLIbook_NetData_s, *PLIbook_NetData_p;

/**********************************************************************
 * misctf routine
 **********************************************************************/
int PLIbook_DisplayNets_misctf(int user_data, int reason, int paramvc)
{
  PLIbook_NetData_p  net_data;
  handle  mod_h;

  acc_initialize();
  acc_configure(accDisplayWarnings, "true");

  switch (reason) {
    case REASON_ENDOFCOMPILE:
      acc_initialize();
      net_data = (PLIbook_NetData_p)malloc(sizeof(PLIbook_NetData_s));
      net_data->module_h  = acc_handle_tfarg(1);
      net_data->net_array = acc_collect(acc_next_net,
                                        net_data->module_h,
                                        &net_data->net_count);
      io_printf("Total nets collected in misctf routine = %d\n",
                net_data->net_count);
      tf_setworkarea((PLI_BYTE8 *)net_data);
      break;
    case REASON_FINISH:
      net_data = (PLIbook_NetData_p)tf_getworkarea();
      acc_free(net_data->net_array);
      break;
  }
  acc_close();
  return(0);
}
```

```
/******************************************************************
 * calltf routine
 ******************************************************************/
int PLIbook_DisplayNets_calltf(int user_data, int reason)
{
  PLIbook_NetData_p  net_data;
  int i;

  acc_initialize();
  acc_configure(accDisplayWarnings, "true");

  net_data = (PLIbook_NetData_p)tf_getworkarea();

  io_printf("\nAt time %s, nets in module %s (%s):\n",
            tf_strgettime(),
            acc_fetch_fullname(net_data->module_h),
            acc_fetch_defname(net_data->module_h));
  for (i=0; i<net_data->net_count; i++) {
    io_printf("  %-13s  value is  %s (hex)\n",
              acc_fetch_name(net_data->net_array[i]),
              acc_fetch_value(net_data->net_array[i], "%h", null));
  }

  acc_close();
  return(0);
}
```

15.14 Obtaining object handles using an object's name

The ACC library contains two routines which can obtain a handle for an object, using the name of the object as the reference point.

 TIP   Obtaining an object handle from the name of the object is an expensive process for simulation performance, and should be used judiciously. It is much more efficient to obtain a handle for an object based on its relationship to some other object.

handle **acc_handle_by_name (**
    PLI_BYTE8 *** obj_name,**   /* name of an object */
    handle   **scope)**   /* handle for a scope, or **null** */

The `acc_handle_by_name()` routine is used to obtain a handle for an object, using the name of the object. The routine requires two inputs, a string containing the object's name, and a handle for a Verilog hierarchy scope. A scope in the Verilog hierarchy can be a top-level module, a module instance, a named statement group, a Verilog task or a Verilog function. The scope argument can also be set to `null`, which represents the top level of Verilog hierarchy.

The name provided can be the local name of the object, a relative hierarchical path name, or a full hierarchical path name. The PLI will search for the object, using the same search rules as the Verilog language. The full search rules are outside the scope of this book, but briefly, Verilog searches for a name in the hierarchy scope specified by the scope handle first, then as a relative path, then as a full path. If the object cannot be found, then `acc_handle_by_name()` will return `null`.

The objects for which `acc_handle_by_name()` routine can obtain a handle are listed in Table 15-4.

Table 15-4: Objects supported by acc_handle_by_name()

Modules	Parameters
Primitives	Specparams
Nets	Named blocks
Regs	Verilog HDL tasks
Integer, time and real variables	Verilog HDL functions
Named events	

The ACC library provides a second routine which obtains the handle for an object using the object's name.

handle **acc_handle_object (**
 PLI_BYTE8 *** obj_name)**   /* name of an object */

`acc_handle_object()` is similar to `acc_handle_by_name()`, but differs in where it will search for the object. The `acc_handle_object()` routine will begin searching for the object in the design hierarchy scope in which the PLI is executing. By default, the PLI scope is the Verilog design hierarchy scope from which the PLI application was called. The `acc_set_scope()` routine can be used to change the PLI scope to another point in the Verilog design hierarchy. The `acc_handle_object()` routine will search for the object, using the search rules of the Verilog language. If the object cannot be found by searching from the PLI scope, then a `null` is returned.

The `acc_handle_object()` routine can obtain the handle for the objects listed in Table 15-5. Note that there are three objects for which handles can be obtained using this routine which cannot be obtained with the `acc_handle_by_name()` routine. These are module ports, module paths and data paths.

Table 15-5: Objects supported by acc_handle_object()

Modules	Named events
Module ports	Parameters
Data paths	Specparams
Primitives	Named blocks
Nets	Verilog HDL tasks
Regs	Verilog HDL functions
Integer, time and real variables	

The C code fragment listed below illustrates obtaining an object handle using the object's name. The name of a primitive instance and a delay value are read from a file, and then the handle for the primitive is obtained, using `acc_handle_by_name()`.

```
char      prim_name[1024];
handle    prim_handle;
double    new_delay;

fscanf(file_p, "%s %f", prim_name, &new_delay);
prim_handle = acc_handle_by_name((PLI_BYTE8 *)prim_name,
                                              null);
  if (prim_handle)
    /* add new delay value to the primitive object */
  else
    /* error: primitive not found */
```

15.14.1 Obtaining handles for ports using a port name

The `acc_handle_object()` routine can obtain handles for module ports. However, there is a potential risk when using port names for which a PLI application must make provisions. In most Verilog designs, a module port will be connected to an internal signal of the same name. For example:

```
module my_chip (in, .out(result));
  input  in;
  output result;
  wire   in, result;
  ...
endmodule
```

In this example, there are two objects with the same name of **in** within the module my_chip, the port and the net. If acc_handle_object() were called to obtain a handle for the name **"in"**, a handle for the internal signal (the loconn) will be returned. On the other hand, if a handle for the name **"out"** is requested, there is only one object with that name, the port, and so a port handle will be returned. Therefore, when obtaining handles for either a signal or a module port using the object's name, the PLI application should always check the type of object for which a handle was returned. If the handle references the wrong object, the appropriate ACC handle or next routine should be used to traverse from the object which was returned to the object desired.

15.14.2 Obtaining handles for module paths and data paths using a name

The acc_handle_object() routine can obtain handles for two objects which do not have names. This is done by creating a derived name for the objects. These objects are module paths and data paths. To create the derived name, the name of the signal connected to the input and the name signal connected to the output of the path are concatenated together, with a dollar sign ($) character between the two names.

In the following example, the derived name of the module path is *"a$b"*.

```
(a *> b) = 2.5;
```

15.14.3 Changing the PLI hierarchy scope

The acc_handle_object() routine will search for an object within a specific hierarchy scope of the Verilog design hierarchy. This routine will use the *PLI scope* as the reference point for where in the Verilog design hierarchy to search for objects.

When a PLI application is called, the PLI scope is the Verilog HDL hierarchy scope from which the PLI application was called. A Verilog hierarchy scope may be: a top-level module, a module instance, a Verilog HDL task, a Verilog HDL function, a named begin—end statement group, or a named fork—join statement group. The acc_set_scope() routine is used to change the PLI scope to another point in the Verilog design hierarchy. The syntax of acc_set_scope() is:

PLI_BYTE8 ***acc_set_scope (**
    handle   **module,**   /* handle for a module, or **null** */
    PLI_BYTE8 *** name)**   /* name of a module, or **null** */

`acc_set_scope()` uses the `acc_configure()` routine to control which arguments are used to describe the target hierarchy scope. The syntax for `acc_configure()` was presented earlier in this chapter, in section 15.2 on page 489.

- `acc_configure(accEnableArgs, "no_acc_set_scope")` configures the `acc_set_scope()` routine to only use the `module` argument. The `name` argument is ignored. This is the default configuration.

  - If a handle is specified in the `module` argument, the PLI scope will be set to the hierarchy scope of the module handle.

  - If the `module` argument is set to `null`, the PLI scope will be set to the first top-level module found in the Verilog design hierarchy.

- `acc_configure(accEnableArgs, "acc_set_scope")` configures the `acc_set_scope()` routine to use the `module` argument as the first choice, and the `name` argument as the second choice. This is the default configuration.

  - If a handle is specified in the `module` argument, the PLI scope will be set to the hierarchy scope of the module handle.

  - If the `module` argument is set to `null`, and a literal string or pointer to a string is specified in the `name` argument, the PLI scope will be set to the hierarchy scope specified in the `name` argument.

  - If the `module` argument is set to `null`, and the `name` argument is `null`, the PLI scope will be set to the first top-level module found in the Verilog design hierarchy.

The return from `acc_set_scope()` is a pointer to a string containing the full hierarchical path name of the new PLI scope. The return will be a null if an error occurred.

15.15 Comparing ACC handles

Handles for Verilog objects can be obtained several different ways. Occasionally, a PLI application might need to test to see if two handles reference the same object.

acc_compare_handles() returns a boolean TRUE if two handles reference the same object, and a FALSE if they do not. An ACC handle is an abstraction used to reference an object within simulation, and all ACC routines use this abstraction to access information about the object. This layer of abstraction allows PLI applications to be portable to any number of Verilog simulators. Since the handle for an object is an abstraction, it is not possible to determine handle equivalence using the C '==' operator.

15.16 Summary

The ACC routines in the PLI standard provide dynamic access to what is happening within a Verilog simulation. This access is achieved using an object oriented method, where most Verilog HDL constructs that can exist in a simulation data structure are treated as objects. The ACC routines use *handles* to reference these objects. A large number of ACC routines are provided to obtain handles for the various types of Verilog HDL objects, and other ACC routines retrieve information about the objects, such as the name of an object or the vector size of a net. This chapter has presented many of the 103 ACC routines in the PLI standard, and has shown how these routines are used. The following chapters include several complete examples of PLI applications which use many of the ACC routines that have been introduced in this chapter.

CHAPTER 16   *Reading and Modifying Values Using ACC Routines*

*T*he ACC routines provide access to the simulation values of Verilog objects. These values include information about the object, logic values and delay values. This access allows a PLI application to both read and modify what is happening during a simulation. This chapter presents how to use the ACC routines which read and modify values.

The concepts presented in this chapter are:

- Reading the type, full type and special-type properties of objects
- Using ACC *fetch* routines
- Reading an object's source code file location
- Reading the simulation invocation commands
- Reading the values of system task and system function arguments
- Reading net, reg, variable and UDP logic values
- Modifying logic values
- Reading delay values
- Modifying delay values
- Reading parameter and specparam values
- Reading specparam attribute values
- Reading and modifying pulse control attribute values

16.1 Using ACC fetch routines

Once a handle for an object has been obtained, a number of different properties for the object can be retrieved. The ACC routines which retrieve properties of an object are referred to as ACC *fetch* routines. These routines are property-specific, so there is a different fetch routine for each type of property which can be accessed.

The object diagrams show what properties can be accessed for each object, and which fetch routine is used to access a property. This information is listed in a table below the diagram of the object. For example, the object diagram for modules contains the following table:

 These ACC object diagrams and property tables are not part of the IEEE 1364 standard. They are the invention of the book's author.

Figure 16-1: Partial ACC object diagram for Verilog modules

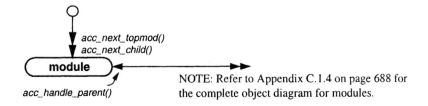

NOTE: Refer to Appendix C.1.4 on page 688 for the complete object diagram for modules.

Related Routines:

acc_fetch_type()	returns **accModule**
acc_fetch_fulltype()	returns **accTopModule, accModuleInstance, accCellInstance**
acc_fetch_name()	returns the instance name of a module
acc_fetch_fullname()	returns the hierarchical path name of a module
acc_fetch_defname()	returns the definition name of a module
acc_fetch_delay_mode()	returns **accDelayModeNone, accDelayModeZero, accDelayModeUnit, accDelayModePath, accDelayModeDistrib, accDelayModeMTM**
acc_fetch_timescale_info()	returns the timescale of a module
acc_fetch_location()	returns source file name & line no. containing module instance

The ACC fetch routines are property-type specific, so there is a different ACC routines for most types of Verilog properties. There are 33 ACC fetch routines:

```
acc_fetch_argc()                    acc_fetch_location()
acc_fetch_argv()                    acc_fetch_name()
acc_fetch_attribute()               acc_fetch_paramtype()
acc_fetch_attribute_int()           acc_fetch_paramval()
acc_fetch_attribute_str()           acc_fetch_polarity()
acc_fetch_defname()                 acc_fetch_precision()
acc_fetch_delay_mode()              acc_fetch_pulsere()
acc_fetch_delays()                  acc_fetch_range()
acc_fetch_delays()                  acc_fetch_size()
acc_fetch_direction()               acc_fetch_tfarg()
acc_fetch_edge()                    acc_fetch_tfarg_int()
acc_fetch_fullname()                acc_fetch_tfarg_str()
acc_fetch_fulltype()                acc_fetch_timescale_info()
acc_fetch_index()                   acc_fetch_type()
acc_fetch_itfarg()                  acc_fetch_type_str()
acc_fetch_itfarg_int()              acc_fetch_value()
acc_fetch_itfarg_str()
```

Most ACC fetch routines require as an input, a handle for the object for which the property is to be accessed. For example, to retrieve the name of a module, a handle for the module must be provided as an input to acc_fetch_name().

```
handle module_handle;
char *module_name;
/* obtain a handle for a module */
module_name = acc_fetch_name(module_handle);
```

The full syntax for each ACC fetch routine is listed in Appendix C, and this information is not duplicated in this chapter. The objective of this chapter is to show how several of these ACC fetch routines can be used in PLI applications.

16.2 Reading object type properties

Every Verilog object which can be accessed by ACC routines has a *type* property and a *fulltype* property. In addition, certain Verilog objects have a *special-type* property, such as a flag to indicate if a net is scalar (1-bit wide) or vector (multiple bits wide).

These properties identify what Verilog object is referenced by a Verilog handle.

- The *type* property identifies the general type of an object. For example, all net data types have the same type property of accNet.

- The *fulltype* property identifies the specific type of an object. For example, a net data type will have one of the fulltype properties: accWire, accWand, accWor, accTri, accTriand, accTrior, accTrireg, accTri0, accTri1, accSupply0 or accSupply1.

- The *special-type* property identifies special attributes an object might have. A net, for example, can have the special-type properties of accScalar, accVector, accExpandedVector and accCollapsedNet.

The ACC object diagrams for each Verilog object list the name of the type, fulltype and special-type constants for that object. Examples of these diagrams are shown in this chapter, and the complete set of diagrams are contained in Appendix C.

acc_fetch_type() retrieves the *type* property of an object, which identifies the general type of an object. The syntax of this routine is:

PLI_INT32 **acc_fetch_type (**
    handle   **object)**   /* handle for an object */

The type property is an integer constant, such as accModule, accPort, accNet, accPrimitive, etc. This property can be used many different ways. One common usage is to verify that a handle which was obtained references the type of object expected. For example, the *$show_all_nets* application requires that the first task/ function argument be a module instance. The PLI application could verify that the argument is correct, using the following code fragment:

```
handle tfarg_handle;
tfarg_handle = acc_handle_tfarg(1);
if (acc_fetch_type(tfarg_handle) != accModule)
  /* report error that argument is not correct */
```

`acc_fetch_fulltype()` retrieves the *fulltype* property of an object. The syntax is:

PLI_INT32 **acc_fetch_fulltype (**
   handle   **object)**   /* handle for an object */

The fulltype property provides more detailed information about an object. For example, if a PLI application has obtained a handle for a Verilog net, the constant returned for `acc_fetch_type(net_handle)` is **accNet**, while the constants returned for `acc_fetch_fulltype(net_handle)` include **accWire, accWor, accWand,** etc. For a Verilog module, `acc_fetch_type(module_handle)` returns the constant **accModule,** and `acc_fetch_fulltype(module_handle)` returns one of the constants: **accTopModule, accModuleInstance, accCellInstance.**

The type and fulltype properties are represented by constants with integer values. Printing the type of an object directly would print the integer value of the constant, rather than the name of the constant. A useful routine for debugging problems in a PLI application is **acc_fetch_type_str()**. This routine takes a type or fulltype constant value as its input and returns a pointer to a string which contains the actual name of the constant. The syntax of this routine is:

PLI_BYTE8 ***acc_fetch_type_str (**
   PLI_INT32   **type)**   /* any type or fulltype constant */

An example of using the `acc_fetch_type_str()` routine to print the name of a type constant is:

```
PLI_INT32 object_type;
object_type = acc_fetch_type(tfarg_handle);
if (object_type != accModule) {
  tf_error("Tfarg type of %s is illegal.\n",
           acc_fetch_type_str(object_type));
}
```

Testing for special-type properties

There are no ACC routines to fetch the special-type properties of objects. Instead, two ACC routines test to see if an object has a special-type property, and return true if the property exists and false if it does not. These true/false tests can be used with any of the type, fulltype or special-type constants.

PLI_INT32 **acc_object_of_type (**
   handle   **object,**   /* handle for an object */
   PLI_INT32   **type)**   /* type, fulltype or special-type property constant */

The `acc_object_of_type()` routine returns true if an object has a specific type, fulltype or special-type property. For example, the following test could be used to perform different operations if an object handle is referencing a module that is at top-level of the design hierarchy or a cell module at the bottom of the design hierarchy (the properties `accTopModule` and `accCellInstance` are fulltypes).

```
handle mod_handle;

/* add code to get a module handle */

if (acc_object_of_type(mod_handle, accTopModule))
  /* process top-level modules */
else if (acc_object_of_type(mod_handle, accCellInstance))
  /* process cell-level modules */
```

PLI_INT32 **acc_object_in_typelist (**
   handle   **object,**   /* handle for an object */
   PLI_INT32 *** type_list)**   /* pointer to a static array of type, fulltype and special-type
                      property constants */

The `acc_object_in_typelist()` routine returns true if an object has any of a list of type, fulltype or special-type properties. The list of types must declared as a static integer array which contains the type, fulltype or special-type constants. The last element in the array must be 0. The following example tests to see if a net is any type of wired logic, using the fulltype property constants which represent the Verilog net data types:

```
handle net_handle;

static PLI_INT32 valid_types[5] = {accWand, accWor,
                                   accTriand, accTrior, 0};

/* add code to get a net handle */

if (acc_object_in_typelist(net_handle, valid_types))
  /* process wired-logic nets */
```

The special-type properties

Certain Verilog objects have special-type properties which can be useful in a PLI application. For example, Verilog nets have the special-type properties `accScalar` and `accVector`, which indicate if the net is 1-bit wide or multiple bits wide. The special-type properties are represented with the following constants:

- **accScope** indicates that an object has it own hierarchy scope. Objects which can have hierarchy scope are modules, named begin—end statement groups, named fork—join statement groups, Verilog HDL tasks and Verilog HDL functions. Objects which have scope can have local reg, variables and parameters declared.

- **accModPathHasIfnone** indicates that a module path object has an ifnone condition. This Verilog language keyword is used to indicate a default delay for modules with conditional path delays.

- **accScalar** indicates that a net or reg object is scalar (1-bit wide)

- **accVector** indicates that a net or reg object is 2 or more bits wide

- **accExpandedVector** indicates a vector net for which the simulator allows access to individual bits within the vector. Access to all bits of a vector is the default in the Verilog language, and can also be explicitly declared in the Verilog source code, using the scalared keyword. For example:

  ```
  wire scalared [63:0] data_bus;
  ```

- **accUnExpandedVector** indicates a vector net for which the simulator prohibits access to individual bits within the vector. This is a feature which some simulators use to optimize run-time performance. If a simulator has flagged a vector as unexpanded, then some ACC routines will only be able to obtain a handle for the complete vector. Vector only access must be explicitly declared in the Verilog source code, using the vectored keyword. For example:

  ```
  wire vectored [31:0] address_bus;
  ```

 A net will always test true for at least one of the accExpandedVector and accUnExpandedVector properties. These properties are not mutually exclusive, and a simulator can return true for both properties. When both properties are true, the rules for expanded vectors will take precedence.

- **accCollapsedNet** indicates that a net object has been removed from the simulation by collapsing the net onto an equivalent net. The equivalent net is referred to as the simulated net. Net collapsing may or may not occur, depending on the optimizations performed by a simulator. An example of where net collapsing might occur is when two modules are connected together using an intermediate net.

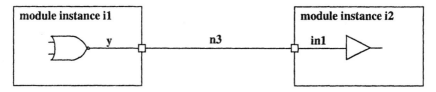

In this example, the nets **y**, **n3** and **in1** are all tied together, effectively becoming the same net. A Verilog simulator might maintain all three nets in its simulation data structure, or it might collapse two of the nets into the third net in order to optimize the simulation performance. When net collapsing occurs, handles for the collapsed nets can still be obtained (using `acc_next_net()`, for example). However, the collapsed nets may not reflect the actual logic value of the design if the simulator is only propagating value changes on the equivalent simulated net. If a net has been collapsed onto another net, the routine **`acc_handle_simulated_net()`** will retrieve the handle for the equivalent net.

The following example uses the `acc_object_of_type()` routine to determine if a net is a scalar net or a vector net, so it can process scalar nets differently than vector nets:

```
handle net_handle;

/* add code to get a net handle */

if (acc_object_of_type(net_handle, accScalar))
  /* process scalar nets */
else if (acc_object_of_type(net_handle, accVector))
  /* process vector nets */
```

16.3 Accessing an object's source code location

The ACC library can access the Verilog HDL source code location for an object.

PLI_INT32 **acc_fetch_location (**

p_location	**location,**	/* pointer to an application-allocated *s_location* structure to receive the location */
handle	**object)**	/* handle for an object */

`acc_fetch_location()` is passed an object handle and a pointer to an **s_location** structure. The Verilog HDL source code location for that object is then retrieved into the structure fields by the simulator. The s_location structure us defined in acc_user.h, as follows:

```
typedef struct t_location {
  PLI_INT32  line_no;
  PLI_BYTE8 *filename;
} s_location, *p_location;
```

The `acc_fetch_location()` routine retrieves the file name into the PLI string buffer, and places a pointer to the string in the `s_location` structure. The string is stored in the temporary ACC string buffer, and should be used immediately, or copied into application-allocated storage.

Not all Verilog objects are supported with `acc_fetch_location()`. Table 16-1 lists the types of objects for which the source code location can be accessed, and what location will be returned.

Table 16-1: Objects supported with `acc_fetch_location()`

Object Type	Source Location Returned
Modules	Module instantiation line
Module ports	Module definition
Module paths	Module path line
Data paths	Module path line
Primitives	Instantiation line
Explicit nets	Definition line
Implicit nets	Line where first used
Regs	Definition line
Integer, time and real variables	Definition line
Named events	Definition line
Parameters	Definition line
Specparams	Definition line
Named blocks	Definition line
Verilog HDL tasks	Definition line
Verilog HDL functions	Definition line

The following code fragment illustrates using `acc_fetch_location()` to print the file name and location of an object.

```
handle obj_h;
s_location source_location;

/* get handle for some object */

acc_fetch_location(&source_location, obj_h);
io_printf ("%s is defined in file %s at line %d.\n",
          acc_fetch_fullname(obj_h),
          source_location->filename,
          source_location->line_no);
}
```

16.4 Reading the simulation invocation commands

The ACC library provides a means for PLI application developers to create user-defined invocation options. This capability makes it possible to configure PLI applications or to pass data to an application from the invocation command line of a Verilog simulator. Two ACC routines are used to read the simulation invocation commands:

PLI_INT32 **acc_fetch_argc ( )**

Returns the number of command line arguments given on the command line used to invoke the simulation.

PLI_BYTE8 ****acc_fetch_argv ( )**

Returns a pointer to an array of character string pointers containing the command line arguments used to invoke the simulation.

The `acc_fetch_argc()` routine returns the number of command line arguments given on the command line used to invoke a Verilog simulator.

The `acc_fetch_argv()` routine returns a pointer to an array of string pointers containing the command line arguments used to invoke a Verilog simulator.

The **argc** and **argv** values are the same command line values defined in the C language. `argc` is the number of invocation command arguments, and `argv` is a pointer to an array of strings, where each string is one argument from the command line.

The ability to check the simulator command line options makes it possible to create user-defined invocation options. Two applications for user-defined options are:

- To enable debug messages when debugging a PLI application. A verbose mode invocation option could be used to specify that debug messages should be printed.

- To pass file names or other values from the command line to a PLI application.

Parsing the -f command file invocation option

The IEEE 1364 Verilog standard does not specify any invocation options for Verilog simulators. Every Verilog simulator, however, has adopted a small number of de facto standard invocation options. One of these is the **-f** option. This invocation option specifies that the file name which follows the option contains additional command line invocation arguments.

When the `argv` value is -f, the next `argv` will be a pointer to a NULL terminated array of pointers to strings. Element 0 in the array will contain the name of the file specified with the -f option, and the remaining elements in the array will be invocation commands contained in the file. Comments are not included.

For example, assume that a command file named *run.f* contained the following:

```
my_chip.v
my_test.v
+my_debug
```

If simulation were invoked with the command:

```
verilog -f run.f -s
```

Then `argv` would point to the following array of strings:

```
argv[0] -> "verilog"
    [1] -> "-f"
    [2] -------> [0] -> "run.f"
                 [1] -> "my_chip.v"
                 [2] -> "my_test"
                 [3] -> "+my_debug"
                 [5] -> NULL
    [3] -> "-s"
```

Example 16-1 illustrates a PLI application called *$print_invoke_commands*. This application uses `acc_fetch_argc()` and `acc_fetch_argc()` routines to print all command line arguments, including commands from within -f command files.

CD The source code for this example is on the CD accompanying this book.

- Application source file: `Chapter.16/invoke_commands_acc.c`
- Verilog test bench: `Chapter.16/invoke_commands_test.v`
- Expected results log: `Chapter.16/invoke_commands_test.log`

Example 16-1: *$print_invoke_commands* — printing invocation commands

```
#include <stdio.h>       /* ANSI C standard input/output library */

/* prototypes of subroutines used by calltf routine */
void PLIbook_ScanCommandFile(PLI_BYTE8 **arg);

int PLIbook_InvokeCommands_calltf(int user_data, int reason)
{
  int        argc, i;
  PLI_BYTE8 **argv;
```

```
   acc_initialize();
   acc_configure(accDisplayWarnings, "true");

   argc = acc_fetch_argc();
   argv = acc_fetch_argv();

   io_printf("\nSimulation invocation commands:\n");
   for (i=0; i<argc; i++) {
     io_printf("  %s\n", *argv);
     if (strcmp(*argv, "-f") == 0) {
       argv++;   /* next arg is address to array of strings */
       i++;
       PLIbook_ScanCommandFile((PLI_BYTE8 **)*argv);
     }
     argv++; /* increment to next argument */
   }
   io_printf("\n\n");
   acc_close();
   return(0);
}

int PLIbook_indent = 0;   /* global variable to format text indenting */

void PLIbook_ScanCommandFile(PLI_BYTE8 **arg)
{
   int i;

   PLIbook_indent += 4; /* increase text indentation */
   while ( *arg != NULL ) { /* loop until null termination */
     for (i=0; i<=PLIbook_indent; i++)
       io_printf(" ");
     io_printf("%s\n", *arg);
     if (strcmp(*arg, "-f") == 0) {
       arg++;   /* next arg is address to array of strings */
       PLIbook_ScanCommandFile((PLI_BYTE8 **)*arg);
     }
     arg++;
   }
   PLIbook_indent -= 4; /* decrease text indentation */
   return;
}
```

16.5 Accessing objects in simulation which have logic values

In order to read an object's logic value or write a new value into an object, a handle
for the object must first be obtained. The ACC routines can read the values of several
different types of objects, but can only modify the values of certain object types. The
Verilog objects for which the ACC routines can access logic values are:

- A *parameter* constant (read only)
- A *specparam* constant (read only)
- A *specparam attribute* constant (read only)
- Any *net* data type: scalar, vector, part-selects and bit-selects of vectors
- The *reg* data type: scalar, vector, part-selects and bit-selects of vectors
- The *integer, time* and *real* variable data types
- A *memory* word select
- A literal *integer* value (read only)
- A literal *real* value (read only)
- A *string* value (read only)
- A *function call* (read only)
- A *system function call*
- A *sequential user-defined primitive*

The routines to obtain handles for the various types of Verilog objects were presented in the previous chapter.

16.6 Reading the values of system task/function arguments

Many of the objects which have logic values can be used as arguments to a system task or system function. The ACC routines offer two methods to access the values of these arguments:

- Read the value directly, using specific ACC routines.
- Obtain a handle for the argument, and read or modify the value of the object, using the appropriate ACC routine for the type of object.

This section shows how to use the specific ACC routines which can read the value of a task/function argument directly, and other sections in this chapter show how to use the ACC routines which read the values of different types of objects, using a handle for the object.

16.6.1 System task/function arguments

A user-defined system task or system function can have any number of arguments, including none. The arguments are numbered from left to right, starting with argument number 1. In the following example:

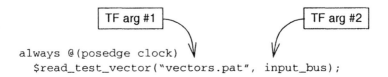

```
always @(posedge clock)
    $read_test_vector("vectors.pat", input_bus);
```

- Task/Function argument number 1 is a string, with the value "vectors.pat".

- Task/Function argument number 2 is a signal, with the name input_bus.

16.6.2 Multiple instances of system tasks and system functions

The Verilog HDL source code can reference the same system task or system function any number of times. For example:

```
always @(posedge clock)
    $read_test_vector("A.dat", data_bus);

always @(negedge clock)
    $read_test_vector("B.dat", data_bus);
```

Just as a Verilog module can be used, or *"instantiated"*, many times in a design, every occurrence of a system task/function is a separate and unique *instance*. Each instance of *$read_test_vector* in the above example has different arguments. The PLI recognizes that each instance is unique, and keeps track of each instance. Therefore, at each positive edge of clock, the *calltf routine* associated with one instance of *$read_test_vector* will be invoked, and at the negative edge of clock, the *calltf routine* associated with a different instance of *$read_test_vector* will be executed. It is important to understand that the Verilog simulator will call the same C functions for each instance of the system task/function, but the inputs and data associated with each call will be unique for each instance.

As an example, *$read_test_vector* might have a *calltf routine* named *gvCall()* associated with it. When the instance of *gvCall()* that is invoked at the positive edge of clock reads the value of task/function argument 1, it will see the string "A.dat". When the instance of *gvCall()* that is invoked at the negative edge of clock read task/function argument number 1, it will see the string "B.dat".

Reading the values of system task/function arguments

The ACC library provides three specific routines to directly read the values of system task/function arguments:

double **acc_fetch_tfarg (**
   PLI_INT32　**n**)   /* index number of a PLI system task/function arg */

PLI_INT32 **acc_fetch_tfarg_int (**
   PLI_INT32　**n**)   /* index number of a PLI system task/function arg */

PLI_BYTE8 ***acc_fetch_tfarg_str (**
   PLI_INT32　**n**)   /* index number of a PLI system task/function arg */

The `acc_fetch_tfarg()` routine returns the value of a task/function argument as a C double. The `acc_fetch_tfarg_int()` routine returns the value of a task/function argument as a 32-bit C integer.

The `acc_fetch_tfarg_str()` routine converts the value of the argument to a string, which is stored in the temporary ACC string buffer. The routine returns a pointer to the string. The primary purpose of `acc_fetch_tfarg_str()` is to read strings from the Verilog language into strings in the C language. Strings are stored differently in Verilog than in C, so this routine converts the string from one format to the other. If the value of the task/function argument is not a string, each 8 bits of the value are converted into an ASCII character, and the value is retrieved as a string.

The *$read_test_vector* system task can be used to illustrate how these ACC routines are used. An example of this system task is:

```
$read_test_vector("vectors.pat", input_bus);
```

The following C code fragment reads the values of the two system task/function arguments. The first argument is read as a string, and the second as an integer.

```
PLI_BYTE8  *file_name;
PLI_INT32  bus_value;

file_name = acc_fetch_tfarg_str(1);
bus_value = acc_fetch_tfarg_int(2);
```

16.6.3 Testing for errors when reading system task/function arguments

An error can occur when reading the value of a system task/function argument if the index number specified is out of range, or if the argument does not have a logic value. The acc_fetch_tfarg() routine will return an exception value of **0.0** if an error occurs. The acc_fetch_tfarg_int() routine will return an exception value of **0**. The acc_fetch_tfarg_str() routine will return an exception value of **null**.

Since the exception values of 0.0 and 0 could be legitimate values, it may not be possible to determine that an error occurred, based on the return value. The **acc_error_flag** will be set if an error occurs, and this flag can be used to test for errors. The acc_error_flag was presented in Chapter 15, section 15.2.2 on page 491.

16.6.4 Instance specific system task/function routines

A PLI *calltf routine*, *misctf routine*, *checktf routine* or *sizetf routine* is directly associated with the name of a system task or system function. This association is part of the PLI interface mechanism, which was presented in Chapter 9. Each of these routines can access the arguments of the system task/function which caused the routine to be called, using acc_handle_tfarg(), acc_handle_tfarg_int() or acc_handle_tfarg_str(). If these routines call another C function, that function can also access the arguments of the system task/function.

The ACC library also provides for another type of PLI routine, called a *consumer routine*. Defining and using *consumer routines* is defined in the next chapter of this book. A *consumer routine* is **not** associated with the name of a system task or system function. Therefore, a *consumer routine* cannot directly access the arguments of system tasks/functions. The PLI also provides a means for PLI applications to indirectly access system task/function arguments. Indirect access is done by obtaining a handle for an instance of a system task/function, and then using a specific set of ACC routines to access the arguments of that system task/function instance.

An instance of a system task/function is an object, and the ACC routines can obtain a handle for that object using, the **acc_handle_tfinst()** routine. The syntax of this routine is:

handle **acc_handle_tfinst ( )**

The acc_handle_tfinst() routine returns a handle for the system task/function which called the PLI application.

There are instance-specific counterparts to the routines which read the values of task/function arguments, or which obtain a handle for an argument:

double **acc_fetch_itfarg (**
PLI_INT32	**n,**	/* index number of a PLI system task/function arg */
handle	**tfinst)**	/* handle for an instance of a PLI system task/function */

PLI_INT32 **acc_fetch_itfarg_int (**
PLI_INT32	**n,**	/* index number of a PLI system task/function arg */
handle	**tfinst)**	/* handle for an instance of a PLI system task/function */

PLI_BYTE8 ***acc_fetch_itfarg_str (**
PLI_INT32	**n,**	/* index number of a PLI system task/function arg */
handle	**tfinst)**	/* handle for an instance of a PLI system task/function */

handle **acc_handle_itfarg (**
PLI_INT32	**n,**	/* position number of a system task/function argument */
handle	**tfinst)**	/* handle for an instance of a system task/function */

Each of these routines require two inputs, the index number of a task/function argument, and a handle for a task/function instance. The typical usage of these routines is for a *calltf routine* or a *misctf routine* to obtain the handle for the system task/function instance which called the routine. This handle is then saved in the user_data field of a *consumer routine*. This gives the consumer routine access to the instance handle, so that the consumer routine can access the arguments of the system task/function. Example 18-6 on page 633 of chapter 18 shows an examples of using these instance specific ACC routines.

 The TF system task/function instance pointer is not the same as the ACC system task/function instance handle. The TF library has instance specific versions of the routines which access the arguments of a system task or system function. However, the instance pointer returned from tf_getinstance() may be different than the ACC handle returned from acc_handle_tfinst(). The instance pointer from the TF library should not be used with ACC routines. Conversely, the instance handle from the ACC library should not be used with TF routines.

16.6.5 Modifying the values of system task/function arguments

The values of system task/function arguments can be modified by obtaining a handle for the argument. The acc_handle_tfarg() routine is used to obtain the argument handle. Note that not all legal arguments to a system task/function have logic values which can be modified. For example, a literal string or a module instance name are valid arguments, but have fixed values which cannot be modified by the PLI. The routines acc_fetch_type() and acc_fetch_fulltype() can be used to determine the type of object which is being used as a task/function argument. The subsequent sections of this chapter present how to read and modify values of different object types.

16.7 Reading object logic values

Using the handle for an object, a PLI application can read the logic value property of the object. The same ACC routine is used to read the logic value of any type of object and any data type of logic value.

16.7.1 Working with a 4-logic value, multiple strength level system

The Verilog HDL supports 4 logic values, **0**, **1**, **z** and **x** and multiple levels of signal strength. There are also two ambiguous logic values, represented by **L** (low) and **H** (high), and many ambiguous strength values. The C programming language does not directly represent the same information. The ACC routine to read logic values provides several ways to automatically translate values between Verilog and C. The translation converts Verilog 4-state logic into the following C types:

- *A 32-bit C integer*: Verilog scalar and vector logic values are converted to a single 32-bit C integer. The Verilog 4-state logic is converted to 2-state logic values of 0 and 1. Logic values of z and x values are converted to 0, and strength levels are ignored.

- *A C double*: Verilog real number values, scalar values and vector values are converted to a C double precision value. Real numbers in Verilog are 2-state decimal values, which convert directly to C doubles. Scalar and vector 4-state logic is converted to 2-state logic, and strength levels are ignored.

- *A C string*: Verilog scalar and vector logic values are converted to a C character string. The Verilog 4-state logic is converted to the letters "0", "1", "z" and "x". Logic strength levels for scalar nets are represented using the 3 character mnemonics defined in the Verilog language. Strength levels are ignored for Verilog vectors and other data types.

- *A C constant*: Verilog scalar values are converted to a C integer constants. The Verilog 4-state logic is converted to the constants **acc0**, **acc1**, **accZ** and **accX**. Logic strength levels are ignored.

- *A C aval/bval structure*: Verilog scalar and vector logic values are converted to a C structure which encodes each bit of a Verilog 4-state value to a pair of bits in C, referred to as an *aval/bval* pair. An array of aval/bval pairs is used to encode Verilog vectors of any size. The logic strength levels are ignored.

16.7.2 The acc_fetch_value() routine

A single ACC routine is used to read the logic value of most Verilog objects.

PLI_BYTE8 *acc_fetch_value (
handle	**object,**	/* handle for a net, reg or variable */
PLI_BYTE8 * **format_str,**		/* character string controlling the radix of the retrieved value: "%b", "%o", "%d", "%h", "%v" or "%%" */
p_acc_value **value)**		/* pointer to an application-allocated s_acc_value structure to receive the value; only used if format_str is "%%" */

The acc_fetch_value() routine converts Verilog logic values into C language representations. The logic value that is retrieved can be passed to the PLI application in one of two ways:

- The retrieved value can be saved in the temporary ACC string buffer, and a pointer to the string is returned by acc_fetch_value().

- The retrieved value can be stored in an **s_acc_value** structure allocated by the PLI application. The structure can receive the Verilog logic value in a variety of C data types.

Retrieving values as character strings

Retrieving values as a C string is the easiest way to represent any size of Verilog vector and any Verilog logic value. This format is also an easy way to retrieve values using acc_fetch_value(). All that is needed is to obtain an object handle and call acc_fetch_value(), saving the string pointer that is returned in a PLI_BYTE8 * variable, if desired.

TIP

The format in which a value is read can have an impact on the run-time performance of a PLI application. The fastest run-time performance will be achieved when a value is retrieved in a format closest to the format in which a value is saved in the simulation structure. For Verilog scalar and vector nets and regs, this format is the aval/bval pair. For Verilog integers, 32-bit C integers are most efficient, and for Verilog reals, C doubles are most efficient. The least efficient method for run-time performance is to retrieve a logic value as a C string.

The acc_fetch_value() routine has three inputs, but only the first two are used when retrieving values as a string. The third input should be set to null. It is not used when retrieving values as C strings.

The first input to acc_fetch_value() is a handle for the object from which the logic value is to be read. The second input to a format·string, which controls how the

Verilog value will be represented in the C string. Table 16-2 shows the formats which are supported for returning string values:

Table 16-2: acc_set_value() format strings

Format String	Return Value Description
"%b"	value is retrieved as a C string with a binary representation ('0', '1', 'z', 'x')
"%o"	value is retrieved as a C string with an octal representation ('0' through '7', 'z', 'Z', 'x', 'X')
"%d"	value is retrieved as a C string with a decimal representation ('0' through '9', 'z', 'Z', 'x', 'X')
"%h"	value is retrieved as a C string with a hexadecimal representation ('0' through 'F', 'z', 'Z', 'x', 'X')
"%v"	value is retrieved as a C string with a 3-character strength representation, such as st0, we1 or 65X
"%%"	value is retrieved into an **s_acc_value** structure, instead of a C string (see section 16.7.7 on page 559)

The formats which return the logic value as a string ("%b", "%o", "%d" and "%h") use the same representation as the Verilog language built-in *$display* system task.

The following example illustrates using acc_set_value() to return the logic values of all nets in a module, represented in a binary notation.

> **CD** The source code for this example is on the CD accompanying this book.
>
> • Application source file: Chapter.16/list_nets_acc.c
> • Verilog test bench: Chapter.16/list_nets_test.v
> • Expected results log: Chapter.16/list_nets_test.log

Example 16-2: *$list_nets* — using acc_fetch_value() to read values as strings

```
int PLIbook_ListNets_calltf(int user_data, int reason)
{
  handle module_h, net_h;
  acc_initialize();
  module_h = acc_handle_tfarg(1);
  io_printf("\nNet values in module %s:\n",
          acc_fetch_fullname(module_h));
```

```
net_h = null;     /* start with known value for target handle */
while (net_h = acc_next_net(module_h, net_h)) {
  io_printf("   %-10s:   %s\n",
            acc_fetch_name(net_h),
            acc_fetch_value(net_h, "%b", null));
}
acc_close();
return(0);
}
```

 The value retrieved from acc_fetch_value() for any of the C string
representations will be stored in the ACC string buffer. A pointer to the string is
returned by acc_fetch_value(). The ACC string buffer storage is temporary.
Therefore, the value should be used immediately, or copied into application-
allocated memory.

Retrieving values into an s_acc_value structure

acc_fetch_value() can convert Verilog logic values into several C data types.
This is done by setting the format string to "%%". The value is then retrieved into an
s_acc_value structure. The PLI application must first allocate this structure, and
pass a pointer to the structure as the third input to acc_fetch_value(). The return
value from acc_fetch_value() is not used, and should be ignored.

The **s_acc_value** structure is defined in acc_user.h, and is listed below.

```
typedef struct t_setval_value {
  PLI_INT32       format;   /* accBinStrVal,   accOctStrVal,
                               accDecStrVal,   accHexStrVal,
                               accScalarVal,   accIntVal,
                               accRealVal,     accStringVal,
                               accVectorVal */
  union {
    PLI_BYTE8   *str;     /* used with string-based formats */
    PLI_INT32   scalar;   /* used with accScalarVal format */
    PLI_INT32   integer;  /* used with accIntVal format */
    double      real;     /* used with accRealVal format */
    p_acc_vecval vector;  /* used with accVectorVal format */
  } value;
} s_setval_value, *p_setval_value,
  s_acc_value, *p_acc_value;
```

This `s_acc_value` structure has two fields, **format** and **value**. The `format` field controls what C language data type should be used to receive the value. The `value` field is a union of C data types which receive the value of the object. The definitions of the format values and the value union fields are listed in the following table.

Table 16-3: The `s_acc_value` structure format constants

Format	Definition
`accBinStrVal`	retrieves the Verilog logic value as a C string, using binary numbers, with 4-state logic
`accOctStrVal`	retrieves the Verilog logic value as a C string, using octal numbers, with 4-state logic
`accDecStrVal`	retrieves the Verilog logic value as a C string, using decimal numbers, with 4-state logic
`accHexStrVal`	retrieves the Verilog logic value as a C string, using hexadecimal numbers, with 4-state logic
`accScalarVal`	retrieves the Verilog scalar logic value as one of the C constants acc0, acc1, accZ or accX
`accIntVal`	retrieves the Verilog logic value as a 32-bit C integer, with 2-state logic
`accRealVal`	retrieves the Verilog logic value as a C double precision number, with 2-state logic
`accStringVal`	retrieves the Verilog string value as a C string
`accVectorVal`	retrieves the Verilog vector value as an array of 32-bit C integers, encoded to represent Verilog 4-state logic

The `value` field of the `s_acc_value` structure is a union of C data types. Which C data type is used to receive the Verilog value is controlled by the format field of the structure:

- The **value.str** field in the value union is used if the format is `accBinStrVal`, `accOctStrVal`, `accDecStrVal`, `accHexStrVal`, or `accStringVal`. The `acc_fetch_value()` routine will fetch the object's logic value, and convert the value into an ASCII string. The routine will place a pointer to the string in the value.str field. The string is stored in the temporary ACC string buffer, and may be overwritten by other ACC routines which return pointers to strings.

- The **value.scalar** field is used if the format is `accScalarVal`. The `acc_fetch_value()` routine will fetch the object's logic value, and place one of the constants: **acci0, acc1, accZ** or **accX** in the value.scalar field.

- The **value.integer** field is used if the format is accIntVal. The acc_fetch_value() routine will fetch the object's logic value, and convert the value into 2-state logic. The routine will place the value in the value.integer field.

- The **value.real** field is used if the format is accRealVal. The acc_fetch_value() routine will fetch the object's logic value, convert the value into 2-state logic, and place the value in the value.real field.

- The **value.vector** field is used if the format is accVectorVal. The acc_fetch_value() routine will fetch the object's logic value into an array of s_acc_vecval structures, and place a pointer to the memory in the value.vector field of the s_acc_value structure. Refer to section 16.7.7 on page 559, for more details on reading vector values.

Before acc_fetch_value() can be called using the "%%" format, the PLI application must first allocate an s_acc_value structure. Following are three ways in which the structure might be allocated:

- Allocate an automatic variable of the s_acc_value type. The storage allocated will automatically be freed when the PLI application exits:

```
s_acc_value  obj_value;
acc_fetch_value(obj_handle, "%%", &obj_value);
```

- Allocate persistent storage, which can be maintained from one call of the PLI application to another. The pointer to the storage can be preserved in the TF work area.

```
p_acc_value  obj_value;
obj_value = (p_acc_value)malloc(sizeof(s_acc_value));
acc_fetch_value(obj_handle, "%%", obj_value);
```

- Allocate static storage, which can be initialized at the time of allocation. The initialization is not repeated for each call to the PLI application, which can improve run-time efficiency.

```
static s_acc_value  obj_value = {accBinStringVal};
```

 NOTE If static storage is used, all instances of the PLI application will share the same static structure. If one call to the application changes the format field of the structure, it will affect other instances of the PLI application.

16.7.3 Reading 4-state logic as C strings, using acc_fetch_value()

The accBinStrVal, accOctStrVal, accDecStrVal and accHexStrVal formats of the s_acc_value structure are nearly identical to the respective "%b", "%o", "%d", and "%h" string formats that can be specified as an argument to acc_fetch_value(). The only difference is that, when using a "%%" format, the pointer to the string will be placed in the **value.str** field of the s_acc_value structure, instead of being returned by the acc_fetch_value() function. The string will be stored in the temporary ACC string buffer.

16.7.4 Reading 2-state logic as a 32-bit C integer, using acc_fetch_value()

The accIntVal format will retrieve Verilog logic values into a 32-bit C integer. The acc_fetch_value() routine will convert Verilog 4-state logic into C 2-state logic. The Verilog value which is read can be any Verilog data type.

 The maximum value which can be read is constrained by the size of a 32-bit C integer. A Verilog can be a vector of any bit width. If the value stored in a Verilog vector is greater than the maximum value which can be stored as a 32-bit integer, then the left-most bits (the most significant bits) of the Verilog vector are truncated.

The steps to read the value of an object as a 32-bit C integer are:

1. Allocate an s_acc_value structure.

2. Set the format field in the structure to **accIntVal**.

3. Call acc_fetch_value(), giving a pointer to the s_acc_value structure as an input, along with a handle for the object from which to read the logic value.

4. Read the logic value of the object from the **value.integer** field of the s_acc_value structure.

The following C code fragment illustrates reading the logic value of a Verilog integer into a 32-bit C integer using acc_fetch_value().

```
handle tfarg_h;
s_acc_value val_s;

tfarg_h = acc_handle_tfarg(1);

val_s.format = accIntVal;

acc_fetch_value(tfarg_h, "%%", &val_s);

io_printf("%s = %d\n",
          acc_fetch_name(tfarg_h),
          val_s.value.integer);
```

16.7.5 Reading 2-state logic as a C double, using acc_fetch_value()

The `accRealVal` format will retrieve Verilog real values into a C double. The `acc_fetch_value()` routine will convert any other Verilog value into a 2-state integer value, and then to a C double.

The steps to read the value of an object as a C double are:

1. Allocate an `s_acc_value` structure.

2. Set the `format` field in the structure to **accRealVal**.

3. Call `acc_fetch_value()`, giving a pointer to the `s_acc_value` structure as an input, along with a handle for the object from which to read the logic value.

4. Read the logic value of the object from the **value.real** field of the `s_acc_value` structure.

16.7.6 Reading Verilog string values into C strings

The `acc_fetch_value()` routine will automatically convert a Verilog string value to a C string, and place a pointer to the string in the **value.str** field of the `s_acc_value` structure. The intended use of the `accStringVal` format is to read Verilog string values. If the value of the object is not a Verilog string, then each 8 bits of the Verilog value will be converted to an ASCII character.

The steps to read the value of an object as a C double are:

1. Allocate an `s_acc_value` structure.

2. Set the `format` field in the structure to **accStringVal**.

3. Call `acc_fetch_value()`, giving a pointer to the `s_acc_value` structure as an input, along with a handle for the object from which to read the logic value.

4. Read the string from the **value.str** field of the `s_acc_value` structure.

 The string is stored in the ACC string buffer, and a pointer to the string is placed in the `value.str` field. The ACC string buffer storage is temporary. Therefore, the value should be used immediately, or copied into application-allocated memory.

16.7.7 Reading Verilog 4-state logic vectors as encoded aval/bval pairs

The `acc_fetch_value()` routine with an **accVectorVal** format will retrieve an object's 4-state logic value as an encoded pair of 32-bit C integers. The encoding uses an *aval/bval* pair of 32-bit integers to represent the 4-state logic values of Verilog. One bit of each aval/bval pair represents a corresponding bit of the Verilog logic value. The TF and VPI libraries use the same encoding, as shown below:

Table 16-4: aval/bval logic value encoding

aval/bval pair	Verilog logic value represented
0/0	0
1/0	1
0/1	Z
1/1	X

The ACC library defines C integers which are 32-bits wide, and therefore uses the aval/bval pair to encode up to 32 bits of a Verilog vector. By using an array of aval/bval integer pairs, vector lengths of any size may be represented. The representation of a 40-bit vector in Verilog can be visualized as:

For the Verilog declaration: `reg [39:0] data;`

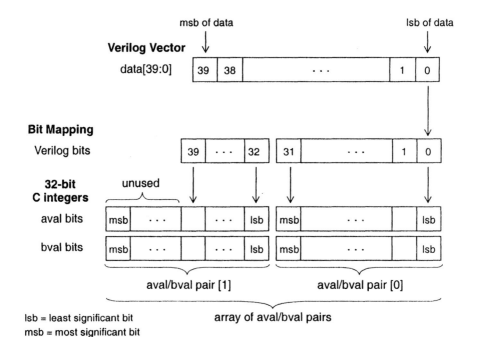

The Verilog language supports any numbering convention for a vector's bit numbers. The least significant bit of the Verilog vector can be the smallest bit number, such as bit 0 (which is referred to as little endian convention). Or, the least significant bit of the Verilog vector can be the largest bit number, such as bit 39 (which is referred to as big endian convention). Verilog does not require that there be a bit zero at all. Each of the following examples are valid vector declarations in Verilog:

```
reg [39:0] data;    /* little endian -- LSB is bit 0  */
reg [0:39] data2;   /* big endian    -- LSB is bit 39 */
reg [40:1] data3;   /* little endian -- LSB is bit 1  */
```

The bit numbering used in Verilog does not affect the aval/bval representation of the Verilog vector. In the array of aval/bval pairs, the LSB of the Verilog vector will always be the LSB of the first 32-bit C integer in the array, and the MSB of the Verilog vector will always be the last bit in the array which is used. The following diagram illustrates the aval/bval array for a Verilog vector declared with a big endian convention.

For the Verilog declaration: **reg [1:40] data;**

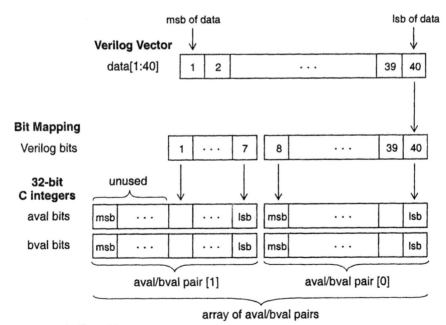

lsb = least significant bit
msb = most significant bit

The aval/bval pair is declared in an **s_acc_vecval** structure, which is defined in the acc_user.h file. The structure definition is:

```
typedef struct t_acc_vecval {
  PLI_INT32  aval;          /* bit encoding: aval/bval:
  PLI_INT32  bval;              0/0==0, 1/0==1, 1/1==X, 0/1==Z */
} s_acc_vecval, *p_acc_vecval;
```

To read a task/function argument's 4-state logic value using acc_fetch_value() involves four basic steps:

1. Allocate an s_acc_value structure.

2. Allocate memory for an array of s_acc_vecval structures. The array must have 1 element for each 32 bits of the Verilog vector.

3. Set the format field in the structure to accVectorVal.

4. Set the value.vector field in the structure to the pointer to the s_acc_vecval array.

5. Call acc_fetch_value(), passing a pointer to the s_acc_value structure as an input, along with a handle for the object from which to read the logic value.

6. Read the logic value of the object from the value.vector field of the s_acc_value structure.

To allocate memory for the s_acc_vecval array requires first determining the number of elements that will be needed in the array. Since there will be one element for each 32 bits of a Verilog vector, the number of array elements can be calculated using:

```
number_of_array_elements = ((vector_size - 1) / 32 + 1);
```

The vector size of the Verilog vector object can be retrieved using the acc_fetch_size() routine. For example:

```
vector_size = acc_fetch_size(vector_handle);
```

Once the number of elements are known, the value of each 32-bit group of the Verilog vector can be accessed by reading the aval/bval pair of each s_acc_vecval structure in the array. Within each aval/bval pair, an individual bit of the Verilog vector can be accessed by masking out the other bits in the aval/bval pair.

When reading values using the accVectorVal format, the memory to store the array of s_acc_vecval structures must be allocated and maintained by the PLI application. If the PLI application needs to preserve the array for future calls to the application, a pointer to the array must be saved. One way to preserve the pointer is to allocate persistent memory for the s_acc_value structure, and save a pointer to this structure in the TF work area. The pointer to the s_acc_vecval array can then be saved in the value.vector field of the s_acc_value structure.

Example 16-3 illustrates using acc_fetch_value() to read the aval/bval encoded 4-state value of a vector. This example assumes the vector is passed to the PLI application as the first system task/function argument. The value of the vector is then printed one bit at a time. For simplicity in this example, it is assumed that the least-significant bit of the Verilog vector is bit 0.

> **CD** The source code for this example is on the CD accompanying this book.
>
> • Application source file: Chapter.16/read_vecval_acc.c
> • Verilog test bench: Chapter.16/read_vecval_test.v
> • Expected results log: Chapter.16/read_vecval_test.log

Example 16-3: *$read_vector_value* — reading vector values as aval/bval pairs

```
int PLIbook_ReadVecVal_calltf(int user_data, int reason)
{
  handle      vector_h;
  s_acc_value vector_val;    /* structure to receive vector value */
  int         i, vector_size, array_size, avalbit, bvalbit, bit_num;
  char        vlogval;

  vector_h = acc_handle_tfarg(1);

  vector_size = acc_fetch_size(vector_h);    /* determine number of...*/
  array_size  = ((vector_size-1) / 32 + 1); /* ...elements in array  */

  vector_val.value.vector =
             (p_acc_vecval)malloc(array_size * sizeof(s_acc_vecval));

  vector_val.format = accVectorVal;          /* set value format field */

  acc_fetch_value(vector_h,"%%",&vector_val); /* read vector's value */

  io_printf("\nVector %s encoded value:\n",
             acc_fetch_name(vector_h));
  for (i=0; i<array_size; i++) {
    /* the following loop assumes the Verilog LSB is bit 0 */
    for (bit_num=0; bit_num<=31; bit_num++) {
      avalbit=PLIbook_getbit(vector_val.value.vector[i].aval, bit_num);
      bvalbit=PLIbook_getbit(vector_val.value.vector[i].bval, bit_num);
      vlogval=PLIbook_get_4state_val(avalbit, bvalbit);
```

```
        io_printf("  bit[%2d]   aval/bval = %d/%d   4-state value = %c\n",
                  (i*32+bit_num), avalbit, bvalbit, vlogval);
        /* quit when reach last bit of Verilog vector */
        if ((i*32+bit_num) == vector_size-1) break;
    }
  }
  return(0);
}

/*************************************************************************
 * Function to determine if a specific bit is set in a 32-bit word.
 * Sets the least-significant bit of a mask value to 1 and shifts the
 * mask left to the desired bit number.
 *************************************************************************/
int PLIbook_getbit(int word, int bit_num)
{
  int mask;
  mask = 0x00000001 << bit_num;
  return((word & mask)? TRUE: FALSE);
}

/*************************************************************************
 * Function to convert aval/bval encoding to 4-state logic represented
 * as a C character.
 *************************************************************************/
char PLIbook_get_4state_val(int aval, int bval)
{
  if        (!bval && !aval) return('0');
  else if (!bval &&  aval) return('1');
  else if ( bval && !aval) return('z');
  else                      return('x');
}
```

16.8 Writing values into Verilog objects

A single ACC routine is used to write values onto any type of object which the ACC supports modifying values.

PLI_INT32 **acc_set_value (**

handle	**object,**	/* handle for a reg, variable, system function call, sequential UDP or net */
p_setval_value	**value,**	/* pointer to an application-allocated s_setval_value structure containing the value to be set */
p_setval_delay	**delay)**	/* pointer to an application-allocated s_setval_delay structure containing a propagation delay value */

The `acc_set_value()` routine converts a value represented as a C data type into Verilog 4-state logic, and writes the value into a Verilog object. The value to be written can be represented a variety of ways in the C language. These representations are the same as was described in section 16.7.1 on page 552 for reading values.

There are three inputs to `acc_set_value()`:

1. A handle for the object into which the value is to be written.

2. A pointer to an **s_setval_value** structure. This structure must be allocated by the PLI application, and the appropriate field within the structure set to the value to be written into the object.

3. A pointer to an **s_setval_delay** structure. This structure is allocated by the PLI application, and set to a propagation delay value.

The `s_setval_delay` structure is used to specify when the simulator should apply the value which is being written by the PLI application. The structure is defined in acc_user.h, as follows:

```
typedef struct t_setval_delay {
  s_acc_time time;
  PLI_INT32  model;         /* accNoDelay, accInertialDelay,
                               accPureTransportDelay,
                               accTransportDelay, accAssignFlag,
                               accDeassignFlag, accForceFlag,
                               accReleaseFlag */
} s_setval_delay, *p_setval_delay;
```

The **model** field in the `s_setval_delay` structure controls how the value to be written will propagate within the simulation. The `acc_set_value()` routine can either write the value into the object immediately, or use the simulator's event scheduling mechanism. By using the event scheduler, a value can be scheduled to occur in a future simulation time step.

The model field is set to one of the following constants:

- **accNoDelay** indicates no propagation delay is to be used. The object may be a Verilog reg, variable, memory word, sequential UDP or system function. When this flag is used, the time field in the `s_setval_delay` structure is not used, and can be set to null.

- **accInertialDelay** indicates that inertial delay propagation is to be used. Any pending events which are scheduled for the object are cancelled. The object may be a Verilog reg, variable, or memory word.

- **accPureTransportDelay** indicates transport delay propagation is to be used. Any pending events which are scheduled for the object remain scheduled (no events are cancelled). The object may be a Verilog reg, variable, memory word or variable array word.

- **accTransportDelay** indicates a modified transport delay propagation is to be used. Any pending events for the object which are scheduled at a later time than this new event are cancelled. The object must be a Verilog reg, variable, memory word or variable array word.

- **accAssignFlag** indicates that the value is to be continuously assigned into the object, overriding any existing values. The value is written into the object in a similar manner as the Verilog HDL procedural `assign` statement, with one important difference; the value assigned is not re-evaluated when anything in the right-hand side expression changes, as it would be in a Verilog HDL procedural continuous assignment. No propagation delay is used, and the `time` field of the `s_setval_delay` structure can be set to `null`. The object may be a Verilog reg or variable. Only one procedural continuous assign value may exist for an object at a time. Setting another assign value on an object will replace any existing procedural continuous assign value, regardless of whether the assign was set within the PLI or within the Verilog HDL.

- **accDeassignFlag** indicates that any existing procedural continuous assign on the object is to be deassigned. This is the same functionality as the Verilog HDL procedural `deassign` statement.

- **accForceFlag** indicates that the value is to be forced into the object, overriding any existing values. The value is written into the object in the similar manner as the Verilog HDL procedural `force` statement, with one important difference; the value assigned is not re-evaluated when anything in the right-hand side expression changes, as it would be in a Verilog HDL force statement. No propagation delay is used, and the `time` field of the `s_setval_delay` structure may be set to `null`. The object may be a Verilog reg, variable, memory word, variable array word, or net. Only one force value may exist for an object at a time. Setting a force value on an object will replace any existing force value, regardless of whether the force was set within the PLI or within the Verilog HDL.

- **accReleaseFlag** indicates that any existing force on the object is to be released. This is the same functionality as the Verilog HDL procedural `release` statement.

 The IEEE 1364-2001 standard adds multi-dimensional arrays, as well as arrays of net data types and real data types. The ACC library does not support these enhancements. Only the VPI library supports multi-dimensional arrays and arrays of all data types.

The **time** field of the s_setval_delay structure is a pointer to an s_acc_time structure. This structure is used to hold the propagation delay which is to be used by acc_set_value().

The s_acc_time structure is defined in acc_user.h, and is shown below:

```
typedef struct t_acc_time
{
  PLI_INT32 type;        /* accTime, accSimTime, accRealTime */
  PLI_INT32 low, high;   /* if type is accTime or accSimTime*/
  double    real;        /* used if type is accRealTime */
} s_acc_time, *p_acc_time;
```

The **type** field in the s_acc_time structure controls how the time value will be specified. The type must be set to one of the following constants:

- **accRealTime** indicates that the delay is represented as a C double. The delay value will be scaled to the time units and precision of the module containing the object into which the value will be written.

- **accTime** indicates that the delay is represented as a pair of 32-bit C integers, which contain the high order 32 bits and the low order 32 bits of the 64-bit simulation time. The delay value will be scaled to the time units and precision of the module containing the object into which the value will be written.

- **accSimTime** indicates that the delay is represented as a pair of 32-bit C integers, which contain the high order 32 bits and the low order 32 bits of the 64-bit simulation time. The delay value is in the simulator's internal time units, and will *not* be scaled to the object's time scale.

To represent a 64 bit time value, the Verilog PLI uses a pair of C unsigned integers to store the full 64 bits of simulation time. The lower 32 bits of Verilog time are placed in one integer, and the upper 32 bits of the Verilog time are stored in a second integer, as shown in the following illustration:

```
int   high, low;
```

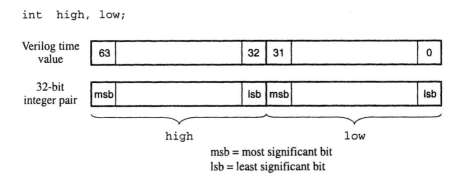

An example of using acc_set_value()

To write a value into an object requires the following steps:

1. Obtain a handle for an object—the object must have a value property, in order to write a value into the object.

2. Allocate an **s_acc_value** structure.

3. Allocate an **s_setval_delay** structure.

4. Allocate an **s_acc_time** structure (if a propagation delay is to be specified).

5. Allocate an array of s_acc_vecval structures, if needed (for the accVectorVal format).

6. Load the value variables or structures with the value to be written.

7. Set the **format** field in the s_acc_value structure to indicate how the logic value is represented in C.

8. Set the appropriate field in the **value** union of the s_acc_value structure to a pointer to the value.

9. Set the **model** field in the s_setval_delay structure to indicate the type of propagation delay which should be used.

10. Set the **time** field of the s_setval_delay structure to a pointer to the s_acc_time structure that was allocated, or null if no delay is to be used.

11. Set the **type** field in the s_acc_time structure to indicate how the delay value is represented.

12. Set the delay value in the appropriate field of the s_acc_time structure.

13. Call acc_set_value(), with pointers to the s_setval_value and s_setval_delay structures.

The following code fragment writes a value represented as a C character string onto the second argument of a system task, using transport delay. The procedure for writing a value represented in other C data types is very similar to this example. The code fragment example assumes a *$read_test_vector* system task, where the second argument is a Verilog reg data type. An example of this system task is:

```
reg [22:0] input_vector;

always @(posedge clock)
  $read_test_vector("vector_file.pat", input_vector);
```

The C code to write a value into the second system task argument is.

```
handle tfarg_h;
char test_vector[1024];
double delay;
s_acc_value value_s;
s_setval_delay delay_s;

/* read test value and delay from file *.

tfarg_h = acc_handle_tfarg(2);

value_s.format    = accStrBinVal;
value_s.value.str = test_vector;
delay_s.model     = accTransportDelay;
delay_s.time.type = accRealTime;
delay_s.time.real = delay;

acc_set_value(tfarg_h, &value_s, &delay_s);
```

No delay versus zero delay

The Verilog PLI standard makes a distinction between putting a value into simulation with no delay and putting a value into simulation with zero delay.

- The **accNoDelay** delay flag indicates that the value will be written into Verilog simulation instantly. When the PLI application returns back to simulation, any values written to an object or system function return using these routines will already be in effect for Verilog HDL statements to use.

- The **accInertialDelay**, **accTransportDelay** and **accPureTransportDelay** flags schedule a value to be written into simulation. If a delay of zero is specified, the value is scheduled to be written into the object later in the current simulation time step. When the system task returns back to simulation, the scheduled value will not yet have taken effect. Other Verilog HDL statements scheduled to be executed in the same simulation time step may or may not see the new value of the object (depending on where the value change which was scheduled by the PLI falls in the simulator's event queue, in relation to other Verilog HDL events).

The following simple Verilog HDL source code illustrates the potential problem of putting a value into simulation using a delay of zero.

```
module test;
  reg [7:0] reg1, reg2;
  initial
    begin
      reg1 = 0; reg2 = 0;
      $put_value(reg1, reg2);
      $display("reg1=%d   reg2=%d", reg1, reg2);
      $strobe ("reg1=%d   reg2=%d", reg1, reg2);
```

```
        #1 $finish;
    end
endmodule
```

If the *calltf routine* for *$put_value* puts a value into reg1 using **accNoDelay**, then when *$put_value* returns to the simulation, and the *$display* statement prints the value of reg1, the <u>new</u> value will be printed.

If, however, the *calltf routine* for *$put_value* writes a value into reg2 using **accInertialDelay** with a delay of zero, then, when *$put_value* returns to the simulation, the *$display* statement may print the <u>old</u> value of reg2. The old value may be printed because the value written by the PLI has been scheduled to take place in the current time step, but will not yet have taken effect. The *$strobe* statement which follows the *$display* will print the new value of both reg1 and reg2, because the definition of *$strobe* is to print its message at the end of the current simulation time step, after all value changes for that moment in time have taken effect.

16.9 Returning logic values of system functions

When a handle for a system function is passed to the acc_set_value() routine, the value will be written as the system function's return value. The acc_handle_tfinst() routine is used to obtain the handle for the system function.

Rules for returning values to system functions

There are two important restrictions on returning values to a system function:

- A value can only be written to the return of a system function from a *calltf routine*, which is when the system function is active. The *calltf routine* is invoked when the system function is encountered by the simulator while simulation is running. System function return values cannot be written from a *misctf routine*, *sizetf routine* or *checktf routine* because the simulation is not executing the statement containing the function at the times these routines are invoked.

- It is illegal to specify a propagation delay when returning a value to a system function. If a delay is specified, the value will not be written. To return a value with zero delay, specify **accNoDelay** for the delay flag.

Types of system functions

The ACC standard allows for two types of system functions:

- Real functions, which will return a Verilog real value. Verilog real variables are double precision floating point values.

- Sized integer functions, which return a Verilog scalar or vector value. Verilog scalars are 1-bit wide, and vectors can be any width. The `acc_set_value()` routine can return a system function value of any vector width.

The type of function is established when the system function is registered through the TF/ACC PLI interface mechanism. Refer to Chapter 9, section 9.11 on page 305, for a full description of registering system functions.

Example 16-4, shown below, implements a system function that returns a 72-bit value. The value is represented using aval/bval pairs to encode 4-state logic in the PLI application. To keep this example simple, the 72-bit value that is returned is hard-coded into the PLI application.

CD The source code for this example is on the CD accompanying this book.

- Application source file: `Chapter.16/func_72bit_acc.c`
- Verilog test bench: `Chapter.16/func_72bit_test.v`
- Expected results log: `Chapter.16/func_72bit_test.log`

Example 16-4: *$func_72bit* — returning 4-state vector values with system functions

```
/*********************************************************************
 * Sizetf application
 *********************************************************************/
int PLIbook_Func72bit_sizetf(int user_data, int reason)
{
  return(72);   /* $pow returns 32-bit values */
}

/*********************************************************************
 * calltf routine
 *********************************************************************/
int PLIbook_Func72bit_calltf(int user_data, int reason)
{
  handle          systf_handle;
  s_setval_value  value;
  s_setval_delay  delay;
  p_acc_vecval    val_array;
  int             array_size, i;

  /* declare an array of aval/bval pairs for the vector size */
  #define VEC_SIZE 72  /* hard coded 72-bit vector for this example */
  array_size = ((VEC_SIZE-1)/32)+1;
  val_array = (p_acc_vecval)malloc(sizeof(s_acc_vecval) * array_size);
```

```
/* set value of vector aval/bval pairs */
for (i=0; i<array_size; i++) {
  val_array[i].aval = 0xAAAAAAAA; /* aval bits encode logic 0 & 1 */
  val_array[i].bval = 0x00000000; /* bval bits encode logic Z & X */
}

systf_handle = acc_handle_tfinst();
value.format = accVectorVal;
value.value.vector = val_array;
delay.model = accNoDelay;

acc_set_value(systf_handle, &value, &delay); /* set sysfunc return */

return(0);
}
```

16.10 Reading module time scale information

Within a Verilog HDL model, delay values can be specified either as an integer or as a floating point real number. Within a Verilog simulation, however, time is represented as a 64 bit unsigned integer. As a Verilog model is compiled or loaded into simulation, delay values in the module are scaled to a simulation time unit. The scaling factor is specified with the `timescale` compiler directive.

The `timescale` directive in Verilog indicates what time units are used in the modules which follow the directive. The directive also indicates a time precision, which is how many decimal points of accuracy are permitted. Time values with more decimal points than the precision are rounded off. Each module in a design can have a different time scale, so one model can specify delays in nanoseconds with two decimal points of precision and another model can specify delays in microseconds with three decimal points. Most Verilog simulators will determine the finest precision of all modules in a simulation, and set that precision as the simulation time units. The simulator then scales the times in all modules to the simulation time units. As an example:

```
`timescale 1ns/10ps  //nanosecond units, 2 decimal points
module A;
   ...
   nand #5.63 n1 (y, a, b);
   ...
endmodule
```

```
`timescale 1us/100ns   //microsecond units, 3 decimal points
module B;
   . . .
   nand #3.581 n1 (y, a, b);
   . . .
endmodule
```

In this example, a simulator will determine that ten-picosecond units of time is the
finest precision used by all modules, and set that as the simulation time units. The
simulator will then scale the 5.63 nanosecond delay in module **A** to 563 10-picosec-
ond units, and it will scale the 3.581 microsecond delay in module **B** to 360,000 10-
picosecond units (the rounding to the module's time precision occurs before the time
value is scaled to the simulator's time units).

16.10.1 Reading time scale factors

The ACC routines can access both the time units and the time precision of a module.

void **acc_fetch_timescale_info (**
   handle   **object,**   /* handle for a module instance, module definition, PLI
                                        system task/function, or **null** */
   p_timescale_info **timescale**) /* pointer to an application-allocated timescale structure */

The `acc_fetch_timescale_info()` routine retrieves the time scale factors of a
module, or a *$timeformat* built-in system task. The information is retrieved into an
application-allocated **s_timescale_info** structure, which is defined in acc_user.h,
as follows:

```
typedef struct t_timescale_info {
  PLI_INT16 unit;
  PLI_INT16 precision;
} s_timescale_info, *p_timescale_info;
```

The object handle passed to `acc_fetch_timescale_info()` can be one of three
types of objects:

- *A handle for a module instance or top-level module* — the time scale factors of that
  module will be retrieved and stored in the `s_timescale_info` structure.

- *A handle for a system task or system function* — the time scale factors of the module containing the instance of the task/function will be retrieved and stored in the
  `s_timescale_info` structure. Handles for system tasks and system functions are
  obtained with the routine `acc_handle_tfinst()`.

- `null` — the time scale factors of the active *$timeformat* built-in Verilog system
  task will be retrieved and stored in the `s_timescale_info` structure.

The time units and time precision are retrieved as integers, which represent the exponent of the scale factor. For example, 1 nanosecond is 1 second times 10^{-9}, so the integer value used to represent nanoseconds is -9. Table 16-5 shows the representations of all time units and precisions supported in the Verilog language.

Table 16-5: Time unit exponents for `acc_fetch_timescale_info()`

Exponent Value	Time Unit Represented
2	100 seconds (1×10^2)
1	10 seconds (1×10^1)
0	1 second (1×10^0)
-1	100 milliseconds (1×10^{-1})
-2	10 milliseconds (1×10^{-2})
-3	1 millisecond (1×10^{-3})
-4	100 microseconds (1×10^{-4})
-5	10 microseconds (1×10^{-5})
-6	1 microsecond (1×10^{-6})
-7	100 nanoseconds (1×10^{-7})
-8	10 nanoseconds (1×10^{-8})
-9	1 nanosecond (1×10^{-9})
-10	100 picoseconds (1×10^{-10})
-11	10 picoseconds (1×10^{-11})
-12	1 picosecond (1×10^{-12})
-13	100 femtoseconds (1×10^{-13})
-14	10 femtoseconds (1×10^{-14})
-15	1 femtosecond (1×10^{-15})

PLI_INT32 acc_fetch_precision ( )

The `acc_fetch_precision()` routine retrieves the internal simulation time units. The internal time units represent the finest increment of time for which the simulator can schedule events. In most Verilog simulators, the internal time unit will be the finest time precision of a Verilog modules which were compiled into the simulation. `acc_fetch_precision()` does not have any inputs, and the return value is an integer which represents the magnitude of 1 second in the same way as with `acc_fetch_timescale_info()`.

The following code fragment retrieves the time scale information of a module instance, the module containing the system task, and the current *$timeformat* system task.

```
s_timescale_info ts_info;
handle module_h, tfinst_h;

/* obtain handle for a module instance */
/* obtain handle for a system task instance */

acc_fetch_timescale_info(module_h, &ts_info);
io_printf("\nModule %s: units = %d  precision = %d\n",
          acc_fetch_fullname(module_h),
          ts_info.unit, ts_info.precision);

acc_fetch_timescale_info(tfinst_h, &ts_info);
io_printf("\nSystem task: units = %d  precision = %d\n",
          ts_info.unit, ts_info.precision);

acc_fetch_timescale_info(null, &ts_info);
io_printf("\n$timeformat units = %d  precision = %d\n\n",
          ts_info.unit, ts_info.precision);
```

16.10.2 Reading the current simulation time

The ACC library does not have a routine to retrieve the current simulation time. The TF routines should be used to read the current simulation time in ACC applications. These routines are: `tf_gettime()`, `tf_getlongtime()`, `tf_getrealtime()` and `tf_strgettime()`.

16.11 Reading delay values

There are several types of Verilog objects which have delays values. The ACC routines can both read and modify the delays of these objects.

16.11.1 Verilog objects which have delay values

Several types of constructs in the Verilog language can have delay values, and each type of object can store a different number of delay values, representing different delays for different output transitions. In addition, each output transition can be represented by a single delay value, referred to as a typical delay, or each output transition can be represented by a set of three delay values for a minimum, typical and maximum delay range. Table 16-6 lists the types of objects for which ACC routines can access delay values. The table shows the number of output transitions for each type of object that can be represented in the Verilog HDL source code, and the number of transitions which will be accessed by the PLI for that object.

Table 16-6: Verilog objects which can have delays

Verilog Object	Verilog HDL Source Code		Delays Accessed by PLI	
	Number of Delay Transitions[1] (and order specified)		**Number of Transitions**[1] (and order accessed)	
2-state primitive	0	zero delay	2	rise, fall
	1	all transitions have same delay		
	2	separate delays for rise, fall		
3-state primitive	0	zero delay	3	rise, fall, toZ
	1	all transitions have same delay		
	2	separate delays for rise, fall		
	3	separate delays for rise, fall, toZ		
module path	(number of values specified is mapped to 12 transitions within simulation data structure)		(number of transitions accessed is set by accPathDelayCount)	
	1	all transitions have same delay	1	all transitions
	2	separate delays for rise, fall	2	rise, fall
	3	separate delays for rise, fall, toZ	3	rise, fall, toZ
	6	separate delays for 0–>1, 1–>0, 0–>Z, Z–>1, 1–>Z, Z–>0	6	0–>1, 1–>0, 0–>Z, Z–>1, 1–>Z, Z–>0
	12	separate delays for 0–>1, 1–>0, 0–>Z, Z–>1, 1–>Z, Z–>0, 0–>X, X–>1, X–>1, 1–>X, X–>0, X–>Z, Z–>X	12	0–>1, 1–>0, 0–>Z, Z–>1, 1–>Z, Z–>0, 0–>X, X–>1, 1–>X, X–>0, X–>Z, Z–>X
module port module port bit	(cannot be represented in Verilog HDL)		3	rise, fall, toZ
module interconnect	(cannot be represented in Verilog HDL)		3	rise, fall, toZ
timing constraint	1	timing limit	1	timing limit
[1]A transition can be a typical delay value, or a min:typ:max set of delay values				

Definitions of terms used in table 16-6

- **2-state primitive:** an instance of a Verilog primitive which does not use high impedance. These are: and, nand, or, nor, xor, xnor, buf, not, pullup, pulldown and *user-defined primitives*.

- **3-state primitive:** an instance of a Verilog primitive which uses high impedance. These are: bufif0, bufif1, notif0, notif1, cmos, rcmos, pmos, rpmos, tran, rtran, tranif0, rtranif0, tranif1, rtranif1.

- **Module paths:** a path delay specified in a Verilog specify block from an input or inout port of a module to an output or inout port of the same module.

- **Module input ports (MIPD):** a delay on all signals which fan-in to an input port of a module instance. There is no construct in the Verilog language to represent module input port delays; but the PLI and SDF can add or modify delays on input ports.

- **Module input port bits:** a delay on all signals which fan-in to a bit of a vector port.

- **Module interconnect paths:** a connection from an output port of one module to the input port of another module. There is no construct in the Verilog language to represent interconnect path delays; but the PLI and SDF can add or modify delays on interconnect paths.

- **Timing constraint checks:** a timing constraint represented in a Verilog specify block, using *$setup*, *$hold*, etc.

There are other objects in the Verilog HDL which can have delay values, but which the ACC library cannot access. These are continuous assignments, and delays within procedural blocks. The VPI library can access the delay values of these objects.

Minimum, typical and maximum delays

Each Verilog delay transition can have:

- A single delay value, representing the typical delay.

- Three delay values, representing the minimum, typical, maximum delay range for that transition.

For example, a bufif1 tri-state buffer gate can represent a propagation delay as:

A tri-state buffer with no delays:

```
bufuf1 g1 (...);
```

A tri-state buffer with delay of 5 for rising, falling, and turn-off transitions:

```
bufuf1 #5 g2 (...);
```

A tri-state buffer with separate delays for rising, falling, and turn-off transitions:

```
bufuf1 #(3, 4, 5) g3(...);
```

A tri-state buffer with separate minimum:typical:maximum delay sets for rising, falling and turn-off transitions:

```
bufuf1 #(2:3:4, 3:4:5, 5:6:7) g4 (...);
```

The Verilog language can represent minimum, typical and maximum delay values. Many Verilog simulators, however, only store a single value in the simulation data structure for each transition. The value that is stored is usually controlled by an invocation option (such as +mindelays). For these types of simulators, the PLI will retrieve the same value for the minimum, typical and maximum delay.

16.11.2 Reading an object's delay values

A single ACC routine, `acc_fetch_delays()`, is used to read the delay values of any Verilog object. This single routine can be configured to read either typical delay values, or to read minimum, typical and maximum delay values. The arguments which are passed to the routine will change according to the configuration.

The syntax for `acc_fetch_delays()` when configured to read just typical delays is:

PLI_INT32 **acc_fetch_delays (**   /* if accMinTypMaxDelays is false (default) */
handle	**object,**	/* handle for a primitive, module path, intermodule path,
		module input port or port bit, or timing check */
double	*** d1...*d12)**	/* pointers to variables to receive the delay values */

The syntax for `acc_fetch_delays()` when configured to read minimum, typical and maximum delays is:

PLI_INT32 **acc_fetch_delays (**   /* if accMinTypMaxDelays is true */
handle	**object,**	/* handle for a primitive, module path, intermodule path,
		module input port or port bit, or timing check */
double	*** dset_array)**	/* pointer to an array to receive the delay values */

The `acc_fetch_delays()` routine retrieves the current delay values of a *primitive instance*, a *module path*, a *module interconnect path*, a *module input port*, a *bit-select of a module input port* or a *timing check*. The routine returns 1 if it was successful, and non-zero if an error occurred.

 The delay values retrieved into the PLI application will automatically be scaled to the time scale of the module containing the object from which the delays are read.

`acc_fetch_delays()` is overloaded in such a way that the number of arguments and the C data types of the arguments used by `acc_fetch_delays()` change, depending on:

- The type of object from which delays are being read.
- The setting of the ACC configuration for **accPathDelayCount**.
- The setting of the ACC configuration for **accMinTypMaxDelays**.

16.11.3 Reading typical delays

`acc_fetch_delays()` can be configured to read either the typical delays or the minimum, typical, maximum delays of an object. The default configuration is to read the typical delays of an object. This configuration can also be explicitly specified, using `acc_configure()`, as follows:

> **acc_configure(accMinTypMaxDelays, "false");**

In the typical delay mode, `acc_fetch_delays()` requires as its arguments:

- A handle for the object from which to read the delay values.
- A pointer to a C double variable for the first delay value to be read.
- A pointer to a C double variable for the second delay value to be read.
- A pointer to a C double variable for the third delay value to be read.

    . . .

- A pointer to a C double variable for the last delay value to be read.

The number of delay values which will be retrieved depends on the type of object. The `acc_fetch_type()` or `acc_object_of_type()` routines can be used to determine the object type.

- 2 delay values if the object is a *2-state primitive instance*.
- 3 delay values if the object is a *3-state primitive instance*.
- 3 delay values if the object is a *module input port*.
- 3 delay values if the object is a *module interconnect path*.
- 1 delay value if the object is a *timing constraint* check.
- **1, 2, 3, 6,** or **12** delay values if the object is a *module path*. The number of delays is controlled by the configuration of `accPathDelayCount`. The default is 6.

The number of *module path* delay transitions to be retrieved by
acc_fetch_delays() can be configured by the PLI application. A module path
will always have 12 output transitions stored in the Verilog simulation data structure.
The acc_fetch_delays() routine will automatically map the 12 delay values of
the module path to the number of transitions requested by the PLI application. The
configuration is set, using acc_configure(), as one of the following:

```
acc_configure(accPathDelayCount, "1");

acc_configure(accPathDelayCount, "2");

acc_configure(accPathDelayCount, "3");

acc_configure(accPathDelayCount, "6");

acc_configure(accPathDelayCount, "12");
```

The default configuration is "6".

It is an error to call acc_fetch_delays() with too few arguments for the number
of delays of an object. If too many arguments are specified, the extra arguments are
ignored. The maximum number of arguments which can be specified when reading
only typical delay values is 13 (the object handle plus pointers to 12 delay variables).

Example 16-5 illustrates using acc_fetch_delays() to read the typical delays of
all primitives in a module. Three delay values are retrieved, which represent the rise,
fall and turn-off delays of the primitive.

CD   The source code for this example is on the CD accompanying this book.

- Application source file: Chapter.16/list_prim_delays_acc.c
- Verilog test bench: Chapter.16/list_prim_delays_test.v
- Expected results log: Chapter.16/list_prim_delays_test.log

Example 16-5: *$list_prim_delays* — reading typical delays

```
int PLIbook_PrimDelays_calltf(int user_data, int reason)
{
  handle module_h, prim_h;
  double rise, fall, toZ;
  static PLI_INT32 three_state[7] = {accBufif0Gate,  accBufif1Gate,
                                     accNotif0Gate,  accNotif1Gate,
                                     accTranif0Gate, accTranif1Gate,
                                     0};

  acc_initialize();
```

```
  acc_configure(accMinTypMaxDelays, "false");

  module_h = acc_handle_tfarg(1);
  io_printf("\nPrimitives in module %s:\n",
           acc_fetch_fullname(module_h));
  prim_h = null;    /* start with known value for target handle */
  while (prim_h = acc_next_primitive(module_h, prim_h)) {
    io_printf("  %-8s instance %-4s:  ",
              acc_fetch_defname(prim_h),
              acc_fetch_name(prim_h));
    if (acc_object_in_typelist(prim_h, three_state)) {
      acc_fetch_delays(prim_h, &rise, &fall, &toZ);
      io_printf("rise=%2.2f, fall=%2.2f, toZ=%2.2f\n", rise, fall, toZ);
    }
    else {
      acc_fetch_delays(prim_h, &rise, &fall);
      io_printf("rise=%2.2f, fall=%2.2f\n", rise, fall);
    }
  }
  acc_close();
  return(0);
}
```

16.11.4 Reading minimum, typical and maximum delays

acc_fetch_delays() can be configured to read the minimum, typical, maximum delays for each output transition of an object. This configuration must be explicitly specified, using acc_configure(), as follows:

acc_configure(accMinTypMaxDelays, "true");

In the min:typ:max delay mode, acc_fetch_delays() requires as its inputs:

- A handle for the object from which to read the delay values.

- A pointer to an array of C double variables, with one element in the array for each delay value to be read.

acc_fetch_delays() will retrieve a minimum, typical and maximum delay value for each delay transition that is read from the object. The total number of delay values which will be retrieved depends on the type of object, as follows:

- **6** delay values if the object is a *2-state primitive instance*.

- **9** delay values if the object is a *3-state primitive instance*.

- **9** delay values if the object is a *module input port*.

- **9** delay values if the object is a *module interconnect path*.

- **3** delay value if the object is a *timing constraint check*.
- **3, 6, 9, 18** or **36** delay values if the object is a module path. The number of delays is controlled by the configuration of `accPathDelayCount`.

Table 16-7 shows the order in which `acc_fetch_delays()` retrieves delay values.

Table 16-7: Number elements and order of delay array

Number of Delay Transitions	Order of retrieved delays		
	array element	object delay	
1 transition	[0] receives	all transitions	min value
	[1]		typ value
	[2]		max value
2 transitions	[0] receives	rise transition	min value
	[1]		typ value
	[2]		max value
	[3] receives	fall transition	min value
	[4]		typ value
	[5]		max value
3 transitions	[0] receives	rise transition	min value
	[1]		typ value
	[2]		max value
	[3] receives	fall transition	min value
	[4]		typ value
	[5]		max value
	[6] receives	turn-off transition	min value
	[7]		typ value
	[8]		max value
6 transitions	[0] receives	1st transition[1]	min value
	[1]		typ value
	[2]		max value
	...		...
	[15] receives	6th transition	min value
	[16]		typ value
	[17]		max value
12 transitions	[0] receives	1st transition[1]	min value
	[1]		typ value
	[2]		max value
	...		...
	[33] receives	12th transition	min value
	[34]		typ value
	[35]		max value

[1]In the preceding table, the order of the transitions for the 6 and 12 transition sets is:
0->1, 1->0, 0->Z, Z->1, 1->Z, Z->0, 0->X, X->1, 1->X, X->0, X->Z, Z->X

> **NOTE** The Verilog language can represent minimum, typical and maximum delay values, but many Verilog simulators only store a single value in the simulation data structure for each transition. The value that is stored is usually controlled by an invocation option, and, by default, will be the typical delay value. If the simulator has not stored the full set of values, then the value that was stored for each transition will be used for all three minimum, typical and maximum fields in the delay array.

The array of C doubles is declared by the PLI application, and must be large enough to receive all the delay values which will be retrieved. Declaring the array with too few elements could potentially crash a PLI application. It is not an error to declare the array with more elements than are needed. The additional elements in the array will be ignored by acc_fetch_delays().

Example 16-6 lists a C function which reads the minimum, typical, maximum rise and fall delays of all module paths in a module. Note that module paths do not have an actual name in the Verilog language. The ACC routines automatically create a derived module path name by concatenating together the names of the nets connected to the first input of the path and the first output of the path, with a '$' in between the two names.

> **CD** The source code for this example is on the CD accompanying this book.
>
> - Application source file: Chapter.16/list_path_delays_acc.c
> - Verilog test bench: Chapter.16/list_path_delays_test.v
> - Expected results log: Chapter.16/list_path_delays_test.log

Example 16-6: *$list_path_delays* — reading min:typ:max delay values

```
int PLIbook_PathDelays_calltf(int user_data, int reason)
{
  handle module_h, path_h;
  double delay_set[18];
  int i;

  acc_initialize();
  acc_configure(accDisplayWarnings, "true");

  acc_configure(accMinTypMaxDelays, "true");
  acc_configure(accPathDelayCount,  "6");

  module_h = acc_handle_tfarg(1);
```

```
io_printf("\nPath delays in module %s:\n",
          acc_fetch_fullname(module_h));
path_h = null;      /* start with known value for target handle */
while (path_h = acc_next_modpath(module_h, path_h)) {
  io_printf("  %-12s :  ",
            acc_fetch_name(path_h));
  acc_fetch_delays(path_h, delay_set);
  for (i=0; i<18; i++) {
    if ( i == 0 )  /* format output like Verilog syntax */
      io_printf("(");
    else if ( (i % 3) )
      io_printf(":");
    else
      io_printf(", ");
    io_printf("%1.1f", delay_set[i]);
  }
  io_printf(")\n\n");
}
acc_close();
return(0);
}
```

16.12 Writing delay values into an object

A pair of ACC routines, acc_append_delays() and acc_replace_delays(), are used to modify the delay values of Verilog objects. These routines work the same as when reading delays values. The routines can be configured to modify either typical delay values, or to modify minimum, typical and maximum delay values. The arguments which are passed to the routines change according to the configuration.

The acc_append_delays() routine adds delays to the existing delays of an object. Delays can be appended to a *primitive instance*, a *module path*, a *module interconnect path*, a *module input port*, a *bit-select of a module input port* and a *timing check*.

The acc_replace_delays() routine replaces any existing delays of an object. Delays can be replaced on a *primitive instance*, a *module path*, a *module interconnect path*, a *module input port*, a *bit-select of a module input port*, and a *timing check*.

The syntax for these routines, when configured to modify just typical delays, is:

PLI_INT32 **acc_append_delays (**   /* if accMinTypMaxDelays is false (default) */
      handle   **object,**   /* handle for a primitive, module path, intermodule path,
                                 module input port or port bit, or timing check */
      double   **d1, ... d12)** /* transition delay values */

PLI_INT32 **acc_replace_delays(**   /* if accMinTypMaxDelays is false (default) */
    handle   **object,**   /* handle for a primitive, module path, intermodule path,
                                            module input port or port bit, or timing check */
    double   **d1,...d12)**   /* transition delay values */

The syntax for these routines, when configured to modify minimum, typical and maximum delays, is:

PLI_INT32 **acc_append_delays (**   /* if accMinTypMaxDelays is true */
    handle   **object,**   /* handle for a primitive, module path, intermodule path,
                                            module input port or port bit, or timing check */
    double   *** dset_array)** /* pointer to an array of transition delay values.

PLI_INT32 **acc_replace_delays(**   /* if accMinTypMaxDelays is true */
    handle   **object,**   /* handle for a primitive, module path, intermodule path,
                                            module input port or port bit, or timing check */
    double   *** dset_array)** /* pointer to an array of transition delay values */

 NOTE The delay values specified in the PLI application will automatically be scaled to the time scale of the module containing the object which is being modified.

The number of arguments and the types of arguments used by `acc_append_delays()` and `acc_replace_delays()` will vary, based on the type of the object and settings of the ACC configurations. The factors which affect the arguments used by these routines are:

- The type of object for which delays are being modified.
- The setting of the ACC configuration for **accPathDelayCount**.
- The setting of the ACC configuration for **accMinTypMaxDelays**.
- The setting of the ACC configuration for **accToHiZDelay**.

16.12.1 Setting the number of module path delays to be modified

When the PLI modifies the delays of a module path, the number of transitions to be modified can be configured by the PLI application. The configuration is set by `acc_configure()`, as one of the following configurations:

```
acc_configure(accPathDelayCount, "1");

acc_configure(accPathDelayCount, "2");

acc_configure(accPathDelayCount, "3");

acc_configure(accPathDelayCount, "6");
```

```
acc_configure(accPathDelayCount, "12");
```

The default configuration is "6".

16.12.2 Setting the calculation of turn-off delays

Verilog three-state primitives, module input ports, and module interconnect paths can have three transitions: the *rise time*, *fall time* and *turn-off time*. The turn-off time is the amount of time it takes to transition from any logic value to high impedance.

The **accToHiZDelay** configuration determines whether a turn-off transition delay should be calculated by the PLI or specified as an input to acc_append_delays() or acc_replace_delays(). The configuration is specified as follows:

```
acc_configure(accToHiZDelay, "min");

acc_configure(accToHiZDelay, "max");

acc_configure(accToHiZDelay, "average");

acc_configure(accToHiZDelay, "from_user");
```

The first three configurations enable the PLI to automatically calculate the turn-off delay time, based on the rise and fall transition times. The calculation can use the shortest, the longest or the average of the rise and fall times. The "from_user" configuration indicates that the turn-off time for objects with three transitions will be specified as an input to acc_append_delays() or acc_replace_delays().

The default setting of accToHiZDelay is "from_user".

 The setting of accToHiZDelay is ignored if the object is a module path.

16.12.3 Modifying typical delays

acc_append_delays() and acc_replace_delays() can be configured to modify either the typical delays or the minimum, typical, maximum delays of an object. The default configuration is to modify the typical delays of an object. This configuration can also be explicitly specified, using acc_configure(), as follows:

```
acc_configure(accMinTypMaxDelays, "false");
```

In the typical delay mode, `acc_append_delays()` and `acc_replace_delays()` require as inputs:

- A handle for the object from which to modify the delay values.
- A C double variable or literal value for the first delay value to be modified.
- A C double variable or literal value for the second delay value to be modified.
- A C double variable or literal value for the third delay value to be modified.

    ...

- A C double variable or literal value for the last delay value to be modified.

The number of delay inputs used will be based on the type of object and the ACC configurations:

- **2** delay values if the object is a *two-state primitive instance*.
- **2** delay values if the object is a *three-state primitive instance*, *module input port* or *module interconnect path* and `accToHiZDelay` is configured for "min", "max" or "average".
- **3** delay values if the object is a *three-state primitive instance*, *module input port* or *module interconnect path* and `accToHiZDelay` is configured for "from_user".
- **1** delay value if the object is a timing constraint check.
- **1, 2, 3, 6,** or **12** delay values if the object is a module path. The number of delays is controlled by the configuration of `accPathDelayCount` (refer back to section 16.12.1 on page 585).

It is an error to call `acc_append_delays()` and `acc_replace_delays()` with too few arguments for the type of object. If too many arguments are specified, the unnecessary arguments are ignored. The maximum number of arguments which can be specified when modifying typical delay values is 13 (the object handle plus 12 delay values).

The following C code fragment illustrates how to append typical rise and fall delays onto a Verilog primitive.

```
handle prim_h;
double add_rise, add_fall;
s_acc_value val_s;

acc_configure(accMinTypMaxDelays, "false");
acc_configure(accToHiZDelay, "average");
...
acc_append_delays(prim_h, add_rise, add_fall);
```

16.12.4 Modifying minimum, typical and maximum delays

`acc_append_delays()` and `acc_replace_delays()` can be configured to modify the minimum, typical, maximum delays for each output transition of an object. This configuration must be explicitly specified using `acc_configure()`, as follows:

`acc_configure(accMinTypMaxDelays, "true");`

In the min:typ:max delay mode, `acc_append_delays()` and `acc_replace_delays()` require as inputs:

- A handle for the object from which to read the delay values.

- A pointer to an array of C double variables, with one element in the array for each delay value to be modified.

The total number of delay values which will be modified depends on the type of object and the ACC configurations, as follows:

- **2** delay values if the object is a *two-state primitive instance*.

- **2** delay values if the object is a *three-state primitive instance*, *module input port* or *module interconnect path* and `accToHiZDelay` is configured for "min", "max" or "average".

- **3** delay values if the object is a *three-state primitive instance*, *module input port* or *module interconnect path* and `accToHiZDelay` is configured for "from_user".

- **1** delay value if the object is a timing constraint check.

- **1, 2, 3, 6,** or **12** delay values if the object is a module path. The number of delays is controlled by the configuration of `accPathDelayCount`.

The delay values to be written onto the object are loaded into the array by the PLI application prior to calling `acc_append_delays()` or `acc_replace_delays()`. The order of the delay values in the array is the same as with `acc_fetch_delays()`, which was shown earlier, in Table 16-7 on page 582.

 The Verilog language can represent minimum, typical and maximum delay values, but many Verilog simulators only store a single value in the simulation data structure for each transition. The value that is stored is usually controlled by an invocation option, and, by default, will be the typical delay value. If the simulator does not store the full set of values, then only the appropriate minimum, typical or maximum fields in the delay array will be used by the simulator.

The array of C doubles is declared by the PLI application, and must be large enough to hold all delay values which will be modified. Declaring the array without enough elements may result in an error, but could potentially crash a PLI application. It is not an error to declare the array with more elements than are needed.

Example 16-7 shows a useful PLI application called *$mipd_delays*. The Verilog HDL does not have a construct to represent module input port delays, but these delays can be added through the PLI. *$mipd_delays* provides a means for a Verilog model to add delays to the input port of a module. The usage of this application is:

```
$mipd_delays(<port_name>, <d1>, <d2>, ... <d9>)
```

Example:

```
$mipd_delays(in1, 1.4, 1.6, 1.9, 1.1, 1.3, 1.5, 0.6, 0.8, 0.9);
```

CD   The source code for this example is on the CD accompanying this book.

- Application source file: `Chapter.16/mipd_delays_acc.c`
- Verilog test bench: `Chapter.16/mipd_delays_test.v`
- Expected results log: `Chapter.16/mipd_delays_test.log`

Example 16-7: *$mipd_delays* — modifying min:typ:max delays

```
int PLIbook_MipdDelays_calltf(int user_data, int reason)
{
  double delay_array[9];
  double rise, fall, toZ;
  handle port_h;
  int i;

  acc_initialize();
  acc_configure(accDisplayWarnings, "true");

  acc_configure(accMinTypMaxDelays, "true");
  port_h = acc_handle_tfarg(1);

  /* most simulators return loconn handle, not port handle */
  if (   (acc_fetch_type(port_h) != accPort)
      && (acc_fetch_type(port_h) != accPortBit) )
    port_h = acc_next_port(port_h, null);
  if (   (acc_fetch_type(port_h) != accPort)
      && (acc_fetch_type(port_h) != accPortBit) ) {
    io_printf("ERR: $mipd_delays could not obtain port handle\n");
    return(0);
  }
```

```
for (i = 0; i < 9; i++)
  delay_array[i] = acc_fetch_tfarg(i+2);

acc_replace_delays(port_h, delay_array);

/* verify new delays took affect */
acc_configure(accMinTypMaxDelays, "false");
rise = fall = toZ = 0.0;
acc_fetch_delays(port_h, &rise, &fall, &toZ);
io_printf("Port %s new delays: (%1.2f, %1.2f, %1.2f)\n\n",
            acc_fetch_name(port_h),
            rise, fall, toZ);

acc_close();
return(0);
}
```

16.13 Reading parameter constant values

The Verilog HDL has two types of constants, **parameter** and **specparam**. Each type of constant can store integer values, real values or string values. There are specific ACC routines provided to determine what type of value is stored in a constant, and to read the value of the constant.

 The IEEE 1364-2001 standard adds a third type of constant, **localparam**. Only the VPI library can access this new constant type.

PLI_INT32 **acc_fetch_paramtype (**
    handle   **object**)   /* handle for a parameter or specparam */

double **acc_fetch_paramval (**
    handle   **object**)   /* handle for a parameter or specparam */

The acc_fetch_paramtype() routine returns an integer constant which represents what type of data that is stored in a Verilog parameter or specparam constant. The constant will be: accIntegerParam for integer values, accRealParam for floating point values, or accStringParam for ASCII string values.

acc_fetch_paramval() returns the value of what is stored in a parameter or specparam constant. *The value is always returned as a C double.* If the value stored in the parameter is an integer, the return from acc_fetch_paramval() can be cast to an int. A Verilog parameter can store 4-state logic. The acc_fetch_paramval() routine returns 2-state logic by converting logic X and Z to 0.

If the value stored in a constant is an ASCII string, acc_fetch_paramval() will retrieve the string into the ACC string buffer, and return a pointer to the string as a C double. Strings stored in the string buffer are temporary, and should be used immediately, or copied to application-allocated storage.

 Some C compilers do not allow casting a C double to a string pointer. Therefore, this example first casts the double to an int, and then casts the int to a char *.

The casting of a pointer to a double may not be portable to all operating systems, in particular, 64-bit operating systems. To work around this issue, some simulators add two special ACC routines which are not part of the IEEE 1364 standard: **acc_fetch_paramval_int()** and **acc_fetch_paramval_str()**.

The following example illustrates accessing and printing the value of all parameter constants in a Verilog model.

CD The source code for this example is on the CD accompanying this book.

- Application source file: Chapter.16/list_parameters_acc.c
- Verilog test bench: Chapter.16/list_parameters_test.v
- Expected results log: Chapter.16/list_parameters_test.log

Example 16-8: *$list_params* — reading parameter and specparam values

```
int PLIbook_ListParams_calltf(int user_data, int reason)
{
  handle module_h, param_h;

  acc_initialize();
  acc_configure(accDisplayWarnings, "true");

  module_h = acc_handle_tfarg(1);
  io_printf("\nConstants in module %s:\n",
            acc_fetch_fullname(module_h));
  param_h = null;
  while(param_h = acc_next_parameter(module_h, param_h)) {
    io_printf("  Parameter %s is: ", acc_fetch_fullname(param_h));
    switch(acc_fetch_paramtype(param_h) ) {
      case accRealParam:
```

```
      io_printf("%f\n", acc_fetch_paramval(param_h));
      break;
    case accIntegerParam:
      io_printf("%d\n", (int)acc_fetch_paramval(param_h));
      break;
    case accStringParam:
      io_printf("%s\n", (PLI_BYTE8 *)(int)acc_fetch_paramval(param_h));
    }
  }
  acc_close();
  return(0);
}
```

16.14 Using constants as model attributes

Often a PLI application needs information about a model which is not part of the model functionality. A standard cell delay calculator, for example, might need the rise, slope and load factors of the cells used in a design. Information that is needed by the PLI application can be stored within the Verilog model, using Verilog `parameter` constants or `specparam` constants. The PLI application can then obtain a handle for the constant and read its value.

The ACC library supports a special usage of the Verilog `parameter` or `specparam` constant, called an ***attribute***. This special usage requires adding a dollar sign ($) to the end of the specparam constant name. Almost without exception, the `specparam` constant type is used to represent an attribute, because this constant type cannot be redefined.

An attribute constant can be associated with all objects within a module, or with specific objects in a module.

- A ***general attribute*** is a constant with a name which ends with a dollar sign. Every object within a Verilog module will be associated with a general attribute.

- An ***object-specific attribute*** is a constant with a base name which ends with a dollar sign, followed by the name of some object in the module. Only the object which is named will be associated with that attribute.

The following example illustrates three specparam attributes:

```
module AN2 (o, a, b); // 2-input AND gate standard cell
  output o;
  input  a, b;
```

```
   and (o, a, b);
   specify
     specparam BaseDelay$  = 2.2;  //general attribute
     specparam InputLoad$a = 0.2;  //object-specific attribute
     specparam InputLoad$b = 0.3;  //object-specific attribute
   endspecify
 endmodule
```

A specific set of ACC routines read the value of the attribute constants. These routines use the attribute name and a handle for the object associated with that attribute. These ACC routines by-pass the need to first obtain a handle for the constant. The routines will search first for an object-specific attribute, then for a general attribute, and finally return a default value.

double **acc_fetch_attribute (**

handle	**object,**	/* handle for a named object */
PLI_BYTE8	*** attribute,**	/* string containing the name of the parameter or specparam attribute associated with the object */
double	**default) ·**	/* default value to be returned if the parameter or specparam does not exist */

PLI_INT32 **acc_fetch_attribute_int (**

handle	**object,**	/* handle for a named object */
PLI_BYTE8	*** attribute,**	/* string with name of attribute associated with object */
PLI_INT32	**default)**	/* default value to be returned */

PLI_BYTE8 ***acc_fetch_attribute_str (**

handle	**object,**	/* handle for a named object */
PLI_BYTE8	*** attribute,**	/* string with name of attribute associated with object */
PLI_BYTE8	*** default)**	/* default string to be returned */

The `acc_fetch_attribute()` routine returns the value of an attribute constant as a C double. `acc_fetch_attribute_int()` returns the value as a 32-bit C integer. `acc_fetch_attribute_str()` returns the value as a pointer to a C string. The string is stored in the temporary ACC string buffer.

These ACC routines access a specparam constant, using the name of the attribute (including the dollar sign) and a handle for the object with which the attribute is associated. For example, to read the value of one of the following attributes:

```
   specparam InputLoad$a = 0.2;  //object-specific attribute
   specparam InputLoad$b = 0.3;  //object-specific attribute
```

The following C code can be used to read the value of the attribute `InputLoad$b`:

```
double load_in;
handle port_h;

/* add code to obtain a handle for module input port b */

load_in = acc_fetch_attribute(port_h, "InputLoad$", 1.0);
```

The ACC routines will search first for an object-specific attribute, then for a general attribute, and finally return a default value.

The default value can be specified in two different ways:

- By setting the ACC configuration to return a default of zero.

- By providing a default value as a third argument to the fetch attribute routine.

The three fetch attribute routines use the `acc_configure()` routine to determine the source of the default return value. The syntax for `acc_configure()` was presented in Chapter 15, in section 15.2 on page 489.

- `acc_configure(accDefaultAttr0, "true")`
  configures the fetch attribute routines to return a default value of zero (represented as 0.0, 0 or "0"). With this configuration, the third argument to the fetch attribute routines is not needed, and will be ignored. This is the default configuration.

- `acc_configure(accDefaultAttr0, "false")`
  configures the fetch attribute routines to use the third argument of the routine as the default value.

 The IEEE 1364-2001 standard adds a new attribute construct to the Verilog language, which uses the **(*** and ***)** tokens. This new attribute token can be associated with any type of Verilog object, including module ports, net, reg and variables, continuous assignments, and procedural statements. Only the VPI library can access these new Verilog HDL attributes and their values.

16.15 Reading and modifying path pulse controls

The Verilog HDL defines a special `specparam` attribute to control how glitches propagate across a module path. The attribute name is **PATHPULSE$**. The ACC library provides specific routines to read and modify the pulse control values.

A *pulse* is two transitions on the same module path that occur in a shorter period of time than the path delay. Pulse control values determine whether a pulse of a certain width can pass through a module path. The pulse control values consist of a ***reject limit*** and an ***error limit*** pair of values, where:

- The *reject limit* sets the threshold for when a pulse is rejected. Any pulse less than the *reject limit* will not propagate to the output of the path.

- The *error limit* sets the threshold for when a pulse generates an error. Any pulse less than the *error limit* and greater than or equal to the *reject limit* will propagate a logic X to the path output.

- A pulse that is greater than or equal to the *error limit* will propagate to the path output.

As with other attribute constants, the PATHPULSE$ attribute can be an object-specific attribute, which is associated with one specific module path, or it can be a general attribute which is associated with all module paths. The following Verilog source code fragment shows both types of pulse control attributes.

```
module and3 (out, in1, in2, in3);
   output out;
   input  in1, in2, in3;
   . . .
   specify
     (in1 => out) = (4, 6);
     (in2 => out) = (4, 5);
     (in3 => out) = (3, 4);

     specparam PATHPULSE$in1$out = 2, 3;
     specparam PATHPULSE$ = 0, 4;

   endspecify
endmodule
```

In this example, the module path from in1 to out has a *reject limit* of 2 and an *error limit* of 3. All other paths in the module have a *reject limit* of 0 and an *error limit* of 4.

There are two important facts to note about the syntax of PATHPULSE$ attributes:

1. The attribute is assigned two values, the *reject limit* and the *error limit*. This usage does not conform to the normal syntax for specparam definitions. Normally, a specparam is assigned a single value. Because of this exception, the ACC routines to read specparam values (acc_fetch_paramval() and acc_fetch_attribute()) cannot be used to read the PATHPULSE$ attribute values. These routines would only return the first value of the pulse control limits.

2. The object specific attribute appends the name of a module path to the end of the PATHPULSE$ attribute name. However, module paths do not have a name in the Verilog language. Therefore, the PATHPULSE$ attribute creates a path name by concatenating the path input name and the path output name together with a **$** (dollar sign) between the names. If the module path specifies multiple inputs or outputs, the names of the first input and output in the path are used to represent the path name.

The ACC library provides a specific set of routines to read and modify pulse control values.

PLI_INT32 **acc_fetch_pulsere (**

handle	**object,**	/* handle for a module path, intermodule path or module input port */
double	***r1, * e1,**	/* pointers to a pulse reject limit / error limit value pair */
	...	
double	*** r12, * e12)**	/* pointers to a pulse reject limit / error limit value pair */

PLI_INT32 **acc_append_pulsere (**

handle	**object,**	/* handle for a module path, intermodule path or module input port */
double	**r1, e1,**	/* pulse reject limit / error limit value pair */
	...	
double	**r12, e12)**	/* pulse reject limit / error limit value pair */

PLI_INT32 **acc_replace_pulsere (**

handle	**object,**	/* handle for a module path, intermodule path or module input port */
double	**r1, e1,**	/* pulse reject limit / error limit value pair */
	...	
double	**r12, e12)**	/* pulse reject limit / error limit value pair */

PLI_INT32 **acc_set_pulsere (**

handle	**object,**	/* handle for a module path, intermodule path or module input port */
double	**r_percent,**	/* pulse reject limit */
double	**e_percent)**	/* pulse error limit */

The acc_fetch_pulsere() routine retrieves the current pulse control limits of a module path, intermodule path or module input port. The inputs to this routine are a handle for a module path or intermodule path, and pointers to two C double variables for each pulse control value to be retrieved.

The `acc_append_pulsere()` routine adds to the current pulse control limits of a module path, intermodule path or module input port.

The `acc_replace_pulsere()` routine replaces the current pulse control limits of a module path, intermodule path or module input port. The values specified are relative to the delays of the path.

The `acc_set_pulsere()` routine replaces the current pulse control limits of a module path, intermodule path or module input port. The values specified are a percentage of the current delays of the path. The return value is not used, and should be ignored.

The inputs to these routines are a handle for a module path, intermodule path or module input port, and a pair of C double variables for each pulse control value to be read or modified. The number of reject/error pairs required is controlled by the configuration of `accPathDelayCount`.

```
acc_configure(accPathDelayCount, "1");

acc_configure(accPathDelayCount, "2");

acc_configure(accPathDelayCount, "3");

acc_configure(accPathDelayCount, "6");

acc_configure(accPathDelayCount, "12");
```

The default configuration is "6".

It is an error to specify too few input arguments. If too many arguments are specified, the unnecessary arguments are ignored. The maximum number of arguments is 25 (a module path handle plus 12 pairs of reject limit and error limit variables).

 The path control values specified in the PLI application will automatically be scaled to the time scale of the module containing the path which is being modified.

Within the Verilog language, a single *reject limit* and an *error limit* is specified for each module path. This single pulse control set is applied to all output transitions of the path. The PLI provides a much greater level of control over pulse control values, by allowing different reject and error limits to be specified for each output transition that can occur for a path delay. Table 16-8 shows how the number of paths specified are mapped to the pulse control values for path output transitions.

Table 16-8: Configuring the number of pulse control values

accPathDelayCount configuration	path delay output transitions represented
"1"	One reject/error pair of values for all transitions
"2"	One reject/error set of values for rising transitions One reject/error set of values for falling transitions
"3"	One reject/error set of values for rising transitions One reject/error set of values for falling transitions One reject/error set of values for toZ transitions
"6" (the default)	One reject/error set of values for 0->1 transitions One reject/error set of values for 1->0 transitions One reject/error set of values for 0->Z transitions One reject/error set of values for Z->1 transitions One reject/error set of values for 1->Z transitions One reject/error set of values for Z->0 transitions
"12"	One reject/error set of values for 0->1 transitions One reject/error set of values for 1->0 transitions One reject/error set of values for 0->Z transitions One reject/error set of values for Z->1 transitions One reject/error set of values for 1->Z transitions One reject/error set of values for Z->0 transitions One reject/error set of values for 0->X transitions One reject/error set of values for X->1 transitions One reject/error set of values for 1->X transitions One reject/error set of values for X->0 transitions One reject/error set of values for X->Z transitions One reject/error set of values for Z->X transitions

When the delay of a module path or interconnect path are changed using `acc_replace_delays()` or `acc_replace_delays()`, the value of the reject and error regions will not be effected unless they exceed the value of the delay. If the reject or error limits exceed the delay they will be truncated down to the new delay limit.

16.16 Summary

This chapter has presented the ACC routines which read property values, read/modify logic values and read/modify delay values in Verilog simulations. The ability to access objects anywhere in a Verilog design hierarchy, and to read and modify the values of those objects, provides a great deal of power for PLI applications.

The next chapter presents another powerful aspect of the ACC library, the ability to synchronize PLI application activity with simulation logic value changes. When the concepts presented in this chapter are combined with those presented in the next chapter, another useful capability of the PLI is made possible—interfacing C language models to Verilog simulations. Examples of how this is done are presented in Chapter 18.

CHAPTER 17 *Using the Value Change Link (VCL)*

*O*ne of the important capabilities provided by the ACC library is the ability to have a Verilog simulator call a PLI application whenever specific objects in the simulation change logic value. This chapter presents how to use the *Value Change Link (VCL)* routines of the ACC library.

The concepts presented in this chapter are:

- An overview of the VCL routines
- Adding VCL flags
- Removing VCL flags
- Using the VCL *consumer routine*
- Synchronizing PLI applications to simulation activity
- Interfacing C language models to Verilog simulations

17.1 An overview of the VCL routines

The *Value Change Link (VCL)* routines in the ACC library schedule with a Verilog simulator to have a PLI application called whenever a specific object changes logic value or strength value. The application that is called is a VCL *consumer routine*.

There are many ways a PLI application can utilize the Value Change Link capability. Just a few possibilities are:

- A graphical display, such as a waveform display. Selected signals in a design can be monitored and each logic change on the signal recorded in a data base which can then be displayed graphically.

- To monitor test vector coverage of a design. All critical signals in a design can be monitored and a count maintained for the number of times each signal changes value. A summary report can be generated to show how often different parts of a design changed value during a simulation.

- To create an interface to a C language model. When an input to the model changes during a Verilog simulation, the PLI *consumer routine* is called and can communicate the value change to the C model.

The typical steps of using the Value Change Link routines in a PLI application are:

1. A *calltf routine* or *misctf routine* obtains a handle for an object that the routine wishes to monitor for value changes.

2. The *calltf routine* or *misctf routine* adds a VCL flag to the object. The flag indicates the name of the *consumer routine* that should be called for logic value changes.

3. When the object changes logic value during simulation, the simulator calls the *consumer routine*.

4. The *consumer routine* processes what ever the PLI application needs to do, such as reading the new logic value of the object that changed.

17.2 Adding and removing VCL flags on Verilog objects

Value Change Link flags are added to an object using **acc_vcl_add()**, and are removed using **acc_vcl_delete()**.

17.2.1 Adding VCL flags

A single routine is used to add VCL flags to objects:

void **acc_vcl_add (**
handle	**object,**	/* handle for a reg, variable, net, event, port, primitive output terminal or primitive inout terminal */
PLI_INT32	*** calback_rtn,**	/* unquoted name of a C function, without parenthesis */
PLI_BYTE8	*** user_data,**	/* user-defined data value */
PLI_INT32	**vcl_flag)**	/* constant **vcl_verilog_logic**, **vcl_verilog_strength** */

The `acc_vcl_add()` routine adds a Value Change Link flag to a Verilog object. From that point on, until the flag is removed, whenever a value change occurs on that object, the specified *consumer routine* will be called.

Most objects which have logic values can have a Value Change Link flag added to the object. These objects are:

- *net* data types: The net must be scalar, an unexpanded vector, or a bit-select of an expanded vector
- *reg* data types: The reg must be scalar, an unexpanded vector, or a bit-select of an expanded vector
- The *integer* and *time* variable data types
- The *real* and *realtime* variable data types
- The *event* data type
- A *primitive output or inout terminal*
- A *scalar module port* or a *bit-select of a module port*

 NOTE When a handle for a module port or port bit is specified, the VCL flag is placed on the loconn of the port. The loconn is the signal inside the module which is connected to that port. When the *consumer routine* is called by the simulator, the information passed to the *consumer routine* about the change will be based on the loconn signal.

The **consumer** argument passed to `acc_vcl_add()` is a pointer to the C function which the simulator should call when a change occurs on the object. A pointer to the function is the unquoted name of the function, without the parentheses after the function name. The PLI standard expects the *consumer routine* to be a C function of type `PLI_INT32`.

The **reason_flag** indicates what types of value changes should cause the *consumer routine* to be called. The *reason_flag* is one of the following two constants:

- **vcl_verilog_logic** indicates the *consumer routine* should be called for logic value changes on the object.
- **vcl_verilog_strength** indicates the *consumer routine* should be called for both logic value changes and strength level changes on the object.

The **user_data** value specified as an argument to `acc_vcl_add()` will be passed to the *consumer routine* whenever a callback occurs. Examples of using the user_data value are presented later in this chapter. The `user_data` is a pointer, which can store a single value, or a pointer to a block of data. If no `user_data` value is needed, the argument should be set to `null`.

The user_data value can be used to pass the handle for an object to the *consumer routine*. Because the user_data is a pointer, several values can be stored in an application-allocated block of memory, and a pointer to the memory block placed in the user_data field. When a pointer to a memory block is placed in the user_data field, the PLI application must ensure that the memory is persistent, and will be available when the *consumer routine* is called. Since the VCL flag will be added from a C function that was called by the simulator, such as a *calltf routine*, local automatic storage cannot be used to store data that is to be accessed from the *consumer routine*. Automatic storage will automatically be freed when the C function exits.

The following code fragment places a VCL flag on to a net, and stores a pointer to the name of the net in the user_data field. The net name is stored in memory that was allocated using malloc. Note that the value stored in user_data must be cast to a PLI_BYTE8* pointer.

```
handle   net_handle;
char     *net_name;
char     *net_name_keep;

/* obtain a handle for a net from the first task arg */
net_handle = acc_handle_tfarg(1);

/* allocate memory for the name of the net */
net_name = (char *)acc_fetch_name(net_handle);
net_name_keep = malloc(strlen(net_name)+1);
strcpy(net_name_keep, net_name);  /* save net_name */

/* add a VCL flag to net--user_data is pointer to net name */
acc_vcl_add(net_handle, my_consumer_routine,
            (PLI_BYTE8 *)net_name_keep, vcl_verilog_logic);
```

An object can have any number of VCL flags added to it, as long as each flag is unique. Either the name of the *consumer routine* or the value of the *user_data* must be different in order to make the flag unique. Using a different reason constant does not make the VCL flag unique.

17.2.2 Removing VCL flags from Verilog objects

A VCL flag can be removed from an object with the following routine:

void **acc_vcl_delete (**
handle	**object,**	/* handle for an object with a VCL flag */
PLI_INT32	*** calback_rtn,**	/* unquoted name of a C function */
PLI_BYTE8	*** user_data,**	/* user-defined data value */
PLI_INT32	**vcl_flag)**	/* constant: **vcl_verilog** */

`acc_vcl_delete()` removes a Value Change Link flag from a Verilog object. The inputs to `acc_vcl_add()` are:

- A *handle for* an object which has a VCL flag.
- The *consumer routine name* which was specified when the VCL flag was added.
- The *user_data* value which was specified when the VCL flag was added.
- A *reason_flag* which must be constant **vcl_verilog**.

An object can have several VCL flags attached to it. Therefore, the arguments to `acc_vcl_delete()` must specify the same *consumer routine name* and the same *user_data* value as in the `acc_vcl_add()` which created the VCL flag.

17.3 Using the VCL consumer routine

When the simulator calls the *consumer routine*, the simulator allocates an **s_vc_record** structure and passes a pointer to the structure as an input to the *consumer routine*. The structure contains information about the value change that occurred. The s_vc_record structure is defined in acc_user.h.

```
typedef struct t_vc_record {
  PLI_INT32 vc_reason;              /* one of the constants:
                                       logic_value_change,
                                       strength_value_change,
                                       vector_value_change,
                                       sregister_value_change,
                                       vregister_value_change,
                                       integer_value_change,
                                       real_value_change,
                                       time_value_change,
                                       event_value_change */
  PLI_INT32  vc_hightime;    /* upper 32-bits of time */
  PLI_INT32  vc_lowtime;     /* lower 32-bits of time */
                       /* (time is in simulator time units) */
  PLI_BYTE8 *user_data;      /* value passed to acc_vcl_add*/
  union {
    PLI_UBYTE8 logic_value;  /* for logic_value_change, one
                                of: vcl0, vcl1, vclZ, vclX */
    double     real_value;   /* for real_value_change */
    handle     vector_handle; /* for vector_value_change,
                                  vregister_value_change,
                                  integer_value_change,
                                  time_value_change */
    s_strengths strengths_s; /* for strength_value_change */
  } out_value;
```

The fields of the s_vc_record structure are filled in by the simulator. The **vc_reason** is a constant which represents the type of Verilog object which changed value, as listed in table 17-1.

Table 17-1: vc_reason constants

vc_reason Constant	Description
logic_value_change	scalar net or bit-select of a vector net changed logic value
strength_value_change	scalar net or bit-select of a vector net changed strength
vector_value_change	vector net or part-select of a vector net changed logic value
sregister_value_change	scalar register or bit-select of a vector register changed logic value
vregister_value_change	vector register or part-select of vector register changed logic value
integer_value_change	integer variable changed value
real_value_change	real variable changed value
time_value_change	time variable changed value
event_value_change	event data type had an event

vc_hightime and **vc_lowtime** contain the simulation in which the value change occurred (which is the current simulation time). The 64-bit simulation time is stored as two 32-bit C integers, as shown in the following diagram.

```
unsigned int   vc_hightime, vc_lowtime;
```

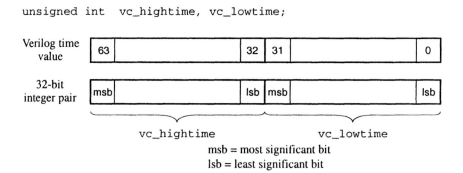

The simulation time is represented in the internal simulation time units. It is *not* scaled to the time units of the module containing the object with the VCL flag.

The **out_value** field contains information about the new logic value of the object which has the VCL flag. Because there are many types of objects which can have VCL flags, the out_value is a union of C data types. The field within the union that contains the value information is based on the **vc_reason** constant. Table 17-2 shows which `out_value` union field is used for the different reason constants.

Table 17-2: s_vc_record out_value member fields

vc_reason	out_value Field With New Value	The New Logic Value Representation
logic_value_change	**logic_value**	one of the constants: **vcl0 vcl1 vclX vclZ**
strength_value_change	**strengths_s**	a structure with the logic and strength
vector_value_change	**vector_handle**	a handle for a vector net or part-select of a vector net
sregister_value_change	**logic_value**	one of the constants: **vcl0 vcl1 vclX vclZ**
vregister_value_change	**vector_handle**	a handle for a vector reg or part-select of a vector reg
integer_value_change	**vector_handle**	a handle for an integer variable
real_value_change	**real_value**	the value of a real variable
time_value_change	**vector_handle**	a handle for a time variable
event_value_change	**none**	(event types have no logic value)

When the object which changed logic value is a vector signal, the simulator passes a handle for the object to the *consumer routine*. Using the handle, the routine can retrieve the logic value using `acc_fetch_value()`. This allows the routine to retrieve the vector value in a wide variety of formats.

When the object which changed value is a scalar net or bit select of a vector net, and the VCL flag was for `vcl_verilog_logic`, the new value is represented as a constant, which encodes the Verilog 4-state logic. Note that the constants used to represent the logic values are defined as `char` data type. This is unique from most other PLI constants, which are of `int` data types.

When the object which changed is a scalar net or bit select of a vector net, and the VCL flag was for `vcl_verilog_strength`, the new value is represented as an **s_strengths** structure. Within the strength structure, the new logic value is represented with a constant which encodes the Verilog 4-state logic. The constants are the same as with scalar logic value changes: **vcl0, vcl1, vclX** and **vclZ**. The strength is represented as a pair of constants, which represent the 8 strength levels available for a logic 0 and logic 1. The strength constants are: **vclSupply, vclStrong, vclPull, vclLarge, vclWeak, vclMedium, vclSmall** and **vclHighZ**. The s_strengths structure is defined in acc_user.h.

```
typedef struct t_strengths {
  PLI_UBYTE8 logic_value;  /* one of: vcl0, vcl1, vclZ, vclZ */
  PLI_UBYTE8 strength1;    /* each strength byte is one of: */
  PLI_UBYTE8 strength2;    /*    vclSupply, vclStrong, vclPull,
                                 vclLarge,  vclWeak, vclMedium,
                                 vclSmall,  vclHighZ */
} s_strengths, *p_strengths;
```

17.4 An example of using Value Change Link routines

$my_monitor is a complete PLI application which uses the ACC Value Change Link capability. The application monitors value changes of vector nets, regs or variables. When a signal changes, the application prints the current simulation time, the old logic value and the new logic value of the signal. (Chapter 18 contains additional examples of using the Value Change Link routines.)

CD The source code for this example is on the CD accompanying this book.

- Application source file: `Chapter.17/my_monitor_acc.c`
- Verilog test bench: `Chapter.17/my_monitor_test.v`
- Expected results log: `Chapter.17/my_monitor_test.log`

Example 17-1: *$my_monitor* — using the acc_vcl_add() routine

```
/*****************************************************************
 * calltf routine
 *****************************************************************/
typedef struct PLIbook_MyMon_t {
  PLI_BYTE8 signalName[256];  /* signal names--up to 255 characters */
  PLI_BYTE8 lastValue[2];     /* scalar logic value stored as a string */
} PLIbook_MyMon_s, *PLIbook_MyMon_p;

PLIbook_MyMonitor_calltf(int user_data, int reason) {
  handle signal_h;
  int    i, numargs;

  /* allocate memory for an array of p_monitor structures */
  PLIbook_MyMon_p monArray; /* starting address for the array */
  numargs = (int)tf_nump();
  monArray=(PLIbook_MyMon_p)malloc(numargs*(sizeof(PLIbook_MyMon_s)));

  acc_initialize();
  /* save name and current logic value of each signal */
  for (i=0; i<numargs; i++) {
    signal_h = acc_handle_tfarg(i+1);
    strcpy(monArray[i].signalName, acc_fetch_fullname(signal_h));
```

```
    strcpy(monArray[i].lastValue,acc_fetch_value(signal_h,"%b",null));
  /* add a VCL flag to each net--user_data is a pointer to saved info */
    acc_vcl_add(signal_h,
                PLIbook_MyMonitor_consumer,
                (PLI_BYTE8 *)&monArray[i],
                vcl_verilog_logic);
  }
  acc_close();
  return(0);
}

/********************************************************************
 * consumer routine
 ********************************************************************/
int PLIbook_MyMonitor_consumer(p_vc_record vc_record)
{
  char newValue[2];
  /* retrieve pointer to data structure array from user_data field */
  PLIbook_MyMon_p ArrayElem_p = (PLIbook_MyMon_p)vc_record->user_data;
  switch (vc_record->vc_reason) { /* check reason call-back occurred */
    case logic_value_change:            /* scalar net changed */
    case sregister_value_change : {  /* scalar register changed */
      switch (vc_record->out_value.logic_value) { /* convert value */
        case vcl0: strcpy(newValue, "0"); break;  /* to string */
        case vcl1: strcpy(newValue, "1"); break;
        case vclX: strcpy(newValue, "x"); break;
        case vclZ: strcpy(newValue, "z"); break;
      }
      io_printf("At time %4d: %-20s last value=%s    new value=%s\n",
                vc_record->vc_lowtime, ArrayElem_p->signalName,
                ArrayElem_p->lastValue, newValue);
      strcpy(ArrayElem_p->lastValue, newValue); /* save the new value */
    }
  }
  return(0);
}
```

17.5 Obtaining object handles from the consumer routine

The VCL *consumer routine* needs the handles for the system task arguments which represent the C model, in order to read the input values and write the C model output values. Since the *consumer routine* is not associated with the system task, access to the task arguments must be passed to the *consumer routine* through the VCL user_data field. This information can be passed in either of two ways:

- When the *calltf routine* sets up the VCL callbacks, the *calltf routine* can obtain a handle for the system task instance, and pass the handle to the *consumer routine* through the VCL `user_data` field. The *consumer routine* can then obtain the handle for any of the system task arguments using `acc_handle_itfarg()`, which is the instance specific version of `acc_handle_tfarg()`.

- When the *calltf routine* sets up the VCL callbacks, the *calltf routine* can allocate persistent storage. The handles for the system task arguments can be stored in the memory which was allocated. A pointer to the storage can be passed to the *consumer routine* through the VCL `user_data` field. The *consumer routine* can then obtain the handles from the storage block.

Each of these methods has advantages. Saving just the handle for the system task instance is simpler, and makes the instance handle available for other uses, such as scheduling a callback to the *misctf routine* for that instance of the system task. Saving all handles in a block of memory means the handles do not need to be obtained each time a value change occurs. Example 18-3, in the following chapter, illustrates passing the system task instance handle, and example 18-4 illustrates allocating a block of memory to store the handles for all system task arguments, and passing a pointer to the memory to the consumer routine.

17.6 Summary

This chapter has presented one of the more powerful capabilities of the ACC library—the ability to have a Verilog simulator call a PLI application whenever any specific objects change logic value. The ACC routines used for Value Change Links are simple to use.

The next chapter shows additional ways to use the ACC Value Change Link capability. The VCL callbacks will be used to create both combinational and sequential interfaces between a Verilog simulation and hardware designs modeled in the C programming language.

CHAPTER 18 *Interfacing to C Models Using ACC Routines*

*T*he Value Change Link in the ACC routines can be used to create a simple and efficient interface to C language models. The ACC routines can also be used in conjunction with the TF routines for synchronizing C models with changes in simulation time. The power of the ACC and TF libraries together with the flexibility of C programming provide countless ways to represent a C model interface. This chapter shows several of these ways.

The concepts presented in this chapter are:

- Representing hardware models in C
- Verilog HDL shell modules
- Combinational logic interfaces to C models
- Sequential logic interfaces to C models
- Synchronizing with the end of a simulation time step
- Synchronizing with a future simulation time step
- Multiple instances of a C model
- Creating instance specific storage within C models
- Representing propagation delays in C models

TIP

One reason for representing hardware models in the C language is to achieve faster simulation performance. The C programming language allows a very abstract, algorithmic representation of hardware functionality, without representing detailed timing, multi-state logic, hardware concurrency and other hardware specific details offered by the Verilog language.

The PLI can be a means to access the efficiency of a highly abstract C model. However, a poorly written PLI application can become a bottleneck that offsets much of the efficiency gains. Care must be taken to write PLI applications that execute as efficiently as possible. Some guidelines that can help maximize the efficiency and run-time performance of PLI applications are:

- Good C programming practices are essential. General C programming style and technique are not discussed within the scope of this book.

- Consider every call to a PLI routine as expensive, and try to minimize the number of calls.

- Routines which convert logic values from a simulator's internal representation to C strings, and vice-versa, are very expensive in terms of performance. Best efficiency is attained when the value representation in C is as similar as possible to the value representation in Verilog.

- Use the Verilog language to model the things hardware description languages do well, such as representing hardware parallelism and hardware propagation times. Simulator vendors have invested a great deal in optimizing a simulator's algorithms, and that optimization should be utilized.

The objective of this book is to show several ways in which the ACC library can be used to interface to C models. Short examples are presented that are written in a relatively easy to follow C coding style. In order to meet the book's objectives, the examples do not always follow the guidelines of efficient C coding and prudent usage of the PLI routines. It is expected that if these examples are adapted for other applications, the coding style will also be modified to be more efficient and robust.

18.1 How to interface C models with Verilog simulations

The power and flexibility of the C programming language and the Verilog PLI provide a wide variety of methods that can be used to interface a Verilog simulation with a C language model. All methods have three essential concepts in common:

- Value changes which occur in the Verilog simulator must be passed to the C model.

- Value changes within the C model must be passed to the Verilog simulation.

- Simulated time in both the Verilog simulation and the C model must remain synchronized.

This chapter will present some of the more common methods of interfacing a Verilog simulation with a C model. The methods presented are by no means the only ways this interface can be accomplished, and may not always be the most efficient method. However, the methods presented have many advantages, including simplicity to implement, portability to many types of Verilog simulators, and the ability to use the C model any number of times and anywhere in the hierarchy of a Verilog design.

The fundamental steps that are presented in this chapter are:

1. Create the C language model as an independent block of pure C code that does not use the PLI routines in any way. The C model will have inputs and outputs, but it will not know the source of the inputs or the destination of the outputs. The C code to implement the model might be in the form of a C function with no main function, or it might be a complete C program with its own main function.

2. Create a Verilog HDL *shell module*, also called a *wrapper module* or *bus functional module* (BFM), to represent the inputs and outputs of the C language model. This module will be written completely in the Verilog language, but will not contain any functionality. To represent the functionality of the model, the shell module will call a PLI application.

3. Create a PLI application to serve as an interface between the C model and the Verilog shell module. The PLI application is a communication channel, which:

    • Uses PLI routines to retrieve data from the Verilog HDL shell module and pass the data to the C model via standard C programming.

    • Uses standard C programming to receive data from the C model, and pass the data to the Verilog shell module via PLI routines.

The following diagram shows how the blocks which are created in these three steps interact with each other.

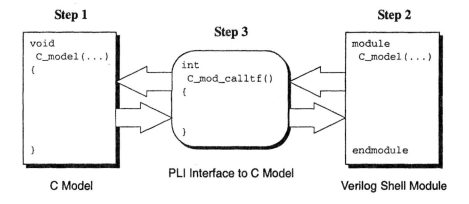

This chapter presents steps 2 and 3 of this interface method in detail. Step 1 is to model some desired functionally or algorithm in the C language. This step is pure C programming, which does not directly involve the Verilog language or the Verilog PLI. This chapter does not cover how to implement ideas in the C language—the focus is on how to interface that implementation with a Verilog simulation. To maintain this focus, the C model example presented in this chapter will be a practical example, but relatively simple to model in C. The C model example used illustrates all of the important concepts of integrating C models into a Verilog simulation.

18.2 Creating the C language model

A hardware model can be represented in the C programming language in two basic forms, either as a C function or as an independent C program.

18.2.1 Using functions to represent the C model

When the C model is represented as a C function, that function can be linked into the Verilog simulator, together with the PLI application that serves as the interface to the model. The PLI application can then call the function when needed, passing inputs to the function, and receiving outputs from the function. One advantage of representing a C model as a function is the simplicity of passing values to and from the model. Another advantage is ease of porting to different operating systems, since the C model is called directly from the PLI application as a C function. A disadvantage of using a function to represent the C model is that the C model must contain additional code to allow a Verilog design to instantiate the C model multiple times. The model needs to specifically create unique storage for each instance.

18.2.2 Using independent programs to represent the C model

When the C model is represented as an independent program, which means it has its own C *main* function, then the Verilog simulation and the C model can be run as parallel processes on the same or on different computers. The PLI application which serves as an interface between the simulation and the model will need to create and maintain some type of communication channel between the two programs. This communication can be accomplished several ways, such as using the exec command in the C standard library. On Unix operating systems, the fork or vfork commands with either Unix pipes or Unix sockets is an efficient method to communicate with the C model program. On PC systems running a DOS or windows operating system, the spawn command can be used to invoke the C model program and establish two-way communications between the PLI application and the C model process.

One of the advantages of representing the C model as an independent model is the
ability to have parallel processes running on the same computer or separate computers. Another advantage is that when a Verilog design instantiates multiple instances of
the C model, each instance will be a separate process with its own memory storage.
The major disadvantage of independent programs when compared to using a C function to represent the C model is that the PLI interface to invoke and communicate
with the separate process is more complex, and might be operating system dependent.

David Roberts, of Cadence Design Systems, who reviewed many of the chapters of
this book, has provided a full example of representing a C model as a separate C program. This example is included with the CD that accompanies this book.

18.3 A C model example

The C model used for the different PLI interfaces shown in this chapter is a scientific
Arithmetic Logic Unit, which utilizes the C math library. The C model is represented
as a C function, which will be called from the PLI interface mechanism. This model is
written entirely with standard C library routines and C data types, without reference
to any PLI routines or PLI data types. This same example is also used in other chapters, to show how a PLI interface to C models can be created using the VPI and TF
libraries of the PLI.

The inputs and outputs of the scientific ALU C model are shown below, and Table 18-
1 shows the operations which the ALU performs.

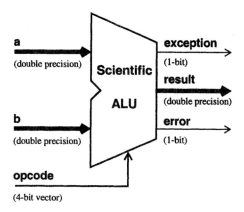

exception is set to 1 whenever an operation results in a value which is out of range
of the double-precision result.

error is set to 1 whenever an input to an operation is out of range for the operation.

Table 18-1: Scientific ALU C model operations

Opcode	C Math Library Operation
0	pow(a,b) — returns a to the power of b
1	sqrt(a) — returns the square root of a
2	exp(a) — returns the natural exponent of a
3	ldexp(a,b) — returns a * (2 to the power of b)
4	fabs(a) — returns the absolute of a
5	fmod(a,b) — returns the floating remainder of a / b
6	ceil(a) — returns smallest whole number not less than a
7	floor(a) — returns largest whole number not more than a
8	log(a) — returns the natural log of a
9	log10(a) — returns the base 10 log of a
A	sin(a) — returns the sine of a
B	cos(a) — returns the cosine of a
C	tan(a) — returns the tangent of a
D	asin(a) — returns the arcsine of a
E	acos(a) — returns the arccosine of a
F	atan(a) — returns the arctangent of a

The source code for the scientific ALU is listed in Example 18-1. This version of the ALU generates a result, but does not store the result. A latched version of the ALU which stores the operation result is listed later in this chapter.

CD The source code for this example is on the CD accompanying this book.

- Application source file: `Chapter.18/sci_alu_combinational_acc.c`
- Verilog shell module: `Chapter.18/sci_alu_combinational_shell.v`
- Verilog test bench: `Chapter.18/sci_alu_combinational_test.v`
- Expected results log: `Chapter.18/sci_alu_combinational_test.log`

Example 18-1: scientific ALU C model — combinational logic version

```
#include <stdlib.h>
#include <stdio.h>
#include <math.h>
#include <errno.h>
void PLIbook_ScientificALU_C_model(
        double  a,        /* input */
        double  b,        /* input */
        int     opcode,   /* input */
        double *result,   /* output from ALU */
        int    *excep,    /* output; set if result is out of range */
        int    *err)      /* output; set if input is out of range */
{
  switch (opcode) {
    case 0x0: *result = pow    (a, b);      break;
    case 0x1: *result = sqrt   (a);         break;
    case 0x2: *result = exp    (a);         break;
    case 0x3: *result = ldexp  (a, (int)b); break;
    case 0x4: *result = fabs   (a);         break;
    case 0x5: *result = fmod   (a, b);      break;
    case 0x6: *result = ceil   (a);         break;
    case 0x7: *result = floor  (a);         break;
    case 0x8: *result = log    (a);         break;
    case 0x9: *result = log10  (a);         break;
    case 0xA: *result = sin    (a);         break;
    case 0xB: *result = cos    (a);         break;
    case 0xC: *result = tan    (a);         break;
    case 0xD: *result = asin   (a);         break;
    case 0xE: *result = acos   (a);         break;
    case 0xF: *result = atan   (a);         break;
  }
  *excep = (errno == ERANGE);  /* result of math func. out-of-range */
#ifdef WIN32  /* for Microsoft Windows compatibility */
  *err   = (_isnan(*result) ||  /* result is not-a-number, or */
            errno == EDOM);      /* arg to math func. is out-of-range */
#else
  *err   = (isnan(*result) ||   /* result is not-a-number, or */
            errno == EDOM);      /* arg to math func. is out-of-range */
#endif
  if (*err) *result = 0.0;      /* set result to 0 if error occurred */
  errno = 0;                    /* clear the error flag */
  return;
}
```

18.4 Creating a Verilog shell module

A *shell module* allows a Verilog design to reference a C model using standard Verilog HDL syntax. The shell module is a Verilog module which has the same input and output ports as the C model, but the module has no functionality modeled within. To represent the module's functionality, the shell module invokes a PLI application, which in turn invokes the C model. A shell module is sometimes referred to as a *wrapper module*, because the module is wrapped around the call to a PLI application.

The shell module for a combinational logic version of the scientific ALU is listed below.

> **CD** The source code for this example is on the CD accompanying this book.
>
> - Application source file: `Chapter.18/sci_alu_combinational_acc.c`
> - Verilog shell module: `Chapter.18/sci_alu_combinational_shell.v`
> - Verilog test bench: `Chapter.18/sci_alu_combinational_test.v`
> - Expected results log: `Chapter.18/sci_alu_combinational_test.log`

Example 18-2: Verilog shell module for the Scientific ALU C model

```
`timescale 1ns / 1ns
module scientific_alu(a_in, b_in, opcode,
                      result_out, exception, error);
  input  [63:0] a_in, b_in;
  input   [3:0] opcode;
  output [63:0] result_out;
  output        exception, error;

  real          a, b, result; // real variables used in this module
  reg           exception, error;

  // convert real numbers to/from 64-bit vector port connections
  assign result_out = $realtobits(result);
  always @(a_in) a  = $bitstoreal(a_in);
  always @(b_in) b  = $bitstoreal(b_in);

  //call the PLI application which interfaces to the C model
  initial
    $scientific_alu(a, b, opcode, result, exception, error);

endmodule
```

In this scientific ALU example, the primary inputs and outputs of the model are double-precision floating point values, represented as Verilog `real` data types. The Verilog language does not permit real numbers to be connected to module ports. However, the language provides built-in system functions which convert real numbers to 64-bit vectors, and vice-versa, so the real values can be passed through a module port connection. These built-in system functions are *$realtobits()* and *$bitstoreal()*.

The Verilog shell module that represents the C model can be instantiated in a design in the same way as any other Verilog module. For example:

```
module chip (...)
    ...
    scientific_alu u1 (a, b, opcode, result, excep, err);
    ...
endmodule
```

Creating a shell module to represent the C model is not mandatory—the PLI application could be called directly from any place in a Verilog design. However, there are important advantages to using a shell module to represent the C model:

- The shell module provides a simple method to encapsulate the C model.
- The shell module can be instantiated anywhere in a Verilog design hierarchy.
- The shell module can be instantiated any number of times in a Verilog design.
- The shell module can add Verilog HDL delays to the C model, which can accurately represent rise and fall delay delays, state-dependent delays, and timing constraints such as setup times.
- Delays within a shell module can be annotated using delay calculators or SDF files for additional delay accuracy for each instance of the shell module.

Section 18.10 later in this chapter discusses how delays can be represented in the Verilog shell module.

18.5 Creating a combinational logic interface to a C model

In a combinational logic model, the outputs of the model continuously reflect the input values of the model. The inputs are asynchronous—when any input changes value, the model outputs are re-evaluated to reflect the input change.

The ACC Value Change Link provides an efficient method of creating a combinational logic interface to a C model. VCL flags can be attached to each input of the C model. Whenever an input changes, the new input values can be passed to the C model. The outputs of the C model are then passed back to the Verilog simulation by writing the results onto the system task arguments in the Verilog shell module. How the ACC Value Change Link is used was presented in Chapter 17.

The basic steps involved with using the Value Change Link to implement a combinational logic interface are:

1. Create a PLI application system task to represent the interface between the Verilog shell module and the C model. The system task is invoked from the shell module, and all of the C model inputs and outputs are listed as arguments to the system task. For example:

    ```
    initial
        $scientific_alu(a, b, opcode, result, exception, error);
    ```

    Note that, in this example, the system task is called from a Verilog **initial** procedure, which means the task will only be invoked one time for each instance of the shell module.

2. In the *calltf routine* associated with the system task, add VCL flags to each system task argument which represents an input to the C model. The same VCL *consumer routine* is listed for each model input.

3. In the VCL *consumer routine*, which is called whenever an input changes value, read the values of all inputs and pass the values to the C model. The output values of the C model are returned to the same *consumer routine*, which then writes the values to the system task arguments that represent the outputs of the C model.

 The VCL *consumer routine* is not directly associated with the system task which represents the C model interface. Therefore, the *consumer routine* cannot obtain handles to the system task arguments using acc_handle_tfarg(), unless a handle to the system task is passed to the *consumer routine* using the user_data field. Using that handle, the system task arguments can be read using the instance specific fetch routines, such as acc_fetch_itfarg(). An alternate method is to allocate storage in the calltf routine, and save the handles for all of the system task/function arguments.A pointer to the storage can then be passed to the consumer routine through the user_data field.

The following example implements a combinational logic interface for the scientific ALU C model.

CD The source code for this example is on the CD accompanying this book.

- Application source file: `Chapter.18/sci_alu_combinational_acc.c`
- Verilog shell module: `Chapter.18/sci_alu_combinational_shell.v`
- Verilog test bench: `Chapter.18/sci_alu_combinational_test.v`
- Expected results log: `Chapter.18/sci_alu_combinational_test.log`

Example 18-3: combinational logic C model interface using ACC routines

```
#define ALU_A       1  /* system task arg 1 is ALU A input      */
#define ALU_B       2  /* system task arg 2 is ALU B input      */
#define ALU_OP      3  /* system task arg 3 is ALU opcode input  */
#define ALU_RESULT  4  /* system task arg 4 is ALU result output */
#define ALU_EXCEPT  5  /* system task arg 5 is ALU exception output */
#define ALU_ERROR   6  /* system task arg 6 is ALU error output  */


/*********************************************************************
 * VCL simulation callback routine: Serves as an interface between
 * Verilog simulation and the C model.  Called whenever the C model
 * inputs change value, passes the values to the C model, and puts
 * the C model outputs into simulation.
 *********************************************************************/
int PLIbook_ScientificALU_interface(p_vc_record vc_record)
{

    double         a, b, result;
    int            opcode, excep, err;
    handle         instance_h, result_h, excep_h, err_h,
                   a_h, b_h, opcode_h;
    s_setval_value value_s;
    s_setval_delay delay_s;
    s_acc_time     time_s;

    acc_initialize();

    /* Retrieve instance handle from VCL user_data field */
    instance_h = (handle)vc_record->user_data;

    /* Obtain handles to all task arguments */
    a_h      = acc_handle_itfarg(ALU_A,      instance_h);
    b_h      = acc_handle_itfarg(ALU_B,      instance_h);
    opcode_h = acc_handle_itfarg(ALU_OP,     instance_h);
    result_h = acc_handle_itfarg(ALU_RESULT, instance_h);
    excep_h  = acc_handle_itfarg(ALU_EXCEPT, instance_h);
    err_h    = acc_handle_itfarg(ALU_ERROR,  instance_h);

    /* Read current values of C model inputs from Verilog simulation */
    value_s.format = accRealVal;
    acc_fetch_value(a_h, "%%", &value_s);
    a = value_s.value.real;
```

```
    acc_fetch_value(b_h, "%%", &value_s);
    b = value_s.value.real;

    value_s.format = accIntVal;
    acc_fetch_value(opcode_h, "%%", &value_s);
    opcode = (int)value_s.value.integer;


    /****** Call C model ******/
    PLIbook_ScientificALU_C_model(a, b, opcode, &result, &excep, &err);


    /* Write the C model outputs onto the Verilog signals */
    delay_s.model      = accNoDelay;
    delay_s.time       = time_s;
    delay_s.time.type  = accRealTime;
    delay_s.time.real  = 0.0;

    value_s.format       = accRealVal;
    value_s.value.real = result;
    acc_set_value(result_h, &value_s, &delay_s);

    value_s.format         = accIntVal;
    value_s.value.integer = (PLI_INT32)excep;
    acc_set_value(excep_h, &value_s, &delay_s);

    value_s.value.integer = (PLI_INT32)err;
    acc_set_value(err_h, &value_s, &delay_s);

    acc_close();
    return(0);
}

/******************************************************************
 * calltf routine: Registers a callback to the C model interface
 * whenever any input to the C model changes value
 ******************************************************************/
int PLIbook_ScientificALU_calltf(int user_data, int reason)
{
    handle  instance_h, a_h, b_h, opcode_h;

    acc_initialize();

    /* get handles for signals in task args which are C model inputs */
    a_h        = acc_handle_tfarg(ALU_A);
    b_h        = acc_handle_tfarg(ALU_B);
    opcode_h   = acc_handle_tfarg(ALU_OP);

    /* get handles for this system task instance to pass to VCL app. */
    instance_h = acc_handle_tfinst();

    /* add VCL flags to all signals which are inputs to the C model  */
```

```
    /* pass handle for task instance as the user_data value          */
    acc_vcl_add(a_h, PLIbook_ScientificALU_interface,
               (PLI_BYTE8 *)instance_h, vcl_verilog_logic);
    acc_vcl_add(b_h, PLIbook_ScientificALU_interface,
               (PLI_BYTE8 *)instance_h, vcl_verilog_logic);
    acc_vcl_add(opcode_h, PLIbook_ScientificALU_interface,
               (PLI_BYTE8 *)instance_h, vcl_verilog_logic);

    acc_close();
    return(0);
}

/**********************************************************************
 * checktf routine: Verifies that $scientific_alu() is used correctly.
 *    Note: For simplicity, only limited data types are allowed for
 *    task arguments.  Could add checks to allow other data types.
 **********************************************************************/
int PLIbook_ScientificALU_checktf(int user_data, int reason)
{
    acc_initialize();

    if (tf_nump() != 6)
        tf_error("$scientific_alu requires 6 arguments");

    else {
        if (!(acc_object_of_type(acc_handle_tfarg(ALU_A), accRealVar)))
            tf_error("$scientific_alu arg 4 must be a real variable\n");

        if (!(acc_object_of_type(acc_handle_tfarg(ALU_B), accRealVar)))
            tf_error("$scientific_alu arg 5 must be a real variable\n");

        if (!(acc_object_of_type(acc_handle_tfarg(ALU_OP), accWire)))
            tf_error("$scientific_alu arg 6 must be a net\n");
        else if (acc_fetch_size(acc_handle_tfarg(ALU_OP)) != 4)
            tf_error("$scientific_alu arg 6 must be a 4-bit vector\n");

        if (!(acc_object_of_type(acc_handle_tfarg(ALU_RESULT),accRealVar)))
            tf_error("$scientific_alu arg 1 must be a real variable\n");

        if (!(acc_object_of_type(acc_handle_tfarg(ALU_EXCEPT), accReg)))
            tf_error("$scientific_alu arg 2 must be a reg\n");
        else if (acc_fetch_size(acc_handle_tfarg(ALU_EXCEPT)) != 1)
            tf_error("$scientific_alu arg 2 must be scalar\n");

        if (!(acc_object_of_type(acc_handle_tfarg(ALU_ERROR), accReg)))
            tf_error("$scientific_alu arg 3 must be a reg\n");
        else if (acc_fetch_size(acc_handle_tfarg(ALU_ERROR)) != 1)
            tf_error("$scientific_alu arg 3 must be scalar\n");
    }

    acc_close();
    return(0);
}
```

18.6 Creating a sequential logic interface to a C model

In a sequential logic model, the outputs of the model change synchronously with an input strobe, such as a positive edge of a clock. There may also be one or more asynchronous inputs, such as a reset signal.

The ACC Value Change Link is a straightforward way to model a sequential logic interface to a C model. VCL flags are attached only to the clock input and any asynchronous inputs of the C model. Only when those inputs change, are the new input values passed to the C model.

The basic steps involved with using the Value Change Link to implement a synchronous sequential logic interface are very similar to implementing a combinational logic interface. The one difference is that VCL flags are only attached to specific C model inputs instead of all inputs.

Regardless of whether the interface is combinational or sequential logic, the VCL *consumer routine* needs the handles for the system task arguments which represents the C model, in order to read the inputs values and write the C model output values. Access to the task arguments must be passed to the *consumer routine*.

Example 18-4, which follows, illustrates allocating a block of memory to store the handles for all system task arguments. This example implements a sequential logic interface for the scientific ALU C model, where all inputs are synchronized to value changes of a clock input.

> **CD** The source code for this example is on the CD accompanying this book.
>
> - Application source file: `Chapter.18/sci_alu_sequential_acc.c`
> - Verilog shell module: `Chapter.18/sci_alu_sequential_shell.v`
> - Verilog test bench: `Chapter.18/sci_alu_sequential_test.v`
> - Expected results log: `Chapter.18/sci_alu_sequential_test.log`

Example 18-4: sequential logic C model interface using ACC routines

```
#define ALU_CLOCK   1  /* system task arg 1 is ALU clock input    */
#define ALU_A       2  /* system task arg 2 is ALU A input        */
#define ALU_B       3  /* system task arg 3 is ALU B input        */
#define ALU_OP      4  /* system task arg 4 is ALU opcode input   */
#define ALU_RESULT  5  /* system task arg 5 is ALU result output  */
#define ALU_EXCEPT  6  /* system task arg 6 is ALU exception output */
#define ALU_ERROR   7  /* system task arg 7 is ALU error output   */

/******************************************************************
```

```
 * Definition for a structure to hold the data to be passed from
 * calltf routine to the ALU interface (a VCL consumer routine).
 ******************************************************************/
typedef struct PLIbook_SciALU_data {
  handle  clock_h, a_h, b_h, opcode_h, result_h, excep_h, err_h;
} PLIbook_SciALU_data_s, *PLIbook_SciALU_data_p;

/******************************************************************
 * VCL simulation callback routine: Serves as an interface between
 * Verilog simulation and the C model.  Called whenever the C model
 * clock input changes value, reads the values of all model inputs,
 * passes the values to the C model, and writes the C model outputs
 * into simulation.
 ******************************************************************/
int PLIbook_ScientificALU_interface(p_vc_record vc_record)
{
  double           a, b, result;
  int              clock, opcode, excep, err;
  s_setval_value   value_s;
  s_setval_delay   delay_s;
  s_acc_time       time_s;

  PLIbook_SciALU_data_p  ALUdata;

  acc_initialize();

  /* Retrieve pointer to ALU data structure from VCL user_data field */
  ALUdata = (PLIbook_SciALU_data_p)vc_record->user_data;

  /* Read current values of C model inputs from Verilog simulation */
  value_s.format = accIntVal;
  acc_fetch_value(ALUdata->clock_h, "%%", &value_s);
  clock = (int)value_s.value.integer;
  if (clock != 1) /* abort if not a positive edge of the clock input */
    return(0);

  value_s.format = accRealVal;
  acc_fetch_value(ALUdata->a_h, "%%", &value_s);
  a = value_s.value.real;

  acc_fetch_value(ALUdata->b_h, "%%", &value_s);
  b = value_s.value.real;

  value_s.format = accIntVal;
  acc_fetch_value(ALUdata->opcode_h, "%%", &value_s);
  opcode = (int)value_s.value.integer;


  /****** Call C model ******/
  PLIbook_ScientificALU_C_model(0, a, b, opcode, &result, &excep, &err);


  /* Write the C model outputs onto the Verilog signals */
```

```
  delay_s.model      = accNoDelay;
  delay_s.time       = time_s;
  delay_s.time.type  = accRealTime;
  delay_s.time.real  = 0.0;

  value_s.format     = accRealVal;
  value_s.value.real = result;
  acc_set_value(ALUdata->result_h, &value_s, &delay_s);

  value_s.format      = accIntVal;
  value_s.value.integer = (PLI_INT32)excep;
  acc_set_value(ALUdata->excep_h, &value_s, &delay_s);

  value_s.value.integer = (PLI_INT32)err;
  acc_set_value(ALUdata->err_h, &value_s, &delay_s);

  acc_close();
  return(0);
}

/***********************************************************************
 * calltf routine: Registers a callback to the C model interface
 * whenever the clock input to the C model changes value
 ***********************************************************************/
int PLIbook_ScientificALU_calltf(int user_data, int reason)
{
  PLIbook_SciALU_data_p  ALUdata;

  acc_initialize();

  ALUdata=(PLIbook_SciALU_data_p)malloc(sizeof(PLIbook_SciALU_data_s));

  /* get handles for all signals in Verilog which connect to C model */
  ALUdata->clock_h  = acc_handle_tfarg(ALU_CLOCK);
  ALUdata->a_h      = acc_handle_tfarg(ALU_A);
  ALUdata->b_h      = acc_handle_tfarg(ALU_B);
  ALUdata->opcode_h = acc_handle_tfarg(ALU_OP);
  ALUdata->result_h = acc_handle_tfarg(ALU_RESULT);
  ALUdata->excep_h  = acc_handle_tfarg(ALU_EXCEPT);
  ALUdata->err_h    = acc_handle_tfarg(ALU_ERROR);

  /* add VCL flag to the clock input to the C model */
  /* pass pointer to storage for handles as user_data value */
  acc_vcl_add(ALUdata->clock_h, PLIbook_ScientificALU_interface,
              (PLI_BYTE8 *)ALUdata, vcl_verilog_logic);

  acc_close();
  return(0);
}
```

18.7 Synchronizing with the end of a simulation time step

Within a simulation, several inputs to the C model might change at the same moment in simulation time. In Verilog simulators, the Value Change Link *consumer routine* will be called for each input change, which means these routine may be called before all input value changes have occurred for that time step.

With a combinational logic interface, the *consumer routine* will be called for every input change. This is the correct functionality for combinational logic—at the completion of a simulation time step, the outputs from the C model represent the most current input values. However, by synchronizing the call to the C model with the end of the simulation time step in which changes occur, the multiple calls to the C model within a time step could be optimized to a single call.

With a sequential logic interface synchronized to a clock, when the *consumer routine* is called at a clock change, other input changes at that moment in simulation time may or may not have occurred. It may be desirable to ensure that the C model is not called until all inputs have their most current value for the time step in which the clock changes.

By using both the ACC and TF libraries, both combinational logic and sequential logic C model interfaces can be synchronized to the end of a current simulation time step. This is done by using the VCL *consumer routine* to schedule a callback to the *misctf routine* for that instance of the system task which represents the C model.

Since the VCL *consumer routine* is not directly associated with a system task, a handle to the system task must be passed to the *consumer routine*. The *consumer routine* can then schedule a callback to the *misctf routine* using `tf_isynchronize()`, which is the instance specific version of `tf_synchronize()`.

Note that `tf_synchronize()` does not guarantee that all changes in the current simulation time step have occurred. The `tf_synchronize()` routine schedules a read/write synchronization callback to the *misctf routine* after all known events in a time step have been processed by the simulator. However, other PLI applications or simulation activity may schedule new events in the same simulation time step. The events caused by other PLI applications could affect the inputs of the C model after the `tf_synchronize()` callback has occurred.

 The IEEE 1364 Verilog standard allows PLI read-write synchronization events to be processed either before or after nonblocking assignments. This ambiguity can lead to different results from different simulators. The reason the Verilog standard allows this latitude in event scheduling is discussed in Chapter 12, on page 221.

If the C model interface requires absolute assurance that no other inputs will change in the current simulation time step, the PLI application must use the `tf_rosynchronize()` read-only synchronization callback, instead of a read/write synchronization. In a read-only synchronization, the *misctf routine* is not allowed to write return values into the simulation using any of the TF put routines or the `acc_set_value()` routine. The VPI routines, however, can return values into simulation at a future time from a read-only simulation callback.

Example 18-5 modifies the combinational logic interface presented in Example 18-3. This modified version schedules a *misctf* synchronize callback at the end of a simulation time step in which an input changed value.

This example uses the TF work area to store a flag to indicate when a callback to the *misctf routine* has already been scheduled for the current simulation time step. This flag is used in the C model interface to prevent more than one callback to the *misctf routine* being requested in the same time step. Since the TF work area is unique for each instance of a system task, each instance of the C model will have a unique flag.

CD The source code for this example is on the CD accompanying this book.

- Application source file: `Chapter.18/sci_alu_synchronized_acc.c`
- Verilog shell module: `Chapter.18/sci_alu_synchronized_shell.v`
- Verilog test bench: `Chapter.18/sci_alu_synchronized_test.v`
- Expected results log: `Chapter.18/sci_alu_synchronized_test.log`

Example 18-5: C model interface synchronized to the end of a time step

```
#define ALU_A      1  /* system task arg 1 is ALU A input        */
#define ALU_B      2  /* system task arg 2 is ALU B input        */
#define ALU_OP     3  /* system task arg 3 is ALU opcode input   */
#define ALU_RESULT 4  /* system task arg 4 is ALU result output  */
#define ALU_EXCEPT 5  /* system task arg 5 is ALU exception output */
#define ALU_ERROR  6  /* system task arg 6 is ALU error output   */

/***********************************************************************
 * VCL simulation callback routine: Schedules a callback to the misctf
 * application synchronized to the end of a time step.  Only schedules
 * one callback for a time step
 **********************************************************************/
int PLIbook_ScientificALU_interface(p_vc_record vc_record)
{
  PLI_BYTE8 *instance_p;

  /* Retrieve instance handle from VCL user_data field */
  instance_p = vc_record->user_data;
```

```
  /* If the TF work area for this instance is NULL, then no misctf   */
  /* synchronize callback has been scheduled for this time step (the */
  /* work area is set to non-null by this routine, and is set to     */
  /* NULL by the misctf after a callback is processed.               */
  if (tf_igetworkarea(instance_p) == NULL) {
    /* Schedule a synchronize callback to misctf for this instance */
    tf_isynchronize(instance_p);
    tf_isetworkarea("1", instance_p); /* set work area to non-null */
  }
  return(0);
}

/**********************************************************************
 * misctf routine: Serves as an interface between Verilog simulation
 * and the C model.  Called by the VCL consumer application whenever
 * the C model inputs change value, reads the values of all inputs,
 * passes the values to the C model, and writes the C model outputs
 * into simulation.
 **********************************************************************/
int PLIbook_ScientificALU_misctf(int user_data, int reason, int pvc)
{
  double       a, b, result;
  int          opcode, excep, err;
  handle       instance_h, result_h, excep_h, err_h,
               a_h, b_h, opcode_h;
  s_setval_value  value_s;
  s_setval_delay  delay_s;
  s_acc_time      time_s;

  if (reason != REASON_SYNCH)
    return(0);  /* abort misctf if not called for synchronize reason */
  acc_initialize();
  /* Set the TF work area for this instance to null (a flag that    */
  /* this callback has been processed)                              */
  tf_setworkarea(null);

  /* Obtain handles to all task arguments */
  a_h      = acc_handle_tfarg(ALU_A);
  b_h      = acc_handle_tfarg(ALU_B);
  opcode_h = acc_handle_tfarg(ALU_OP);
  result_h = acc_handle_tfarg(ALU_RESULT);
  excep_h  = acc_handle_tfarg(ALU_EXCEPT);
  err_h    = acc_handle_tfarg(ALU_ERROR);

  /* Read current values of C model inputs from Verilog simulation */
  value_s.format = accRealVal;
  acc_fetch_value(a_h, "%%", &value_s);
  a = value_s.value.real;

  acc_fetch_value(b_h, "%%", &value_s);
  b = value_s.value.real;

  value_s.format = accIntVal;
  acc_fetch_value(opcode_h, "%%", &value_s);
  opcode = (int)value_s.value.integer;
```

```
/****** Call C model ******/
PLIbook_ScientificALU_C_model(a, b, opcode, &result, &excep, &err);

/* Write the C model outputs onto the Verilog signals */
delay_s.model      = accNoDelay;
delay_s.time       = time_s;
delay_s.time.type  = accRealTime;
delay_s.time.real  = 0.0;

value_s.format     = accRealVal;
value_s.value.real = result;
acc_set_value(result_h, &value_s, &delay_s);

value_s.format        = accIntVal;
value_s.value.integer = (PLI_INT32)excep;
acc_set_value(excep_h, &value_s, &delay_s);

value_s.value.integer = (PLI_INT32)err;
acc_set_value(err_h, &value_s, &delay_s);

acc_close();
return(0);
}

/**********************************************************************
 * calltf routine: Registers a callback to the C model interface
 * whenever any input to the C model changes value
 **********************************************************************/
int PLIbook_ScientificALU_calltf(int user_data, int reason)
{
  handle      a_h, b_h, opcode_h;
  PLI_BYTE8 *instance_p;

  acc_initialize();
  /* get handles for signals in task args which are C model inputs */
  a_h        = acc_handle_tfarg(ALU_A);
  b_h        = acc_handle_tfarg(ALU_B);
  opcode_h   = acc_handle_tfarg(ALU_OP);

  /* get pointer for this system task instance to pass to VCL app. */
  instance_p = tf_getinstance();
  /* set the TF work area for this instance to null */
  tf_setworkarea(NULL);

  /* add VCL flags to all signals which are inputs to the C model  */
  /* pass handle for task instance as the user_data value          */
  acc_vcl_add(a_h, PLIbook_ScientificALU_interface,
              instance_p, vcl_verilog_logic);
  acc_vcl_add(b_h, PLIbook_ScientificALU_interface,
              instance_p, vcl_verilog_logic);
  acc_vcl_add(opcode_h, PLIbook_ScientificALU_interface,
              instance_p, vcl_verilog_logic);
  acc_close();
  return(0);
}
```

18.8 Synchronizing with a future simulation time step

In certain C model applications, it may be necessary to synchronize C model activity with future simulation activity. The `tf_setdelay()` routine and variations of this routine from the TF library can be used to schedule a call to the *misctf routine* for a specific amount of time in the future, relative to the current simulation time.

The `tf_getnextlongtime()` routine returns the future simulation time in which the next simulation event is scheduled to occur. This provides a way for a PLI application to synchronize activity for when the Verilog simulator is processing simulation events.

These TF routines for synchronizing with future simulation times were presented in more detail in Chapter 12.

18.9 Allocating storage within a C model

Special attention and care must be taken when a C model uses static variables or allocates memory.

The Verilog language can instantiate a model any number of times. Each instance of the Verilog shell module creates a unique instance of the system task which invokes the PLI interface to the C model. Therefore, the *calltf routine* which is invoked by a system task instance, the handle for the task instance, and the handles for the arguments for the task instance will all be unique for each system task instance. Any memory which is allocated by the *calltf routine* will also be unique for each instance of the system task.

When a C model is represented as an independent program, multiple instances of the model are not a problem, as each instance will invoke a new process with unique storage for each process. When the C model is represented as a C function, however, multiple instances of the model will share the same function. The model must allow for the possibility of multiple instances, and provide unique storage for each instance.

In addition, the VCL *consumer routine* is not unique to each instance of a system task. Any static variables or memory allocated by a *consumer routine* will be common to all instances of the C model. The *consumer routine* must therefore devise a scheme to allocate unique memory storage for each instance of a system task which can call the routine.

One simple way to create unique storage for each instance is to pass the handle for the system task instance to the *consumer routine* using the user_data field of the VCL callback. The *consumer routine* can then allocate memory for each task instance which calls it, and use the instance handle to associate the memory block with a specific task instance.

Example 18-6 presents a latched version of the scientific ALU, which can store the result of a previous operation indefinitely, and Example 18-7 presents a combinational logic interface to the model.

This example allocates unique storage within the C model for each instance of the C model. The storage is allocated within the C model function the first time the C model is called for a specific instance of the *$scientific_alu* system task. A pointer to the memory allocated is saved in a linked list. The instance handle for the system task is used to identify which block of memory in the linked list is associated with a specific instance of *$scientific_alu*. Each time the C model is called, the instance handle which caused the call is compared to the instance handles saved in the linked list, until a match is found. If no match is found, it is an indication that this is the first time that instance has called the C model. A new block of memory is allocated and added to the linked list.

 Instance handles cannot be compared using the C == equivalence operator. A handle is a pointer to information about an object. Each time a handle for an object is obtained, the pointer value may be different. The routine acc_compare_handles() must be used to determine if two handles reference the same object.

In example 18-6, the instance handle is obtained by the *calltf routine* for a system task instance, passed to the Value Change Link *consumer routine* through the user_data field, and then passed to the C model as an input to the model function.

CD The source code for this example is on the CD accompanying this book.

- Application source file: Chapter.18/sci_alu_latched_acc.c
- Verilog shell module: Chapter.18/sci_alu_latched_shell.v
- Verilog test bench: Chapter.18/sci_alu_latched_test.v
- Expected results log: Chapter.18/sci_alu_latched_test.log

Example 18-6: scientific ALU C model with latched outputs

```
/*********************************************************************
 * Definition for a structure to store output values when the ALU is
 * latched.  When enable is 1, the ALU returns the currently calculated
 * outputs, and when 0, the ALU returns the latched previous results.
 *********************************************************************/
#include <stdio.h>
typedef struct PLIbook_SciALUoutputs  *PLIbook_SciALUoutputs_p;
typedef struct PLIbook_SciALUoutputs {
  PLI_BYTE8 *instance_p; /* shows which task instance owns this space /
  double     result;     /* stored result of previous operation */
  int        excep;
  int        err;
  PLIbook_SciALUoutputs_p next_ALU_outputs; /* next stack location */
} PLIbook_SciALUoutputs_s;

  /* declare global stack pointer */
  static PLIbook_SciALUoutputs_p ALU_outputs_stack = NULL;

/*********************************************************************
 * C model of a Scientific Arithmetic Logic Unit.
 *    Latched outputs version.
 *********************************************************************/
#include <math.h>
#include <errno.h>
void PLIbook_ScientificALU_C_model(
        int     enable,     /* input; 0 = latched */
        double  a,          /* input */
        double  b,          /* input */
        int     opcode,     /* input */
        double *result,     /* output from ALU */
        int     *excep,     /* output; set if result is out of range */
        int     *err,       /* output; set if input is out of range */
        PLI_BYTE8 *instance_p) /* input; pointer to sys task instance */
{
  PLIbook_SciALUoutputs_p ALU_outputs;

  /* Locate the output storage in the stack for this model instance  */
  /* If no storage is found, then allocate a storage block and add    */
  /* the storage to the stack.                                        */
  ALU_outputs = ALU_outputs_stack; /* top-of-stack is in global var. */
  while (ALU_outputs && (ALU_outputs->instance_p != instance_p))
    ALU_outputs = ALU_outputs->next_ALU_outputs;

  /* If no storage area found for this model instance, create one */
  if (ALU_outputs == NULL) {
    ALU_outputs =
      (PLIbook_SciALUoutputs_p)malloc(sizeof(PLIbook_SciALUoutputs_s));
    ALU_outputs->instance_p = instance_p; /* set owner of this space */
    ALU_outputs->next_ALU_outputs = ALU_outputs_stack;
    ALU_outputs_stack = ALU_outputs; /* save new top-of-stack */
  }
```

```
if (enable) { /* ALU is not latched, calculate outputs and store */
   switch (opcode) {
      case 0x0: ALU_outputs->result = pow    (a, b);       break;
      case 0x1: ALU_outputs->result = sqrt   (a);          break;
      case 0x2: ALU_outputs->result = exp    (a);          break;
      case 0x3: ALU_outputs->result = ldexp  (a, (int)b);  break;
      case 0x4: ALU_outputs->result = fabs   (a);          break;
      case 0x5: ALU_outputs->result = fmod   (a, b);       break;
      case 0x6: ALU_outputs->result = ceil   (a);          break;
      case 0x7: ALU_outputs->result = floor  (a);          break;
      case 0x8: ALU_outputs->result = log    (a);          break;
      case 0x9: ALU_outputs->result = log10  (a);          break;
      case 0xA: ALU_outputs->result = sin    (a);          break;
      case 0xB: ALU_outputs->result = cos    (a);          break;
      case 0xC: ALU_outputs->result = tan    (a);          break;
      case 0xD: ALU_outputs->result = asin   (a);          break;
      case 0xE: ALU_outputs->result = acos   (a);          break;
      case 0xF: ALU_outputs->result = atan   (a);          break;
   }
   ALU_outputs->excep = (errno == ERANGE);/* result out-of-range */
   #ifdef WIN32   /* for Microsoft Windows compatibility */
    ALU_outputs->err  = (_isnan(*result) ||  /* not-a-number, or */
                         errno == EDOM);      /* arg out-of-range */
   #else
    ALU_outputs->err  = (isnan(*result) ||   /* not-a-number, or */
                         errno == EDOM);      /* arg out-of-range */
   #endif
   if (ALU_outputs->err) ALU_outputs->result = 0.0;
   errno = 0;                                 /* clear error flag */
 }

 /* return the values stored in the C model */
 *result = ALU_outputs->result;
 *err    = ALU_outputs->err;
 *excep  = ALU_outputs->excep;

 return;
}
```

The C source code listing in Example 18-7 illustrates a combinational logic interface for the latched scientific ALU model.

Example 18-7: combinational logic interface to latched scientific ALU C model

```
#define ALU_ENABLE 1  /* system task arg 1 is ALU enable input   */
#define ALU_A      2  /* system task arg 2 is ALU A input        */
#define ALU_B      3  /* system task arg 3 is ALU B input        */
#define ALU_OP     4  /* system task arg 4 is ALU opcode input   */
#define ALU_RESULT 5  /* system task arg 5 is ALU result output  */
```

```
#define ALU_EXCEPT 6   /* system task arg 6 is ALU exception output */
#define ALU_ERROR  7   /* system task arg 7 is ALU error output     */

/**********************************************************************
 * VCL simulation callback routine: Serves as an interface between
 * Verilog simulation and the C model.  Called whenever the C model
 * inputs change value, passes the values to the C model, and puts
 * the C model outputs into simulation.
 **********************************************************************/
int PLIbook_ScientificALU_interface(p_vc_record vc_record)
{
   double          a, b, result;
   int             opcode, excep, err, enable;
   handle          instance_h, result_h, excep_h, err_h,
                   a_h, b_h, opcode_h, enable_h;
   s_setval_value  value_s;
   s_setval_delay  delay_s;
   s_acc_time      time_s;

   acc_initialize();

   /* Retrieve instance handle from VCL user_data field */
   instance_h = (handle)vc_record->user_data;

   /* Obtain handles to all task arguments */
   enable_h = acc_handle_itfarg(ALU_ENABLE, instance_h);
   a_h      = acc_handle_itfarg(ALU_A,      instance_h);
   b_h      = acc_handle_itfarg(ALU_B,      instance_h);
   opcode_h = acc_handle_itfarg(ALU_OP,     instance_h);
   result_h = acc_handle_itfarg(ALU_RESULT, instance_h);
   excep_h  = acc_handle_itfarg(ALU_EXCEPT, instance_h);
   err_h    = acc_handle_itfarg(ALU_ERROR,  instance_h);

   /* Read current values of C model inputs from Verilog simulation */
   value_s.format = accRealVal;
   acc_fetch_value(a_h, "%%", &value_s);
   a = value_s.value.real;

   acc_fetch_value(b_h, "%%", &value_s);
   b = value_s.value.real;

   value_s.format = accIntVal;
   acc_fetch_value(opcode_h, "%%", &value_s);
   opcode = (int)value_s.value.integer;

   acc_fetch_value(enable_h, "%%", &value_s);
   enable = (int)value_s.value.integer;

   /****** Call C model  ******/
   PLIbook_ScientificALU_C_model(enable, a, b, opcode,
                                 &result, &excep, &err,
                                 (PLI_BYTE8 *)instance_h);

   /* Write the C model outputs onto the Verilog signals */
   delay_s.model      = accNoDelay;
   delay_s.time       = time_s;
```

```
  delay_s.time.type  = accRealTime;
  delay_s.time.real  = 0.0;

  value_s.format      = accRealVal;
  value_s.value.real = result;
  acc_set_value(result_h, &value_s, &delay_s);

  value_s.format        = accIntVal;
  value_s.value.integer = (PLI_INT32)excep;
  acc_set_value(excep_h, &value_s, &delay_s);

  value_s.value.integer = (PLI_INT32)err;
  acc_set_value(err_h, &value_s, &delay_s);

  acc_close();
  return(0);
}

/***********************************************************************
 * calltf routine: Registers a callback to the C model interface
 * whenever any input to the C model changes value
 ***********************************************************************/
int PLIbook_ScientificALU_calltf(int user_data, int reason)
{
  handle  instance_h, enable_h, a_h, b_h, opcode_h;

  acc_initialize();

  /* get handles for signals in task args which are C model inputs */
  enable_h = acc_handle_tfarg(ALU_ENABLE);
  a_h      = acc_handle_tfarg(ALU_A);
  b_h      = acc_handle_tfarg(ALU_B);
  opcode_h = acc_handle_tfarg(ALU_OP);

  /* get handles for this system task instance to pass to VCL app. */
  instance_h = acc_handle_tfinst();

  /* add VCL flags to all signals which are inputs to the C model   */
  /* pass handle for task instance as the user_data value           */
  acc_vcl_add(enable_h, PLIbook_ScientificALU_interface,
              (PLI_BYTE8 *)instance_h, vcl_verilog_logic);
  acc_vcl_add(a_h, PLIbook_ScientificALU_interface,
              (PLI_BYTE8 *)instance_h, vcl_verilog_logic);
  acc_vcl_add(b_h, PLIbook_ScientificALU_interface,
              (PLI_BYTE8 *)instance_h, vcl_verilog_logic);
  acc_vcl_add(opcode_h, PLIbook_ScientificALU_interface,
              (PLI_BYTE8 *)instance_h, vcl_verilog_logic);

  acc_close();
  return(0);
}
```

18.10 Representing propagation delays in a C model

Propagation delays from an input change to an output change in a C model can be represented in two ways:

• Using delays in the PLI interface.

• Using delays in the Verilog shell module.

Delays in the PLI interface are represented by specifying a delay value with the `acc_set_value()` routine which writes values onto the system task arguments. Either inertial or transport event propagation can be used, depending on the requirements of the C model. However, using `acc_set_value()` does not offer a great deal of flexibility on creating delays which are different for each instance of a model. Nor can the `acc_set_value()` routine represent minimum, typical and maximum delays, different delays for rise and fall transitions, or annotation of delays using delay calculators or SDF files.

C model propagation delays can also be represented using the pin-to-pin path delays in the Verilog shell module. This method provides the greatest amount of flexibility and accuracy in modeling propagation delays. All path delays constructs can be used, as well and Verilog timing constraints.

 Some Verilog simulators restrict the use of pin-to-pin path delays and SDF delay back annotation to Verilog models which are represented with Verilog primitives and net data types. To use path delays on a C model with these simulators, buffers must be added to all input and output ports, with net data types connected to the inputs and outputs of these buffers.

In Chapter 13, example 13-9 on page 456 illustrates adding pin-to-pin path delays to the scientific ALU shell module.

18.11 Summary

This chapter has presented just a few ways in which the ACC library can be used to interface a C language model with Verilog simulations. The Value Change Link routines in the ACC library provide an efficient means to pass input changes to a C model. The ACC routines to read and modify logic values allow information to be exchanged in a variety of formats. By creating a shell module which contains the system task that invokes the C model interface, the C model can be used in a Verilog design just as any other Verilog module.

Appendices

A ppendix A provides a description of how PLI applications are linked into several major Verilog simulators.

Appendices B, C and D present the complete IEEE 1364-2001 PLI libraries for the TF, ACC and VPI libraries, respectively. Each library routine is listed with its return value type, input value types, and a brief description of the routines purpose. Object diagrams are provided for the VPI and ACC libraries to show Verilog objects can be accessed using these libraries. The diagrams also show the properties of those objects and the relationships of each object to other objects. The names of all C constants and structure definitions used with the libraries are listed with the routine descriptions or in the object diagrams.

APPENDIX A *Linking PLI Applications to Verilog Simulators*

*P*LI applications must be linked into Verilog simulators. There are many simulators available, and each simulator has a unique method of linking PLI applications. This appendix presents how PLI applications are linked into a few of the simulators available at the time this book was written. The topics presented include:

- The IEEE 1364 interface mechanism for VPI based applications

- The IEEE 1364 interface mechanism for TF and ACC based applications

- Linking – PLI applications to the following simulators:

  - *Verilog-XL*™ and *NC-Verilog*® from Cadence Design Systems, Inc.

  - *VCS*™ from Synopsys, Inc.

  - *ModelSim*™ from Model Technology, Inc.

 There are many excellent Verilog simulators on the market. Due to space limitations, it was not possible to include details on how PLI applications are linked into each and every simulator. It is intended that by showing how PLI applications are linked into a few simulators, the overall process will be understand. Most simulators will use methods similar to the ones shown in this appendix.

Company names and product names listed in this appendix are trademarks or registered trademarks of the respective company with which they are associated. Other names may also be trademarks or registered trademarks of their respective companies.

A.1 The PLI interface mechanism

A Verilog PLI application comprises:

- A system task or system function name
- A set of C functions for:
  - A *calltf routine*
  - A *compiletf routine* or *checktf routine*
  - A *sizetf routine*
  - A *misctf routine* (for PLI applications written with the TF/ACC libraries).

After the system task/function name and PLI application C functions have been defined, two actions are required:

1. Associate the system task/function name with the various application routines.

   The PLI *interface mechanism* is used to make the associations between the system task/function name and the application routines. There are two generations of the interface mechanism, one that was created for the older TF and ACC libraries, and a newer mechanism created for the VPI library.

2. Link the applications into a Verilog simulator, so the simulator can call the appropriate routine when the system task/function name is encountered.

   The PLI standard does not provide any guidelines on how PLI applications should be linked into a Verilog simulator. There are many different C compilers and operating systems available to Verilog users, and each compiler and operating system has unique methods for compiling and linking programs.

A.1.1 Interfacing PLI applications using the VPI library

 Chapter 2 discusses the VPI interface mechanism in detail. This appendix only presents a brief summary of the interface.

The IEEE 1364 standard VPI interface mechanism involves creating a register function, which associates the system task/function name with the application routines. After the register function is defined, the Verilog simulator must be notified about the registration function. The register function uses an **s_vpi_systf_data** structure to specify information about the PLI application. This structure is defined as part of the VPI standard, in the PLI *vpi_user.h* file. Refer to chapter 2 for an explanation of the fields in the s_vpi_systf_data structure. The definition of the structure is:

```
typedef struct t_vpi_systf_data {
  PLI_INT32    type;                    /* vpiSysTask, vpiSysFunc */
  PLI_INT32    sysfunctype;             /* vpiIntFunc, vpiRealFunc,
                                           vpiTimeFunc, vpiSizedFunc,
                                           vpiSizedSignedFunc */
  PLI_BYTE8   *tfname;                  /* quoted task/function name */
                                        first character must be $ */
  PLI_INT32 (*calltf)(PLI_BYTE8 *);     /* name of C func */
  PLI_INT32 (*compiletf)(PLI_BYTE8 *);  /* name of C func */
  PLI_INT32 (*sizetf)(PLI_BYTE8 *);     /* name of C func */
  PLI_BYTE8   *user_data;               /* returned with callback */
} s_vpi_systf_data, *p_vpi_systf_data;
```

The following steps are used to register a system task or system function using the VPI interface mechanism:

1. Create a C function to register the system task/function. The C function name is application-defined and can be any legal C name.

2. Allocate an s_vpi_systf_data C structure.

3. Fill in the fields of the structure with the information about the system task or system function.

4. Register the system task/function by calling the VPI routine **vpi_register_systf()**.

5. Specify the name of the register function to the Verilog simulator.

The following example registers a PLI application called *$hello*.

Example A-1: VPI register function for the *$hello* system function

```
/* prototypes of PLI application routine names */
int hello_calltf(), hello_compiletf();

void hello_register()
{
  s_vpi_systf_data tf_data;

  tf_data.type        = vpiSysTask;
  tf_data.sysfunctype = 0;
  tf_data.tfname      = "$hello";
  tf_data.calltf      = hello_calltf;
  tf_data.compiletf   = hello_compiletf;
  tf_data.sizetf      = NULL;
  tf_data.user_data   = NULL;
  vpi_register_systf(&tf_data);
}
```

Notifying Verilog simulators about the VPI register functions

Once the register function has been defined, the Verilog simulator must be notified of
the name of the register function, so that the simulator can call the function. The VPI
standard requires that all Verilog simulators provide a special array, called
vlog_startup_routines, in order to notify a simulator about the register functions. All
functions listed in this array will be called by the simulator as the simulator is starting
up.

 Simulators may provide other ways to specify the names of the register functions
created in step 1. For example, some simulators can specify the register functions
through invocation options instead of the *vlog_startup_routines array*.

An example *vlog_startup_routines* array is listed below, with entries for the *$hello*
PLI application register function, and a *$pow* register function.

Example A-2: sample *vlog_startup_routines* array

```
/* prototypes of the PLI application register routines */
extern void hello_register(), pow_register();

void (*vlog_startup_routines[])() =
{
    /*** add user entries here ***/
  hello_register,
  pow_register,
  0 /*** final entry must be 0 ***/
};
```

 The IEEE 1364 standard does not define where the *vlog_startup_routines* array
should be located. Consult the reference manual of the simulator for the location of
the start-up array used by that simulator.

 Do not place the vlog_startup_routines array in the same C source file as the PLI
application! In a typical design environment, PLI applications will come from
several sources, such as internally developed applications and 3rd party applications.
If the *vlog_startup_routines* array and a PLI application and are in the same file, then
the source code for the application must be available whenever another PLI
application needs to be added to the start-up array.

A.1.2 Interfacing PLI applications using the TF/ACC libraries

 Chapter 9 discusses the TF/ACC interface mechanism in detail. This appendix only presents a brief summary of the interface.

The TF/ACC interface mechanism is derived from the 1990 OVI PLI 1.0 standard. This older interface mechanism defines what all Verilog simulators should provide for interfacing PLI applications to a simulator, but does not define how the interface should be implemented.

The de facto standard TF/ACC veriusertfs array

Many Verilog simulators have adopted a similar method of specifying the information about a PLI application. This method uses a C array called *veriusertfs* (which stands for *Verilog user tasks and functions*). The veriusertfs array is an array of **s_tfcell** structures. Each structure specifies the information about a PLI application. There can be any number of structures in the array. A structure with the first field set to 0 is used to denote the end of the array.

The s_tfcell structure is not defined in the IEEE standard. However, nearly all simulators use the same structure definition. Refer to chapter 9 for an explanation of the the fields in the typical s_tfcell structure. The typical structure definition is:

```
typedef struct t_tfcell {
    short type;            /* one of the constants: usertask,
                              userfunction, userrealfunction */
    short data;            /* data passed to user routine */
    int (*checktf)();      /* pointer to the checktf routine */
    int (*sizetf)();       /* pointer to the sizetf routine */
    int (*calltf)();       /* pointer to the calltf routine */
    int (*misctf)();       /* pointer to the misctf routine */
    char *tfname;          /* name of the system task/function */
    int  forwref;          /* usually set to 1 */
    char *tfveritool;      /* usually ignored */
    char *tferrmessage;    /* usually ignored */
} s_tfcell, *p_tfcell;
```

Example A-3 illustrates an example veriusertfs array with the information required for the *$hello* system task, and a *$pow* system function. The structure fields for the *$hello* application include comments. The structure fields for the *$pow* application are listed on a single line without comments.

Example A-3: sample veriusertfs array, as used by many Verilog simulators

```
/* prototypes of the PLI application routines */
extern int hello_calltf(), hello_checktf();
extern int pow_check(), pow_call(), pow_size();

/* the veriusertfs table */
s_tfcell veriusertfs[] =
{
    {usertask,              /* type of PLI routine */
     0,                     /* user_data value */
     hello_checktf,         /* checktf routine */
     0,                     /* sizetf routine */
     hello_calltf,          /* calltf routine */
     0,                     /* misctf routine */
     "$hello",              /* system task/function name */
     1                      /* forward reference = true */
    },

    {userfunction, 0, pow_check, pow_size, pow_call, 0, "$pow", 1},

    {0} /*** final entry must be 0 ***/
};
```

 The *veriusertfs array* is not specified in the IEEE 1364 standard. Though many simulators use this array, it is not required, and some simulators use different methods of specifying the PLI application information for the TF/ACC interface.

 Do not specify the veriusertfs array in the same file as the PLI application!

TIP
  • Not all Verilog simulators use the *veriusertfs array* to specify PLI application information.

  • The array is not standardized, and may be different in different simulators.

  • The C language does not allow multiple global arrays with the same name. If two applications both contained a *veriusertfs array* definition, the applications could not be used together.

A.1.3 Should PLI routines be `PLI_INT32`, `int` or `void` functions?

The IEEE 1364 standard defines that VPI *calltf*, *compiletf* and *sizetf* routines should have a return type `PLI_INT32`. On 32-bit operating systems, `PLI_INT32` is typically defined to be a C `int`. On 64-bit operating systems, however, `PLI_INT32` may be defined differently. Therefore, VPI routines should always be defined to of type `PLI_INT32`, in order to be portable to all operating systems.

The IEEE 1364 standard does not define what data type TF/ACC *calltf, compiletf* and *misctf* routines should return. The standard does require that *sizetf* routines return an integer value. Most Verilog simulators use the de facto standard s_tfcell to define TF/ACC PLI applications, as discussed in section A.1.2. The typical s_tfcell structure defines the *calltf, compiletf, misctf* and *sizetf* routines to have a return type **int**.

For both VPI and TF/ACC PLI applications, the simulator only uses the return value from the *sizetf routine*. The return values from the *calltf routine, compiletf routine, checktf routine* and *misctf routine* are ignored. Since the return value is ignored, some PLI application developers will declare these routines as void functions. Declaring the functions as a different type from their prototype may result in a C compiler warning message when the routine is compiled. To prevent this warning message, the pointer to the void function can be cast to a pointer to an int function. For example:

```
(int(*)())Pow_calltf
```

Note, however, that some C compilers may not permit casting a function pointer from one type to another.

A.1.4 C versus C++

Most Verilog simulators expect PLI applications to be written in ANSI C. The PLI standard includes a library of C functions as well as definitions of C constants, special data types and structures. The PLI library is compliant with the ANSI C standard.

Many Verilog simulators were written in the C language, and may or may not support linking C++ applications to the simulator. PLI application developers who wish to work with C++ should first check the limitations of the simulator products to which the applications will be linked. For maximum portability, it is recommended that PLI applications be written in ANSI C.

A.1.5 Static linking versus dynamic linking

There are two general methods in which a PLI application can be linked into a Verilog simulator: static linking and dynamic linking.

Some Verilog simulators statically link PLI applications into the simulator. Static linking requires:

1. The simulator vendor must provide all of the object files for the simulator.

2. The end-user must create a new simulator executable by relinking the simulator's object files with the compiled PLI application object files.

Static linking is simple, but does have some drawbacks. The relinking process often depends on operating system graphical libraries being present (such as the X-11 libraries on some Unix systems). If a required operating system object file is missing, is an incorrect version, or is in a different location than expected, the Verilog simulator will not link properly.

Some Verilog simulators dynamically link PLI applications into the simulator. With this method:

1. Object files are compiled separately as position independent code, to form object libraries (or dll's in Microsoft Windows).

2. The standard simulator executable dynamically loads the separately compiled PLI applications when needed.

With dynamic linking, the simulator vendor does not need to provide the object files for the Verilog simulator. A major advantage of this is that the user does not need to create new, customized simulator executables, and does not need to worry about having all the correct operating system object files. A disadvantage of dynamic linking is a dependency on operating system and compiler versions.

A.2 Linking to the Cadence *Verilog-XL and NC-Verilog* simulators

The Cadence *Verilog-XL*™ simulator is the original Verilog simulator, and was first introduced in 1985. At the time this book was written, Verilog-XL ran on several Unix workstations, and on PC's running Linux operating systems (but not the Windows operating systems).

The Cadence *NC-Verilog*® simulator is Cadence's second generation Verilog simulator. NC-Verilog is a "native compiled" code simulator, which compiles Verilog source code into machine-dependent object files. As part of the compilation, NC-Verilog can highly optimize the object files for a specific operating system. At the time this book was written, NC-Verilog ran on several Unix workstations, and on PCs running the Linux, Windows NT and Windows 2000 operating systems.

A.2.1 IEEE 1364-2001 compliance

The examples in this book were tested with Verilog-XL version 3.2, and NC-Verilog version 3.2, which were the latest versions available at the time the second edition of this book was prepared. These versions support the full IEEE 1364-1995 PLI standard. Only a portion of the new features in the IEEE 1364-2001 standard for the Verilog HDL and PLI have been implemented in these versions. Full support of the IEEE 1364-2001 standard is planned for future releases of both products.

These versions of Verilog-XL and NC-Verilog use the standard IEEE 1364-2001 *vpi_user.h*, *veriuser.h* and *acc_user.h* header files, which define the VPI, TF and ACC PLI libraries. Verilog-XL and NC-Verilog also provide a number of extensions to the PLI standard, which support proprietary features of the Cadence simulators. The extensions to the IEEE standard are specified in three proprietary header files, *vpi_user_cds.h*, *vxl_veriuser.h* and *vxl_acc_user.h*. It is not necessary to include these proprietary header files unless the proprietary extensions to the PLI are being used. The examples in this book all run with just the standard IEEE 1364-2001 header files.

A.2.2 Specifying PLI application information

Verilog-XL and NC-Verilog have two PLI interface mechanisms:

- A VPI interface mechanism
- A TF/ACC interface mechanism

Sections A.2.4 and A.2.5, which follow, present first the VPI interface mechanism and then the older TF/ACC interface mechanism.

A.2.3 Enabling PLI support with NC-Verilog

The NC-Verilog simulator optimizes the run-time performance of PLI applications by restricting the level of access the PLI has into the Verilog simulation data structure. By default, NC-Verilog supports the TF library, and limited functionality from the ACC and VPI libraries. The TF routines can only access the simulation data that is passed to the PLI application through system task or system function arguments. This limited access means that at compile time, NC-Verilog can determine exactly what data will be accessed by the application. The ACC and VPI routines, however, can arbitrarily access information anywhere and at any time in the simulation data structure. This arbitrary access cannot be predicted at compile time, and therefore cannot be optimized as efficiently.

NC-Verilog allows the user to specify ACC and VPI access as read-only, read and write, and/or connectivity tracing. This access can be specified:

- Using one or more access files, which can define the allowable access for each module, or each instance of a module. The names of the access files are passed to the NC-Verilog elaborator (ncelab) using a -afile invocation option. Refer to the Cadence documentation for details on specifying access files.

- Using an invocation option to the NC-Verilog elaborator (ncelab), to globally enable ACC and VPI access to the entire simulation data structure.

While not as optimal as access files, global access is the easiest method to test the examples in this book. Global access is enabled as follows:

```
ncelab -access +rwc
```

or

```
ncverilog +ncaccess+rwc
```

By default, NC-Verilog optimizes its storage of memory arrays modeled in the Verilog HDL. This optimization affects one routine from the TF library, tf_nodeinfo(). In order to use this routine, memory optimization must be disabled in an access file for specific modules or module instances. The optimization can also be disabled globally using an invocation option:

```
ncelab -nomempack
```

or

```
ncverilog +ncnomempack
```

A.2.4 Interfacing VPI applications

The VPI interface mechanism defines a standard method that all simulators should use. This method involves creating an application-defined register function, as discussed in section A.1.1 on page 642. The register function is then added to a *vlog_startup_routines* array, or specified as an invocation option to either Verilog-XL or NC-Verilog. Cadence provides a sample *vlog_startup_routines* array in a file called *vpi_user.c*. To specify the PLI application register function using this array:

1. Copy the sample C source file called *vpi_user.c*, that is provided in the Verilog-XL or NC-Verilog installation directory. The vpi_user.c file is located in the Cadence installation directory, typically in:

    ```
    <cadence_install_directory>/tools/src/vpi_user.c
    ```

2. In the copy of the *vpi_user.c* file, edit the *vlog_startup_routines* array to add the names of the register functions that need to be invoked by the simulator.

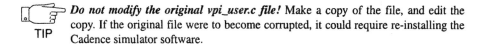

TIP *Do not modify the original vpi_user.c file!* Make a copy of the file, and edit the copy. If the original file were to become corrupted, it could require re-installing the Cadence simulator software.

Example A-1 on page 643 lists the lines that would be added to a vpi_user.c file to register two PLI applications. After the copy of vpi_user.c has been modified, the PLI applications and the vpi_user.c file need to be compiled and linked. Sections A.2.6 and A.2.7 show how this is done with Verilog-XL and NC-Verilog.

An alternate to the vlog_startup_routines array and vpi_user.c file

Verilog-XL and NC-Verilog allow the names of one or more VPI register functions to specified as an invocation option. This alternate method allows applications to be registered without having to list the register functions in the *vlog_startup_routines* array.

The invocation option for Verilog-XL is:

```
+loadvpi=<pli_object_file>:<register_func_name>,<register_func_name>
```

For example:

```
verilog test.v +loadvpi=my_pli_apps:hello_register,pow_register
```

The invocation option for NC-Verilog (as part of the ncelab command) is:

```
-loadvpi <pli_object_file>:<register_func_name>,<register_func_name>
```

For example:

```
ncelab test.v -loadvpi my_pli_apps:hello_register,pow_register
```

`<pli_object_file>` is the name of a shared object file, also called a dynamically linked library file, that contains the compiled and linked PLI applications. The object file is created using the procedure discussed in section A.2.7 on page 655, and will be of type `.so` on the Solaris and Linux operating systems, `.sl` on the HPUX operating system, and `.dll` on the Windows operating systems. This extension does not need to be specified in the `loadvpi` invocation option. A single shared object file can contain any number of PLI applications. Multiple shared object files, each containing different PLI applications, can be specified using multiple `loadvpi` invocation options.

`<register_func_name>` is the name of the function within the PLI application which registers the application. The names of one or more register functions are listed as a comma-separated list as part of the `loadvpi` invocation option. The register functions for the PLI applications must also be compiled into the shared object file. The simulator will execute these register functions as simulation is starting, thereby registering the PLI applications.

A.2.5 Interfacing TF/ACC applications

Verilog-XL and NC-Verilog use the same ***veriusertfs array*** used by most other Verilog simulators to specify PLI application information. Refer to Section A.1.2 on page 645 for a description of the *veriusertfs* array. The *veriusertfs* array, along with other information used by Verilog-XL and NC-Verilog, is specified in a file called ***veriuser.c***. To use the TF/ACC PLI interface:

1. Copy the C source file called ***veriuser.c***, that is provided with in the Verilog-XL or NC-Verilog installation directory. The file is located in the installation directory of the simulator software. The typical location is:

    `<cadence_install_directory>/tools/src/veriuser.c`

2. In the copy of the *veriuser.c* file, edit the *veriusertfs array* to add the information for each PLI application to the array.

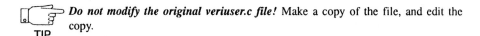
Do not modify the original veriuser.c file! Make a copy of the file, and edit the copy.
TIP

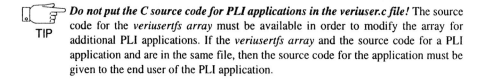
Do not put the C source code for PLI applications in the veriuser.c file! The source code for the *veriusertfs array* must be available in order to modify the array for additional PLI applications. If the *veriusertfs array* and the source code for a PLI application and are in the same file, then the source code for the application must be given to the end user of the PLI application.
TIP

Example A-3 on page 646 lists the lines that would be added to a veriuser.c file to register two PLI applications.

An alternate to the veriusertfs array and veriuser.c file

By default, Verilog-XL and NC-Verilog look in the *veriusertfs array* for the definitions of PLI applications. The modified array must be compiled and linked into the simulator. The array is defined in a global name space, and therefore only one array with the name *veriusertfs* can be used.

Both Verilog-XL and NC-Verilog also allow PLI application information to be specified in one or more alternate arrays. The array must be placed in an applications-supplied or user-supplied function, referred to as a *bootstrap function*. The name of the function is then specified as an invocation option, so that the simulation knows to call the function and register the PLI applications. This alternate method allows PLI applications to be registered without the need to list all PLI applications in a single *veriuser.c* file.

The invocation option for Verilog-XL is:

> **+loadpli1=**<*pli_object_file*>:<*boot_func_name*>,<*bootstrap_func_name*>

>   For example:

```
verilog test.v +loadpli1=my_pli_apps:my_apps_boot,other_apps_boot
```

The invocation option for NC-Verilog (as part of the ncelab command) is:

> **-loadpli1** <*pli_object_file*>:<*boot_func_name*>,<*boot_func_name*>

>   For example:

```
ncelab test.v -loadpli1 my_pli_apps:my_apps_boot,other_apps_boot
```

<*pli_object_file*> is the name of a shared object file, also called a dynamically linked library file, that contains the compiled and linked PLI applications. This shared object file is created using the procedure discussed in section A.2.7 on page 655, and will be of type .so on the Solaris and Linux operating systems, .sl on the HPUX operating system, and .dll on Windows operating systems. Note that this extension does not need to be specified in the loadpli1 invocation option. A single shared object file can contain any number of PLI applications. Multiple shared object files, each containing different PLI applications, can be specified using multiple loadpli1 invocation options.

<*bootstrap_func_name*> is the name of user-supplied C function that returns a pointer to a NULL-terminated array of s_tfcell structures. The array is identical to what is contained within the veriuser.c file discussed on the previous pages. However, the boot function is not the same as a veriuser.c file, and the array can be called any name. The array can contain the registration information for any number of PLI applications. The names of one or more bootstrap functions are listed as a comma-separated list as part of the loadpli1 invocation option. The bootstrap functions for the

PLI applications must also be compiled into the shared object file. The simulator will execute these boot functions as simulation is starting, thereby registering the PLI applications.

Example A-4: Sample Cadence TF/ACC bootstrap file

```
#include "veriuser.h"
#include "vxl_veriuser.h"

/* prototypes of the PLI application routines */
extern int hello_calltf(), hello_checktf();
extern int pow_check(), pow_call(), pow_size();

p_tfcell my_pli_apps_boot ()
{
  /* array size must be at least 1 more than the number of PLI apps */
  static s_tfcell my_pli_apps[3] =
  {
    {usertask,                /* type of PLI routine */
     0,                       /* user_data value */
     hello_checktf,           /* checktf routine */
     0,                       /* sizetf routine */
     hello_calltf,            /* calltf routine */
     0,                       /* misctf routine */
     "$hello",                /* system task/function name */
     1                        /* forward reference = true */
    },

    {userfunction, 0, pow_check, pow_size, pow_call, 0, "$pow", 1},

    {0} /*** final entry must be 0 ***/
  };
  return(my_pli_apps);
}
```

A.2.6 Compiling and linking PLI applications—the old way

The Unix version of Verilog-XL includes a utility program called **vconfig**. This program is used to configure the options to be linked into the Verilog-XL simulator, such as which waveform display to use. The *vconfig* program is a text-based program that asks a series of questions, including whether to link in applications using the older TF/ACC interface mechanism or the newer VPI interface mechanism, or both. The *vconfig* utility program also ask whether to use dynamic or static linking. The utility asks for the name of the modified veriuser.h file (for the TF/ACC interface) and vpi_user.h file (for the VPI interface). Finally, *vconfig* asks for the names of the source or object files containing the PLI applications.

When the *vconfig* program completes, it generates a Unix C-shell script, which by default is called *cr_vlog*. This script contains the C compile and link commands for the PLI application. The script is executed at a Unix prompt. This script does not require additional input from the user. If dynamic linking was selected, the *cr_vlog* script will generate a C object library file. If static linking was selected, then the *cr_vlog* script will generate a new simulator executable program.

A.2.7 Compiling and linking PLI applications—the new way

Later releases of Verilog-XL and NC-Verilog provide a *"PLI Wizard"* utility program, called *pliwiz*. This utility program is available on Unix, Linux and Windows operating systems. *PLI Wizard* is graphical program. It provides forms that allow the user to specify whether using the TF/ACC interface, the VPI interface, or both, and whether to link PLI applications using static linking or dynamic linking. The *PLI Wizard* forms provide places to specify the name of the modified veriuser.h file (for the TF/ACC interface) and vpi_user.h file (for the VPI interface), and the names of the source or object files containing the PLI applications.

Once all the forms are completed, the *pliwiz* program generates a set of C make files. The C make utility can then be run to compile and link the PLI application. The make command can also be directly executed from a button in the *PLI Wizard*.

A.2.8 Running Verilog-XL or NC-Verilog with PLI applications

 The full details on how to invoke Verilog-XL and NC-Verilog with PLI applications are outside the scope of this book. The following section only present the most basic steps to running these simulators with PLI applications. Refer to the manuals included with the product for the full details on using these simulators.

Once the PLI applications have been compiled and linked, they can be executed by the Verilog simulator. If the PLI applications are statically linked into Verilog-XL, then simulations are run using the newly created simulator executable programs. If the PLI applications are dynamically linked, the simulation is run using the standard Verilog-XL or NC-Verilog simulation executable. The simulator will dynamically load the PLI object libraries. In order for the simulator to load a dynamically linked PLI library file, it must be in the operating system's dynamic linker search path. This search path is operating system dependent. For example on the Sun Solaris operating system, the linker uses the LD_LIBRARY_PATH environment variable, and on Windows NT, the linker uses the PATH environment variable. Verilog-XL or NC-Verilog will also search for the dynamically loaded library files in the directory specified with the **loadvpi** invocation option for VPI applications and the **loadpli1** invocation option for TF/ACC applications.

Verilog-XL

Verilog-XL is an interpretive type of simulator. The Verilog source code does not need to be pre-compiled. The Verilog-XL simulator reads in Verilog HDL source code directly, interprets each statement and executes it. Verilog-XL simulations are run with a single command:

verilog <invocation_options> <verilog_source_files>

NC-Verilog

NC-Verilog is a compiled-code type of simulator. Running a simulation requires multiple steps: compile, then elaborate and then run simulation. In addition, a preparation step must be executed one time, to set up the directories used by NC-Verilog. The commands to run NC-Verilog are: ncprep, ncvlog, ncelab and ncsim.

NC-Verilog also provides a one-command method of invocation, ncverilog. This command combines all of the separate commands into one. Only the single command method is listed here. Refer to Cadence manuals for the details on using the separate commands for each phase of compiling and running simulations. The one-command method of invoking NC-Verilog works the same as Verilog-XL.

ncverilog <invocation_options> <verilog_source_files>

To enable full ACC and VPI access to the simulation data structure when using the one-command invocation, use the following options:

+ncaccess+rwc

To enable tf_nodeinfo(), use:

+ncnomempack

A.3 Linking to the Synopsys *VCS* simulator

The *VCS*™ simulator from Synopsys is a high performance Verilog simulator that is well known for its RTL simulation speed. VCS was originally developed by Chronologic, Inc. in 1991. Chronologic was acquired by Viewlogic, Inc., which was later acquired by Synopsys. VCS runs on most Unix operating systems, including Linux.

A.3.1 IEEE 1364-2001 compliance

The examples in this book were tested with VCS version 6.1 beta, which was the latest release available at the time the second edition of this book was prepared. This version of VCS supports the IEEE 1364-2001 TF and ACC routines, and much of the IEEE 1364-1995 VPI routines. Only a portion of the new PLI enhancements in the IEEE 1364-2001 standard were supported in this beta release. Full support for HDL and PLI enhancements in the IEEE 1364-2001 standard is planned for future releases.

VCS version 6.1 uses the standard IEEE 1364-2001 *vpi_user.h*, *veriuser.h* and *acc_user.h* header files, which define the VPI, TF and ACC PLI libraries. VCS also provides a number of extensions to the PLI standard, which support proprietary features of the simulator. The extensions to the IEEE standard are specified in a proprietary header file, *vcs_acc_user.h*. It is not necessary to include this proprietary header file unless the proprietary extensions to the PLI are being used. The examples in this book all run with just the standard IEEE 1364-2001 header files.

 VCS 6.0 and later use the IEEE standard *veriuser.h* file. Previous versions of VCS used a proprietary header file, called *vcsuser.h*, to define the TF library. For backward compatibility, the older proprietary header file is now a link to the standard *veriuser.h* file. This link allows PLI applications that include the older *vcsuser.h* file to still compile correctly.

Deviations from the IEEE 1364 Verilog PLI standard

The IEEE standard defines that the routines `tf_error()` and `tf_message()` should print an error message, and, when called from a *checktf routine*, they should abort simulation before it starts running. In the VCS 6.1 beta version, as well as in earlier versions of VCS, these routines only print an error message. To cause VCS to abort after an error, the `tf_dofinish()` routine must be called after printing the message.

The IEEE standard requires that the return width of system functions that return sized vectors be defined using a *sizetf routine*. This allows the return width to be calculated by the PLI application. VCS does not use sizetf routines, and instead hard codes the system function returns size in the VCS PLI table file (see A.3.4).

The beta version of VCS 6.1 used to test examples in this book did not support the use of IEEE 1364 VPI *register functions* to define VPI-based PLI applications. Several examples in this book will not work correctly without the IEEE standard register functions. Section x on the following page explains some of the capabilities available through the use of VPI register functions.

A.3.2 Enabling PLI support with VCS

The VCS simulator optimizes the run-time performance of the PLI applications by providing PLI application developers a means to configure the level of access the PLI has into the Verilog simulation data structure.

By default, VCS supports the TF library, and some limited functionality from the ACC library. Full support for the ACC library must be enabled using access commands, specified as part of registering the PLI application (see section A.3.4). Support for the VPI library must be enabled with an invocation option, **+vpi**, in conjunction with the same access commands used with the ACC library.

This requirement to specifically enable ACC and VPI access allows VCS to optimize simulation for maximum performance. The TF routines can only access the simulation data that is passed to the PLI application through system task or system function arguments. This limited access means that at compile time, VCS can determine exactly what data will be accessed by the simulator, and can optimize the simulation data structure for faster performance. The ACC and VPI routines, however, can arbitrarily access information anywhere, and at any time, in the simulation data structure. This arbitrary access cannot be predicted at compile time, and therefore cannot be optimized as efficiently. VCS enables support for the ACC and VPI libraries in increments, and can specify the added support for the entire design hierarchy, or for just certain regions of a design. VCS allows the support for ACC and VPI access to be enabled per Verilog module definition or per ACC and VPI capability (for example: read only access, read and write access, etc.).

The fastest simulation performance is achieved with VCS when only the TF library is used in PLI applications. When ACC or VPI routines are used, simulation performance may be impacted. However, by allowing a PLI application developer to control how much access into the simulation data structure that the ACC and VPI routines have the impact on simulation performance can be controlled.

A.3.3 Interfacing VPI applications

VCS version 6.1 beta does *not* adhere to the IEEE 1364 standard for the VPI interface mechanism. Instead, the beta version of VCS 6.1 uses the VCS interface for the older TF/ACC PLI applications. This interface is described in the sections A.3.4, on the following page.

The IEEE standard VPI interface mechanism involves creating an application-defined *register function*, as discussed in section A.1.1 on page 642. After the register function is defined, it should be listed in a *vlog_startup_routines* array, or specified in some other form, such as an invocation option to the simulator.

By not using the IEEE 1364 VPI interface, VCS cannot take advantage of some of the advantages of VPI based applications. Several examples in this book will not work correctly without register functions. Some of the limitations in VCS 6.1 beta are:

- The IEEE standard VPI interface mechanism allows the system task name and the names of the *calltf*, *compiletf* and the *sizetf* routines to be specified within the PLI application, and thus hidden and protected from the end user of the PLI application. All the end user needs to know is the name of the register function. The non-standard The older TF/ACC interface mechanism requires that the end user know and specify all the internal function names of the PLI application.

- The IEEE standard VPI interface mechanism allows for several types of system function returns, including real, integer, unsigned vectors and signed vectors. The VCS interface only supports real and unsigned vector returns.

- The IEEE standard VPI interface mechanism allows the vector size of system function returns to be calculated by the PLI application. The VCS interface requires the vector size of system function returns to be hard coded in the interface.

- The IEEE standard VPI interface mechanism uses a startup array to register system tasks and functions, as well as VPI simulation callbacks for end-of-compile and start-of-simulation. The VCS interface cannot specify register functions that must be called before simulation starts running.

- The IEEE standard VPI interface mechanism allows memory or other information to be dynamically allocated by the interface, and a pointer to the memory or information to be passed to all instances of the system task or function, as the user_data value. The VCS interface only allows a hard-coded integer value to defined as the user_data value.

It should be noted that these limitations in VCS were in the pre-release beta version of VCS version 6.1, which was the latest version of VCS available at the time this book was written. Later versions of VCS may implement the IEEE 1364 standard VPI interface mechanism.

A.3.4 Interfacing TF/ACC applications—the VCS *PLI table* file

VCS does not use the de facto standard *veriusertfs array* of s_tfcell structures that most other Verilog simulators use to specify PLI application information. Instead, VCS uses a proprietary ***PLI table*** file. The table file is much more simple and flexible than the veriusertfs array used by most simulators. The VCS PLI table specifies:

- The PLI application information.

- The configuration of ACC and VPI library access into the simulation data structure.

The PLI table file can contain the specification for any number of PLI applications. Each specification must be on a single line. Comments may be included in the file using a // token to begin the comment and a carriage return to terminate the comment. VCS allows any number of PLI tables to be used. This means a separate PLI table file can be created for each PLI application, and for any given simulation, only the PLI tables which are required need to be specified to the simulator.

The general format of the VCS PLI table is:

```
$<system_task_function_name> <PLI_spec> <access_spec>
```

The <PLI_spec> is zero, one or more of the specifications listed in Table A-1. The specifications can be listed in any order, but must be listed on the same line.

Table A-1: The Synopsys VCS PLI table PLI routine specifications

PLI Specification	Description
call=<routine_name>	Name of the *calltf routine* for the PLI application
check=<routine_name>	Name of the *checktf routine* or *compiletf routine* for the PLI application
misc=<routine_name>	Name of the *misctf routine* for the PLI application
size=<number_or_r>	Size of a system function return, or **r** if the return is a real number; VCS does not support the *sizetf routine* required by the IEEE standard
data=<number>	Integer user_data value to be passed to the TF/ACC *calltf routine*, *checktf routine* and *misctf routine*; does not support the VPI version of user_data, which is a pointer

 VCS infers that the PLI application is a system task if there is no size specification for the system task name. A system function is inferred if there is a size specification.

The <access_spec> defines what type of access within the simulation data structure ACC routines and VPI routines will be allowed. The <access_spec> is zero, one or more of the specifications listed in Table A-2. The specifications can be listed in any order, but must follow the <PLI_spec>, and be listed on the same line as the <PLI_spec>. The format of the <access_spec> is:

acc=<operation><capability>:<design_scopes>+

The **<operation>** is one of the four tokens listed in Table A-2.

Table A-2: The VCS PLI table ACC/VPI specification operations

Access Operation	Description
+=	Add the specified capabilities to the specified scope
=	The same as +=
-=	Remove the specified capabilities from the specified scope
:=	Set the specified capabilities to the specified scope

The **<capability>** of the access specification must be one of commands listed in Table A-3.

Table A-3: The VCS PLI table ACC/VPI capability specification

Access Capability
read (abbreviation: **r**) Allow reading values of nets, regs and variables
read_write (abbreviation: **rw**) Allow reading values of nets, regs and variables, and writing values to regs and variables
callback (abbreviation: **cbk**) Allow VCL (Value Change Link) and VPI simulation callbacks on named objects
callback_all (abbreviation: **cbka**) Allow VCL and VPI simulation callbacks on named and unnamed objects
force (abbreviation: **frc**) Allow forcing values onto nets and regs
timing_check_backannotation (abbreviation: **tchk**) Allow delay back annotation of timing check limits

Table A-3: The VCS PLI table ACC/VPI capability specification (continued)

Access Capability
gate_backannotation (abbreviation: **gate**) Allow delay back annotation of primitives
module_path_backannotation (abbreviation: **mp**) Allow delay back annotation of module paths
module_input_port_backannotation (abbreviation: **mip**) Allow delay back annotation of module input ports (cannot use wild card for scope)
module_input_port_bit_backannotation (abbreviation: **mipb**) Allow delay back annotation of bits of a module input port (cannot use wildcard for scope)

The **<scope>** of the access specification is one of names listed in Table A-4.

Table A-4: The VCS PLI table access specification scope

Access Scope	Description
<module_name>	The definition name of any Verilog module. The specified operation and capability will apply to all instances of that module.
%TASK	Any module which contains the system task/function that is being specified. Applies to all instances of those modules.
%CELL	Any module flagged as a cell (using the `celldefine` compiler directive or the -y or -v invocation options). Applies to all instances of those modules.
*	A wild card which indicates all Verilog modules in the design.

If a plus sign (+) is specified after the <scope>, then the access capability is applied to both the specified module, and all levels of hierarchy below that module. If the plus sign is not specified, then the capability is only applied to the specified module.

Example A-5 lists a VCS PLI table file for a *$hello* application and a *$pow* application.

Example A-5: PLI table file to specify PLI applications for the VCS simulator

```
// Example Synopsys VCS PLI table to register PLI applications

$hello check=hello_checktf call=hello_calltf data=0 acc+=read:*
$pow check=pow_checktf call=pow_calltf misc=pow_misctf size=32 data=0
```

A.3.5 Compiling and running PLI applications with VCS

 The full details on how to run VCS with PLI applications are outside the scope of this book. The following section only present the most basic steps to running VCS with PLI applications. Refer to the manuals included with the product for the full details on using VCS.

Running a simulation with VCS requires two steps:

1. Invoking the VCS compiler.

2. Invoking the simulation executable created by the compiler

To invoke the VCS compiler with a PLI application, the names of the PLI table files are specified using the **-P** invocation option, and the names of the PLI application source or object files are listed as part of the invocation command. For example, a simulation using the *$hello* and *$pow* PLI applications might be invoked using:

```
vcs test.v -P my_pli_apps.tab pow.c hello.c -CC "-I$VCS_HOME/include"
```

The C source files for the PLI applications can be specified as inputs to the VCS compiler, and VCS will automatically compile and link the applications into the VCS simulation executable. Alternatively, the PLI application can be pre-compiled into object files or shared library files, and the object file names specified on the VCS compiler command line. If the PLI applications are pre-compiled, the VCS compiler will just link the applications into the VCS simulation executable.

The -CC option shown in the preceding example is used to specify C compiler options. The VCS compiler will simply pass the string following the -CC to the C compiler. The preceding example uses this option to specify the location of the PLI include files, *vpi_user.h*, *veriuser.h* and *acc_user.h*.

The output from the vcs command is an executable program, called **simv**. Simulation is run by executing the simv program.

A.4 Linking to the Model Technology *ModelSim* simulator

The *ModelSim™* simulator from Model Technology, Inc. (a wholly owned subsidiary of Mentor Graphics, Inc.) is an easy to use Verilog simulator with a well integrated user interface. At the time this book was written, ModelSim ran on several Unix workstations, and on PCs running the Linux, Windows NT and Windows 2000 operating systems.

A.4.1 IEEE 1364-2001 compliance

The examples in this book were tested with ModelSim SE version 5.5e, which was the latest release available at the time the second edition of this book was prepared. This version of ModelSim supports the IEEE 1364-2001 TF and ACC routines, and most of the IEEE 1364-2001 VPI routines. Most the new features in the IEEE 1364-2001 standard are available in this release of ModelSim. Full support of the IEEE 1364-2001 PLI standard is planned for future releases.

ModelSim 5.5e uses the standard IEEE 1364-2001 *vpi_user.h*, *veriuser.h* and *acc_user.h* header files, which define the VPI, TF and ACC PLI libraries. ModelSim also provides a number of extensions to the PLI standard, which support proprietary features of the simulator. The extensions to the IEEE standard are specified in proprietary header files, *acc_vhdl.h* and *mti.h*. It is not necessary to include these proprietary header files unless the proprietary extensions to the PLI are being used. The examples in this book all run with just the standard IEEE 1364-2001 header files.

A.4.2 Interfacing VPI PLI applications

ModelSim uses the IEEE 1364 standard VPI interface mechanism. This method involves creating an application-defined *register function*, as discussed in section A.1.1 on page 642. The register function is then added to a ***vlog_startup_routines*** array. To specify the PLI application register function using this array:

1. Create a new C source file, which can be any name.

2. Within the file, define a global array called *vlog_startup_routines*, which contains an array of register function names.

3. Compile and link the PLI application source files and the file containing the *vlog_startup_routines array* into one or more shared libraries. The compile and link commands are shown in section A.4.4, which follows.

Example A-1 on page 643 lists an example *vlog_startup_routines array* file to register two PLI applications.

A.4.3 Interfacing TF/ACC PLI applications

ModelSim uses the same **_veriusertfs array_** of s_tfcell structures used by most other Verilog simulators to specify PLI application information. Refer to section A.1.2 on page 645 for a description of the _veriusertfs array_ and s_tfcell structure.

To specify the TF/ACC PLI application with ModelSim:

1. Create a new C source file, which can be any name

2. Within the file, define a global array called _veriusertfs_, which contains an array of s_tfcell structures. Each structure contains information about a PLI application.

3. Compile and link the PLI application source files and the file containing the _veriusertfs array_ into one or more shared libraries. The compile and link commands are shown in section A.4.4, which follows.

Example A-3 on page 646 lists the lines that would be added to a veriusertfs array file to register two PLI applications. ModelSim will automatically read the contents of the _veriusertfs array_ as it is loading the simulation.

A.4.4 Compiling and running PLI applications with ModelSim

 The full details on how to run ModelSim with PLI applications are outside the scope of this book. The following section only present the most basic steps to running ModelSim with PLI applications. Refer to the manuals included with the product for the full details on using ModelSim.

With ModelSim, running a simulation that has PLI applications requires three steps:

1. Compile and link the PLI applications source files and the file containing the _veriusertfs array_ as one or more shared object libraries. The compile and link commands are specific to the operating system and C compiler, and are covered in the ModelSim documentation. Three examples are:

    On a Sun system with the Solaris operating system and the Sun C compiler:

    ```
    cc -c -I/<install_dir>/modeltech/include <app_source_files>
    ld -G -B symbolic -o mti_pli_apps.so <pli_app_object_files>
    ```

    On a Linux operating systems with the GNU C compiler:

    ```
    gcc -c -I/<install_dir>/modeltech/include <app_source_files>
    ld -shared -E -o mti_pli_apps.so <pli_app_object_files>
    ```

On a PC system with the Windows NT operating system and the Visual C++ compiler (note: this example lists export commands for both the *veriusertfs array* for TF/ACC applications and the *vlog_startup_routines array* for VPI applications; only the export command for the type of application being used should be specified):

```
cl -c -I<install_dir>\modeltech\include <app_source_files>
link -dll -export:veriusertfs -export:vlog_startup_routines \
  <pli_app_object_files> \
  <install_dir>\modeltech\win32\mtipli.lib out:mti_pli_apps.dll
```

2. Modify the ModelSim **.ini** initialization file and specify the names of the shared object libraries in the **Veriuser** variable. This variable is typically found towards the middle of the initialization file, and by default it is commented out. Any number of shared object files may be listed, separated by a white space. As an example, if the shared library created in step 1 was called mti_pli_apps.so, then the *Veriuser* variable would be set to:

```
; List of dynamically loaded objects for Verilog PLI applications
Veriuser=mti_pli_apps.so
```

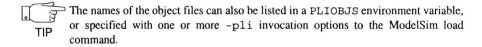 The names of the object files can also be listed in a PLIOBJS environment variable,
TIP or specified with one or more -pli invocation options to the ModelSim load command.

3. Invoke the ModelSim simulator in the usual way. The simulator will automatically load and dynamically link the shared object files referenced in the *Veriuser* variable in the .ini file. The *veriusertfs array* must be contained in one of the object files listed.

A.5 Summary

This appendix has presented how Verilog PLI applications are linked into a few Verilog simulators. The primary objective has been to show enough information to aid in getting started with using the PLI with these simulators. It is *not* the intent to present all the features available with these simulators. Readers should always refer to documentation supplied with a simulator for the most up-to-date information on using the Verilog PLI with a specific simulator.

There are many other excellent Verilog simulators, which were not included in this appendix. Due to space limitations, it was simply not possible to include the details on linking PLI applications into each and every simulator. While each simulator has unique ways of handling the PLI interface, the purpose is the same for all simulators. The few simulators covered in this appendix should serve as a guideline on what the PLI interface must do, and how the interfaces might be implemented in different products.

APPENDIX B

APPENDIX B *The IEEE 1364-2001 TF Routine Library*

Appendix B lists the definitions of the TF routines contained in the IEEE 1364-2001 PLI standard. The TF library is the first generation of the Verilog PLI, and was first introduced about 1985 as a proprietary interface to one simulator. The TF routines provide access to the arguments of user-defined system tasks and system functions, as well several utility routines such as the ability to print messages and to control Verilog simulations.

Many of the TF routines have two versions:

- A standard version, which works with the instance of the system task or system function that called the PLI application.
- An instance-specific version, which requires a pointer to an instance of a system task or system function. The instance pointer is obtained using tf_getinstance().

The TF library is defined in the file **veriuser.h**, which should be one of the files provided with each Verilog simulator. The veriuser.h file defines:

- A lower-case **null** constant, which is used by the TF library.
- Data types with fixed widths, for portability to different operating systems. Simulator vendors may modify these definitions as required for a specific operating system. The default definitions in the IEEE 1364-2001 standard are:

```
typedef int            PLI_INT32;
typedef unsigned int   PLI_UINT32;
typedef short          PLI_INT16;
typedef unsigned short PLI_UINT16;
typedef char           PLI_BYTE8;
typedef unsigned char  PLI_UBYTE8;
```

B.1 TF routine definitions

void **io_mcdprintf** (
PLI_INT32 **mcd,**	/* multi-channel descriptor of open files */
PLI_BYTE8 *** format,**	/* character string with a formatted message */
arg1...arg12)	/* arguments to the formatted message string */

Prints a formatted message to one or more open files. Uses formatting controls similar to the C printf() function, with a maximum of 12 arguments. The **mcd** values are generated in simulation by the Verilog $fopen system function, and can be passed to the PLI application as an argument to a system task or function.

void **io_printf** (
PLI_BYTE8 *** format,**	/* character string with a formatted message */
arg1...arg12)	/* arguments to the formatted message string */

Prints a formatted message to the simulator's output channel and output log file. Uses formatting controls similar to the C printf() function, with a maximum of 12 arguments.

PLI_BYTE8 ***mc_scan_plusargs** (
PLI_BYTE8 *** plusarg**)	/* name of the invocation option */

Tests to see if a string was included as a + option on the simulator product invocation commands. Returns **null** if the string does not exist. Returns a pointer to a null string ("\0") if the string exists exactly as tested. Returns a pointer to a string with any suffix characters if the string exists with suffix characters (e.g.: if the invocation option is +size64 and **plusarg** is "size", then a pointer to the string "64" is returned. The plus sign used on the command line is not included as part of the string that is tested.

PLI_INT32 **tf_add_long** (
PLI_INT32 *** low1,**	/* pointer to lower 32 bits of first operand */
PLI_INT32 *** high1,**	/* pointer to upper 32 bits of first operand */
PLI_INT32 **low2,**	/* lower 32 bits of second operand */
PLI_INT32 **high2**)	/* upper 32 bits of second operand */

Adds two 64-bit values, represented as pairs of 32-bit integers, and puts the result back into the first pair of integers. The return value is not used.

PLI_INT32 **tf_asynchoff** ( )
PLI_INT32 **tf_iasynchoff** (
PLI_BYTE8 *** tfinst**)	/* pointer to an instance of a system task/function */

Disables calling of the *misctf* routine with **reason_paramvc** for the calling or specific instance of a system task/function. The return value is not used.

PLI_INT32 **tf_asynchon** ( )
PLI_INT32 **tf_iasynchon** (
PLI_BYTE8 *** tfinst**)	/* pointer to an instance of a system task/function */

Enables asynchronous calling of the *misctf* routine with **reason_paramvc** whenever any argument of the calling or specific instance of a system task/function changes value or strength. Returns **1** if successful, and **0** if an error occurred.

PLI_INT32 **tf_clearalldelays ( )**
PLI_INT32 **tf_iclearalldelays (**
   PLI_BYTE8 *** tfinst)**   /* pointer to an instance of a system task/function */

Clears callbacks to the *misctf* routine scheduled by tf_setdelay() (or its related routines) for the calling or specific instance of a system task/function. The return value is not used.

PLI_INT32 **tf_compare_long (**
PLI_UINT32	**low1,**	/* lower 32 bits of the first operand */
PLI_UINT32	**high1,**	/* upper 32 bits of the first operand */
PLI_UINT32	**low2,**	/* lower 32 bits of the second operand */
PLI_UINT32	**high2)**	/* upper 32 bits of the second operand */

Compares two 64-bit integers, represented as pairs of 32-bit integers. Returns **0** if equal, **1** if operand 1 is greater than operand 2, and **–1** if operand 1 is less than operand 2.

PLI_INT32 **tf_copypvc_flag (**
   PLI_INT32   **n)**   /* index number of a system task/function arg, or –1 */

PLI_INT32 **tf_icopypvc_flag (**
PLI_INT32	**n,**	/* index number of a system task/function arg, or –1 */
PLI_BYTE8	*** tfinst)**	/* pointer to an instance of a system task/function */

Copies the current parameter value change (PVC) flag of argument **n** of the calling or specific instance of a system task/function to the saved PVC flag. Returns the value of the copied flag. If **n** is **-1**, then the PVC flags of all arguments are copied, and the logical OR of the saved flags is returned.

void **tf_divide_long (**
PLI_INT32	*** low1,**	/* pointer to lower 32 bits of first operand */
PLI_INT32	*** high1,**	/* pointer to upper 32 bits of first operand */
PLI_INT32	**low2,**	/* lower 32 bits of second operand */
PLI_INT32	**high2)**	/* upper 32 bits of second operand */

Divides two 64-bit values, represented as pairs of 32-bit integers, and puts the result into the first pair of integers.

PLI_INT32 **tf_dofinish ( )**

Executes the same functionality as the Verilog $finish task. The return value is not used.

PLI_INT32 **tf_dostop ( )**

Executes the same functionality as the Verilog $stop task. The return value is not used.

PLI_INT32 **tf_error (**
PLI_BYTE8	*** format,**	/* character string with a formatted message */
	arg1...arg5)	/* arguments to the formatted message string */

Prints a formatted message to the simulator's output channel and output log file. Uses formatting controls similar to the C printf() function, with a maximum of 5 arguments. The return value is not used.

PLI_INT32 **tf_evaluatep (**
    PLI_INT32   **n)**   /* index number of a system task/function argument */

PLI_INT32 **tf_ievaluatep (**
    PLI_INT32   **n,**   /* index number of a system task/function argument */
    PLI_BYTE8 *** tfinst)**   /* pointer to an instance of a system task/function */

Evaluates the value of argument **n** of the calling or specific instance of a system task/function and places it into an application-allocated s_tfexprinfo structure. The tf_exprinfo() routine must have been previously called to setup the structure. Returns **1** if successful, and **0** if an error occurred.

p_tfexprinfo **tf_exprinfo (**
    PLI_INT32   **n,**   /* index number of a system task/function argument */
    p_tfexprinfo **info)**   /* pointer to application-allocated s_tfexprinfo structure */

p_tfexprinfo **tf_iexprinfo (**
    PLI_INT32   **n,**   /* index number of a system task/function argument */
    p_tfexprinfo **info,**   /* pointer to application-allocated s_tfexprinfo structure */
    PLI_BYTE8 *** tfinst)**   /* pointer to an instance of a system task/function */

Retrieves information about argument **n** of the calling or specific instance of a system task/function argument, and places the information into an s_tfexprinfo structure pointed to by **info**. Returns the value of **info** if successful and **0** if an error occurred.

```
typedef struct t_tfexprinfo {
  PLI_INT16  expr_type;    /* tf_nullparam,
                              tf_readonly,     tf_readonlyreal,
                              tf_readwrite,    tf_readwritereal,
                              tf_rwbitselect, tf_rwpartselect,
                              tf_rwmemselect, tf_string  */
  PLI_INT16  padding;
  struct t_vecval *expr_value_p;
  double     real_value;
  PLI_BYTE8 *expr_string;
  PLI_INT32  expr_ngroups;
  PLI_INT32  expr_vec_size;
  PLI_INT32  expr_sign;        /* may not be supported */
  PLI_INT32  expr_lhs_select;  /* may not be supported */
  PLI_INT32  expr_rhs_select;  /* may not be supported */
} s_tfexprinfo, *p_tfexprinfo;
```

```
typedef struct t_vecval {
  PLI_INT32  avalbits;  /* aval/bval encoding: 0/0 == 0, */
  PLI_INT32  bvalbits;  /* 1/0 == 1, 0/1 == Z, 1/1 == X  */
} s_vecval, *p_vecval;
```

PLI_BYTE8 ***tf_getcstringp (**
    PLI_INT32   **n**)   /* index number of a system task/function argument */

PLI_BYTE8 ***tf_igetcstringp (**
    PLI_INT32   **n,**   /* index number of a system task/function argument */
    PLI_BYTE8 *** tfinst**)   /* pointer to an instance of a system task/function */

Returns a pointer to a string containing the value argument **n** of the calling or specific instance of a system task/function argument. Returns **null** if an error occurs. Verilog strings are converted to C strings. Each 8 bits of a Verilog vector is interpreted as an ASCII character, starting at the right-most bit. Real values cannot be converted.

PLI_BYTE8 ***tf_getinstance ( )**

Returns a pointer to the instance of the system task/function that called the PLI application. This is the **tfinst** pointer used with the instance-specific TF routines.

PLI_INT32 **tf_getlongp (**
    PLI_INT32 *** highvalue,**   /* pointer to the upper 32-bits of the value */
    PLI_INT32   **n**)   /* index number of a system task/function argument */

PLI_INT32 **tf_igetlongp (**
    PLI_INT32 *** highvalue,**   /* pointer to the upper 32-bits of the value */
    PLI_INT32   **n,**   /* index number of a system task/function argument */
    PLI_BYTE8 *** tfinst**)   /* pointer to an instance of a system task/function */

Retrieves the value of argument **n** of the calling or specific instance of a system task/function argument as a 64-bit integer represented as a pair of 32-bit integers. The lower 32-bits are returned, and the upper 32-bits are placed into **highvalue**.

PLI_INT32 **tf_getlongtime (**
    PLI_INT32 *** hightime**)   /* pointer to the upper 32-bits of the time */

PLI_INT32 **tf_igetlongtime (**
    PLI_INT32 *** hightime,**   /* pointer to the upper 32-bits of the time */
    PLI_BYTE8 *** tfinst**)   /* pointer to an instance of a system task/function */

Retrieves the current 64-bit simulation time. The lower 32-bits are returned, the upper 32-bits are loaded into **hightime**. Time is scaled to the time scale of the module containing the calling or specific instance of the system task/function.

PLI_INT32 **tf_getnextlongtime (**
    PLI_INT32 *** lowtime,**   /* pointer to the lower 32-bits of the time */
    PLI_INT32 *** hightime**)   /* pointer to the upper 32-bits of the time */

Retrieves the 64-bit simulation time of the next scheduled simulation event when called from a *misctf* routine that was called with **reason_rosynch**. The lower 32-bits are placed into **lowtime** and the upper 32-bits are placed into **hightime**. Time is scaled to the time scale of the module containing the instance of the system task/function. The return value is determined by when the routine is called:

• Returns **0** if called from a *misctf* routine that was called with **reason_rosynch**.

• Returns **1** if there are no more simulation events; A time of 0 is retrieved.

• Returns **2** if not called from a *misctf* routine that was called with **reason_rosynch**; The current simulation time is retrieved as the time of the next event.

PLI_INT32 **tf_getp (**
    PLI_INT32   **n**)   /* index number of a system task/function argument */

PLI_INT32 **tf_igetp (**
    PLI_INT32   **n,**   /* index number of a system task/function argument */
    PLI_BYTE8 *** tfinst**)   /* pointer to an instance of a system task/function */

Returns the current value of argument **n** of the calling or specific instance of a system task/function. If the argument is a scalar, vector or real value, it is converted to a 32-bit C integer (X and Z values are converted to 0). If the argument is a string, a pointer, cast to a *PLI_INT32* integer, to a string containing the value is returned. (Note: the casting of a string pointer to an integer may not work correctly on some operating systems—use tf_getcstringp() instead).

PLI_INT32 **tf_getpchange (**
    PLI_INT32   **n**)   /* index number of a system task/function argument */

PLI_INT32 **tf_igetpchange (**
    PLI_INT32   **n,**   /* index number of a system task/function argument */
    PLI_BYTE8 *** tfinst**)   /* pointer to an instance of a system task/function */

Returns the index number of the next system task/function argument greater than **n** that changed value at the current simulation time step. A **0** is returned if no other arguments changed value or an error occurred. tf_movepvc_flag(-1) must first be called in the current call to the PLI application to store the PVC (parameter value change) flags. Index **n** must be **0** the first time the routine is called. tf_asynchon() or tf_iasynchon() must be active to enable PVC flags.

double **tf_getrealp (**
    PLI_INT32   **n**)   /* index number of a system task/function argument */

double **tf_igetrealp (**
    PLI_INT32   **n,**   /* index number of a system task/function argument */
    PLI_BYTE8 *** tfinst**)   /* pointer to an instance of a system task/function */

Returns the current value of argument **n** of the calling or specific instance of a system task/function as a double. The argument must be a real, scalar or vector value (X and Z values are converted to 0.0). Returns **0.0** if an error occurred.

double **tf_getrealtime ( )**

double **tf_igetrealtime (**
    PLI_BYTE8 *** tfinst**)   /* pointer to an instance of a system task/function */

Retrieves the current simulation time as a real number. Time is scaled to the time scale of the module containing the calling or specific instance of a system task/function.

PLI_INT32 **tf_gettime ( )**

PLI_INT32 **tf_igettime(**
    PLI_BYTE8 *** tfinst**)   /* pointer to an instance of a system task/function */

Retrieves the lower 32-bits of current simulation time as an integer. Time is scaled to the time scale of the module containing the calling or specific instance of a system task/ function.

PLI_INT32 **tf_gettimeprecision ( )**

PLI_INT32 **tf_igettimeprecision (**
    PLI_BYTE8 *** tfinst)**   /* pointer to an instance of a system task/func, or **null** */

Returns an integer representing the time scale precision of the module containing the calling or specific instance of a system task/function. If **tfinst** is **null**, the internal simulation time unit is returned (usually the smallest simulation time precision of all modules in the design). Time is represented as: 2==100 seconds, 1==10 s, 0==1 s, -1==100 ms, -2 ==10 ms, -3==1 ms,... -6==1 us,... -9==1 ns,... -12==1 ps,... -15==1 fs.

PLI_INT32 **tf_gettimeunit ( )**

PLI_INT32 **tf_igettimeunit (**
    PLI_BYTE8 *** tfinst)**   /* pointer to an instance of a system task/func, or **null** */

Returns an integer representing the time scale unit of the module containing the calling or specific instance of a system task/function. If **tfinst** is **null**, the internal simulation time unit is returned (usually the smallest simulation time precision of all modules in the design). Time is represented as: 2==100 seconds, 1==10 s, 0==1 s, -1==100 ms, -2 ==10 ms, -3==1 ms,... -6==1 us,... -9==1 ns,... -12==1 ps,... -15==1 fs.

PLI_BYTE8 ***tf_getworkarea ( )**

PLI_BYTE8 ***tf_igetworkarea (**
    PLI_BYTE8 *** tfinst)**   /* pointer to an instance of a system task/function */

Returns the pointer stored in the built-in, instance-specific storage for the calling or specific instance of a system task/function. A pointer is stored in the work area using tf_setworkarea().

void **tf_long_to_real (**
    PLI_INT32   **low,**   /* lower (right-most) 32-bits of a 64-bit integer */
    PLI_INT32   **high,**   /* upper (left-most) 32-bits of a 64-bit integer */
    double   *** real)**   /* pointer to a double precision variable */

Converts a 64-bit integer represented as a pair of 32-bit integers into a double precision number.

PLI_BYTE8 ***tf_longtime_tostr (**
    PLI_INT32   **lowtime,**   /* lower (right-most) 32-bits of a 64-bit integer */
    PLI_INT32   **hightime)**   /* upper (left-most) 32-bits of a 64-bit integer */

Converts simulation time (a 64-bit unsigned integer) represented as a pair of 32-bit integers to a character string. Returns a pointer to the string.

PLI_INT32 **tf_message (**
   PLI_INT32   **level,**   /* a constant representing the error severity level; one of:
                                   ERR_ERROR, ERR_SYSTEM, ERR_INTERNAL,
                                   ERR_WARNING, ERR_MESSAGE */
   PLI_BYTE8 *** facility,**   /* quoted string to be appended to the output message;
                                   must be 10 or less characters */
   PLI_BYTE8 *** code,**   /* quoted string to be appended to the output message after
                                   facility; must be 10 or less characters */
   PLI_BYTE8 *** format,**   /* quoted string of the formatted message */
              arg1...arg5) /* arguments to the formatted message string */

Prints a formatted message to the simulator's output channel and output log file. Uses formatting controls similar to the C printf() function, with a maximum of 5 arguments. If called by a *checktf* routine during compiling or loading the Verilog code, and if **level** is **ERR_ERROR, ERR_SYSTEM** or **ERR_INTERNAL**, then the elaboration or loading of simulation is aborted. The return value is not used.

PLI_BYTE8 ***tf_mipname ( )**

PLI_BYTE8 ***tf_imipname (**
   PLI_BYTE8 *** tfinst)**   /* pointer to an instance of a system task/function */

Returns a pointer to a string containing the full hierarchical path name of the module containing the calling or specific instance of a system task/function.

PLI_INT32 **tf_movepvc_flag (**
   PLI_INT32   **n)**   /* index number of a system task/function arg, or −1 */

PLI_INT32 **tf_imovepvc_flag (**
   PLI_INT32   **n,**   /* index number of a system task/function arg, or −1 */
   PLI_BYTE8 *** tfinst)**   /* pointer to an instance of a system task/function */

Moves the current Parameter Value Change (PVC) flag of argument **n** of the calling or specific instance of a system task/function to the saved PVC flag and clears the current flag. Returns the value of the flag that was moved. If **n** is −1, then all the task/function argument PVC flags are moved, and a logical OR of all the saved flags is returned.

void **tf_multiply_long (**
   PLI_INT32 *** low1,**   /* pointer to lower 32 bits of first operand */
   PLI_INT32 *** high1,**   /* pointer to upper 32 bits of first operand */
   PLI_INT32 **low2,**   /* lower 32 bits of second operand */
   PLI_INT32 **high2)**   /* upper 32 bits of second operand */

Multiplies two 64-bit values, represented as pairs of 32-bit integers, and puts the result back into the first pair of integers.

p_tfnodeinfo **tf_nodeinfo (**
   PLI_INT32   **n,**   /* index number of a system task/function argument */
   p_tfnodeinfo **info)**   /* pointer to application-allocated s_tfnodeinfo structure */

p_tfnodeinfo **tf_inodeinfo ()**
   PLI_INT32   **n,**   /* index number of a system task/function argument */
   p_tfnodeinfo **info,**   /* pointer to application-allocated s_tfnodeinfo structure */
   PLI_BYTE8 ***tfinst)**   /* pointer to an instance of a system task/function */

Retrieves node information about argument **n** of the calling or specific instance of a system task/function argument. Places the information in an s_tfnodeinfo structure pointed to by **info**. Returns the value of **info** if successful and **0** if an error occurred. tf_nodeinfo() supports the following Verilog data types: scalar or vector nets, scalar or vector regs, integer, time or real variables, a word select of a one-dimensional array of reg, integer or time. Some simulators may support additional types, such as bit or part selects of vectors, but different simulators may not return the same results for these additional types.

```
typedef struct t_tfnodeinfo {
  PLI_INT16 node_type; /* tf_null_node, tf_reg_node,
                          tf_integer_node, tf_real_node,
                          tf_time_node, tf_netvector_node,
                          tf_netscalar_node, tf_memory_node */
  PLI_INT16 padding;
  union {
    struct t_vecval       *vecval_p;
    struct t_strengthval  *strengthval_p;
    PLI_BYTE8             *memoryval_p;
    double                *real_val_p;
  } node_value;
  PLI_BYTE8 *node_symbol;      /* signal name */
  PLI_INT32  node_ngroups;
  PLI_INT32  node_vec_size;
  PLI_INT32  node_sign;        /* may not be supported */
  PLI_INT32  node_ms_index;    /* may not be supported */
  PLI_INT32  node_ls_index;    /* may not be supported */
  PLI_INT32  node_mem_size;
  PLI_INT32  node_lhs_element; /* may not be supported */
  PLI_INT32  node_rhs_element; /* may not be supported */
  PLI_INT32 *node_handle;      /* not used */
} s_tfnodeinfo, *p_tfnodeinfo;
```

```
typedef struct t_vecval {
  PLI_INT32 avalbits;   /* aval/bval encoding: 0/0 == 0, */
  PLI_INT32 bvalbits;   /* 1/0 == 1, 0/1 == Z, 1/1 == X  */
} s_vecval, *p_vecval;
```

```
typedef struct t_strengthval {
  PLI_INT32 strength0;
  PLI_INT32 strength1;
} s_strengthval, *p_strengthval;
```

PLI_INT32 **tf_nump ( )**

PLI_INT32 **tf_inump (**
 PLI_BYTE8 *** tfinst)**　　　　/* pointer to an instance of a system task/function */

Returns the number of arguments in the calling or a specific system task/function.

PLI_INT32 **tf_propagatep (**
 PLI_INT32 **n)**　　　　/* index number of a system task/function argument */

PLI_INT32 **tf_ipropagatep (**
 PLI_INT32 **n,**　　　　/* index number of a system task/function argument */
 PLI_BYTE8 *** tfinst)**　　　　/* pointer to an instance of a system task/function */

Writes a value stored in an **s_tfexprinfo** structure onto argument **n** of the calling or specific instance of a system task/function, and propagates the value to any continuous assignments that read the argument's value. tf_exprinfo() must have been previously called to set up the value structure. Returns **0** if successful, and **1** if an error occurred.

```
typedef struct t_tfexprinfo {
  PLI_INT16  expr_type;    /* tf_nullparam,
                             tf_readonly,    tf_readonlyreal,
                             tf_readwrite,   tf_readwritereal,
                             tf_rwbitselect, tf_rwpartselect,
                             tf_rwmemselect, tf_string   */
  PLI_INT16  padding;
  struct t_vecval *expr_value_p;
  double     real_value;
  PLI_BYTE8 *expr_string;
  PLI_INT32  expr_ngroups;
  PLI_INT32  expr_vec_size;
  PLI_INT32  expr_sign;          /* may not be supported */
  PLI_INT32  expr_lhs_select;    /* may not be supported */
  PLI_INT32  expr_rhs_select;    /* may not be supported */
} s_tfexprinfo, *p_tfexprinfo;
```

```
typedef struct t_vecval {
  PLI_INT32  avalbits;    /* aval/bval encoding: 0/0 == 0, */
  PLI_INT32  bvalbits;    /* 1/0 == 1, 0/1 == Z, 1/1 == X   */
} s_vecval, *p_vecval;
```

PLI_INT32 **tf_putlongp (**
 PLI_INT32 **n,**　　　　/* index number of a system task/function arg., or 0 */
 PLI_INT32 **lowvalue,**　　　　/* lower (right-most) 32-bits of a 64-bit integer */
 PLI_INT32 **highvalue)**　　　　/* upper (left-most) 32-bits of a 64-bit integer */

PLI_INT32 **tf_iputlongp (**
 PLI_INT32 **n,**　　　　/* index number of a system task/function arg., or 0 */
 PLI_INT32 **lowvalue,**　　　　/* lower (right-most) 32-bits of a 64-bit integer */
 PLI_INT32 **highvalue,**　　　　/* upper (left-most) 32-bits of a 64-bit integer */
 PLI_BYTE8 *** tfinst)**　　　　/* pointer to an instance of a system task/function */

Writes a 64-bit integer value, represented as a pair of 32-bit integers, onto argument **n** of the calling or specific instance of a system task/function. If **n** is **0**, the value is written as the return of a system function. Returns **0** if successful, and **1** if an error occurred.

PLI_INT32 **tf_putp (**
    PLI_INT32   **n,**   /* index number of a system task/function arg., or 0 */
    PLI_INT32   **value)**   /* a 32-bit integer value */

PLI_INT32 **tf_iputp (**
    PLI_INT32   **n,**   /* index number of a system task/function arg., or 0 */
    PLI_INT32   **value,**   /* a 32-bit integer value */
    PLI_BYTE8 *** tfinst)**   /* pointer to an instance of a system task/function */

Writes a 32-bit integer value onto argument **n** of the calling or specific instance of a system task/function. If **n** is **0**, then the value is written as the return of a system function. Returns **0** if successful, and **1** if an error occurred.

PLI_INT32 **tf_putrealp (n, value)**
    PLI_INT32   **n,**   /* index number of a system task/function arg., or 0 */
    double   **value)**   /* a double precision real number */

PLI_INT32 **tf_iputrealp (n, value, tfinst)**
    PLI_INT32   **n,**   /* index number of a system task/function arg., or 0 */
    double   **value,**   /* a double precision real number */
    PLI_BYTE8 *** tfinst)**   /* pointer to an instance of a system task/function */

Writes a real value onto argument **n** of the calling or specific instance of a system task/function. If **n** is **0**, then the value is written as the return of a system function. Returns **0** if successful, and **1** if an error occurred.

PLI_INT32 **tf_read_restart (**
    PLI_BYTE8 *** blockptr,**   /* pointer to a block of memory */
    PLI_INT32   **blocklength)** /* length of the block of memory in bytes */

Loads a block of memory with data that was saved with tf_write_save(). This routine must be called from a *misctf* routine that was called with **reason_restart**. Returns **1** if successful and **0** if an error occurred.

void **tf_real_to_long (**
    double   **real,**   /* a double precision variable */
    PLI_INT32   *** low,**   /* pointer to the lower 32-bits of a 64-bit integer */
    PLI_INT32   *** high)**   /* pointer to upper 32-bits of a 64-bit integer */

Converts a real number to a 64-bit long integer, represented as a pair of 32-bit integers.

PLI_INT32 **tf_rosynchronize ( )**

PLI_INT32 **tf_irosynchronize (**
    PLI_BYTE8 *** tfinst)**   /* pointer to an instance of a system task/function */

Schedules a callback to the *misctf* routine with **reason_rosynch** for the calling or specific instance of a system task/function. The callback occurs at the end of all events in the current simulation time step. The PLI is not allowed to schedule any additional events at the current or future simulation time. Returns **0** if successful, and **1** if an error occurred.

void **tf_scale_longdelay (**

PLI_BYTE8	*** tfinst,**	/* pointer to an instance of a system task/function */
PLI_INT32	**lowdelay1,**	/* lower (right-most) 32-bits of first 64-bit time value */
PLI_INT32	**highdelay1,**	/* upper (left-most) 32-bits of first 64-bit time value */
PLI_INT32	**lowdelay2,**	/* lower (right-most) 32-bits of second 64-bit time value */
PLI_INT32	**highdelay2)**	/* upper (left-most) 32-bits of second 64-bit time value */

Scales a 64-bit time value, represented as a pair of 32-bit integers, to the time scale of the module containing a specific instance of a system task/function. The value in arguments **lowdelay1** and **highdelay1** are scaled and placed into **lowdelay2** and **highdelay2**.

void **tf_scale_realdelay (**

PLI_BYTE8	*** tfinst,**	/* pointer to an instance of a system task/function */
double	**realdelay1,**	/* first time value */
double	*** realdelay2)**	/* pointer to second time value */

Scales a double precision real number time value to the time scale of the module containing a specific instance of a system task/function. The value in **realdelay1** is scaled and placed into **realdelay2**.

PLI_INT32 **tf_setdelay (**

PLI_INT32	**delay**	/* a 32-bit time value greater than or equal to 0 */

PLI_INT32 **tf_isetdelay (**

PLI_INT32	**delay,**	/* a 32-bit time value greater than or equal to 0 */
PLI_BYTE8	*** tfinst)**	/* pointer to an instance of a system task/function */

Schedules a callback to the *misctf* routine for the calling or specific instance of a system task/function. The callback occurs with **reason_reactivate** after the amount of time specified by **delay**. The 32-bit integer delay value is scaled to the time scale of module containing the calling or specific instance of a system task/function. Returns **1** if successful and **0** if an error occurred.

PLI_INT32 **tf_setlongdelay (**

PLI_INT32	**lowdelay,**	/* lower (right-most) 32-bits of a 64-bit time value */
PLI_INT32	**highdelay)**	/* upper (left-most) 32-bits of a 64-bit time value */

PLI_INT32 **tf_isetlongdelay (**

PLI_INT32	**lowdelay,**	/* lower (right-most) 32-bits of a 64-bit time value */
PLI_INT32	**highdelay,**	/* upper (left-most) 32-bits of a 64-bit time value */
PLI_BYTE8	*** tfinst)**	/* pointer to an instance of a system task/function */

Same as tf_setdelay() except that the delay is a 64-bit integer value, represented as a pair of 32-bit integers.

PLI_INT32 **tf_setrealdelay (**

double	**realdelay)**	/* a real time value greater than or equal to 0.0 */

PLI_INT32 **tf_isetrealdelay (**

double	**realdelay,**	/* a real time value greater than or equal to 0.0 */
PLI_BYTE8	*** tfinst)**	/* pointer to an instance of a system task/function */

Same as tf_setdelay() except that the delay is a double precision value.

PLI_INT32 **tf_setworkarea (**
    PLI_BYTE8 *** workarea**)   /* pointer to application-allocated storage */

PLI_INT32 **tf_isetworkarea (**
    PLI_BYTE8 *** workarea,**   /* pointer to application-allocated storage */
    PLI_BYTE8 *** tfinst**)   /* pointer to an instance of a system task/function */

Writes a pointer value into the built-in, instance-specific storage area of the calling or specific instance of a system task/function. The return value is not used.

PLI_INT32 **tf_sizep (**
    PLI_INT32   **n**)   /* index number of a system task/function argument */

PLI_INT32 **tf_isizep (**
    PLI_INT32   **n,**   /* index number of a system task/function argument */
    PLI_BYTE8 *** tfinst**)   /* pointer to an instance of a system task/function */

Returns the size in number of bits of argument **n** of the calling or specific instance of a system task/function.

PLI_BYTE8 ***tf_spname ( )**

PLI_BYTE8 ***tf_ispname (**
    PLI_BYTE8 *** tfinst**)   /* pointer to an instance of a system task/function */

Returns a pointer to a string containing the full hierarchical path name of the scope containing the calling or specific instance of a system task/function.

PLI_INT32 **tf_strdelputp (**
    PLI_INT32   **n,**   /* index number of a system task/function argument */
    PLI_INT32   **bit_length,**   /* number of bits to be written */
    PLI_INT32   **format_char,** /* character in single quotes indicating the value format, as
                                   'b', **'B'**, **'o'**, **'O'**, **'d'**, **'D'**, **'h'** or **'H'** */
    PLI_BYTE8 *** value,**   /* string representing the value to be written */
    PLI_INT32   **delay,**   /* 32-bit integer delay before value is written */
    PLI_INT32   **delay_type**) /* code indicating delay method: **0** for inertial, **1** for
                                   modified transport, **2** for pure transport */

PLI_INT32 **tf_istrdelputp (**
             . . . ,   /* same arguments as tf_strdelputp() */
    PLI_BYTE8 *** tfinst**)   /* pointer to an instance of a system task/function */

Writes a value to argument **n** of the calling or specific instance of a system task. The value to be written is represented as a string, with binary, octal, decimal or hexadecimal characters, and can include z, Z, x or X characters. The value representation is specified in the **format_char** field, which must be set to **'b'** or **'B'** for binary, **'o'** or **'O'** for octal, **'d'** or **'D'** for decimal, **'h'** or **'H'** for hexadecimal. The value change is scheduled in simulation after a delay, which must be greater than or equal to 0. The simulation event scheduling method is specified in the **delay_type** field, which must be set to **0** for inertial delay, **1** for modified transport delay, **2** for pure transport delay. Time is scaled to the time scale of the module containing the calling or specific instance of the system task. Returns **1** if successful and **0** if an error occurred.

PLI_BYTE8 ***tf_strgetp (**
   PLI_INT32   **n,**   /* index number of a system task/function argument */
   PLI_INT32   **format_char**)/* character in single quotes representing value format */

PLI_BYTE8 ***tf_istrgetp (**
   PLI_INT32   **n,**   /* index number of a system task/function argument */
   PLI_INT32   **format_char,** /* character in single quotes representing value format */
   PLI_BYTE8 *** tfinst**)   /* pointer to an instance of a system task/function */

Returns a pointer to a string containing the value of argument **n** of the calling or specific instance of a system task/function. The value is converted to a string, with binary, octal, decimal or hexadecimal characters, and can include z, Z, x or X characters. How the value is converted is represented by format_char, which must be **'b'** or **'B'** for binary, **'o'** or **'O'** for octal, **'d'** or **'D'** for decimal, **'h'** or **'H'** for hexadecimal.

PLI_BYTE8 ***tf_strgettime ( )**

Retrieves a pointer to a string containing the current simulation time. Time is returned in the simulator's time units, which is the smallest time precision of all modules which make up the simulation.

PLI_INT32 **tf_strlongdelputp (**
   PLI_INT32   **n,**   /* index number of a system task/function argument */
   PLI_INT32   **length,**   /* number of bits to be written */
   PLI_INT32   **format_char,** /* character in single quotes indicating the value format, as
                             'b', 'B', 'o', 'O', 'd', 'D', 'h' or **'H'** */
   PLI_BYTE8 *** value,**   /* string representing the value to be written */
   PLI_INT32   **lowdelay,**   /* lower (right-most) 32-bits of a 64-bit time value */
   PLI_INT32   **highdelay,**   /* upper (left-most) 32-bits of a 64-bit time value */
   PLI_INT32   **mode**)   /* code indicating delay method: **0** for inertial, **1** for
                             modified transport, **2** for pure transport */

PLI_INT32 **tf_istrlongdelputp (**
             . . . ,   /* same arguments as tf_strlongdelputp() */
   PLI_BYTE8 *** tfinst**)   /* pointer to an instance of a system task/function */

Same as tf_strdelputp() except that the delay is specified as a 64-bit integer value, represented as a pair of 32-bit integers.

PLI_INT32 **tf_strrealdelputp (**
   PLI_INT32   **n,**   /* index number of a system task/function argument */
   PLI_INT32   **length,**   /* number of bits to be written */
   PLI_INT32   **format_char,** /* character in single quotes indicating the value format, as
                             'b', 'B', 'o', 'O', 'd', 'D', 'h' or **'H'** */
   PLI_BYTE8 *** value,**   /* string representing the value to be written */
   double   **delay,**   /* 32-bit integer delay before value is written */
   PLI_INT32   **mode**)   /* code indicating delay method: **0** for inertial, **1** for
                             modified transport, **2** for pure transport */

PLI_INT32 **tf_istrrealdelputp (**
             . . . ,   /* same arguments as tf_strrealdelputp() */
   PLI_BYTE8 *** tfinst**)   /* pointer to an instance of a system task/function */

Same as tf_strdelputp() except that the delay is specified as a double precision value.

PLI_INT32 **tf_subtract_long (**
    PLI_INT32   *** low1,**   /* pointer to lower 32 bits of first operand */
    PLI_INT32   *** high1,**   /* pointer to upper 32 bits of first operand */
    PLI_INT32   **low2,**   /* lower 32 bits of second operand */
    PLI_INT32   **high2)**   /* upper 32 bits of second operand */

Subtracts two 64-bit values, represented as pairs of 32-bit integers, and puts the result back into the first pair of integers. The return value is not used.

PLI_INT32 **tf_synchronize ( )**

PLI_INT32 **tf_isynchronize (**
    PLI_BYTE8   *** tfinst)**   /* pointer to an instance of a system task/function */

Schedules a callback to the *misctf* routine with **reason_synch** for the calling or specific instance of a system task/function. The callback is scheduled at the end of the list of events in the current simulation time step. The PLI is allowed to schedule additional events at the same or a later simulation time. Returns **0** if successful and **1** if an error occurred.

PLI_INT32 **tf_testpvc_flag (**
    PLI_INT32   **n)**   /* index number of a system task/function arg, or **−1** */

PLI_INT32 **tf_itestpvc_flag (**
    PLI_INT32   **n,**   /* index number of a system task/function arg, or **−1** */
    PLI_BYTE8   *** tfinst)**   /* pointer to an instance of a system task/function */

Returns the value of the saved PVC (parameter value change) flag of argument **n** of the calling or specific instance of a system task/function. If **n** is **−1**, a logical OR of all saved PVC flags is returned, which can be used to determine if any saved flag is set.

PLI_INT32 **tf_text (**
    PLI_BYTE8   *** format,**   /* character string with a formatted message */
                   arg1...arg5) /* arguments to the formatted message string */

Queues a formatted message into a text buffer, which will be printed when tf_message() is called. The return value is not used.

PLI_INT32 **tf_typep(n)**
    PLI_INT32   **n)**   /* index number of a system task/function argument */

PLI_INT32 **tf_itypep(n, tfinst)**
    PLI_INT32   **n,**   /* index number of a system task/function argument */
    PLI_BYTE8   *** tfinst)**   /* pointer to an instance of a system task/function */

Returns a constant representing the data type of argument **n** of the calling or specific instance of a system task/function. The type is one of: **tf_nullparam, tf_string, tf_readonly, tf_readwrite, tf_readonlyreal, tf_readwritereal**. A read/write argument is a Verilog data type that is legal on the left-hand side of a procedural assignment (a reg or variable data type). A read-only argument is a Verilog net data type, a literal value or a name.

void **tf_unscale_longdelay (**

PLI_BYTE8	*** tfinst,**	/* pointer to an instance of a system task/function */
PLI_INT32	**lowdelay1,**	/* lower 32-bits of first 64-bit time value */
PLI_INT32	**highdelay1,**	/* upper 32-bits of first 64-bit time value */
PLI_INT32	***lowdelay2,**	/* pointer to lower 32-bits of second 64-bit time value */
PLI_INT32	***highdelay2)**	/* pointer to upper 32-bits of second 64-bit time value */

Converts a 64-bit time value, represented as a pair of 32-bit integers, expressed in simulation time units to the time scale of a specific instance of a system task/function. The value in arguments **lowdelay1** and **highdelay1** are converted and placed into **lowdelay2** and **highdelay2**.

void **tf_unscale_realdelay (**

PLI_BYTE8	*** tfinst,**	/* pointer to an instance of a system task/function */
double	**realdelay1,**	/* real number value of first delay value */
double	*** realdelay2)**	/* pointer to real number value of second delay value */

Converts a real number value expressed in simulation time units to the time scale of a specific instance of a system task/function. The value in **realdelay1** is converted and placed into **realdelay2**.

PLI_INT32 **tf_warning (**

PLI_BYTE8	*** format,**	/* character string with a formatted message */
	arg1...arg5)	/* arguments to the formatted message string */

Prints a formatted message to the simulator's output channel and output log file. Uses formatting controls similar to the C printf() function, with a maximum of 5 arguments. Does not abort elaboration or loading. The return value is not used.

PLI_INT32 **tf_write_save (**

PLI_BYTE8	*** blockptr,**	/* pointer to a block of application-allocated memory */
PLI_INT32	**blocklength)**	/* length of the block of memory in bytes */

Causes a block of memory to be included in a data file created by the $save system task. This routine may only be called by a *misctf* routine that was called with **reason_save**. Returns non-zero if successful and **0** if an error occurred.

APPENDIX C *The IEEE 1364-2001 ACC Routine Library*

T he ACC is library is the second generation of the PLI, and was first introduced about 1988 as an extension to the TF library. The ACC routines provide access to the structural constructs within a simulation data structure, such as modules, primitives and nets. This appendix lists the ACC routines in the IEEE 1364-2001 standard.

This appendix also includes **ACC object diagrams**, which show what objects can be accessed by ACC routines, and how to traverse the simulation data structure in order to access one object from another object. These diagrams are not part of the IEEE 1364 standard.

The ACC library is defined in the file **acc_user.h**, which should be one of the files provided with each Verilog simulator. The acc_user.h file defines:

- A lower-case **handle** data type, which is a pointer to information about an object.

- A lower-case **null** constant, which is used by the ACC library (*note*: this lower-case **null** is not the same as the ANSI C upper-case **NULL**).

- Data types with fixed widths, for portability to different operating systems. Simulator vendors may modify this section of acc_user.h as required for a specific operating system. The default definitions in the IEEE 1364-2001 standard are:

```
typedef int             PLI_INT32;
typedef unsigned int    PLI_UINT32;
typedef short           PLI_INT16;
typedef unsigned short  PLI_UINT16;
typedef char            PLI_BYTE8;
typedef unsigned char   PLI_UBYTE8;
```

C.1 ACC object relationships

C.1.1 Object diagram legend

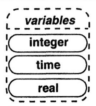

A *solid enclosure with bold font* represents an *object definition*. The diagram shows all relationships which can be traversed to and from the object.

A *dotted enclosure with bold-italic font* represents an *object class definition*. The name of the object class is listed at the top of the enclosure (a class may be unnamed). The diagram shows all relationships which can be traversed to and from the object class.

A *solid enclosure with standard font* represents an *object reference*. Another diagram defines the referenced object.

A *dotted enclosure with italic font* represents an *object class reference*. Another diagram defines the referenced object class.

A *single arrow* represents a *one-to-one relationship* which is traversed using the routine shown, usually an **acc_handle_*** routine. Refer to the description of the routine for the arguments required.

A *double arrow* represents a *one-to-many relationship* which is traversed using the routine shown, usually an **acc_next_*** routine. Refer to the description of the routine for the arguments required.

A *small circle* represents either the *top of hierarchy* or the *current PLI hierarchy scope*. Objects are accessed with the acc_handle_* or acc_next_* routine shown.

C.1.2 system task/function call objects

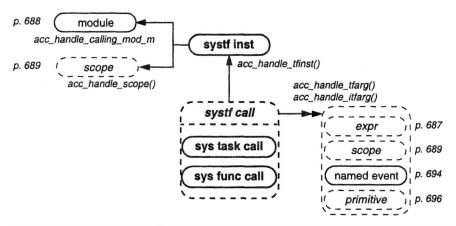

acc_fetch_type()	returns **accUserTask, accUserFunction, accUserRealFunction**
acc_fetch_fulltype()	returns **accUserTask, accUserFunction, accUserRealFunction**
acc_fetch_tfarg()	returns the value of a system task/function argument as a double
acc_fetch_tfarg_int()	returns the value of a system task/function argument as an integer
acc_fetch_tfarg_str()	returns the value of a system task/function argument as a string
acc_set_value()	writes a value onto a system task/function argument writes the return of a system function instance
acc_fetch_value()	retrieves the current value of a system function instance
acc_fetch_location()	returns source file name and line number containing a system task/function instance

C.1.3 expression objects

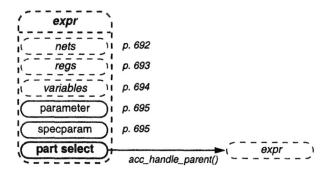

acc_fetch_type()	returns **accNet, accReg** (or **accRegister**), **accIntegerVar, accTimeVar, accRealVar, accConstant, accParameter, accSpecparam, accPartSelect**
acc_fetch_fulltype()	refer to the full type constants listed in the diagrams for each object

C.1.4 module objects

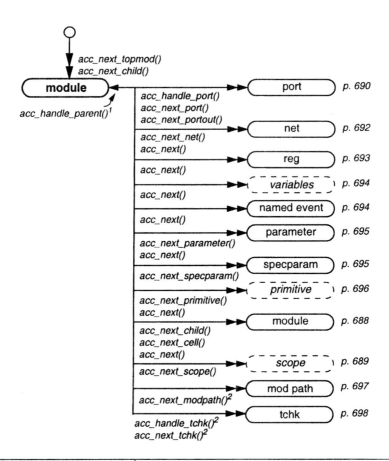

acc_fetch_type()	returns **accModule**
acc_fetch_fulltype()	returns **accTopModule, accModuleInstance, accCellInstance**
acc_fetch_name()	returns the instance name of a module
acc_fetch_fullname()	returns the hierarchical path name of a module
acc_fetch_defname()	returns the definition name of a module
acc_fetch_delay_mode()	returns **accDelayModeNone, accDelayModeZero, accDelayModeUnit, accDelayModePath, accDelayModeDistrib, accDelayModeMTM**
acc_fetch_timescale_info()	returns the timescale of a module
acc_fetch_location()	returns source file name & line no. of module instance

1. **acc_handle_parent()** does not work with scope types of **accTask, accFunction, accNamedBeginStat,** and **accNamedForkStat**.

2. In some simulators, **acc_next_modpath()**, **acc_next_tchk()** and **acc_handle_tchk()** only work if the module fulltype is **accCellInstance**.

C.1.5 scope objects

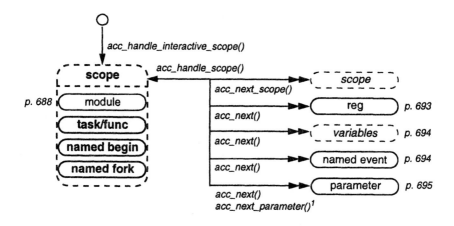

acc_fetch_type()	returns **accModule, accFunction, accTask, accStatement**
acc_fetch_fulltype()	returns **accTopModule, accModuleInstance, accCellInstance, accFunction, accNamedBeginStat, accNamedForkStat**
acc_object_of_type()	returns 1 (true) for special-type of **accScope**
acc_fetch_name()	returns the instance name of a scope
acc_fetch_fullname()	returns the hierarchical path name of a scope
acc_fetch_location()	returns source file name and line number containing a scope
acc_set_interactive_scope()	sets the simulator's interactive debug scope
acc_set_scope()	sets the PLI search scope for routines which search for objects by name
acc_configure()	**accEnableArgs** configures the arguments required by the acc_set_scope() routine

1. Some simulators cannot access parameters declared within scope types of **accNamedBeginStat** and **accNamedForkStat** using **acc_next_parameter()**.

C.1.6 port objects

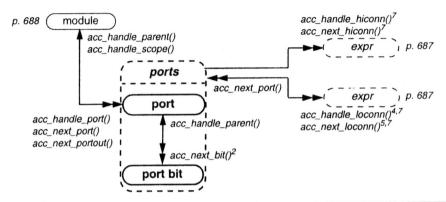

p. 688 module
acc_handle_parent()
acc_handle_scope()

acc_handle_hiconn()[7]
acc_next_hiconn()[7]

expr p. 687

ports

acc_next_port()

expr p. 687

port

acc_handle_port()
acc_next_port()
acc_next_portout()

acc_handle_parent()

acc_handle_loconn()[4,7]
acc_next_loconn()[5,7]

acc_next_bit()[2]

port bit

acc_fetch_type()	returns **accPort** or **accPortBit**
acc_fetch_fulltype()	returns **accScalarPort**, **accVectorPort**, **accBitSelectPort**[1], **accPartSelectPort**[1], **accConcatPort**[1]
acc_fetch_name()	returns the instance name of a port or port bit
acc_fetch_fullname()	returns the hierarchical path name of a port or bit
acc_fetch_direction()	returns **accInput**, **accOutput**, **accInout**, **accMixedIo**
acc_fetch_index()	returns the port's index position in the module port list (beginning with 0 for the left-most port)
acc_fetch_size()	returns the number of bits in a port
acc_fetch_location()	returns the source file name and line number containing a port
acc_object_of_type()	returns 1 (true) or 0 (false); use to verify a port is of special type **accExpandedVector** before calling **acc_next_bit()**
acc_fetch_delays()	returns delays of a scalar input port or input port bit
acc_append_delays()	appends delays to a scalar input port or input port bit
acc_replace_delays()	replaces delays of a scalar input port or input port bit
acc_fetch_pulsere()	returns the pulse control values of an intermodule path
acc_append_pulsere()	adds to the pulse control values of an input port or input port bit
acc_replace_pulsere()	replaces the pulse control values of an input port or input port bit
acc_set_pulsere()	sets the pulse control values of an input port or input port bit
acc_vcl_add()	adds a VCL flag to a scalar port or port bit
acc_vcl_delete()	removes a VCL flag on a scalar port or port bit

1. **accBitSelectPort** represents "module m1 (a[0])",
   accPartSelectPort represents "module m1 (a[15:8])",
   accConcatPort represents "module m1 (.a({n,m}))".

2. **acc_next_bit()** requires the **accExpandedVector** property to be true.

3. When the port fulltype is **accConcatPort**, **acc_next_bit()** will return handles for each vector in the concatenation, not each bit of each vector.

4. **acc_handle_loconn()** does not work with ports of fulltype **accConcatPort**.

(notes continued on next page)

5. **acc_next_loconn()**, **acc_fetch_name()** and **acc_fetch_fullname()** do not work with ports of fulltype **accPartSelectPort** and **accConcatPort**.

6. If a port name is specified as a system task/function argument, **acc_handle_tfarg()** will return a handle for the loconn signal connected to the port.

7. The loconn and hiconn handle returned is determined by the port properties, as follows:

routine	port property	connection handle
acc_handle_loconn() **acc_handle_hiconn()**	scalar port	scalar connection
	expanded vector port	vector connection
	unexpanded vector port	vector connection
	bit select of a port	bit select of connection
acc_next_loconn()	scalar port	scalar connection
	expanded vector port	bit select of connection
	unexpanded vector port	vector connection
	bit select of a port	illegal
acc_next_hiconn()	scalar port	scalar connection
	expanded vector port	vector connection
	unexpanded vector port	vector connection
	bit select of a port	illegal

C.1.7 net objects

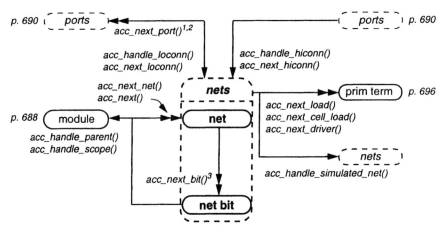

acc_fetch_type()	returns **accNet**
acc_fetch_fulltype()	returns **accSupply0, accSupply1, accTri, accTriand, accTrior, accTrireg, accTri0, accTri1, accWand, accWire, accWor**
acc_object_of_type()	returns 1 (true) if an object has a special-type property; use to test for: **accScalar, accVector, accCollapsedNet, accExpandedVector**
acc_fetch_name()	returns the instance name of the net or bit
acc_fetch_fullname()	returns the hierarchical path name of the net or bit
acc_fetch_value()	returns the logic value of a net or net bit
acc_set_value()	sets the logic value of a net or net bit (force only)
acc_fetch_size()	returns the number of bits in a net
acc_fetch_range()[3]	returns the most significant bit and least significant bit declaration values of a vector net
acc_fetch_location()	returns the source file name and line number containing a net
acc_handle_object() acc_handle_by_name()	obtains a handle for a net using the net's name
acc_vcl_add()[4]	adds a value change link flag to a net or net bit
acc_vcl_delete()	removes a value change link flag from a net or bit

1. **acc_next_port()** requires the special property **accScalar** be true.

2. **acc_next_port()** with a net handle should return a handle for a module port, but some simulators return a handle for a **port bit**.

3. **acc_next_bit()** and **acc_fetch_range()** require that the special property **accExpandedVector** be true.

4. Some simulators restrict the net types supported by **acc_vcl_add()** to scalar nets, unexpanded vector nets and bit selects of expanded vector nets.

5. Verilog-2001 adds arrays of nets. Each net within the array is a net object, and can be accessed if a task/function argument selects a single net from the array. The ACC library cannot access the complete array.

C.1.8 reg objects

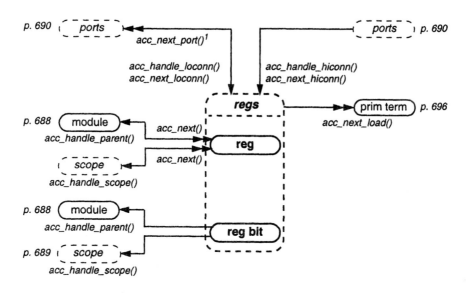

acc_fetch_type()	returns **accReg** (or **accRegister**), **accRegBit**
acc_fetch_fulltype()	returns **accReg** (or **accRegister**), **accRegBit**
acc_object_of_type()	returns 1 (true) or 0 (false); use to test if a reg has a special-type property of **accScalar** or **accVector**
acc_fetch_name()	returns the instance name of reg or reg bit
acc_fetch_fullname()	returns the hierarchical path name of reg or bit
acc_fetch_value()	returns the logic value of a reg or reg bit
acc_set_value()	sets the logic value of a reg or reg bit
acc_fetch_range()[2]	returns the most significant bit and least significant bit declaration values
acc_fetch_size()	returns the number of bits in a reg
acc_fetch_location()	returns the source file name and line number containing a reg
acc_handle_object() acc_handle_by_name()	obtains a handle for a reg using the name of a reg
acc_vcl_add()	adds a value change link flag to a reg or reg bit
acc_vcl_delete()	removes a value change link flag from an object

1. **acc_next_port()** requires the special property **accScalar** to be true.

2. **acc_fetch_range()** requires the special property **accVector** to be true.

3. Verilog-1995 allows one-dimensional arrays of regs. Verilog-2001 adds multi-dimensional arrays of regs. Each reg within the array is a reg object, and can be accessed if a task/function argument selects a single reg from the array. The ACC library cannot access the complete array, but **tf_nodeinfo()** can be used to access all values in a one-dimensional reg array.

C.1.9 variable and named event objects

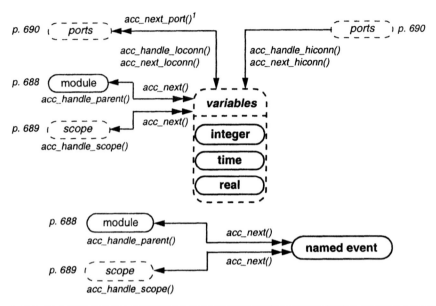

acc_fetch_type()	returns **accIntegerVar, accTimeVar, accRealVar, accNamedEvent**
acc_fetch_fulltype()	returns **accIntegerVar, accTimeVar, accRealVar, accNamedEvent**
acc_object_of_type()	returns 1 (true) or 0 (false); use to test if a reg has a special-type property of **accScalar** or **accVector**
acc_fetch_name()	returns the instance name of the object
acc_fetch_fullname()	returns the hierarchical path name of the object
acc_fetch_value()	returns the logic value of a variable
acc_set_value()	sets the logic value of a variable
acc_fetch_size()[2]	returns the number of bits in a reg or variable
acc_fetch_location()	returns the source file name and line number containing a variable or named event
acc_handle_object() acc_handle_by_name()	obtains a handle for a variable or named event using the name of the object
acc_vcl_add()	adds a value change link flag to an object
acc_vcl_delete()	removes a value change link flag from an object

1. **acc_next_port()** requires the special property **accScalar** to be true.

2. **acc_fetch_size()** does not apply to real variables or named events.

3. Verilog-1995 allows one-dimensional arrays of integer and time variables. Verilog-2001 adds multi-dimensional arrays of variables. Each variable within the array is a variable object, and can be accessed if a task/function argument selects a single variable from the array. The ACC library cannot access the complete array, but **tf_nodeinfo()** can be used to access all values in a one-dimensional integer or time array.

C.1.10 parameter and specparam objects

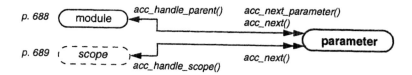

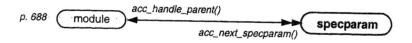

acc_fetch_type()	returns **accParameter** (or **accStatement**[1]) or **accSpecparam**
acc_fetch_fulltype()	returns **accIntegerParam, accRealParam, accStringParam**
acc_fetch_paramtype()	returns **accIntegerParam, accRealParam, accStringParam**
acc_fetch_name()	returns the instance name of a parameter or specparam
acc_fetch_fullname()	returns the hierarchical path name of a parameter or specparam
acc_fetch_paramval()	returns the logic value of a parameter or specparam
acc_fetch_attribute() acc_fetch_attribute_int() acc_fetch_attribute_str()	returns the logic value of a parameter or specparam attribute associated with specific objects or all objects within a module
acc_configure()	**accDefaultAttr0** configures the default return value of acc_fetch_attribute() routines. **accPathDelimStr** configures the delimiter string between path inputs and outputs used by the acc_fetch_attribute() routines.
acc_fetch_location()	returns source file name and line number containing a parameter or specparam
acc_handle_object() acc_handle_by_name()	obtains a handle for a parameter or specparam using the name of the object

1. When the parameter is defined in a Verilog task or function, some simulators may return a type of **accStatement** instead of **accParameter**.

C.1.11 primitive objects

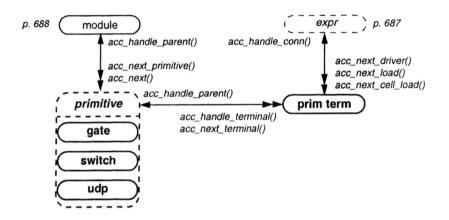

acc_fetch_type()	returns **accPrimitive** or **accTerminal**
acc_fetch_fulltype()	for primitive objects, returns **accAndGate, accBufGate, accBufif0Gate, accBufif1Gate, accCmosGate, accNandGate, accNmosGate, accNorGate, accNotGate, accNotif0Gate, accNotif1Gate, accOrGate, accPmosGate, accPulldownGate, accPullupGate, accRcmosGate, accRnmosGate, accRpmosGate, accRtranGate, accRtranif0Gate, accRtranif1Gate, accTranGate, accTranif0Gate, accTranif1Gate, accXnorGate, accXorGate, accCombPrim** (UDP), **accSeqPrim** (UDP)
	for prim term objects, returns **accInputTerminal, accOutputTerminal** or **accInoutTerminal**
acc_fetch_name()	returns the instance name of a primitive
acc_fetch_fullname()	returns the hierarchical path name of the primitive
acc_fetch_defname()	returns the definition name of the primitive
acc_fetch_direction()	for a primitive terminal, returns **accInput, accOutput, accInout**
acc_fetch_index()	returns the primitive terminal's index position in a primitive (beginning with 0 as the left-most terminal)
acc_fetch_location()	returns source file name and line number containing a primitive
acc_handle_object() acc_handle_by_name()	obtains a handle for a primitive using the primitive's instance name or full hierarchical name
acc_fetch_delays()	retrieves the delay values of a primitive
acc_append_delays()	adds to the delay values of a primitive
acc_replace_delays()	sets the delay values of a primitive
acc_set_value()	sets the logic value of a sequential UDP
acc_vcl_add()	adds Value Change Link flag to primitive input or inout terminal
acc_vcl_delete()	removes VCL flag from a primitive input or inout terminal

C.1.12 module path delay objects

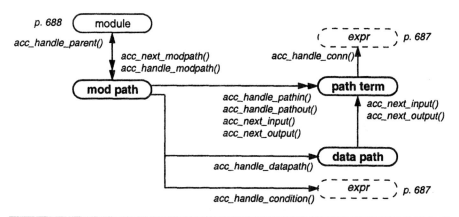

acc_fetch_type()	returns **accModPath, accDataPath** or **accPathTerminal**
acc_fetch_fulltype()	returns **accModPath, accDataPath, accPathInput, accPathOutput**
acc_fetch_polarity()	for a module path or data path, returns **accPositive, accNegative, accUnknown**
acc_fetch_edge()	for a path term, returns **accNoedge, accPosedge, accNegedge, accEdge01, accEdge10, accEdge0x, accEdgex1, accEdge1x, accEdgex0**
acc_fetch_name()	returns a derived name of a module path
acc_fetch_fullname()	returns a derived hierarchical name of a module path
acc_fetch_delays()	retrieves the delay values of a module path
acc_append_delays()	adds to the delay values of a module path
acc_replace_delays()	replaces the delay values of a module path
acc_fetch_pulsere()	returns the pulse control values of a module path
acc_append_pulsere()	adds to the pulse control values of a module path
acc_replace_pulsere()	replaces the pulse control values of a module path
acc_set_pulsere()	sets the pulse control values of a module path
acc_configure()	**accMinTypeMaxDelays** sets type of delays to be accessed. **accPathDelayCount** sets number of delays accessed by delay and pulsere routines. **accEnableArgs** configures the arguments required by acc_handle_modpath(). **accPathDelimStr** sets delimiter string between path inputs and outputs used by acc_fetch_name() and acc_fetch_fullname().
acc_object_of_type()	returns 1 (true) or 0 (false); use with **accModPathHasIfnone** to test if a path delay has an ifnone condition
acc_fetch_location()	returns source file name and line number of a module path
acc_handle_object() acc_handle_by_name()	obtain a handle for a module path using the module path's name
acc_release_object()	frees memory used by acc_next_input() and acc_next_output()

C.1.13 intermodule path delay objects

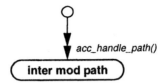

acc_fetch_type()	returns **accIntermodPath** or **accWirePath**
acc_fetch_fulltype()	returns **accIntermodPath**
acc_fetch_delays()	retrieves the delay values of an intermodule path
acc_replace_delays()	sets the delay values of an intermodule path
acc_fetch_pulsere()	returns the pulse control values of an intermodule path
acc_append_pulsere()	adds to the pulse control values of an intermodule path
acc_replace_pulsere()	replaces the pulse control values of an intermodule path
acc_set_pulsere()	sets the pulse control values of an intermodule path
acc_configure()	**accPathDelayCount** sets number of delays accessed by delay and pulsere routines. **accMapToMipd** sets how intermodule path delays are mapped.

C.1.14 timing check objects

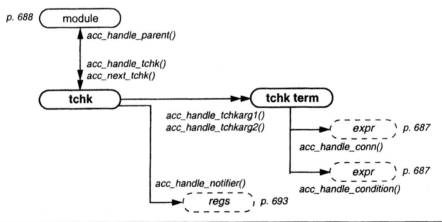

acc_fetch_type()	returns **accTchk** or **accTchkTerminal**
acc_fetch_fulltype()	returns **accSetup, accHold, accSetuphold, accPeriod, accSkew, accRecovery, accWidth, accNochange, accTchkTerminal**
acc_fetch_edge()	returns **accNoedge, accPosedge, accNegedge, accEdge01, accEdge10, accEdge0x, accEdgex1, accEdge1x, accEdgex0**
acc_fetch_delays()	retrieves the timing limit values of a timing check
acc_append_delays()	adds to the timing limit values of a timing check
acc_replace_delays()	sets the timing limit values of a timing check
acc_configure()	**accEnableArgs** sets arguments required by acc_handle_tchk()
acc_fetch_location()	returns source file name and line number containing timing check

C.2 ACC routine definitions

PLI_INT32 **acc_append_delays (** /* if accMinTypMaxDelays is false (default) */
| *handle* | **object,** | /* handle for a primitive, module path, intermodule path, module input port or port bit, or timing check */ |
| *double* | **d1, ... d12)** | /* transition delay values */ |

PLI_INT32 **acc_append_delays (** /* if accMinTypMaxDelays is true */
| *handle* | **object,** | /* handle for a primitive, module path, intermodule path, module input port or port bit, or timing check */ |
| *double* | ***dset_array)** | /* pointer to an array of transition delay values. |

Appends delays onto an object. Returns **1** if successful, and **0** if an error occurred. acc_configure(accMinTypMaxDelays, ...) controls how values are passed:

• If "false" (default), values are passed as arguments, with one delay for each transition.

• If "true", values are passed as a pointer to an array. Three delay values are used for each transition, representing a min:typ:max delay set.

The number of delay transitions required depends on the type of object. At least the number of delay values for the type object must be provided. Any additional delay values or delay sets beyond the object's number of transitions are ignored.

• 2-state primitives require 2 transitions or 2 transition sets.

• 3-state primitives and module ports require 3 transitions or 3 transition sets if acc_configure(accToHiZDelay, ...) is "from_user" (the default), and require 2 transitions or 2 transition sets if acc_configure(accToHiZDelay, ...) is "average", "min" or "max".

• Module paths require 1, 2, 3, 6 or 12 transitions or transition sets, depending on the setting of acc_configure(accPathDelayCount, ...).

• Timing checks require 1 transition or transition set; representing the check limit.

PLI_INT32 **acc_append_pulsere (**
handle	**object,**	/* handle for a module path, intermodule path or module input port */
double	**r1, e1,**	/* pulse reject limit / error limit value pair */
	...	
double	**r12, e12)**	/* pulse reject limit / error limit value pair */

Appends pulse control values to a module path, intermodule path or module input port. May have 1, 2, 3, 6 or 12 reject and error pairs, depending on the setting of acc_configure(accPathDelayCount, ...). Returns 1 if successful, 0 if an error occurred.

void **acc_close ( )**

Frees any memory allocated by acc_initialize() and sets ACC configurations to defaults.

handle ***acc_collect (**
handle	***next_routine,**	/* name, without parenthesis, of any ACC *next* routine, except acc_next_topmod */
handle	**object,**	/* reference object for the ACC *next* routine */
PLI_INT32	***count)**	/* the number of objects collected */

Returns a pointer of an array of handles found by a *next* routine. Automatically allocates memory for the array, which must be released using acc_free() when no longer needed.

PLI_INT32 **acc_configure (**
    PLI_INT32   **config,**   */* one of the constants listed in following table */*
    PLI_BYTE8 ***value)**   */* configuration value as a character string */*

Configures specific ACC routines as shown in the following table. Returns **1** if successful and **0** if an error occurred.

config constant	Value	Description
accDisplayErrors affects all ACC routines	"true" "false"	*default*—display run-time errors from ACC routines do not display run-time errors
accDisplayWarnings affects all ACC routines	"true" "false"	display run-time warnings generated by ACC routines *default*—do not display run-time warnings
accMinTypMaxDelays affects acc_append_delays(), acc_fetch_delays(), acc_replace_delays()	"false" "true"	*default*—access typical delays only access min, typ and max delay sets
accPathDelayCount affects acc_append_delays(), acc_fetch_delays(), acc_replace_delays(), acc_append_pulsere(), acc_fetch_pulsere(), acc_replace_pulsere()	"1" "2" "3" "6" "12"	access a single delay value for all output transitions access delay values for rise, fall transitions access delay values for rise, fall, toZ transitions *default*—access 0->1, 1->0, 0->Z, Z->1, 1->Z, Z->0 access 0->1, 1->0, 0->Z, Z->1, 1->Z, Z->0, 0->X, X->1, 1->X, X->0, X->Z, Z->X
accToHiZDelay affects acc_append_delays() and acc_replace_delays()	"min" "max" "average" "from_user"	set toZ delay as minimum of rise and fall set toZ delay as maximum of rise and fall set toZ delay as average of rise and fall *default*—specify toZ delay as input to ACC routine
accMapToMipd affects acc_append_delays() and acc_replace_delays()	"min" "max" "latest"	map shortest intermodule path delay to module input port delay (mipd) *default*—map longest intermodule path delay to mipd map most recent intermodule path delay to mipd
accPathDelimStr affects acc_fetch_name(), acc_fetch_fullname(), acc_fetch_attribute(), acc_fetch_attribute_int(), acc_fetch_attribute_str()	string	define delimiter string between source and destination in path names; default is "$"
accDefaultAttr0 affects acc_fetch_attribute(), acc_fetch_attribute_int(), acc_fetch_attribute_str()	"true" "false"	default return is 0 if attribute not found *default*—default return value is an input to routine
accEnableArgs affects acc_handle_modpath(), acc_handle_tchk(), acc_set_scope()		enables certain inputs to specific ACC routines—see the description of the affected routine for values

PLI_INT32 **acc_compare_handles (**
handle	**object1,**	/* handle for an object */
handle	**object2)**	/* handle for an object */

Returns **1** if two handles refer to the same object, and **0** if they do not.

PLI_INT32 **acc_count (**
handle	*** next_routine,**	/* name, without parenthesis, of any ACC *next* routine, except acc_next_topmod */
handle	**object)**	/* reference object for the ACC *next* routine */

Returns the number of objects found by an ACC *next* routine.

PLI_INT32 **acc_fetch_argc ( )**

Returns the number of command line arguments given on the command line used to invoke the simulation.

PLI_BYTE8 ****acc_fetch_argv ( )**

Returns a pointer to an array of character string pointers containing the command line arguments used to invoke the simulation.

double **acc_fetch_attribute (**
handle	**object,**	/* handle for a named object */
PLI_BYTE8	*** attribute,**	/* string containing the name of the parameter or specparam attribute associated with the object */
double	**default)**	/* default value to be returned if the parameter or specparam does not exist */

Returns the value as a double of an object's attribute parameter or specparam (X and Z values are converted to 0.0). The default argument can be dropped by setting acc_configure(accDefaultAttr0, "true"). *An "attribute" for this routine is not the same as the HDL (* *) attribute.* This pseudo-attribute is a parameter or specparam name that ends with a dollar sign (**$**). Optionally, the name of an object can be appended to the attribute name, making the attribute object specific. The object name is not included in the **attribute** string. A module path name is the name of the path input and path output, separated by a delimiter string specified with acc_configure(accPathDelimStr).

PLI_INT32 **acc_fetch_attribute_int (**
handle	**object,**	/* handle for a named object */
PLI_BYTE8	*** attribute,**	/* string with name of attribute associated with object */
PLI_INT32	**default)**	/* default value to be returned */

Returns the value as an integer of an object's attribute parameter or specparam (X and Z values are converted to 0). Refer to acc_fetch_attribute() for more details.

PLI_BYTE8 ***acc_fetch_attribute_str (**
handle	**object,**	/* handle for a named object */
PLI_BYTE8	*** attribute,**	/* string with name of attribute associated with object */
PLI_BYTE8	*** default)**	/* default string to be returned */

Returns the value as a string of an object's attribute parameter or specparam. Refer to acc_fetch_attribute() for more details.

PLI_BYTE8 ***acc_fetch_defname (**
   handle   **object)**   /* handle for a module or primitive */

Returns a pointer to a character string containing the definition name of a module instance or primitive instance.

PLI_INT32 **acc_fetch_delay_mode (**
   handle   **object)**   /* handle for a module */

Returns a constant indicating the delay mode of a module. The constant is one of: **accDelayModeNone**, **accDelayModeZero**, **accDelayModeUnit**, **accDelayModePath**, **accDelayModeDistrib** or **accDelayModeMTM**.

PLI_INT32 **acc_fetch_delays (**   /* if accMinTypMaxDelays is false (default) */
   handle   **object,**   /* handle for a primitive, module path, intermodule path, module input port or port bit, or timing check */
   double   *** d1...*d12)**   /* pointers to variables to receive the delay values */
PLI_INT32 **acc_fetch_delays (**   /* if accMinTypMaxDelays is true */
   handle   **object,**   /* handle for a primitive, module path, intermodule path, module input port or port bit, or timing check */
   double   *** dset_array)** /* pointer to an array to receive the delay values */

Retrieves the delays of an object. Returns **1** if successful, and **0** if an error occurred. The setting of acc_configure(accMinTypMaxDelays,...) determines where the delay values are placed:

- If "false" (default), then values are placed into pointers passed as arguments. A single delay value is used for each transition.
- If "true", then values are placed into an array passed as a pointer. Three delay values are used for each transition, representing a min:typ:max delay set.

The number of delay transitions retrieved depends on the type of object. At least the number of pointers or delay sets to receive values for the type of object must be specified. Any additional pointers or delay sets beyond the object's number of transitions are ignored.

- 2-state primitives retrieve 2 transitions or 2 transition sets.
- 3-state primitives and module ports retrieve 3 transitions or 3 transition sets if acc_configure(accToHiZDelay, ...) is "from_user" (the default), and 2 transitions or 2 transition sets if acc_configure(accToHiZDelay, ...) is "average", "min" or "max".
- Module paths may retrieve 1, 2, 3, 6 or 12 transitions or transition sets, depending on the setting of acc_configure(accPathDelayCount, ...).
- Timing checks retrieve 1 transition or 1 transition set; representing the check limit.

PLI_INT32 **acc_fetch_direction (**
   handle   **object)**   /* handle for a module port or primitive terminal */

Returns an integer constant representing the direction of a module port or primitive terminal. The constant returned is one of **accInput**, **accOutput**, **accInout**, or **accMixedIo**.

PLI_INT32 **acc_fetch_edge (**
    handle   **object**)   /* handle for a module path input or output terminal or a timing check terminal */

Returns an integer value representing the edge specifier of a module path terminal or timing check terminal. The value returned is either:

- One of the constants: **accNoedge, accPosedge** or **accNegedge**.
- A logical OR of one or more of the constants: **accEdge01, accEdge10, accEdge0x, accEdgex1, accEdge1x** or **accEdgex0** (each constant has a single bit set).

PLI_BYTE8 ***acc_fetch_fullname (**
    handle   **object**)   /* handle for an object */

Returns a pointer to a string containing the full hierarchical path name of an object. Not all objects have a full name—refer to the object diagram to determine if this property exists. Returns **null** (not a null string) if an error occurs.

PLI_INT32 **acc_fetch_fulltype (**
    handle   **object**)   /* handle for an object */

Returns a constant representing the full type of an object. Refer to the object diagrams for a list of full type constants.

PLI_INT32 **acc_fetch_index (**
    handle   **object**)   /* handle for a module port or primitive terminal */

Returns the index number of a module port or primitive terminal, starting with the left-most index as 0. A port index is its position in the module definition. A primitive terminal index is its position in the primitive instance.

PLI_INT32 **acc_fetch_location (**
    p_location   **location,**   /* pointer to an application-allocated *s_location* structure to receive the location */
    handle   **object**)   /* handle for an object */

Retrieves the file name and line number in the Verilog source code for an object into an *s_location* structure. Returns **1** if successful and **0** if an error occurred.

```
typedef struct t_location {
  PLI_INT32  line_no;
  PLI_BYTE8 *filename;
} s_location, *p_location;
```

PLI_BYTE8 ***acc_fetch_name (**
    handle   **object**)   /* handle for an object */

Returns a pointer to a string containing the name of an object. Not all objects have a name—refer to the object diagram to determine if this property exists. Returns **null** (not a null string) if an error occurs.

PLI_INT32 **acc_fetch_paramtype (**
    handle   **object)**   /* handle for a parameter or specparam */

Returns a constant representing the data type of a parameter or specparam. One of: **accIntegerParam, accRealParam,** or **accStringParam.**

double **acc_fetch_paramval (**
    handle   **object)**   /* handle for a parameter or specparam */

Returns the value of a parameter or specparam as a double. If the parameter or specparam contains a string, a pointer, cast to a double, to a string containing the value is returned. Note: the casting of a string pointer to a double may not work correctly on some operating systems. Most Verilog simulators include a non IEEE standard routine called acc_fetch_paramval_str() for reading string values.

PLI_INT32 **acc_fetch_polarity (**
    handle   **object)**   /* handle for a module path or data path */

Returns an integer constant representing the polarity of a module path or data path. One of: **accPositive, accNegative,** or **accUnknown.**

PLI_INT32 **acc_fetch_precision ( )**

Returns an integer representing the smallest simulation time unit of an instantiated design: 2==100 seconds, 1==10 s, 0==1 s, -1==100 ms, -2 ==10 ms, -3==1 ms,... -6==1 us,... -9==1 ns,... -12==1 ps,... -15==1 fs.

PLI_INT32 **acc_fetch_pulsere (**
    handle   **object,**   /* handle for a module path, intermodule path or module
                                 input port */
    double   ***r1, *e1,**   /* pointers to a pulse reject limit / error limit value pair */
                  ...
    double   ***r12, *e12)**   /* pointers to a pulse reject limit / error limit value pair */

Retrieves pulse control values of a module path, intermodule path or module input port. The number of reject/error value pairs retrieved is set by acc_configure(accPathDelayCount,...). Returns **1** if successful, and **0** if an error occurred.

PLI_INT32 **acc_fetch_range (**
    handle   **object,**   /* handle for a net, reg, integer or time vector */
    PLI_INT32 ***msb,**   /* pointer to a variable to receive the value of the most-
                                 significant (left-most) bit of a vector */
    PLI_INT32 ***lsb)**   /* pointer to a variable to receive the value of the least-
                                 significant (right-most) bit of a vector */

Retrieves the most-significant bit and least-significant bit declaration values of a net, reg, integer or time vector.

PLI_INT32 **acc_fetch_size (**
    handle   **object)**   /* handle for a net, reg, integer variable, time variable or
                               module port */

Returns the number of bits in a vector.

double **acc_fetch_tfarg (**
   PLI_INT32   **n)**   /* index number of a PLI system task/function arg */

double **acc_fetch_itfarg (**
   PLI_INT32   **n,**   /* index number of a PLI system task/function arg */
   handle   **tfinst)**   /* handle for an instance of a PLI system task/function */

Returns the value as a double of an argument in the calling, or a specific instance, of a system task/function (X and Z values are converted to 0.0). Arguments are numbered from left to right, beginning with 1.

PLI_INT32 **acc_fetch_tfarg_int (**
   PLI_INT32   **n)**   /* index number of a PLI system task/function arg */

PLI_INT32 **acc_fetch_itfarg_int (**
   PLI_INT32   **n,**   /* index number of a PLI system task/function arg */
   handle   **tfinst)**   /* handle for an instance of a PLI system task/function */

Returns the value as a 32-bit integer of an argument in the calling, or a specific instance, of a system task/function (X and Z values are converted to 0). Arguments are numbered from left to right, beginning with 1.

PLI_BYTE8 ***acc_fetch_tfarg_str (**
   PLI_INT32   **n)**   /* index number of a PLI system task/function arg */

PLI_BYTE8 ***acc_fetch_itfarg_str (**
   PLI_INT32   **n,**   /* index number of a PLI system task/function arg */
   handle   **tfinst)**   /* handle for an instance of a PLI system task/function */

Returns a pointer to a string that contains the value of an argument in the calling, or a specific instance, of a system task/function. Arguments are numbered from left to right, beginning with 1. Verilog strings are converted to C strings. Verilog vectors are converted by treating each 8 bits as an ASCII value, beginning with the right-most bit. Real values cannot be converted. Returns **null** (not a null string) if an error occurs.

void **acc_fetch_timescale_info (**
   handle   **object,**   /* handle for a module instance, module definition, PLI
                           system task/function, or **null** */
   p_timescale_info **timescale)** /* pointer to an application-allocated timescale structure */

Retrieves the timescale defined for a Verilog module into an s_timescale_info structure. If **object** is **null**, the timescale for the active $timeformat system task is returned. Time is represented as an integer, where: 2==100 seconds, 1==10 s, 0==1s, -1==100 ms, -2==10 ms, -3==1 ms, ... -6== 1 us, ... -9==1 ns, ... -12==1 ps, ... -15==1 fs.

```
typedef struct t_timescale_info {
  PLI_INT16 unit;
  PLI_INT16 precision;
} s_timescale_info, *p_timescale_info;
```

PLI_INT32 **acc_fetch_type (**
   handle   **object)**   /* handle for an object */

Returns a constant representing the type of an object. Refer to the object diagrams for the type constant names for each object.

PLI_BYTE8 ***acc_fetch_type_str (**

   PLI_INT32   **type)**   /* any type or fulltype constant */

Returns a pointer to a string containing the name of the type or fulltype integer constant returned by acc_fetch_type() or acc_fetch_full_type(). Returns **null** (not a null string) if an error occurs.

PLI_BYTE8 ***acc_fetch_value (**

   handle   **object,**   /* handle for a net, reg or variable */
   PLI_BYTE8 * **format_str,**   /* character string controlling the radix of the retrieved
                                  value: **"%b", "%o", "%d", "%h", "%v"** or **"%%"** */
   p_acc_value   **value)**   /* pointer to an application-allocated s_acc_value
                                  structure to receive the value; only used if format_str is
                                  "%%" */

Retrieves the logic value of an object. If the format string is "%b", "%o", "%d", "%h" or "%v", the value is returned as a pointer to a string. The format of the string is the same as with the return from the Verilog $display system task. If format is "%%", the value is retrieved into an s_acc_value structure, based on the structure **format** field.

```
typedef struct t_setval_value {
   PLI_INT32         format;  /* accBinStrVal,    accOctStrVal,
                                 accDecStrVal,    accHexStrVal,
                                 accScalarVal,    accIntVal,
                                 accRealVal,      accStringVal,
                                 accVectorVal */
   union {
      PLI_BYTE8    *str;     /* used with string-based formats */
      PLI_INT32    scalar;   /* used with accScalarVal format */
      PLI_INT32    integer;  /* used with accIntVal format */
      double       real;     /* used with accRealVal format */
      p_acc_vecval vector;   /* used with accVectorVal format */
   } value;
} s_setval_value, *p_setval_value,
  s_acc_value, *p_acc_value;
```
```
typedef struct t_acc_vecval {
   PLI_INT32  aval;     /* bit encoding: aval/bval:
   PLI_INT32  bval;              0/0==0, 1/0==1, 1/1==X, 0/1==Z */
} s_acc_vecval, *p_acc_vecval;
```

void **acc_free (**

   handle   * **array_ptr)**   /* pointer to an array of handles */

Frees memory that was allocated by acc_collect().

handle **acc_handle_by_name (**

   PLI_BYTE8 * **obj_name,**   /* name of an object */
   handle   **scope)**   /* handle for a scope, or **null** */

Returns a handle for an object using its name. Begins searching for the object in the scope specified, and follows the Verilog HDL search rules. If **scope** is **null**, begins searching in the scope of the system task/function that called the PLI application. Cannot obtain handles for ports, module paths, data paths, or intermodule paths.

handle **acc_handle_calling_mod_m**

Returns a handle for the module from which the PLI application was called. *Note:* This routine is defined as a macro, not a function. Therefore, it should not be called with parenthesis at the end of the name.

handle **acc_handle_condition (**

    handle   **object)**   /* handle for a module path, data path or timing check */

Returns a handle for a conditional expression of a module path, data path, or timing check terminal. Returns **null** if there is no condition or the condition is ifnone.

handle **acc_handle_conn (**

    handle   **object)**   /* handle for a primitive terminal, path terminal or timing check terminal */

Returns a handle for the expression connected to the terminal.

handle **acc_handle_datapath (**

    handle   **object)**   /* handle for a module path */

Returns a handle for the data path associated with an edge sensitive module path.

handle **acc_handle_hiconn (**

    handle   **object)**   /* handle for a scalar port or bit select of vector port */

Returns a handle for the net connected externally to a port (hierarchically higher).

handle **acc_handle_interactive_scope ( )**

Returns a handle for the current interactive debug scope of the simulator.

handle **acc_handle_loconn (**

    handle   **object)**   /* handle for a scalar port or bit select of vector port */

Returns a handle for the expression connected internally to a port (hierarchically lower).

handle **acc_handle_modpath (**

    handle   **object,**   /* handle for a module */
    PLI_BYTE8 *** src_name,**   /* name of net connected to path source, or **null** */
    PLI_BYTE8 *** dest_name,**   /* name of net connected to path destination, or **null** */
    handle   **src_handle,**   /* handle for net connected to path source, or **null** */
    handle   **dest_handle)** /*handle for net connected to path destination, or **null** */

Returns a handle for a module path using either path input/output net names or net handles. acc_configure(accEnableArgs, ...) controls which arguments are used. If **accEnableArgs** is "**no_acc_handle_modpath**" (the default), then only the **src_name** and **dest_name** arguments are used (the **src_handle** and **dest_handle** arguments are ignored, and can be omitted). If **accEnableArgs** is "**acc_handle_modpath**", then the name arguments are used if specified, and the **src_handle** and **dest_handle** arguments are used if the name pointers are **null**.

handle **acc_handle_notifier (**

    handle   **object)**   /* handle for a timing check */

Returns a handle for the notifier reg of a timing check.

handle **acc_handle_object (**
  _PLI_BYTE8_ *** obj_name)**   /* name of an object */

Returns a handle for a named object using either a local name or a hierarchical path name. Uses the Verilog HDL search rules to find the object, beginning in the current PLI search scope. The PLI search scope is set using acc_set_scope(). By default, the PLI search scope is the scope from with the PLI application was called.

handle **acc_handle_parent (**
  handle   **object)**   /* handle for any object */

For most objects, returns the handle for the module containing an object. There are three exceptions: the parent of a primitive terminal is a primitive, the parent of a bit-select or part-select of a port is a port, and the parent of a part-select of a vector is a vector (but the parent of a bit-select of a vector is a module).

handle **acc_handle_path (**
  handle   **source,**   /* handle for a scalar output or inout port, or bit-select of a
                                vector output or inout port */
  handle   **destination)** /* handle for a scalar input or inout port, or bit-select of a
                                vector input or inout port */

Returns a handle for an intermodule path (the output of one module to the input of another module).

handle **acc_handle_pathin (**
  handle   **object)**   /* handle for a module path */
Returns a handle for the net connected to the first source in a module path.

handle **acc_handle_pathout (**
  handle   **object)**   /* handle for a module path */
Returns handle for the net connected to the first destination in a module path.

handle **acc_handle_port (**
  handle   **object,**   /* handle for a module */
  _PLI_INT32_   **index)**   /* port position index */

Returns a handle for a specific module port based on the port position in the module definition. Ports are numbered from left to right starting with 0.

handle **acc_handle_scope (**
  handle   **object)**   /* handle for an object */
Returns the handle for the scope containing an object.

handle **acc_handle_simulated_net (**
  handle   **object)**   /* handle for a net */

Returns the handle for the net being simulated after equivalent nets to the handle provided have been collapsed. The simulated net returned may not be the same on all simulators, because simulators can perform net collapsing differently.

handle **acc_handle_tchk (**
handle	**object,**	/* handle for a module */
PLI_INT32	**type,**	/* constant representing timing check type. One of: **accHold, accNochange, accPeriod, accRecovery, accSetup, accSkew** or **accWidth** */
PLI_BYTE8	*** name1,**	/* name of net connected to the 1st tchk arg, or **null** */
PLI_INT32	**edge1,**	/* constant representing the edge of the 1st timing check argument (constant names are listed below) */
PLI_BYTE8	*** name2,**	/* name of net connected to the 2nd tchk arg, or **null** */
PLI_INT32	**edge2,**	/* constant representing the edge of the 2nd tchk arg (constant names are listed below) */
handle	**conn1,**	/* handle for net connected to 1st tchk arg, or **null** */
handle	**conn2)**	/* handle for net connected to 2nd tchk arg, or **null** */

Returns the handle for a timing check using a description of the check. The **edge1** and **edge2** identifiers are one of the following:

• One of the constants: **accNoedge, accPosedge, accNegedge**

• List of constants separated by a plus sign (+): **accEdge01, accEdge0x, accEdgex1**

‧ List of constants separated by a plus sign (+): **accEdge10, accEdge1x, accEdgex0**

The routine acc_configure(accEnableArgs, ...) controls which arguments are used. If **accEnableArgs** is "no_acc_handle_tchk" (the default), then only the **name1** and **name2** arguments are used (the **conn1** and **conn2** arguments are ignored, and can be omitted). If **accEnableArgs** is "acc_handle_tchk", then the name arguments are used if specified, and the **conn1** and **conn2** arguments are used if the name pointers are **null**.

handle **acc_handle_tchkarg1 (**
handle	**object)**	/* handle for a timing check */

Returns the handle for the timing check terminal connected to the first argument of a timing check.

handle **acc_handle_tchkarg2 (**
handle	**object)**	/* handle for a timing check */

Returns the handle for the timing check terminal connected to the second argument of a timing check.

handle **acc_handle_terminal (**
handle	**object,**	/* handle for a primitive */
PLI_INT32	**index)**	/* primitive terminal position */

Returns the handle for a specific primitive terminal based on the terminal position in the primitive instance. Terminals are numbered from left to right starting with 0.

handle **acc_handle_tfarg (**
   PLI_INT32   **n)**   /* position number of a system task/function argument */

handle **acc_handle_itfarg (**
   PLI_INT32   **n,**   /* position number of a system task/function argument */
   handle   **tfinst)**   /* handle for an instance of a system task/function */

Returns a handle for the object named as an argument in the calling system task/function or a specific instance of a system task/function. Arguments are numbered from left to right beginning with 1. If the argument is the name of a port, a handle for the loconn of the port (the internal signal connected to the port) is returned.

handle **acc_handle_tfinst ( )**

Returns a handle for the system task/function which called the PLI application.

PLI_INT32 **acc_initialize ( )**

Initializes the ACC environment. Resets configurations to default values. Returns **1** if successful and **0** if an error occurred.

handle **acc_next (**
   PLI_INT32 *** type_list,**   /* pointer to a static array with a list of type or fulltype
                                 constants */
   handle   **scope,**   /* handle for the scope in which to scan for objects */
   handle   **prev_object)** /* handle for the previous object found; initially **null** */

Returns the handle for the next object of the types listed in an array within the scope specified. Returns **null** when there are no more objects to be found. The type and fulltype constants in the list must be for modules, primitives, nets, regs, variables or parameters. Refer to the object diagrams for these objects for the constant names.

handle **acc_next_bit (**
   handle   **object,**   /* handle for a vector or path terminal */
   handle   **prev_bit)** /* handle for the previous bit found; initially **null** */

Returns the handle for the next bit within a vector or path terminal. Returns **null** when there are no more bits. For vectors, acc_object_of_type(**accExpandedVector**) must return **1** (true), and acc_object_of_type(**accCollapsedNet**) must return **0** (false).

handle **acc_next_cell (**
   handle   **module,**   /* handle for a module */
   handle   **prev_cell)** /* handle for the previous cell found; initially **null** */

Returns the handle for the next cell module at or below the specified scope. Returns **null** when there are no more cell modules. A cell module is a module which is flagged with the 'cell_define compiler directive or was loaded using a library scan option.

handle **acc_next_cell_load (**
   handle   **net,**   /* handle for a scalar net or bit-select of a vector net */
   handle   **prev_load)** /* handle for the previous load found; initially **null** */

Returns the handle for the next primitive input or inout terminal loading a net. Returns **null** when there are no more loads. Loads that are not in cell modules are not returned. Only the first load within a cell will be returned if there are multiple loads in that cell.

handle **acc_next_child (**
| *handle* | **module,** | /* handle for a module, or **null** */ |
| *handle* | **prev_child)** | /* handle for the previous child found; initially **null** */ |

Returns the handle for the next module instantiated within the specified module. Returns **null** when there are no more child modules. If the **module** handle is **null**, a child will be the next top-level module in the simulation.

handle **acc_next_driver (**
| *handle* | **net,** | /* handle for a scalar net or bit-select of a vector net */ |
| *handle* | **prev_driver)** | /* handle for the previous driver found; initially **null** */ |

Returns the handle for the next primitive output or inout terminal driving a net. Returns **null** when there are no more drivers.

handle **acc_next_hiconn (**
| *handle* | **port,** | /* handle for a port */ |
| *handle* | **prev_hiconn)** | /* handle for the previous hiconn found; initially **null** */ |

Returns the handle for the next externally connected net (hierarchically higher) on a module port. Returns **null** when there are no more hiconns. Vectored ports are scanned one bit at a time, beginning with the most-significant bit.

handle **acc_next_input (**
| *handle* | **object,** | /* handle for a module path, data path or timing check */ |
| *handle* | **prev_input)** | /* handle for the previous input found; initially **null** */ |

Returns handle for the next input terminal of a mod path, source of a data path, or input terminal of a timing check. Returns **null** when there are no more objects.

handle **acc_next_load (**
| *handle* | **net,** | /* handle for a scalar net or bit-select of a vector net */ |
| *handle* | **prev_load)** | /* handle for the previous load found; initially **null** */ |

Returns the handle for the next primitive input or inout terminal loading a net. Returns **null** when there are no more loads. If there are multiple loads within a module, all loads will be returned .

handle **acc_next_loconn (**
| *handle* | **port,** | /* handle for a port */ |
| *handle* | **prev_loconn)** | /* handle for the previous loconn found; initially **null** */ |

Returns the handle for the next internally connected expression (hierarchically lower) on a module port. Returns **null** when there are no more loconns. Vectored ports are scanned one bit at a time, beginning with the most-significant bit.

handle **acc_next_modpath (**
| *handle* | **module,** | /* handle for the module in which to scan for paths */ |
| *handle* | **prev_path)** | /* handle for the previous path found; initially **null** */ |

Returns the handle for the next module pin-to-pin timing path specified within a module. Returns **null** when there are no more paths.

handle **acc_next_net (**
 handle   **module,** /* handle for a module */
 handle   **prev_net)** /* handle for the previous net found; initially **null** */

Returns the handle for the next net within a module. Returns **null** when there are no more nets. Both explicitly and implicitly declared nets are returned.

handle **acc_next_output (**
 handle   **path,**   /* handle for a module path or data path */
 handle   **prev_output)** /* handle for the previous output found; initially **null** */

Returns handle for next output terminal of a module path or the destination of a data path. Returns **null** when there are no more outputs.

handle **acc_next_parameter (**
 handle   **scope,** /* handle for a scope */
 handle   **prev_param)** /* handle for previous parameter found; initially **null** */

Returns the handle for the next parameter within a scope (a module, Verilog task, Verilog function or named statement group). Returns **null** when there are no more parameters.

handle **acc_next_port (**
 handle   **object,** /* handle for a module, scalar net, bit-select of a vector
              net, scalar reg or bit-select of a vector reg */
 handle   **prev_port)** /* handle for the previous port found; initially **null** */

If object is a module handle, returns the handle for the next port in the module definition. If object is a net or reg handle, returns the handle for the next port in the module definition connected to the net or reg. (Note: some simulators return a port bit handle instead of a port handle when the object is a scalar net or scalar reg). Returns **null** when there are no more ports.

handle **acc_next_portout (**
 handle   **module,** /* handle for a module */
 handle   **prev_port)** /* handle for the previous port found; initially **null** */

Returns the handle for the next output or inout port in a module definition. Returns **null** when there are no more ports.

handle **acc_next_primitive (**
 handle   **module,** /* handle for a module */
 handle   **prev_prim)** /* handle for previous primitive found; initially **null** */

Returns the handle for the next built-in or user-defined primitive within a module. Returns **null** when there are no more primitives.

handle **acc_next_scope (**
 handle   **ref_scope,** /* handle for a scope */
 handle   **prev_scope)** /* handle for the previous scope found; initially **null** */

Returns the handle for the next scope within a scope. A scope is a module, Verilog task, Verilog function or a named statement group. Returns **null** when there are no more scopes.

handle **acc_next_specparam (**
    handle   **module,**   /* handle for a module */
    handle   **prev_sparam**) /*handle for previous specparam found; initially **null** */

Returns the handle for the next specparam within a module. Returns **null** when there are no more spceparams.

handle **acc_next_tchk (**
    handle   **module,**   /* handle for a module */
    handle   **prev_tchk**)   /* handle for the previous tchk found; initially **null** */

Returns the handle for the next timing check task within a module. Returns **null** when there are no more timing checks.

handle **acc_next_terminal (**
    handle   **primitive,**   /* handle for a primitive */
    handle   **prev_term**)   /* handle for previous terminal found; initially **null** */

Returns the handle for the next terminal of a primitive. Returns **null** when there are no more terminals. Terminals are returned in the order of the primitive instance, starting with the left-most terminal (index 0). The terminals of a primitive are analogous to the ports of a module.

handle **acc_next_topmod (**
    handle   **prev_topmod**) /* handle for the previous top level module found;
                                 initially **null** */

Returns the handle for the next top-level module within simulation. Returns **null** when there are no more top-level modules. acc_next_topmod() cannot be used as an argument to acc_collect() or acc_count()—use acc_next_child(null, ...) instead.

PLI_INT32 **acc_object_in_typelist (**
    handle   **object,**   /* handle for an object */
    PLI_INT32   *** type_list**)   /* pointer to a static array of type, fulltype and special-type
                                 property constants */

Returns **1** if object matches any of a list of type, fulltype or special-type properties, and **0** if it does not match. Refer to the ACC object diagrams for the type and fulltype constants, and acc_object_of_type() for special-type constants.

PLI_INT32 **acc_object_of_type (**
    handle   **object,**   /* handle for an object */
    PLI_INT32   **type**)   /* type, fulltype or special-type property constant */

Returns **1** if an object matches a type, fulltype or special-type property, and **0** if it does not match. Refer the object diagrams for the names type and fulltype constants. The special-type property constants are **accScalar, accVector, accCollapsedNet, accExpandedVector, accUnExpandedVector, accScope** and **accModPathHasIfnone**.

PLI_INT32 **acc_product_type ( )**

Returns a constant representing the type of simulator that called the PLI application. Product types are **accSimulator, accTimingAnalyzer, accFaultSimulator** and **accOther**.

PLI_BYTE8 *acc_product_version ( )

Returns a string with the version of the simulator that called the PLI application.

PLI_INT32 acc_release_object (

    handle   **object**)   /* handle for a module path or data path terminal */

Frees memory allocated by calls to acc_next_input() and acc_next_output() for the module path or data path. Returns **1** if successful and **0** if an error occurred.

PLI_INT32 acc_replace_delays(/* if accMinTypMaxDelays is false (default) */

    handle   **object,**   /* handle for a primitive, module path, intermodule path, module input port or port bit, or timing check */

    double   **d1,...d12**)   /* transition delay values */

PLI_INT32 acc_replace_delays(/* if accMinTypMaxDelays is true */

    handle   **object,**   /* handle for a primitive, module path, intermodule path, module input port or port bit, or timing check */

    double   *** dset_array**) /* pointer to an array of transition delay values */

Appends delays onto an object. Returns **1** if successful, and **0** if an error occurred. acc_configure(accMinTypMaxDelays, ...) controls how values are passed:

- If "false" (default), then values are listed as arguments, with a single delay value for each transition.
- If "true", then values are passed as a pointer to an array. Three delay values are used for each transition, representing a min:typ:max delay set.

The number of delay transitions required depends on the type of object. At least the number of delay values for the type object must be provided. Any additional delay values or delay sets beyond the object's number of transitions are ignored.

- 2-state primitives require 2 transitions or 2 transition sets.
- 3-state primitives and module ports require 3 transitions or 3 transition sets if acc_configure(accToHiZDelay, ...) is "from_user" (the default) and require 2 transitions or 2 transition sets if acc_configure(accToHiZDelay, ...) is "average", "min" or "max".
- Module paths require 1, 2, 3, 6 or 12 transitions or transition sets, depending on the setting of acc_configure(accPathDelayCount, ...).
- Timing checks require 1 transition or 1 transition set; representing the check limit.

PLI_INT32 acc_replace_pulsere (

    handle   **object,**   /* handle for a module path, intermodule path or module input port */

    double   **r1, e1,**   /* pulse reject limit / error limit value pair */

                 ...

    double   **r12, e12**) /* pulse reject limit / error limit value pair */

Replaces pulse control values of a module path, intermodule path or module input port. The number of reject and error value pairs used depends on the setting of acc_configure(accPathDelayCount,...). Returns **1** if successful, **0** if an error occurred.

void acc_reset_buffer ( )

Resets the pointer to the ACC string buffer to the beginning of the buffer. The string buffer is used by many of the ACC routines that return pointers to strings.

PLI_INT32 acc_set_interactive_scope (

handle	**scope,**	/* handle for the scope to be the simulator's new interactive debug scope */
PLI_INT32	**callback_flag)**	/* call *misctf* routine if **1** (true) */

Sets the scope of the simulator's interactive debug mode. If **callback_flag** is **1** (true), the *misctf* routine will be called with **reason_scope** after the current application returns. Returns **1** if successful and **0** if an error occurred (*note*: the descriptions in the IEEE 1364-1995 and 1364-2001 standards state that this routine returns a handle for the new interactive scope if successful, but this is an errata).

PLI_INT32 acc_set_pulsere (

handle	**object,**	/* handle for a module path, intermodule path or module input port */
double	**r_percent,**	/* pulse reject limit */
double	**e_percent)**	/* pulse error limit */

Sets pulse control values of a path as a percentage of the path delays. The return value is not used.

PLI_BYTE8 *acc_set_scope (

handle	**module,**	/* handle for a module, or **null** */
PLI_BYTE8	* **name)**	/* name of a module, or **null** */

Sets the PLI search scope, which affects where routines such as acc_handle_object() and acc_fetch_attribute() search for named objects. The arguments supplied to acc_set_scope() are based on the configuration of acc_configure(accEnableArgs, ...):

• If "**no_acc_set_scope**" (default), the module handle is used (the name argument is ignored and can be omitted). If **module** is **null**, the PLI search scope is set to the first top-level module.

• If "**acc_set_scope**", the **module** argument used if a valid module handle is provided, and the **name** argument is used if the module handle is invalid or **null**. If both module and name are **null**, then the PLI search scope is set to the first top-level module.

Returns a pointer to a string containing the full hierarchical path name of the new PLI search scope, or **null** if an error occurred.

PLI_INT32 **acc_set_value (**
   handle   **object,**   /* handle for a reg, variable, system function call,
                                        sequential UDP or net */

   p_setval_value **value,**   /* pointer to an application-allocated s_setval_value
                                          structure containing the value to be set */

   p_setval_delay **delay)**   /* pointer to an application-allocated s_setval_delay
                                          structure containing a propagation delay value */

Schedules a logic value change on an object at the current or future simulation time. The value and delay information are stored in structures. Returns **0** if successful and non-zero if an error occurred. The delay model constant controls how events propagate:

- **accNoDelay**, **accInertialDelay**, **accTransportDelay** or **accPureTransportDelay** writes and propagates a value on a reg, variable or sequential UDP.

- **accAssignFlag** continuously assigns a value on a reg or variable; equivalent to the assign Verilog HDL statement.

- **accDeassignFlag** removes a continuous assign; similar to deassign HDL statement.

- **accForceFlag** forces a value on a net, reg or variable; similar to force HDL statement.

- **accReleaseFlag** removes a force; equivalent to the release Verilog HDL statement.

- **accNoDelay** writes the return value of a system function call.

```
typedef struct t_setval_value {
  PLI_INT32 format; /* accBinStrVal, accOctStrVal, accDecStrVal,
                        accHexStrVal, accScalarVal, accIntVal,
                        accRealVal, accStringVal, accVectorVal */
  union {
      PLI_BYTE8   *str;      /* used with string-based formats */
      PLI_INT32   scalar;   /* used with accScalarVal format */
      PLI_INT32   integer;  /* used with accIntVal format */
      double      real;     /* used with accRealVal format */
      p_acc_vecval vector;  /* used with accVectorVal format */
  } value;
} s_setval_value, *p_setval_value, s_acc_value, *p_acc_value;
```

```
typedef struct t_acc_vecval {
  PLI_INT32 aval;          /* bit encoding: aval/bval:
  PLI_INT32 bval;             0/0==0, 1/0==1, 1/1==X, 0/1==Z */
} s_acc_vecval, *p_acc_vecval;
```

```
typedef struct t_setval_delay {
  s_acc_time time;
  PLI_INT32 model; /* accNoDelay, accInertialDelay,
                      accPureTransportDelay, accTransportDelay,
                      accAssignFlag, accDeassignFlag,
                      accForceFlag, accReleaseFlag */
} s_setval_delay, *p_setval_delay;
```

```
typedef struct t_acc_time
{
  PLI_INT32 type;        /* accTime, accSimTime, accRealTime */
  PLI_INT32 low, high; /* if type is accTime or accSimTime*/
  double    real;        /* used if type is accRealTime */
} s_acc_time, *p_acc_time;
```

void **acc_vcl_add (**

> | *handle* | **object,** | /* handle for a reg, variable, net, event, port, primitive output terminal or primitive inout terminal */ |
> | *PLI_INT32* | *** calback_rtn,** | /* unquoted name of a C function, without parenthesis */ |
> | *PLI_BYTE8* | *** user_data,** | /* user-defined data value */ |
> | *PLI_INT32* | **vcl_flag)** | /* constant **vcl_verilog_logic, vcl_verilog_strength** */ |

Adds a value change link monitor to an object. Each time the object changes value (for **vcl_verilog_logic**) or changes either value or strength (for **vcl_verilog_strength**), the callback routine (referred to as the *"consumer routine"*) is called and passed a pointer to a simulation-allocated vc_record structure, which contains information about the change. For nets, regs and variables, the object must be scalar, an unexpanded vector, or a bit-select of an expanded vector (use acc_object_of_type() to determine if an object is expanded or unexpanded)

```
typedef struct t_vc_record {
  PLI_INT32 vc_reason;              /* one of the constants:
                                        logic_value_change,
                                        strength_value_change,
                                        vector_value_change,
                                        sregister_value_change,
                                        vregister_value_change,
                                        integer_value_change,
                                        real_value_change,
                                        time_value_change,
                                        event_value_change */

  PLI_INT32  vc_hightime;     /* upper 32-bits of time */
  PLI_INT32  vc_lowtime;      /* lower 32-bits of time */
                       /* (time is in simulator time units) */

  PLI_BYTE8 *user_data;          /* value passed to acc_vcl_add*/

  union {
    PLI_UBYTE8 logic_value;    /* for logic_value_change, one
                                  of: vcl0, vcl1, vclZ, vclX */
    double     real_value;     /* for real_value_change */
    handle     vector_handle;  /* for vector_value_change,
                                  vregister_value_change,
                                  integer_value_change,
                                  time_value_change */
    s_strengths strengths_s;   /* for strength_value_change */
  } out_value;
```
```
typedef struct t_strengths {
  PLI_UBYTE8 logic_value; /* one of: vcl0, vcl1, vclZ, vclZ */
  PLI_UBYTE8 strength1;   /* each strength byte is one of: */
  PLI_UBYTE8 strength2;   /*   vclSupply, vclStrong, vclPull,
                               vclLarge,  vclWeak,  vclMedium,
                               vclSmall,  vclHighZ */
} s_strengths, *p_strengths;
```

void **acc_vcl_delete (**
    handle   **object,**   /* handle for an object with a VCL flag */
    PLI_INT32   *** calback_rtn,** /* unquoted name of a C function */
    PLI_BYTE8 *** user_data,** /* user-defined data value */
    PLI_INT32   **vcl_flag)**   /* constant: **vcl_verilog** */

Removes a value change link monitor on an object. (Note: some simulators allow the vcl_flag constant to be **vcl_verilog_logic** or **vcl_verilog_strength**).

PLI_BYTE8 ***acc_version ( )**

Returns a pointer to a character string with the version of ACC routines being used by the simulator.

APPENDIX D *The IEEE 1364-2001 VPI Routine Library*

*T*his section lists the definitions of the VPI routines contained in the IEEE 1364-2001 Verilog PLI standard. The VPI library is the third generation of the PLI, and is a full super set of the older TF and ACC PLI libraries. The VPI library provides full access to all parts of the simulation data structure, including both structural and RTL modeling constructs. The VPI library was introduced in 1993 by Open Verilog International under the name "PLI 2.0" (the IEEE standard does not use this label).

The VPI object diagrams in this guide are similar to the object diagrams contained in the IEEE 1364 standard, but are drawn in a different format and with additional information not contained in the IEEE standard. These diagrams document the properties that can be accessed for each object, and the relationships of an object to other objects.

The VPI library is defined in the file **vpi_user.h**, which should be one of the files provided with each Verilog simulator. This file defines data types with fixed widths, for portability to different operating systems. Simulator vendors may modify this section of vpi_user.h as required for an operating system. The default definitions in the IEEE 1364 standard are:

```
typedef int              PLI_INT32;
typedef unsigned int     PLI_UINT32;
typedef short            PLI_INT16;
typedef unsigned short   PLI_UINT16;
typedef char             PLI_BYTE8;
typedef unsigned char    PLI_UBYTE8;
```

Note: the vpi_user.h file does not define a **NULL** constant, as this is defined in the standard ANSI C libraries. Therefore, it is necessary to include either the **<stdlib.h>** or **<stdio.h>** C header files (or both) in all PLI applications that use VPI routines.

D.1 VPI object relationships

D.1.1 Object diagram legend

> (**module**)
>
> A *solid enclosure with bold font* represents an *object definition*. The diagram shows all relationships which can be traversed to and from the object.

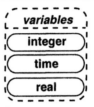

A *dotted enclosure with bold-italic font* represents an *object class definition*. The name of the object class is listed at the top of the enclosure (a class may be unnamed). The diagram shows all relationships which can be traversed to and from the object class.

> (port)
>
> A *solid enclosure with standard font* represents an *object reference*. Another diagram defines the referenced object.

> (*variables*)
>
> A *dotted enclosure with italic font* represents an *object class reference*. Another diagram defines the referenced object class.

A *single arrow* represents a *one-to-one relationship* which is traversed using **vpi_handle()** with the constant shown. For example:

```
expression_handle = vpi_handle(vpiExpr, prim_term_handle);
```

> vpiNet (net)
>
> A *double arrow* represents a *one-to-many relationship* which is traversed using **vpi_iterate()** with the constant shown. **vpi_iterate()** returns an iterator handle, which is then passed to **vpi_scan()** to access all objects of that type. For example:

```
net_iterator = vpi_iterate(vpiNet, module_handle);
while (net_handle = vpi_scan(net_iterator) != NULL ) {...}
```

A *small circle* indicates a **NULL** is used for the reference handle supplied to **vpi_handle()** or **vpilterate()**.

D.1.2 task/function call objects

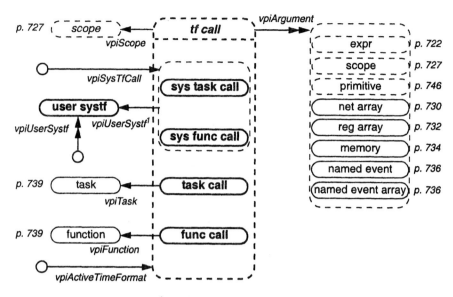

int / str	vpiType	returns **vpiTaskCall**, **vpiFuncCall**, **vpiSysTaskCall**, **vpiSysFuncCall** or **vpiUserSystf** (a system task/function instance)
int	vpiFuncType	for **vpiFuncCall** or **vpiSysFuncCall** objects, returns **vpiIntFunc**, **vpiTimeFunc, vpiRealFunc, vpiSizedFunc, vpiSizedSignedFunc**
bool	vpiUserDefn	returns 1 (true) if the system task/function call is user-defined (a PLI system task or system function)
str	vpiName	returns the name of the task, function, system task or system function
str	vpiDecompile	for **vpiSysTaskCall** or **vpiSysFuncCall** objects, returns a string with a functionally equivalent call as what was in the original Verilog source
str	vpiFile	returns the file name containing the tf call or user systf instance
int	vpiLineNo	returns the file line number of the tf call or user systf instance

1. **vpi_handle(vpiSysTfCall, NULL)** will obtain a handle for the system task/function call which invoked the PLI application.

2. **vpi_iterate(vpiUserSystf, NULL)** will obtain an iterator for all user-defined system task/ function instances in the simulation.

3. **vpi_get_systf_info**(user_systf_handle) will access the registration information for a user-defined system task/function (**vpiUserDefn** must be true).

4. A null task/function argument in the Verilog HDL will be an expr object with a **vpiType** of **vpiOperation** and a **vpiOpType** of **vpiNullOp**.

5. **vpi_handle(vpiActiveTimeFormat, NULL)** obtains a handle for the last instance of the $timeformat system task called by the simulator. If $timeformat has not been called, then **NULL** is returned.

6. **vpi_get_value()** returns the current value of a function call or system function call.

7. **vpi_put_value()** writes the return value of a system function call into simulation.

D.1.3 expression objects

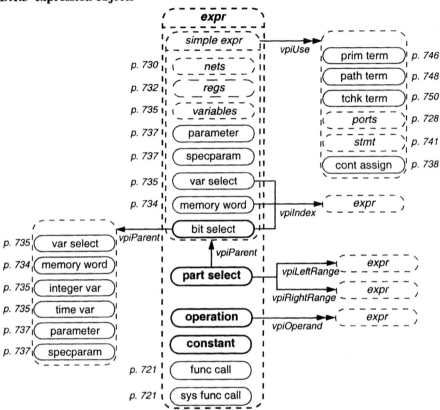

int / *str*	**vpiType**	returns **vpiNet, vpiReg, vpiIntegerVar, vpiTimeVar, vpiRealVar, vpiParameter, vpiSpecparam, vpiVarSelect, vpiMemoryWord, vpiBitSelect, vpiPartSelect, vpiOperation, vpiConstant, vpiFuncCall, vpiSysFuncCall**
int / *str*	**vpiOpType**	if type is **vpiOperation**, returns one of the operator constants shown in table on the following page
int	**vpiConstType**	if type is **vpiConstant**, returns **vpiIntConst, vpiRealConst, vpiBinaryConst, vpiOctConst, vpiDecConst, vpiHexConst, vpiStringConst**
bool	**vpiConstantSelect**	returns 1 (true) if a part select or bit select expression is a constant
int	**vpiSize**	returns the bit size of the expression result
str	**vpiName**	returns the name of a simple expression
str	**vpiFullName**	returns the full hierarchical path name of a simple expression
str	**vpiDecompile**	returns a decompilation of the expression
str	**vpiFile**	returns the file name containing the expression
int	**vpiLineNo**	returns the file line number of the expression

1. The value of an expression is accessed with **vpi_get_value()**.

2. Expressions with multiple operands will result in a handle of type **vpiOperation**.

vpiOpType Constants		
vpiAddOp	+	arithmetic addition operator
vpiArithLShiftOp	<<<	arithmetic left shift operator
vpiArithRShiftOp	>>>	arithmetic right shift operator
vpiBitAndOp	&	bitwise and operator
vpiBitNegOp	-	bitwise negation operator
vpiBitOrOp	|	bitwise or operator
vpiBitXNorOp	~^ ^~	bitwise xnor operator (can be either operator token)
vpiBitXorOp	^	bitwise xor operator
vpiCaseEqOp	===	case equality operator
vpiCaseNeqOp	!==	case inequality operator
vpiConcatOp	{ }	concatenation operator
vpiConditionOp	? :	ternary conditional operator
vpiDivOp	/	arithmetic division operator
vpiEqOp	==	relational equality operator
vpiEventOrOp	or ,	event or operator (can be either operator token)
vpiGeOp	>=	relational greater-than-or-equal operator
vpiGtOp	>	relational greater-than operator
vpiLShiftOp	<<	binary left shift operator
vpiLeOp	<=	binary less-than-or-equal operator
vpiListOp	,	list of expressions
vpiLogAndOp	&&	binary logical and operator
vpiLogOrOp	||	binary logical or operator
vpiLtOp	<	relational less-than operator
vpiMinTypMaxOp	:	min:typ:max: delay expression
vpiMinusOp	-	unary minus operator
vpiModOp	%	binary modulus operator
vpiMultOp	*	binary multiplication operator
vpiMultiConcatOp	{{ }}	replication operator (repeated concatenation)
vpiNegedgeOp	negedge	negedge operator
vpiNeqOp	!=	relational inequality operator
vpiNotOp	!	unary not operator
vpiNullOp	;	null operation
vpiPlusOp	+	unary plus operator
vpiPosedgeOp	posedge	posedge operator
vpiPowerOp	**	arithmetic power operator
vpiRShiftOp	>>	binary right shift operator
vpiSubOp	-	binary subtraction operator
vpiUnaryAndOp	&	unary reduction and operator
vpiUnaryNandOp	!&	unary reduction nand operator
vpiUnaryNorOp	!|	unary reduction nor operator
vpiUnaryOrOp	|	unary reduction or operator
vpiUnaryXNorOp	~^ ^~	unary reduction xnor operator (can be either token)
vpiUnaryXorOp	^	unary reduction xor operator

D.1.4 module objects

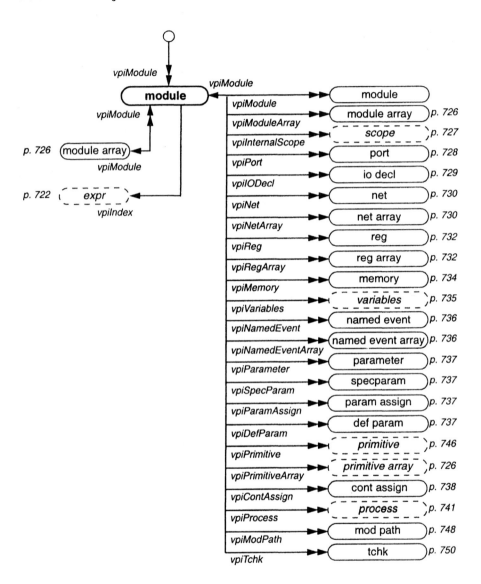

int / str	**vpiType**	returns **vpiModule**
bool	**vpiTopModule**	returns 1 (true) if a module is a top-level module
bool	**vpiCellInstance**	returns 1 (true) if a module is tagged as a cell (e.g. by `celldefine)
bool	**vpiArray**	returns 1 (true) if a module is part of an instance array
bool	**vpiProtected**	returns 1 (true) if the module source is protected
		(table continued on next page)

int	**vpiTimeUnit**	returns the module time unit as 2 down to -15, where 2==100 seconds, 1==10s, 0==1s, -1==100ms, -2 ==10ms, -3==1ms, ... -6==1us,... -9==1ns, ... -12==1ps, ... -15==1fs
int	**vpiTimePrecision**	returns module time precision as 2 to -15, as with time units
int	**vpiDefNetType**	returns the default net type as **vpiWire, vpiWand, vpiWor, vpiTri, vpiTri0, vpiTri1, vpiTriReg, vpiTriAnd, vpiTriOr, vpiSupply1, vpiSupply0** or **vpiNone** (no default net type)
int	**vpiUnconnDrive**	returns the unconnected port drive strength as **vpiHighZ, vpiPull1, vpiPull0**
int	**vpiDefDelayMode**	returns the delay mode of the module as: **vpiDelayModeNone, vpiDelayModePath, vpiDelayModeDistrib, vpiDelayModeUnit, vpiDelayModeZero, vpiDelayModeMTM**
int	**vpiDefDecayTime**	returns the default decay time for trireg nets
str	**vpiName**	returns the instance name of the module
str	**vpiFullName**	returns the full hierarchical path name of the module
str	**vpiDefName**	returns the definition name of the module
str	**vpiFile**	returns the file name containing the module instance
int	**vpiLineNo**	returns the file line number of the module instance
str	**vpiDefFile**	returns the file name where the module is defined
int	**vpiDefLineNo**	returns the file line number where module is defined
str	**vpiConfig**	returns the library and cell names of the config statement which controlled the binding of this particular module instance; the :config portion of the statement is not included in the string
str	**vpiLibrary**	returns the name of the configuration library to which the module instance was bound
str	**vpiCell**	returns the name of the configuration cell to which the module instance was bound

1. **vpi_iterate(vpiModule, NULL)** returns an iterator for all top-level modules.

2. **vpi_handle(vpiExpr,** module_handle) returns the module's index within a module instance array.. If the module is not part of an array, a **NULL** will be returned.

3. **vpi_get(vpiTimeUnit, NULL)** or **vpi_get(vpiTimePrecision, NULL)** will return the simulator's time units, which is the smallest time precision of all modules in the instantiated design.

D.1.5 instance array objects

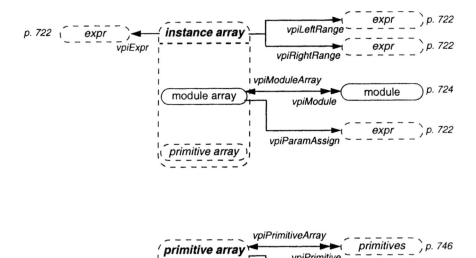

int / str	**vpiType**	returns **vpiModuleArray, vpiGateArray, vpiSwitchArray** or **vpiUdpArray**
int	**vpiSize**	returns the number of instances in the array
str	**vpiName**	returns the declaration name of the array
str	**vpiFullName**	returns the full hierarchical path name of the array
str	**vpiDefFile**	returns the file name containing the array instance
int	**vpiLineNo**	returns the file line number of the array instance

1. **vpi_handle(vpiExpr,** array_handle) will return an expression object of type **vpiOperation** with a **vpiOpType** of **vpiListOp**. This expression can be used to access the actual list of connections to the module or primitive instance array in the Verilog source code.

D.1.6 scope objects

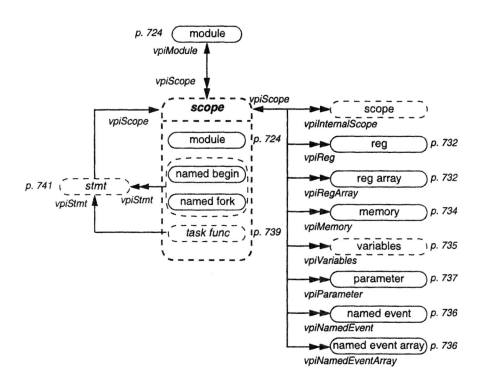

int / str	**vpiType**	returns **vpiModule, vpiNamedBegin, vpiNamedFork, vpiTask** or **vpiFunction**
str	**vpiName**	returns the name of the scope
str	**vpiFullName**	returns the full hierarchical path name of the scope

D.1.7 port objects

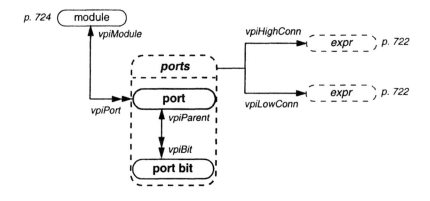

int / str	vpiType	returns **vpiPort** or **vpiPortBit**
bool	**vpiScalar**	returns 1 (true) if the port is scalar (1 bit wide)
bool	**vpiVector**	returns 1 (true) if the port is a vector
bool	**vpiExplicitName**	returns 1 (true) if the port or port bit is explicitly named
bool	**vpiConnByName**	returns 1 (true) if the port or port bit is connected by name
int	**vpiSize**	returns the size of the port
int	**vpiPortIndex**	returns the index position of the port (left-most port is 0); does not apply to port bits
int	**vpiDirection**	returns **vpiInput, vpiOutput, vpiInout, vpiMixedIO, vpiNoDirection**
str	**vpiName**	returns the name of the port; does not apply to port bits
str	**vpiFile**	returns the file name containing the port or port bit declaration
int	**vpiLineNo**	returns the file line number of the port or port bit declaration

1. Module input port delays (MIPD's) are accessed with **vpi_get_delays()** and **vpi_put_delays()**.

2. To access the port connected to an expression, determine the type of the expression and refer to the object diagram for that object type.

D.1.8 I/O declaration objects

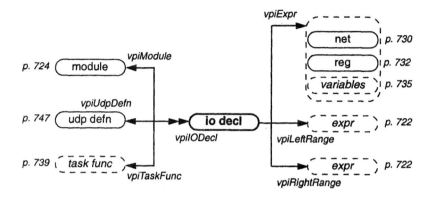

int / str	vpiType	returns **vpiIODecl**
bool	**vpiScalar**	returns 1 (true) if the I/O declaration is scalar
bool	**vpiVector**	returns 1 (true) if the I/O declaration is a vector
bool	**vpiSigned**	returns 1 (true) if the I/O declaration is a signed
int	**vpiSize**	returns the bit size of the I/O declaration
int	**vpiDirection**	returns **vpiInput, vpiOutput, vpiInout, vpiMixedIO** or **vpiNoDirection**
str	**vpiName**	returns the name of the I/O declaration
str	**vpiFile**	returns the file name containing the I/O declaration
int	**vpiLineNo**	returns the file line number of the I/O declaration

D.1.9 range objects

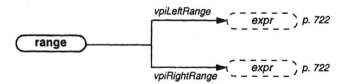

int / str	vpiType	returns **vpiRange**
int	**vpiSize**	returns the number of elements in the range
str	**vpiDefFile**	returns the file name containing the range declaration
int	**vpiLineNo**	returns the file line number of the range declaration

D.1.10 net objects

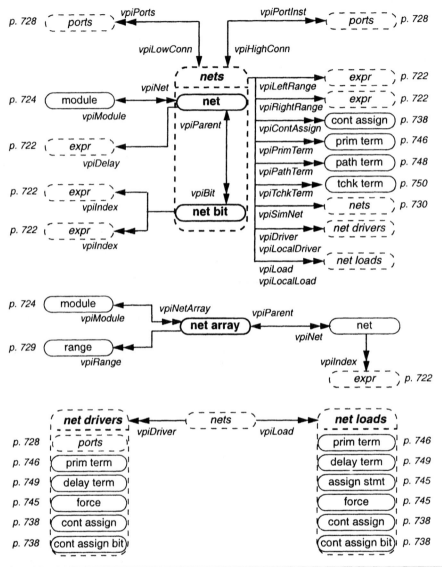

int / str	**vpiType**	returns **vpiNet**, **vpiNetBit** or **vpiNetArray**
int / str	**vpiNetType** **vpiResolvedNetType**	returns **vpiWire**, **vpiWand**, **vpiWor**, **vpiTri**, **vpiTri0**, **vpiTri1**, **vpiTriReg**, **vpiTriAnd**, **vpiTriOr**, **vpiSupply1**, **vpiSupply0**
bool	**vpiScalar**	for net objects, returns 1 (true) if the net is scalar (1-bit wide) for net array objects, returns 1 if nets in the array are scalar
bool	**vpiVector**	for net objects, returns 1 (true) if the net is a vector for net array objects, returns 1 if nets in the array are vectors
		(table continued on next page)

bool	vpiSigned	returns 1 (true) if the net is declared as signed
bool	vpiArray	returns 1 (true) if the net is a member of a net array
bool	vpiExplicitScalared	returns 1 (true) if the net is explicitly declared scalared
bool	vpiExplicitVectored	returns 1 (true) if the net is explicitly declared vectored
bool	vpiExpanded	returns 1 (true) if the net is expanded; bits of a net can be accessed regardless vector expansion
bool	vpiNetDeclAssign	returns 1 (true) if the net decl. is combined with a cont assign
bool	vpiImplicitDecl	returns 1 (true) if the net is implicitly declared
bool	vpiConstantSelect	returns 1 (true) if the net bit select expression is a constant
int	vpiSize	for net objects, returns the number of bits in the net; for net array objects, returns the total number of nets in the array
int	vpiStrength0	returns the logic 0 strength of the net
int	vpiStrength1	returns the logic 1 strength of the net
int	vpiChargeStrength	returns the capacitance strength of the net
str	vpiName	returns the declaration name of the net or net array
str	vpiFullName	returns full hierarchical path name of the net or net array
str	vpiFile	returns the file name containing the net or net array declaration
int	vpiLineNo	returns the file line number of the net or net array declaration

1. The value of a net or net bit is accessed with **vpi_get_value()** and **vpi_put_value()**. A net array does not have a value.

2. Cont assigns and prim terms can only be accessed from scalar nets or bits of vector nets, and will be accessed across hierarchical boundaries.

3. **vpi_iterate(vpiIndex,** net_handle) returns an iterator for the set of indices within an array, starting with the index for the net and working outward. If the net is not part of an array, then **NULL** is returned.

4. **vpi_handle(vpiIndex,** net_bit_handle) returns the bit index for the net bit.
vpi_iterate(vpiIndex, net_bit_handle) returns an iterator for the set of indices of a net bit select in an array, starting with the index for the net bit and working outward.

5. **vpi_iterate(vpiPorts,** net_bit_handle) returns an iterator for handles for port bits.

6. **vpi_iterate(vpiPortInst,** net_bit_handle) returns an iterator for handles for port bits or scalar ports. It is possible that a net bit is not connected to any bit of the port, if there is a mismatch in the sizes.

7. **vpiLoad** and **vpiDriver** return iterators for all loads or drivers, both within and without the module. Only active force or assign objects are considered loads or drivers. Different simulators may not return the same loads or drivers because net collapsing can be different.

8. **vpiLocalLoad** and **vpiLocalDriver** return only the loads or drivers that within the module, including any ports (output and inout ports are loads, input and inout ports are drivers).

9. **vpiSimNet** returns a handle for the simulated net after any net collapsing.

10. A **cbForce, cbRelease** or **cbDisable** callback cannot be placed on a variable bit select.

11. For implicit nets, **vpiLineNo** is 0, and **vpiFile** is the file where the net is first referenced.

D.1.11 reg objects

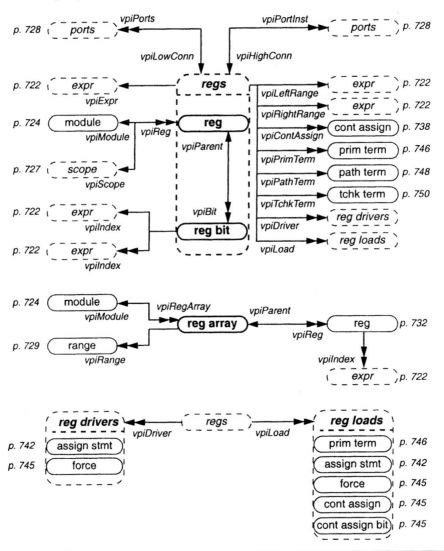

int / str	**vpiType**	returns **vpiReg, vpiRegBit** or **vpiRegArray**
bool	**vpiScalar**	for reg objects, returns 1 (true) if the reg is scalar (1-bit wide) for reg array objects, returns 1 (true) if regs in the array are scalar
bool	**vpiVector**	for reg objects, returns 1 (true) if the reg is a vector for reg array objects, returns 1 (true) if regs in the array are vectors
bool	**vpiSigned**	returns 1 (true) if the reg is declared as signed
bool	**vpiArray**	returns 1 (true) if the reg is a member of a reg array
		(table continued on next page)

bool	**vpiConstantSelect**	returns 1 (true) if a reg bit select expression is a constant
bool	**vpiAutomatic**	returns 1 (true) if a reg or reg array is in an automatic task or function
bool	**vpiValid**	returns 1 (true) if a handle for an automatic reg or reg array accessed from a frame handle is still active in simulation
int	**vpiSize**	for reg objects, returns the bit size of the reg; for reg array objects, returns the total number of regs in the array
str	**vpiName**	returns the declaration name of the reg or reg array
str	**vpiFullName**	returns the full hierarchical path name of the reg or reg array
str	**vpiFile**	returns the file name containing the reg or reg array declaration
int	**vpiLineNo**	returns the file line number of the reg or reg array declaration

1. The value of a reg or reg bit is accessed with **vpi_get_value()** and **vpi_put_value()**. A value with a delay cannot be placed on a reg or reg bit when **vpiAutomatic** is true. A reg array does not have a value.

2. **vpi_handle(vpiExpr**, reg_or_reg_bit_handle) returns a handle for the default initialization assignment expression, if any (e.g. reg r1 = 0;).

3. Continuous assignments and primitive terminals can only be accessed from scalar reg or bit selects of regs, and are accessed across hierarchical boundaries.

4. **vpi_iterate(vpiIndex**, reg_handle) returns the set of indices for a reg within an array, starting with the index for the reg and working outward. If the reg is not part of an array, then **NULL** is returned.

5. **vpi_handle(vpiIndex**, reg_bit_handle) returns the bit index for the reg bit. **vpi_iterate(vpiIndex**, reg_bit_handle) returns the set of indices for a reg bit select in an array, starting with the index for the reg bit and working outward.

6. **vpi_iterate(vpiPorts**, reg_bit_handle) returns an iterator for handles for port bits.

7. **vpi_iterate(vpiPortInst**, reg_bit_handle) returns an iterator for handles for port bits or scalar ports. It is possible that a reg bit is not connected to any bit of the port, if there is a mismatch in the sizes.

8. **vpiLoad** and **vpiDriver** return iterators for all loads or drivers, both within and without the module. Only active force or assign objects are considered loads or drivers. Different simulators may not return the same loads or drivers because net collapsing can be different.

9. **vpiLocalLoad** and **vpiLocalDriver** return only the loads or drivers that within the module, including any ports (output and inout ports are loads, input and inout ports are drivers).

10.A **cbForce**, **cbRelease** or **cbDisable** callback cannot be placed on a variable bit select.

11.A **cbValueChange** callback cannot be put on a reg or reg bit if **vpiAutomatic** is true.

D.1.12 memory objects

A memory is a 1-dimensional array of reg (e.g.: reg [7:0] RAM [0:1023];)

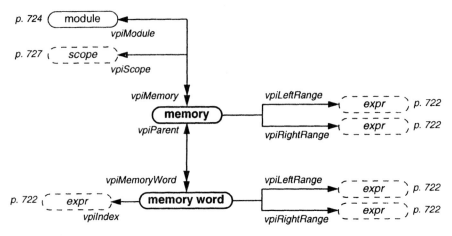

int / str	vpiType	returns **vpiMemory** or **vpiMemoryWord**
bool	**vpiSigned**	returns 1 (true) if the memory is declared as signed
bool	**vpiConstantSelect**	returns 1 (true) if the memory word index expression is a constant
bool	**vpiAutomatic**	returns 1 (true) if a memory or memory word is in an automatic task or function
bool	**vpiValid**	returns 1 (true) if a handle for an automatic memory or memory word accessed from a frame handle is still active in simulation
int	**vpiSize**	returns the number of words in a memory; returns the number of bits in a memory word
str	**vpiName**	returns the name of a memory or memory word
str	**vpiFullName**	returns the full hierarchical path name of a memory or memory word
str	**vpiFile**	returns the file name containing the memory declaration
int	**vpiLineNo**	returns file line number of the memory declaration

1. The access methods in this diagram are obsolete, but are included for backward compatibility with Verilog-1995. The Verilog-2001 standard adds multi-dimensional reg arrays, which is a super set of memory arrays. The preferred method of accessing words within a memory array is to use the methods shown in reg diagram in section D.1.11.

2. The value of a memory word is accessed with **vpi_get_value()** and **vpi_put_value()**. A value with a delay cannot be placed on a memory word when **vpiAutomatic** is true. A memory object does not have a value.

3. For a memory object, **vpiLeftRange** and **vpiRightRange** refer to the starting and ending address of a memory declaration. For a memory word object, **vpiLeftRange** and **vpiRightRange** refer to the msb and lsb of the word.

4. A **cbValueChange** callback cannot be put on an object if **vpiAutomatic** is true.

D.1.13 variable objects

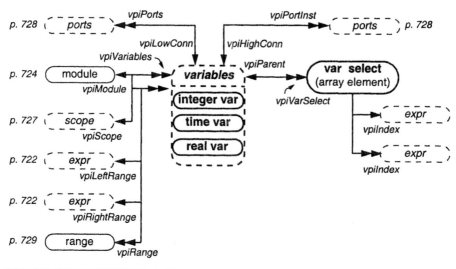

int / str	vpiType	returns **vpiIntegerVar, vpiTimeVar, vpiRealVar** or **vpiVarSelect** (a variable selection from within a variable array)
bool	**vpiArray**	returns 1 (true) if the variable object is an array of variables
bool	**vpiSigned**	returns 1 (true) if the variable object is declared as signed
bool	**vpiConstantSelect**	returns 1 (true) if the var select index expression is a constant
bool	**vpiAutomatic**	returns 1 (true) if the variable is in an automatic task or function
bool	**vpiValid**	returns 1 (true) if a handle for an automatic variable accessed from a frame handle is still active in simulation
int	**vpiSize**	for a variable object when **vpiArray** is false, returns the bit size of the variable; when **vpiArray** is true, returns the number of words in the array; for a var select object, returns the bit size of the select
str	**vpiName**	returns the name of a variable or var select
str	**vpiFullName**	returns the full hierarchical path name of a variable or var select
str	**vpiFile**	returns file name containing the variable definition or var select
int	**vpiLineNo**	returns file line number of the variable definition or var select

1. The value of a variable or var select is accessed with **vpi_get_value()** and **vpi_put_value()**. A value with a delay cannot be placed on a variable or var select when **vpiAutomatic** is true. Variables where the property **vpiArray** is true do not have a value.

2. If **vpiArray** is false, **vpiLeftRange** and **vpiRightRange** refer to the msb and lsb of the variable. If **vpiArray** is true, **vpiLeftRange** and **vpiRightRange** refer to the starting and ending address of the variable array declaration.

3. **vpi_handle(vpiIndex,** var_select_handle**)** returns the index of a var select in a 1-dimensional array. **vpi_iterate(vpiIndex,** var_select_handle**)** returns an iterator for the set of indices for the var select in a multi-dimensional array, starting with the index for the var select, and working outward.

4. A **cbValueChange** callback cannot be put on a variable if **vpiAutomatic** is true.

D.1.14 named event objects

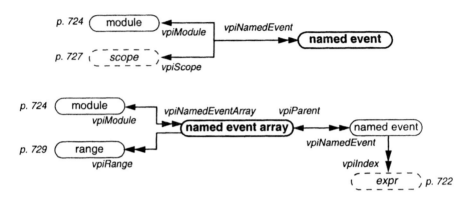

int / str	**vpiType**	returns the type as **vpiNamedEvent** or **vpiNamedEventArray**
bool	**vpiArray**	returns 1 (true) if the named event object is a member of an array of named events
bool	**vpiAutomatic**	returns 1 (true) if a named event or named event array is in an automatic task or function
bool	**vpiValid**	returns 1 (true) if a handle for an automatic named event or named event array accessed from a frame handle is still active in simulation
str	**vpiName**	returns the declaration name of the named event or named event array
str	**vpiFullName**	returns the full hierarchical path name of the named event or named event array
str	**vpiFile**	returns the file name containing the named event declaration
int	**vpiLineNo**	returns the file line number of the named event declaration

1. **vpi_put_value()** can be used to trigger an event on a named event. The value structure is not required; the pointer to the value structure can be set to **NULL**. An event trigger with a delay cannot be placed on a named event when **vpiAutomatic** is true.

2. **vpi_iterate(vpiIndex,** named_event_handle) returns an iterator for the set of indices within an array, starting with the index for the named event and working outward. If the named event is not part of an array, then **NULL** is returned.

3. A **cbValueChange** callback cannot be put on a named event if **vpiAutomatic** is true.

D.1.15 parameter, specparam, param assign and defparam objects

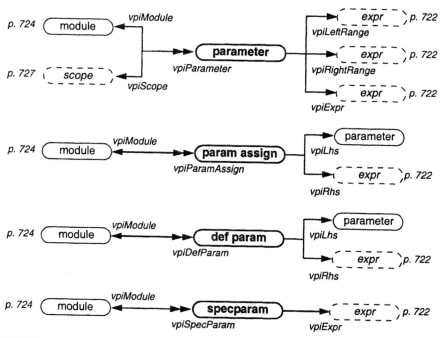

int / str	vpiType	returns **vpiParameter, vpiParamAssign, vpiDefParam** or **vpiSpecParam**
int	vpiConstType	for parameter or specparam, returns **vpiDecConst, vpiRealConst, vpiBinaryConst, vpiOctConst, vpiHexConst, vpiStringConst**
bool	vpiLocalParam	returns 1 (true) if a parameter is declared as a **localparam**
bool	vpiSigned	returns 1 (true) if a parameter is declared as signed
bool	vpiAutomatic	returns 1 (true) if a parameter is in an automatic task or function
bool	vpiValid	returns 1 (true) if a handle for an automatic parameter accessed from a frame handle is still active in simulation
bool	vpiConnByName	returns 1 (true) if a param assign uses explicit in-line parameter redefinition; e.g. mod u1 #(.width(64), .size(1024)) (...); returns 0 (false) if a param assign uses the implicit in-line parameter redefinition; e.g. mod u1 #(64,1024) (...);
int	vpiSize	returns the number of bits in a parameter or specparam
str	vpiName	returns the name of a parameter or specparam
str	vpiFullName	returns the full hierarchical path name of a parameter or specparam
str	vpiFile	returns the file name containing the object
int	vpiLineNo	returns the file line number of the object

1. The value of a parameter or specparam is accessed with **vpi_get_value()**. The value will be the final value after any parameter redefinitions. If **vpiConstType** is **vpiStringVal**, the value must be retrieved using a format of **vpiStringVal** or **vpiVectorVal** (the vector will be treated as a string of 8-bit characters, starting from the right-most bit).

D.1.16 continuous assignment objects

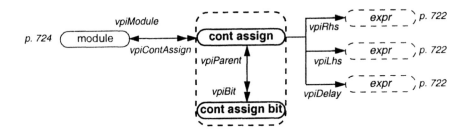

$\frac{int\,/}{str}$	**vpiType**	returns **vpiContAssign**
bool	**vpiNetDeclAssign**	returns 1 (true) if the cont assign is combined with a net declaration
int	**vpiStrength0**	returns the logic 0 strength of a combined cont assign and net declaration
int	**vpiStrength1**	returns the logic 1 strength of a combined cont assign and net declaration
int	**vpiOffset**	returns a cont assign bit's position relative to the least-significant bit. The lsb offset is 0.
str	**vpiFile**	returns the file name containing the cont assign
int	**vpiLineNo**	returns the file line number containing the cont assign

1. The logic value of a cont assign or cont assign bit is accessed with **vpi_get_value()**.

2. The delay value of a cont assign or cont assign bit is accessed with **vpi_get_delays()**.

D.1.17 task/function declaration objects

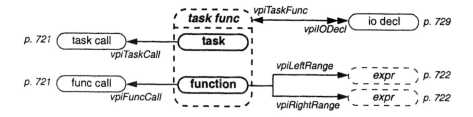

int / str	**vpiType**	returns **vpiTask** or **vpiFunction**
int	**vpiFuncType**	if the object is a function, returns **vpiIntFunc**, **vpiTimeFunc**, **vpiRealFunc**, **vpiSizedFunc** or **vpiSizedSignedFunc**
bool	**vpiSigned**	returns 1 (true) if the function is declared as signed
int	**vpiSize**	returns the number of bits that the function returns
str	**vpiFile**	returns the file name containing the task or function declaration
int	**vpiLineNo**	returns the file line number containing the task or function declaration

1. A Verilog function contains an object with the same name, size and type as the function (this
   is the object to which the output of the function is assigned).

D.1.18 re-entrant task and recursive function frame objects

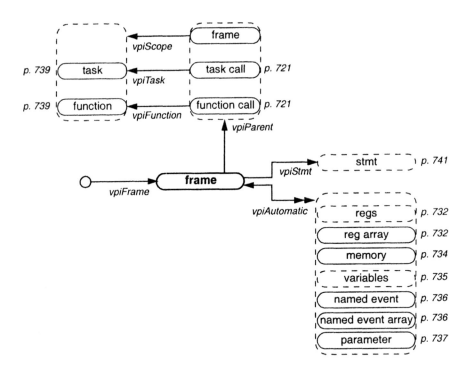

int / str	vpiType	returns **vpiFrame**
bool	**vpiValid**	returns 1 (true) if the frame handle is still active in simulation (this test can also be used with a handle for any of the objects accessed with vpiAutomatic)
bool	**vpiActive**	returns 1 (true) if the frame is currently active in simulation
bool	**vpiAutomatic**	returns 1 (true) if a reg, reg array, memory, variable, named event, named event array or parameter is automatic

1. **vpi_handle(vpiFrame, NULL)** obtains a handle for the currently active frame. There is at most only one active frame at any time.

2. Frame handles must be freed using **vpi_free_object()** once the application no longer needs the handle. If the handle is not freed, it will continue to exist, even after the frame has completed execution.

3. **vpi_handle(vpiStmt,** frame_handle) returns a handle to the currently active statement within the frame.

4. It is illegal to place value change callbacks on an object if **vpiAutomatic** is true.

5. It is illegal to put a value with a delay on an object if **vpiAutomatic** is true.

D.1.19 process (procedure) and statement group objects

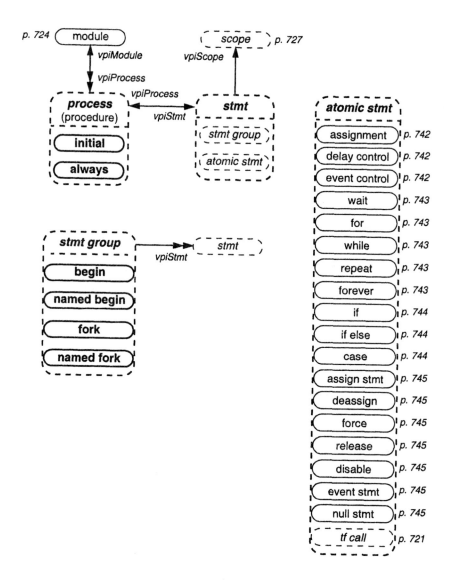

int / str	vpiType	for process objects, returns **vpiInitial** or **vpiAlways**
		for statement group objects, returns **vpiBegin**, **vpiNamedBegin**, **vpiFork** or **vpiNamedFork**
str	vpiFile	returns the file name containing the object
int	vpiLineNo	returns the file line number of the object

D.1.20 assignment statement objects

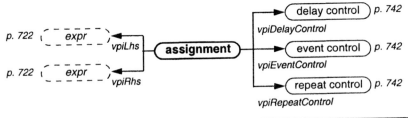

int / str	vpiType	returns **vpiAssignment**
bool	vpiBlocking	returns 1 (true) if the assignment is blocking (=); returns 0 (false) if the assignment is nonblocking (<=)
str	vpiFile	returns the file name containing the assignment
int	vpiLineNo	returns the file line number of the assignment

1. The delay, event or repeat control transition is to an intra-assignment time control (e.g. q <= #5 d;).

D.1.21 delay control, event control and repeat control objects

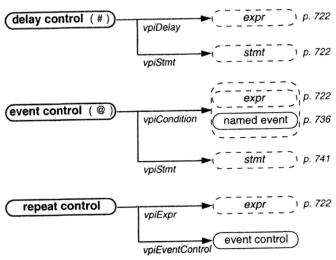

int / str	vpiType	returns **vpiDelayControl**, **vpiEventControl** or **vpiRepeatControl**
str	vpiFile	returns the file name containing the object
int	vpiLineNo	return the file line number of the object

1. The delay value of a delay control is accessed with **vpi_get_delays()**.

2. There is no transition from an intra-assignment time control (e.g. q <= #5 d;) back to the statement; therefore **vpi_handle(vpiStmt,** control_handle) will return **NULL**.

D.1.22 wait, for, while, repeat and forever statement objects

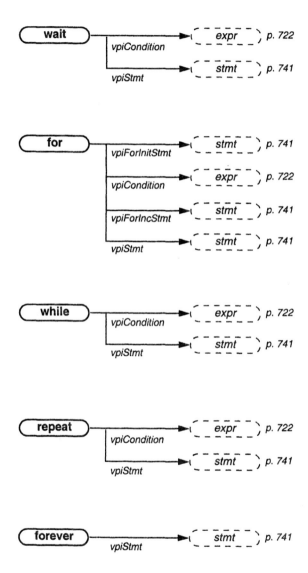

int / str	vpiType	returns **vpiWait, vpiFor, vpiWhile, vpiRepeat, vpiForever**
str	vpiFile	returns the file name containing the statement
int	vpiLineNo	returns the file line number of the statement

D.1.23 if, if–else and case statement objects

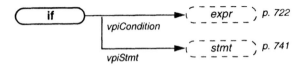

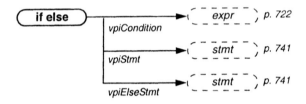

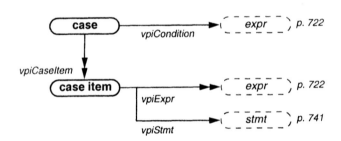

int / str	vpiType	returns **vpiIf, vpiIfElse, vpiCase** or **vpiCaseItem**
int	vpiCaseType	returns **vpiCaseExact, vpiCaseX** or **vpiCaseZ**
str	vpiFile	returns the file name containing the statement
int	vpiLineNo	returns the file line number of the statement

1. The case item groups all case conditions which branch to the same statement.

2. **vpi_iterate(vpiExpr,** case_item_handle) returns **NULL** for a default case item (there is no expression for the default case).

D.1.24 assign, deassign, force, release, disable, event and null statement objects

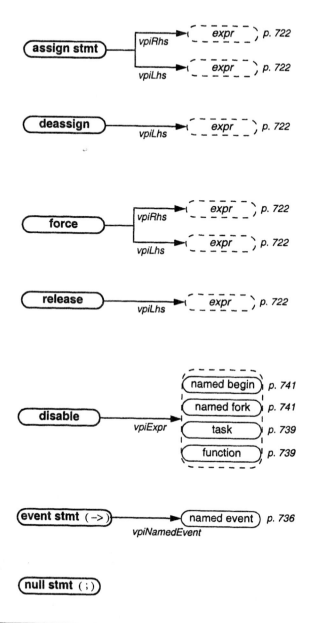

int / str	vpiType	returns **vpiAssignStmt, vpiDeassign, vpiForce, vpiRelease,** **vpiDisable, vpiEventStmt** or **vpiNullStmt**
str	vpiFile	returns the file name containing the statement
int	vpiLineNo	returns the file line number of the statement

D.1.25 primitive and prim term objects

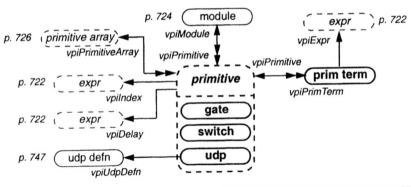

Primitive Object Properties		
int / *str*	**vpiType**	returns **vpiGate** or **vpiSwitch, vpiUDP**
int / *str*	**vpiPrimType**	returns **vpiAndPrim, vpiNandPrim, vpiNorPrim, vpiOrPrim, vpiXorPrim, vpiXnorPrim, vpiBufPrim, vpiNotPrim, vpiBufif0Prim, vpiBufif1Prim, vpiNotif0Prim, vpiNotif1Prim, vpiNmosPrim, vpiPmosPrim, vpiCmosPrim, vpiRnmosPrim, vpiRpmosPrim, vpiRcmosPrim, vpiRtranPrim, vpiRtranif0Prim, vpiRtranif1Prim, vpiTranPrim, vpiTranif0Prim, vpiTranif1Prim, vpiPullupPrim, vpiPulldownPrim, vpiCombPrim** (UDP), **vpiSeqPrim** (UDP)
bool	**vpiArray**	returns 1 (true) if a primitive is a member of an instance array
int	**vpiSize**	returns the number of inputs for a primitive
int	**vpiStrength0**	returns the logic 0 strength of the primitive
int	**vpiStrength1**	returns the logic 1 strength of the primitive
str	**vpiName**	returns the instance name of the primitive
str	**vpiFullName**	returns the full hierarchical path name of the primitive
str	**vpiDefName**	returns the definition name of the primitive
str	**vpiFile**	returns the file name containing the primitive
int	**vpiLineNo**	returns the file line number containing the primitive

Prim Term Object Properties		
int / *str*	**vpiType**	returns **vpiPrimTerm**
int	**vpiTermIndex**	returns the terminal index number (left-most terminal is 0)
int	**vpiDirection**	returns the terminal direction as **vpiInput, vpiOutput** or **vpiInout**

1. **vpi_get_value**(primitive_handle,...) retrieves the current output value of the primitive. **vpi_get_value**(prim_term_handle,...) retrieves the current value of that terminal.

2. Primitive delays are accessed with **vpi_get_delays()** and **vpi_put_delays()**.

3. The output value of a sequential UDP primitive can be set using **vpi_put_value()**.

4. **vpi_handle(vpiIndex,** primitive_handle) returns the primitive's index within an instance array. If the primitive is not a member of an array, a **NULL** is returned.

D.1.26 UDP definition objects

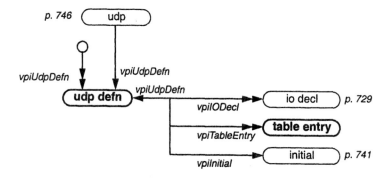

int / str	vpiType	returns **vpiUDPDefn** or **vpiTableEntry**
int / str	vpiPrimType	returns **vpiSeqPrim** or **vpiCombPrim**
bool	vpiProtected	returns 1 (true) if the UDP definition source code is protected
int	vpiSize	for UDP definition objects, returns the number of inputs for table entry objects, returns the number of symbol entries
str	vpiDefName	returns the name of a UDP definition
str	vpiFile	returns the file name containing the udp defn or table entry
int	vpiLineNo	returns the file line number containing the udp defn or table entry

1. UDP Table entry values can be accessed with **vpi_get_value()**; the values must be read as a
   string or as a vector (the vector will contain 8-bit ASCII values).

D.1.27 module path delay objects

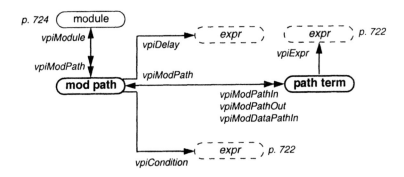

int / str	**vpiType**	returns **vpiModPath** or **vpiPathTerm**
bool	**vpiModPathHasIfNone**	returns 1 (true) if the module path has an ifnone condition
int	**vpiPathType**	for mod path objects, returns **vpiPathFull** or **vpiPathParallel**
int	**vpiPolarity**	for mod path objects, returns **vpiPositive, vpiNegative** or **vpiUnknown**
int	**vpiDataPolarity**	for mod path objects, returns **vpiPositive, vpiNegative** or **vpiUnknown**
int	**vpiDirection**	for path term objects, returns **vpiInput, vpiOutput** or **vpiInout**
int	**vpiEdge**	for path term objects, returns **vpiNoEdge, vpiPosedge, vpiNegedge, vpiAnyEdge, vpiEdge01, vpiEdge10, vpiEdge0x, vpiEdgex1, vpiEdge1x** or **vpiEdgex0**
str	**vpiFile**	returns the file name containing the mod path or path term
int	**vpiLineNo**	returns the file line number of the mod path or path term

1. Module paths delays are accessed with **vpi_get_delays()** and **vpi_put_delays()**.

D.1.28 intermodule path delay objects

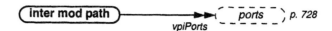

int / str	vpiType	returns **vpiInterModPath**

1. Use **vpi_handle_multi(vpiInterModPath,** output_port_handle, input_port_handle) to obtain inter mod path handles.

2. intermodule path delays are accessed with **vpi_get_delays()** and **vpi_put_delays()**.

D.1.29 delay device and delay term objects

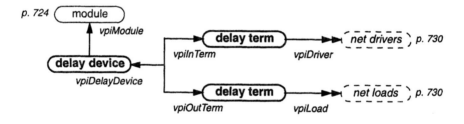

int / str	vpiType	returns **vpiDelayDevice** or **vpiDelayTerm**
int	vpiDelayType	returns **vpiModPathDelay, vpiInterModPathDelay** or **vpiMIPDelay** for either a delay device or a delay term

1. A delay device handle may be returned when iterating on the drivers or loads of a net.

2. A delay device is an internal simulation object which *may* be added by a simulator to propagate delayed events across nets, such as for an intermodule path delay from the output of one module to the input of another. Different simulators may have different delay device objects.

3. The value of the input delay term changes before the delay associated with the delay device. The value of the output delay term changes after the delay associated with the delay device. The current logic value of a delay term can be accessed with **vpi_get_value()**

4. The delay device and delay terms do not have delay values. The delays are properties of the drivers or loads of a delay term..

D.1.30 timing check objects

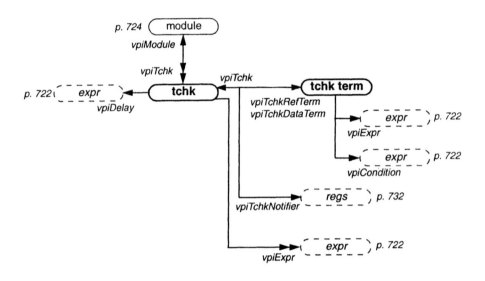

int / str	vpiType	returns **vpiTchk** or **vpiTchkTerm**
int	vpiTchkType	for tchk objects, returns **vpiSetup, vpiHold, vpiPeriod, vpiWidth, vpiSkew, vpiRecovery, vpiNoChange** or **vpiSetupHold**
int	vpiEdge	for tchk term objects, returns **vpiNoEdge, vpiPosedge, vpiNegedge, vpiAnyEdge, vpiEdge01, vpiEdge10, vpiEdge0x, vpiEdgex1, vpiEdge1x** or **vpiEdgex0**
str	vpiFile	returns the file name containing the timing check or timing terminal
int	vpiLineNo	returns the file line number containing the timing check or timing terminal

1. Timing check limit values are accessed with **vpi_get_delays()** and **vpi_put_delays()**.

2. The **vpiTchkRefTerm** is the first terminal for all timing checks except $setup, where **vpiTchkDataTerm** is the first terminal.

3. **vpi_iterate(vpiExpr,** tcheck_handle) will return an iterator for handles for all the arguments of a timing check, including the optional arguments used with negative timing checks. The reference, data and notifier terminals will have a **vpiType** property of **vpiTchkTerm**. All other arguments will have **vpiType** properties matching the expression.

D.1.31 simulation time queue objects

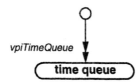

int / str	vpiType	returns **vpiTimeQueue**

1. The time value of the queue is accessed with **vpi_get_time()**.

2. Time queue objects are returned in increasing order of simulation time.

D.1.32 scheduled event objects

sched event

int / str	vpiType	returns **vpiSchedEvent**
bool	vpiScheduled	returns 1 (true) if the event is still scheduled (it has not transpired)

1. A scheduled event handle is returned from **vpi_put_value()** when the event is scheduled.

2. A scheduled event can be cancelled with **vpi_put_value(event_handle, NULL, NULL, vpiCancelFlag)**.

D.1.33 simulation callback objects

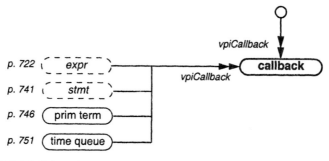

int / str	vpiType	returns **vpiCallback**

1. Simulation callbacks are registered with **vpi_register_cb**.

2. **vpi_iterate(vpiCallback, NULL)** returns an iterator for all active callbacks which have been registered using **vpi_register_cb()**.

3. Information about the callback object is accessed with **vpi_get_cb_info()**.

D.1.34 iterator objects

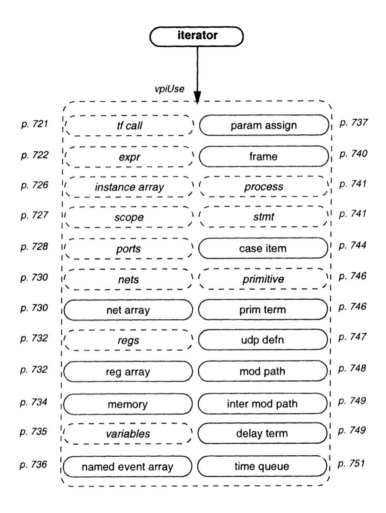

int / str	vpiType	returns **vpiIterator**
int	vpiIteratorType	returns the **vpiType** constant representing the object used to create the iterator (see the diagram for the object)

1. **vpi_handle(vpiUse,** iterator_handle) returns a handle for the reference object which use used with **vpi_iterate()** to create the iterator object. A **NULL** is returned if the reference handle supplied to **vpi_iterate()** was **NULL.**

D.1.35 attribute objects

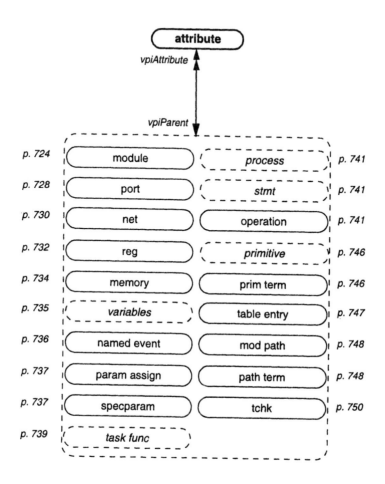

int / str	vpiType	returns **vpiAttribute**
bool	vpiDefAttribute	returns 1 (true) if the attribute was specified for the module definition, and therefore applies to all instances of the object; returns 0 (false) if the attribute was specified for that instance of the object
str	vpiName	returns the attribute name

1. An attribute is defined with (* *) in the Verilog HDL source code.

2. The value of an attribute is accessed with **vpi_get_value()**.

3. This diagram lists the objects shown in the IEEE 1364-2001 standard for the VPI. The Verilog language may allow attributes on other objects that are not listed in this diagram.

D.2 VPI routine definitions

PLI_INT32 **vpi_chk_error (**
   p_vpi_error_info **info)**   /* pointer to an application-allocated s_vpi_error_info structure, or **NULL** */

Returns **0** if the previous call to a VPI routine was successful, and an error severity code if the last VPI routine called resulted in an error. The severity codes are **vpiNotice**, **vpiWarning**, **vpiError**, **vpiSystem** and **vpiInternal**. If an error occurred, information about the error is retrieved into an s_vpi_error structure pointed to by **info**. If the information is not required, a **NULL** can be passed to the routine.

```
typedef struct t_vpi_error_info {
  PLI_INT32  state;      /* vpiCompile, vpiPLI, vpiRun */
  PLI_INT32  level;      /* vpiNotice, vpiWarning, vpiError,
                            vpiSystem, vpiInternal */
  PLI_BYTE8 *message;
  PLI_BYTE8 *product;
  PLI_BYTE8 *code;
  PLI_BYTE8 *file;
  PLI_INT32  line;
} s_vpi_error_info, *p_vpi_error_info;
```

PLI_INT32 **vpi_compare_objects (**
   vpiHandle   **object1,**   /* handle for an object */
   vpiHandle   **object2)**   /* handle for an object */

Returns **1** (true) if **object1** and **object2** handles reference the same object and **0** (false) if not.

PLI_INT32 **vpi_control (**
   PLI_INT32   **operation,**   /* constant representing the operation to perform */
      ...)   /* variable number of arguments, as required by the operation */

Allows a PLI application to control certain aspects of simulation. Returns **1** if successful and **0** if an error occurred. Several operation constants are defined in the IEEE 1364 standard (simulators may add additional flags specific to that product):

- **vpiStop** causes the $stop() built-in Verilog system task to be executed upon return of the PLI application. Requires one additional argument of type *PLI_INT32*, which is the same as the diagnostic message level argument passed to $stop().
- **vpiFinish** causes the $finish() built-in Verilog system task to be executed upon return of the PLI application. Requires one additional argument of type *PLI_INT32*, which is the same as the diagnostic message level argument passed to $finish().
- **vpiReset** causes the $reset() built-in Verilog system task to be executed upon return of the PLI application Requires three additional arguments of type *PLI_INT32*: **stop_value**, **reset_value** and **diagnostic_level**, which are the same values passed to the $reset() system task.
- **vpiSetInteractiveScope** causes a simulator's interactive debug scope to be immediately changed to a new scope. Requires one additional argument of type *vpiHandle*, which is a handle to an object in the scope class (see page 727).

PLI_INT32 **vpi_flush ( )**

Flushes the output buffer for the simulator's output channel and the simulator's log file. Returns **0** if successful, and **1** if an error occurs.

PLI_INT32 **vpi_free_object (**

| vpiHandle | **handle)** | /* handle for an object */ |

Frees memory allocated by the PLI for the specified handle. Returns **1** if successful, and **0** if an error occurred. Should be called to free an iterator handle whenever vpi_scan() is terminated before it returns **NULL**. Should also be called to free a handle returned from vpi_register_cb() or vpi_put_value() when the handle is no longer needed.

PLI_INT32 **vpi_get (**

| PLI_INT32 | **property,** | /* constant representing an object's property */ |
| vpiHandle | **object)** | /* handle for an object */ |

Returns the value associated with integer and boolean properties of an object. Boolean properties return **1** for true and **0** for false. Returns the constant **vpiUndefined** if an error occurs. For properties **vpiTimeUnit** and **vpiTimePrecision**, if object is **NULL**, then the simulation time unit is returned.

void **vpi_get_cb_info (**

| vpiHandle | **object,** | /* handle for an object */ |
| p_cb_data | **data)** | /* pointer to application-allocated s_cb_data structure */ |

Retrieves information about a simulation callback and places in an s_cb_data structure.

```
typedef struct t_cb_data {
  PLI_INT32    reason;    /* callback reason */
  PLI_INT32    (*cb_rtn)(struct t_cb_data *); /* routine name */
  vpiHandle    obj;       /* trigger object */
  p_vpi_time   *time;     /* callback time */
  p_vpi_value  *value;    /* trigger object value */
  PLI_INT32    index;     /* index of memory word or var select
                             that changed value */
  PLI_BYTE8    *user_data;
} s_cb_data, *p_cb_data;
```

PLI_INT32 **vpi_get_data (**
   PLI_INT32   **id,**   /* a save/restart ID */
   PLI_BYTE8 *** data_addr,**   /* address of application allocated storage */
   PLI_INT32   **byte_count**) /* number of bytes to be retrieved from the simulator's
                                save/restart file; must be greater than 0 */

Retrieves **byte_count** bytes of data from a simulation's save/restart file into the location pointed to by **data_addr**. Returns the actual number of bytes retrieved, or **0** if an error occurred. May only be called from a PLI application that has been called for reason **cbStartOfRestart** or **cbEndOfRestart**. **id** is the value created when the data was saved (refer to vpi_put_data()). The **id** from the save application can be passed to the restart application as the **user_data** value when the callback for **cbStartOfRestart** or **cbEndOfRestart** is registered. The first call to vpi_get_data() for a given **id** will retrieve data starting at the beginning of what was saved for that **id**. Each subsequent call for that **id** will start retrieving data where the last call left off. The name of the save/restart file can be accessed with vpi_get_str(vpiSaveRestartLocation, NULL).

void **vpi_get_delays (**
   vpiHandle   **object,**   /* handle for an object */
   p_vpi_delay **delay**)   /* pointer to application-allocated s_vpi_delay structure */

Retrieves an object's delays or pulse control values into an array of s_vpi_time structures onto an object. An s_vpi_delay structure controls how the delays are read.

```
typedef struct t_vpi_delay {
  struct t_vpi_time  *da; /* pointer to application allocated
                             array of delay values */
  PLI_INT32 no_of_delays; /* number of delay transitions */
  PLI_INT32 time_type;    /* vpiScaledRealTime, vpiSimTime */
  PLI_INT32 mtm_flag;     /* set to 0 (false) to retrieve a
                             single delay value for each
                             transition; set to 1 (true) to
                             retrieve min:typ:max delay sets
                             for each transition */
  PLI_INT32 append_flag;  /* not used by vpi_get_delays() */
  PLI_INT32 pulsere_flag; /* set to 0 (false) to retrieve only
                             the delays of the object; set to
                             1 (true) to read both delays and
                             pulse control values */
} s_vpi_delay, *p_vpi_delay;
```

```
typedef struct t_vpi_time {
  PLI_INT32  type;   /* not used by vpi_get_delays() */
  PLI_UINT32 high;   /* when using vpiSimTime */
  PLI_UINT32 low;    /* when using vpiSimTime */
  double     real;   /* when using vpiScaledRealTime */
} s_vpi_time, *p_vpi_time;
```

PLI_BYTE8 ***vpi_get_str (**
   PLI_INT32   **property,**   /* constant representing an object property */
   vpiHandle   **object**)   /* handle for an object */

Returns a pointer to a string containing the value of string properties of an object.

void vpi_get_systf_info (

vpiHandle	object,	/* handle for an object */
p_vpi_systf_data	data)	/* pointer to an application-allocated s_vpi_systf_data structure */

Retrieves information about a system task/function callback and places into an s_vpi_systf_data structure.

```
typedef struct t_vpi_systf_data {
    PLI_INT32  type;              /* vpiSysTask, vpiSysFunc */
    PLI_INT32  sysfunctype;       /* vpiIntFunc, vpiRealFunc,
                                     vpiTimeFunc, vpiSizedFunc,
                                     vpiSizedSignedFunc */
    PLI_BYTE8  *tfname;           /* quoted task/function name */
    PLI_INT32  (*calltf)(PLI_BYTE8 *);    /* name of C func */
    PLI_INT32  (*compiletf)(PLI_BYTE8 *); /* name of C func */
    PLI_INT32  (*sizetf)(PLI_BYTE8 *);    /* name of C func */
    PLI_BYTE8  *user_data;        /* returned with callback */
} s_vpi_systf_data, *p_vpi_systf_data;
```

Note: The Verilog-1995 standard used different constants for system function types: **vpiSysFuncInt, vpiSysFuncTime, vpiSysFuncReal** and **vpiSysFuncSized**. These constants are aliased to the Verilog-2001 constants for backward compatibility.

void vpi_get_time (

vpiHandle	object,	/* handle for an object, or **NULL** */
p_vpi_time	time)	/* pointer to application-allocated s_vpi_time structure */

Retrieves simulation time and places into an s_vpi_time structure. If retrieved as a *double*, the simulation time is scaled to the timescale of the module containing the object. If object is **NULL**, the simulation time is returned in simulation time units. If retrieved as a pair of 32-bit integers, the simulation time is always returned in simulation time units (the **object** argument is not used, and can be set to **NULL**.

```
typedef struct t_vpi_time {
    PLI_INT32  type;     /* vpiScaledRealTime or vpiSimTime */
    PLI_UINT32 high;     /* when using vpiSimTime */
    PLI_UINT32 low;      /* when using vpiSimTime */
    double     real;     /* when using vpiScaledRealTime */
} s_vpi_time, *p_vpi_time;
```

void *vpi_get_userdata (

vpiHandle	tfcall)	/* handle for a system task or system function call */

Retrieves a pointer to application-allocated information that was stored using vpi_put_userdata(). Returns **NULL** if there no user data has been stored. Each instance of a system task or system function has a unique user data storage area, which persists from one application call to another.

void **vpi_get_value (**
   vpiHandle   **object,**   /* handle for an object */
   p_vpi_value   **value)**   /* pointer to application-allocated s_vpi_value structure */

Retrieves the simulation value of an object into an s_vpi_value structure. The value is converted into the type specified by the **format** constant in the **value** structure, and placed into the appropriate field in the **value.value** union. When the **format** is **vpiObjTypeVal**, the routine will convert the value to its most optimal form, and then change the **format** field to show how the value was converted. *Note:* the simulator will allocate a time, vector or strength structure if required for the format, and place a pointer to the structure in the appropriate field in the s_vpi_value structure. These simulator-allocated structures will only be valid until the next call to vpi_get_value() or when the PLI application exits.

```
typedef struct t_vpi_value {
  PLI_INT32     format;   /* vpiBinStrVal,    vpiOctStrVal,
                             vpiDecStrVal,    vpiHexStrVal,
                             vpiScalarVal,    vpiIntVal,
                             vpiRealVal,      vpiStringVal,
                             vpiVectorVal,    vpiTimeVal,
                             vpiStrengthVal, vpiSuppressVal,
                             vpiObjTypeVal */
  union {
    PLI_BYTE8 *str;       /* if any string format */
    PLI_INT32  scalar;    /* if vpiScalarVal: one of vpi0, vpi1,
                             vpiX,vpiZ,vpiH,vpiL,vpiDontCare */
    PLI_INT32  integer;  /* if vpiIntVal format */
    double     real;     /* if vpiRealVal format */
    struct t_vpi_time      *time;    /* if vpiTimeVal */
    struct t_vpi_vecval    *vector;  /* if vpiVectorVal */
    struct t_vpi_strengthval *strength; /* if vpiStrengthVal */
    PLI_BYTE8 *misc;                 /* not used */
  } value;
} s_vpi_value, *p_vpi_value;
```

```
typedef struct t_vpi_time {
  PLI_INT32  type;      /* not used by vpi_get_value() */
  PLI_UINT32 high;      /* upper 32-bits of time value */
  PLI_UINT32 low;       /* lower 32-bits of time value */
  double     real;      /* not used by vpi_get_value() */
} s_vpi_time, *p_vpi_time;
```

```
typedef struct t_vpi_vecval {
  PLI_INT32 aval, bval;  /* bit encoding: a/b:
                            0/0==0, 1/0==1, 1/1==X, 0/1==Z */
} s_vpi_vecval, *p_vpi_vecval;
```

```
typedef struct t_vpi_strengthval {
  PLI_INT32 logic;  /* one of: vpi0, vpi1, vpiX, vpiZ */
  PLI_INT32 s0, s1; /* Logical-OR of the constants:
                       vpiSupplyDrive, vpiStrongDrive,
                       vpiPullDrive, vpiWeakDrive, vpiLargeCharge,
                       vpiMediumCharge, vpiSmallCharge, vpiHiZ */
} s_vpi_strengthval, *p_vpi_strengthval;
```

PLI_INT32 **vpi_get_vlog_info (**
    p_vpi_vlog_info **info)**   /* pointer to an application-allocated s_vpi_vlog_info
                              structure */

Retrieves the simulator's invocation option information into an s_vpi_vlog_info
structure. **argc** is the number of command line arguments given on the command line.
argv is a pointer to an array of character string pointers containing the command line
arguments. Returns **1** if successful and **0** if an error occurred.

```
typedef struct t_vpi_vlog_info {
  PLI_INT32    argc;
  PLI_BYTE8  **argv;
  PLI_BYTE8   *product;
  PLI_BYTE8   *version;
} s_vpi_vlog_info, *p_vpi_vlog_info;
```

vpiHandle **vpi_handle (**
    PLI_INT32   **type,**   /* constant representing an object type */
    vpiHandle   **reference)**   /* handle for an object */

Returns a handle for an object with a one-to-one relationship to a reference object
(single arrow relationships in the VPI object diagrams).

vpiHandle **vpi_handle_by_index (**
    vpiHandle   **parent,**   /* handle for an object with a vpiIndex relationship */
    PLI_INT32   **index)**   /* index number of an object */

Returns a handle for an object based on its index number within a parent object. The
return object must have a **vpiIndex** relationship to the parent object, such as a bit of a
vector.

vpiHandle **vpi_handle_by_multi_index (**
    vpiHandle   **object,**   /* handle for an array object */
    PLI_INT32   **array_size,** /* number or elements in the index array */
    PLI_INT32 *** index_array)** /* pointer to an array of indices */

Returns a handle for a word, or bit of a word, out of an array object.
vpi_get(vpiArray,object) must return true in order to use this routine. num_indices must
contain the number of elements in **index_array**. The order of the indices provided is the
left most select first, progressing to the right most select last.

vpiHandle **vpi_handle_by_name (**
    PLI_BYTE8 *** name,**   /* name of an object */
    vpiHandle   **scope)**   /* handle for a scope object, or **NULL** */

Returns a handle for an object using the name of the object. Objects which can be
searched for are those with a **vpiFullname** property. If the name does not contain a full
hierarchy path then the object is searched for in the specified scope only. If scope is
NULL, the object is searched for in the first top level module found. Returns a **NULL** if
the object cannot be found.

vpiHandle **vpi_handle_multi (**

PLI_INT32	**type,**	/* constant of **vpiInterModPath** */
vpiHandle	**reference1,**	/* handle for an output or inout port */
vpiHandle	**reference2)**	/* handle for an input or inout port */

Returns a handle for an intermodule path, from the output of one module to the input of another module. The output and input ports must be the same size.

vpiHandle **vpi_iterate (**

PLI_INT32	**type,**	/* constant representing an object type */
vpiHandle	**reference)**	/* handle for an object */

Returns a handle of type **vpiIterator** for objects with a one-to-many relationship to a reference object (double arrows in the VPI relationship diagrams). The iterator handle is then passed to **vpi_scan()** to access handles to all objects found by vpi_iterate().

PLI_UINT32 **vpi_mcd_close (**

    PLI_UINT32   **mcd_or_fd)** /* multi-channel descriptor or file descriptor */

Closes one or more files that were opened with either vpi_mcd_open() or $fopen. Returns **0** if successful, and the **mcd** or **fd** of files not closed if an error occurs.

PLI_INT32 **vpi_mcd_flush (**

    PLI_UINT32   **mcd_or_fd)** /* multi-channel descriptor or file descriptor */

Flushes the output buffer for one or more files that were opened with either vpi_mcd_open() or $fopen. Returns **0** if successful, and non-zero if an error occurs.

PLI_BYTE8 ***vpi_mcd_name (**

    PLI_UINT32   **mcd_or_fd)** /* multi-channel descriptor or file descriptor */

Returns a pointer to a string containing the name of a file that was opened with either vpi_mcd_open() or $fopen. The **mcd_or_fd** must reference a single open file.

PLI_UINT32 **vpi_mcd_open (**

    PLI_BYTE8 ***file_name)** /* name of a file to be opened */

Opens a file for writing and returns a multi-channel descriptor number (**mcd**). If the file is already open, the **mcd** or **fd** representing that file is returned. Returns **0** if an error occurred. This routine can only open an **mcd** type of file. It cannot open an **fd** (single-file descriptor) type of file.

PLI_INT32 **vpi_mcd_printf (**

PLI_INT32	**mcd,**	/* multi-channel descriptor */
PLI_BYTE8	*** format,**	/* character string with formatted message */
	...)	/* arguments to the formatted message string */

Prints a formatted message to one or more open files. Returns the number of characters written, or **EOF** if an error occurred. Uses a format string with formatting controls similar to the C printf() function. The **mcd** can be generated by either vpi_mcd_open() or $fopen. Multiple **mcd** values can be ORed together to write to multiple files at the same time. An **mcd** value of **1** represents the simulator's output window and log file (if open). This routine cannot write to a file represented by an **fd** file descriptor returned from $fopen (indicated by the most significant bit being set).

PLI_INT32 **vpi_mcd_vprintf (**

PLI_INT32	**mcd_or_fd,**	/* multi-channel descriptor of open files */
PLI_BYTE8	*** format,**	/* quoted character string of formatted message */
va_list	**arg_list)**	/* an already started list of variable arguments */

Prints a formatted message similar to vpi_mcd_printf(), except that the list of variable arguments has already been started. Returns the number of characters written, or **EOF** if an error occurred. *Note*: the va_list data type is defined in the ANSI C stdargs.h library.

PLI_INT32 **vpi_printf (**

PLI_BYTE8	*** format,**	/* character string containing a formatted message */
	...)	/* arguments to the formatted message string */

Prints a formatted message to the simulation output channel and the current simulation output log file. Uses one or more format strings with formatting controls similar to the C printf() function. Returns the number of characters written, or EOF if an error occurred.

PLI_INT32 **vpi_put_data (**

PLI_INT32	**id,**	/* a save/restart ID */
PLI_BYTE8	*** data_addr,**	/* address of application-allocated storage */
PLI_INT32	**byte_count)**	/* number of bytes to be added to the simulator's save/ restart file; must be greater than 0 */

Adds **byte_count** number of bytes of data located at **data_addr** into a simulator's save/restart file. Returns the number of bytes written, or **0** if an error occurred. Can only be called from an application that has been called for the reason **cbStartOfSave** or **cbEndOfSave**. The **id** is obtained by calling vpi_get(vpiSaveRestartID, NULL). vpi_put_data() can be called any number of times, with different **id**'s or with the same **id**. When the same **id** is used, vpi_get_data() will retrieve data from the save/restart file in the same order in which data was added. The **id** value can be passed to the application that will call vpi_get_data() using the **user_data** field when a callback is registered for **cbStartOfRestart** or **cbEndOfRestart**. Note that handles and pointers that are saved in a save/restart file may not be valid after a restart occurs. The name of the save/restart file can be obtained using vpi_get_str(vpiSaveRestartLocation, NULL).

void **vpi_put_delays (**
  vpiHandle   **object,**   /* handle for an object */
  p_vpi_delay **delay)**   /* pointer to an application-allocated s_vpi_delay structure
                      containing delay information */

Deposits delays or pulse control values stored in an array of application-allocated
s_vpi_time structures onto an object. An s_vpi_delay structure controls how the delays
are placed onto the object.

```
typedef struct t_vpi_delay {
  struct t_vpi_time  *da;    /* pointer to application allocated
                                array of delay values */
  PLI_INT32  no_of_delays;  /* number of delay transitions */
  PLI_INT32  time_type;     /* vpiScaledRealTime, vpiSimTime */
  PLI_INT32  mtm_flag;      /* set to 0 (false) to write a
                                single delay value for each
                                transition; set to 1 (true) to
                                write min:typ:max delay sets
                                for each transition */
  PLI_INT32  append_flag;   /* set to 0 (false) to replace the
                                delays of the object; set to
                                1 (true) to append the delays
                                to any existing delay */
  PLI_INT32  pulsere_flag;  /* set to 0 (false) to write only
                                the delays of the object; set to
                                1 (true) to write both delays
                                and pulse control values */
} s_vpi_delay, *p_vpi_delay;
```

```
typedef struct t_vpi_time {
  PLI_INT32   type;    /* not used by vpi_put_delays() */
  PLI_UINT32 high;     /* when using vpiSimTime */
  PLI_UINT32 low;      /* when using vpiSimTime */
  double      real;    /* when using vpiScaledRealTime */
} s_vpi_time, *p_vpi_time;
```

PLI_INT32 **vpi_put_userdata (**
  vpiHandle   **tfcall,**   /* handle for a system task or system function call */
  void   *** data)**   /* pointer to application-allocated storage */

Stores a pointer to application-allocated storage into simulator-allocated storage for that
instance of a system task or function. Returns **1** if successful and **0** if an error occurred.
Each instance of a system task or system function has a unique user data storage area.
The storage will persist from one call to another. The handle is retrieved from the
tfcall's user data space with *vpi_get_userdata()*.

vpiHandle **vpi_put_value (**
vpiHandle	**object,**	/* handle for an object */
p_vpi_value	**value,**	/* pointer to application-allocated s_vpi_value structure */
p_vpi_time	**time,**	/* pointer to application-allocated s_vpi_time structure */
PLI_INT32	**flag)**	/* constant representing the delay propagation method */

Writes a value specified in an s_vpi_value structure onto an object. The **flag** constant controls how the change is scheduled in simulation. Returns a handle of type **vpiSchedEvent** if the **flag** is ORed with **vpiReturnEvent**, otherwise, returns **NULL**.

- A **vpiNoDelay** flag immediately deposits the value onto the object. The object can be a net, reg, variable, memory word, array word, system function or sequential UDP (values put onto a net only remain in affect until another driver on the net changes).

- A **vpiInertialDelay**, **vpiTransportDelay** or **vpiPureTransportDelay** flag deposits a value at a future simulation time (a delay of 0 is equivalent to #0). The object can be a net, reg, variable, memory word or array word (values put onto a net will only remain in affect until another driver on the net changes).

- **vpiForceFlag** forces a value onto the object after any specified delay. The object can be a net, reg or variable. The force remains in effect until released.

- **vpiReleaseFlag** releases forced values after any specified delay.

- **vpiCancelFlag** cancels an event that was scheduled by a previous call to vpi_put_value(), using the handle returned by vpi_put_value() as the object.

```
typedef struct t_vpi_value {
  PLI_INT32    format;  /* vpiBinStrVal,    vpiOctStrVal,
                           vpiDecStrVal,    vpiHexStrVal,
                           vpiScalarVal,    vpiIntVal,
                           vpiRealVal,      vpiStringVal,
                           vpiVectorVal,    vpiTimeVal,
                           vpiStrengthVal, vpiSuppressVal,
                           vpiObjTypeVal */
  union {
    PLI_BYTE8 *str;     /* if any string format */
    PLI_INT32  scalar;  /* if vpiScalarVal: one of vpi0, vpi1,
                           vpiX, vpiZ, vpiH, vpiL, vpiDontCare */
    PLI_INT32  integer; /* if vpiIntVal format */
    double     real;    /* if vpiRealVal format */
    struct t_vpi_time         *time;    /* if vpiTimeVal */
    struct t_vpi_vecval       *vector;  /* if vpiVectorVal */
    struct t_vpi_strengthval *strength; /* if vpiStrengthVal */
    PLI_BYTE8 *misc;                     /* not used */
  } value;
} s_vpi_value, *p_vpi_value;
```

```
typedef struct t_vpi_time {
  PLI_INT32  type;     /* not used by vpi_get_value() */
  PLI_UINT32 high;     /* upper 32-bits of time value */
  PLI_UINT32 low;      /* lower 32-bits of time value */
  double     real;     /* not used by vpi_get_value() */
} s_vpi_time, *p_vpi_time;
```

(structure definitions continued on next page)

```
typedef struct t_vpi_vecval {
  PLI_INT32 aval, bval;    /* bit encoding: a/b:
                              0/0==0, 1/0==1, 1/1==X, 0/1==Z */
} s_vpi_vecval, *p_vpi_vecval;
```
```
typedef struct t_vpi_strengthval {
  PLI_INT32 logic;  /* one of: vpi0, vpi1, vpiX, vpiZ */
  PLI_INT32 s0, s1; /* Logical-OR of the constants:
                       vpiSupplyDrive, vpiStrongDrive,
                       vpiPullDrive, vpiWeakDrive, vpiLargeCharge,
                       vpiMediumCharge, vpiSmallCharge, vpiHiZ */
} s_vpi_strengthval, *p_vpi_strengthval;
```

vpiHandle **vpi_register_cb (**

   p_cb_data   **data)**   /* pointer to an application-allocated s_cb_data structure
                               containing callback information */

Registers a simulation callback to a PLI application for specific reasons during simulation. Returns a handle of type **vpiCallback** for a callback object. A scheduled callback can be cancelled using vpi_remove_cb(callback_handle). *Note*: the callback handle must be freed by the PLI application when no longer needed, using vpi_free_object(callback_handle).

The **reason** field in the s_cb_data structure is set to one of the constants listed on the following page (simulator's may add additional product-specific reason constants):

- For event-related callbacks, application-allocated s_vpi_value and s_vpi_time structures must be provided. The value structure specifies the format in which the object's new value will be passed to the simulation callback; use **vpiSuppressVal** if the callback does not need the value. The time structure specifies the format in which the simulation time will be passed to the simulation callback; use **vpiSuppressTime** if the callback does not need the time value.

- For all time-related callbacks, an application-allocated s_vpi_time structure must be provided (the value structure is not used, and the **value** pointer can be set to **NULL**). The time structure specifies the format in which the simulation time will be passed to the simulation callback; **vpiSuppressTime** is not allowed.

- For action-related callbacks, the value and time structures are not used; the **value** and **time** pointers can be set to **NULL**.

When the callback occurs, the simulator passes in a simulation-allocated s_cb_data structure, with information about the callback (e.g., for a cbValueChange callback, the structure has the object handle, its new logic value and the current simulation time).

```
typedef struct t_cb_data {
  PLI_INT32      reason;      /* callback reason */
  PLI_INT32      (*cb_rtn)(struct t_cb_data *); /* routine name */
  vpiHandle      obj;         /* trigger object */
  p_vpi_time     *time;       /* callback time */
  p_vpi_value    *value;      /* trigger object value */
  PLI_INT32      index;       /* index of memory word or var
                                 select that changed value */
  PLI_BYTE8      *user_data;
} s_cb_data, *p_cb_data;
```

Event-related Callbacks	
cbValueChange[1]	after a value change occurs on an object
cbStmt[1]	before the execution of a statement object; if object is a module, then before any statement in the module is executed
cbForce[1]	when a force occurs on the object; if object is **NULL**, then when a force occurs on any object in simulation
cbRelease[1]	when a release occurs on the object; if object is **NULL**, then a release occurs on any object in simulation
cbAssign[1]	when a procedural assign statement is executed
cbDeassign[1]	when a procedural deassign statement is executed
cbDisable[1]	when a disable occurs on a scope containing a system task/function
Time-related Callbacks	
cbAtStartOfSimTime[2]	before execution of events at the specified simulation time
cbReadWriteSynch[2]	after execution of events at the specified simulation time; the PLI application can cause additional events at that time
cbReadOnlySynch[2]	after execution of events at the specified simulation time; the PLI application cannot cause additional events at that time
cbNextSimTime[2]	before execution of events in the next event queue.
cbAfterDelay[2]	after a specified amount of time, and before the execution of events in the new time queue
Action-related Callbacks	
cbEndOfCompile[2]	end of the simulation data structure elaboration or build
cbStartOfSimulation[2]	start of simulation (beginning of the time 0 simulation cycle)
cbEndOfSimulation[2]	end of simulation (e.g., $finish system task executed)
cbEnterInteractive[1]	simulation entered interactive debug mode (e.g., by $stop)
cbExitInteractive[1]	simulation is exiting interactive debug mode
cbInteractiveScopeChange[1]	simulation has changed its interactive debug scope
cbError[1]	a simulation run-time error occurred
cbPLIError[1]	a simulation run-time error occurred in a PLI application
cbTchkViolation[1]	a timing check error occurred
cbSignal[1]	an operating system signal occurred
cbStartOfSave[1]	a $save command was invoked
cbEndOfSave[1]	a $save command has completed
cbStartOfRestart[1]	a $restart command was invoked
cbEndOfRestart[1]	a $restart command has completed
cbStartOfReset[1]	a $reset command was invoked
cbEndOfReset[1]	a $reset command has completed
cbUnresolvedSystf[1]	an unknown user-defined system task or function encountered

1. Callback can occur multiple times. The PLI application must unregister the callback using vpi_remove_cb() when no longer needed (vpi_remove_cb() also frees the callback handle).

2. Callback is a one-time event. The simulator will automaticdally remove the callback after it occurs (*note*: the callback handle will still be valid, and must be freed by the PLI application using vpi_free_object() to avoid a memory leak).

vpiHandle **vpi_register_systf (**

p_vpi_systf_data **data)**   /* pointer to the s_vpi_systf_data structure containing
system task/function information */

Registers a PLI system task/function **$** name and associates the name with PLI C
functions. Returns a handle for the callback object. Callbacks can be registered for:

- **calltf** applications—called when the system task/function name is encountered during
simulation.
- **compiletf** applications—called when the system task/function name is encountered
during elaboration or loading.
- **sizetf** applications—called when a system task/function name is encountered during
elaboartion or load, and the function type is vpiSizedFunc or vpiSizedSignedFunc.

If a calltf, compiletf or sizetf application is not used, that field should be set to **NULL**.

```
typedef struct t_vpi_systf_data {
    PLI_INT32   type;                /* vpiSysTask, vpiSysFunc */
    PLI_INT32   sysfunctype;         /* vpiIntFunc, vpiRealFunc,
                                        vpiTimeFunc, vpiSizedFunc,
                                        vpiSizedSignedFunc */
    PLI_BYTE8   *tfname;             /* quoted task/function name */
                                     /* first character must be $  */
    PLI_INT32   (*calltf)(PLI_BYTE8 *);    /* name of C func */
    PLI_INT32   (*compiletf)(PLI_BYTE8 *); /* name of C func */
    PLI_INT32   (*sizetf)(PLI_BYTE8 *);    /* name of C func */
    PLI_BYTE8   *user_data;          /* returned with callback */
} s_vpi_systf_data, *p_vpi_systf_data;
```

Note: The Verilog-1995 standard used different constants for system function types:
vpiSysFuncInt, vpiSysFuncTime, vpiSysFuncReal and **vpiSysFuncSized**. These
constants are aliased to the Verilog-2001 constants for backward compatibility.

PLI_INT32 **vpi_remove_cb (**

vpiHandle   **cb_object)**   /* handle for a callback object */

Removes callbacks to PLI applications which were registered with vpi_register_cb().
Returns **1** if successful, and **0** if an error occurred. Memory associated with the callback
handle is freed, making the handle invalid.

vpiHandle **vpi_scan (**

vpiHandle   **iterator)**   /* handle for an iterator object */

Returns a handle to the next object referenced by an iterator object, which was obtained
by vpi_iterate(). Returns **NULL** when there are no more objects referenced by the
iterator. After returning **NULL**, memory associated with the iterator handle is freed,
making the handle invalid.

PLI_INT32 **vpi_vprintf (**

PLI_BYTE8 *** format,**   /* quoted character string of formatted message */
va_list   **arg_list)**   /* an already started list of variable arguments */

Prints a formatted message. Similar to vpi_printf(), except that the list of var args is
already started. Returns the number of characters printed, or **EOF** if an error occurred.
Note: the *va_list* data type is defined in the ANSI C stdargs.h library.

Index

B

C

About the CD

The CD which accompanies this book contains:

- Over 85 examples of PLI applications, which are the same examples that are listed in the text of the book.

- A comprehensive example of using C sockets to establish communications between a Verilog Simulation and an independent C program. This example is provided courtesy of David Roberts, of Cadence Design Systems.

The examples are organized in chapter subdirectories, which correspond to the chapters of this book. Each chapter includes sample files for registering the PLI examples with the Cadence *Verilog-XL*™ and *NC-Verilog*® simulators, the Synopsys *VCS*™ simulator and the Model Technology *ModelSim*™ simulator. Appendix A of this book presents the basic steps required to compile and link PLI applications into each of these products.

Each example has three parts:

1. The C source file for the PLI example (the file names end in `.c`)

2. The Verilog HDL source code to test the PLI application (these files end in `.v`)

3. The simulation results from simulating the test (these files end with `.log`)

There are two complete directory sets, one for Unix and Linux operating systems, and the other for Windows operating system. The two directories are compressed, as:

- plibook_examples_unix.tar

- plibook_examples_pc.zip

This CD does not include a simulator software. Instead, the author recommends that reader use the latest version of their preferred simulator. That software may be available at the workplace or University. Some Electronic Design Automation companies have PLI-enabled demo versions of their software products available for download from their web sites. The sites for the companies referenced above are: *www.cadence.com*, *www.synopsys.com*, and *www.model.com*.

Note: The examples from the PLI book were tested using the Cadence Verilog-NC simulator. Every effort was made to ensure that the examples are accurate and correct. I am very interested in any suggestions on improving the examples, or about any bugs that are found. Please e-mail me at stuart@sutherland-hdl.com. I will maintain an up-to-date version of these examples on my web page, at www.sutherland-hdl.com. Please refer to this page for a list of errata in the book and updates to examples.

I hope you find these examples useful in learning and understanding the very complex topic of the Verilog Programming Language Interface!

Stuart Sutherland

===
COPYRIGHTS, OWNERSHIP AND DISCLAIMERS:

The electronic files on this CD are distributed by Kluwer Academic Publishers with *ABSOLUTELY NO SUPPORT* and *NO WARRANTY* from Kluwer Academic Publishers and/or Sutherland HDL, Inc. and/or the example authors.

The electronic text files in the directories "plibook_examples_unix" and "plibook_examples_pc" are the copyrighted property of Sutherland HDL, Inc. All rights reserved. Use or reproduction of the information provided in these files for commercial gain is strictly prohibited, with the exceptions noted here. Explicit permission is given for the reproduction and use of this information in an instructional setting, provided proper credit is given to the original source. Express permission is also given to use this information in the development of larger applications, commercial or otherwise, so long as the text files are modified from their original content to fit the context of the larger application. Stuart Sutherland, Sutherland HDL and Kluwer Academic Publishers shall not be liable in any way for damage in connection with, or arising out of, the furnishing, performance or use of this information.

The electronic text files in the directory "pli_socket_example_unix" and "pli_socket_example_pc" are provided by David Roberts. These files are provided with no restrictions on usage, copying or distribution. David Roberts, Stuart Sutherland, Sutherland HDL, Inc., and Kluwer Academic Publishers shall not be liable in any for damage in connection with, or arising out of, the furnishing, performance or use of this information.